2024 IEEE 33rd Conference on Electrical Performance of Electronic Packaging and Systems (EPEPS 2024)

Toronto, Ontario, Canada
6-9 October 2024

IEEE Catalog Number: CFP24EPP-POD
ISBN: 979-8-3503-5124-8

Copyright © 2024 by the Institute of Electrical and Electronics Engineers, Inc. All Rights Reserved

Copyright and Reprint Permissions: Abstracting is permitted with credit to the source. Libraries are permitted to photocopy beyond the limit of U.S. copyright law for private use of patrons those articles in this volume that carry a code at the bottom of the first page, provided the per-copy fee indicated in the code is paid through Copyright Clearance Center, 222 Rosewood Drive, Danvers, MA 01923.

For other copying, reprint or republication permission, write to IEEE Copyrights Manager, IEEE Service Center, 445 Hoes Lane, Piscataway, NJ 08854. All rights reserved.

*** This is a print representation of what appears in the IEEE Digital Library. Some format issues inherent in the e-media version may also appear in this print version.

IEEE Catalog Number:	CFP24EPP-POD
ISBN (Print-On-Demand):	979-8-3503-5124-8
ISBN (Online):	979-8-3503-5123-1
ISSN:	2165-4107

Additional Copies of This Publication Are Available From:

Curran Associates, Inc
57 Morehouse Lane
Red Hook, NY 12571 USA
Phone: (845) 758-0400
Fax: (845) 758-2633
E-mail: curran@proceedings.com
Web: www.proceedings.com

2024 IEEE 33rd Conference on Electrical Performance of Electronic Packaging and Systems (EPEPS 2024)

Toronto, Ontario, Canada
6-9 October 2024

IEEE Catalog Number: CFP24EPP-POD
ISBN: 979-8-3503-5124-8

TABLE OF CONTENTS

Optimizing Power and Power Delivery for Data Center GPUs.. 1
Tawfik Rahal-Arabi, Paul Van Der Arend, Ashish Jain, Mehdi Saidi, Rashad Oreifej, Sriram Sundaram, Rajit Seahra

Electrothermal Co-Design Modeling and Analysis of an Ultra-Low On-Resistance Power Switch 4
Sylvester Ankamah-Kusi, Blake Travis, Swathi Kamath, Rajen Murugan, Tom Kronenberg

Signal Integrity of Die-To-Die Interface with Advanced Packages for Co-Packaged Optics.............................. 7
Jongchul Shin, Hamid Eslampour, Sangnam Jeong, Woopoung Kim, Seokbeom Yong, Sung-Oh Ahn, Eunkyeong Park, Sangsub Song

Delay Rational Macromodelling of Noisy Tabulated Frequency Responses.. 10
Alexander Kirchberger, Anestis Dounavis

Fan-Out Region Crosstalk Optimization of High-Density PCIe 6.0 SMT Connectors 13
Dan Liu, Yangfan Zhong, Minzheng Tian, Mengmeng Guo, Bing Wei, Weizhe Li, Jingbo Li, Tina Bao

Optimization of TSV Array Based on Mathematical Model for HBM3... 16
Yen-Tung Chen, Yu-Ying Cheng, Tzong-Lin Wu

Application of the Reverse Pulse Technique for Worst Case Transient Analysis in HPC PDN Design... 19
Chad M. Smutzer, Jordan R. Keuseman, Alexander P. Hickman, Clifton R. Haider

A DDR5 Interposer De-Embedding Method Based on Transfer Function 22
Aobo Li, Jun Wang, Yan Xu, Kangkang Zhang, Xiuqin Chu

A Hybrid Polynomial Chaos Expansion and Gaussian Process Regression Method for Forward Uncertainty Quantification of Integrated Circuits 25
Paolo Manfredi, Riccardo Trinchero

Parametric S-Parameter Prediction Using Deep Learning.. 28
Vinayak Bansal, Lihong Feng, Valentin De La Rubia, Peter Benner

Modeling Multiplexed Qubit Readout with a Josephson Traveling-Wave Parametric Amplifier....................... 31
Samuel T. Elkin, Michael Haider, Thomas E. Roth

Comparative Evaluation of 100G-PAM4 Ethernet Link Performance in Air and Immersion Cooling Conditions 34
Dan Liu, Oluwafemi Akinwale, Yangfan Zhong, Kai Wang, Cesar Mendez-Ruiz, Kusuma Matta

Cascading of 2D and 3D Simulations of ASIC Substrate Interconnect Up to 100 GHz................................... 37
Zhekun Peng, Junyong Park, Sathvika Bandi, Santosh Pappu, Srinivas Venkataraman, Xu Wang, Granthana Ranzaswamy, Donghyun Kim

Multiphysics Simulation and Measurement Correlation of a Multichip Module IC Package Current Sensor 40
Rajen Murugan, Jie Chen, Guangxu Li, Yutaka Suzuki, Sylvester Ankamah-Kusi

Impact of Non-Functional Pads Location on Eye Diagram Performance.. 43
  Mehdi Mousavi, Kevin Cai, Junyong Park, Chaofeng Li, Manish K. Mathew, Reza Asadi,
  Shameem Ahmed, Donghyun Bill Kim, Bidyut Sen

High-Speed Interconnect Design of Silicon Interposer Based Heterogeneous Integration for AI
Computing.. 46
  Keeyoung Son, Seonguk Choi, Keunwoo Kim, Jiwon Yoon, Junghyun Lee, Haeseok Suh,
  Hyunjun An, Joungho Kim

Design and Analysis of L3 Cache Embedded-GPU-High Bandwidth Memory Architecture with
Reduced Energy and Latency for AI Computing... 49
  Haeseok Suh, Jiwon Yoon, Keeyoung Son, Seonguk Choi, Keunwoo Kim, Junghyun Lee, Taein
  Shin, Hyunjun An, Taesoo Kim, Jungmin Ahn, Hyunah Park, Hyunsik Kim, Taeil Bae,
  Haekang Jung, Joungho Kim

Analysis of Nonlinear Phase Interactions of a Differential Line in the Presence of a Signal Skew 52
  Byung Cheol Min, Mun Ju Kim, Hyun Chul Choi, Kang Wook Kim

A 155 MHz Low-Jitter PLL for Enhanced Signal Integrity in High-Speed Interconnects 55
  Tiruye Mulat Ayinet, Gerba Olani Baissa, Teo T. Hui

Single-Layer Wiring Design in UCIe to Realize Low-Cost Interposer Substrate 58
  Soshi Shimomura, Yutaka Uematsu, Katsuya Kikuchi, Haruo Shimamoto, Yuuki Araga,
  Shinichi Ouchi

PCIe Gen 6.0 SSD Receiver PAM-4 SI Analysis Based on End-Port Time-Domain Measurements
for Unknown System Channel... 61
  Jinwook Song, Jinan Lee, Jonghee Jeong, Seokwoo Hong, Sungwoo Jin, Hyunwoo Kim,
  Sungwon Roh, Chorom Jang, Youngjun Ko, Taehyun Shim, Juneyoung Kim, Dongho Choi,
  Kyungsuk Kim, Sunghoon Chun

Using Generative AI to Predict DC Electrical Performance.. 64
  J. Eric Bracken

PCIe Gen 6.0 SSD PSIJ Estimation Based on Early Design Stage Jitter Sensitivity Measurements 67
  Youngjun Ko, Jinwook Song, Seokwoo Hong, Jinan Lee, Jonghee Jeong, Sungwoo Jin,
  Hyunwoo Kim, Chorom Jang, Sungwon Roh, Dongho Choi, Kyungsuk Kim, Sunghoon Chun

Application of CAMM2 Connector on PCIe Gen 6.0 SSD Host Interface for Low Near-End
Crosstalk.. 70
  Sungwoo Jin, Jinwook Song, Seokwoo Hong, Youngjun Ko, Hyunwoo Kim, Sungwon Roh,
  Chorom Jang, Dongho Choi, Kyungsuk Kim, Sunghoon Chun

An Efficient SPICE-Compatible Model for Fast Co-Simulation of Signal and Power Integrity on
Multilayer PCB with Arbitrary Shape 73
  Hyunwoo Kim, Dongryul Park, Seunghun Ryu, Seonghi Lee, Sanguk Lee, Jinwook Lee,
  Dongkyun Kim, Seungyoung Ahn

Latency Insertion Method for Fast Electro-Thermal Simulation of FinFET with Self-Heating Effect 76
  Yi Zhou, José E. Schutt-Ainé

Eye-Diagram Edge Estimation (EEE) Network for Through Silicon Via Design in Next-Generation
High Bandwidth Memory 79
  Hyunjun An, Junghyun Lee, Keeyoung Son, Seonguk Choi, Taein Shin, Keunwoo Kim, Jiwon
  Yoon, Taesoo Kim, Jungmin Ahn, Hyunah Park, Haeseok Suh, Joungho Kim

Automated Accurate Quadratic Formulation of Nonlinear Circuits 82
Germin Ghaly, Emad Gad, Michel Nakhla

Crosstalk Analysis in Add-In Card Structure for High-Speed SerDes Channels with PCIe Gen6 85
Sungjin Yoon, Manho Lee, Kwangho Kim, Hyeongi Lee, Chulhee Cho, Youngjae Lee, Wooshin Choi, Young-Chul Cho, Jung-Hwan Choi, Young-Soo Sohn

Agile Analysis for Worst-Case Eye-Diagrams in Multi-Line Links of CoWoS Packaging 88
Zhu-Chen Chang, Chien-Min Lin, Ruey-Beei Wu

Operator Inference for Rigid-Flex Printed Circuit Boards Subject to Large Deformations 91
Pascal Den Boef, Diana Manvelyan, Wil Schilders, Joseph Maubach, Nathan Van De Wouw

A Signal Integrity Comparison of VIPPO Technology for PCIe 5.0 DC Blocking Capacitors 94
Andrew Page, Matteo Cocchini

Analysis of Interconnects in Multilayer SIW Bandpass Filters Design 97
Lu Qiu, Xiao-Wei Zhu, Xian-Long Yang

Yield-Aware Interposer Design for UCIe Interconnects 100
Ram Krishna, Ashita Victor, Srujan Penta, Xu Chen, Muhannad S. Bakir, Nam Sung Kim, Elyse Rosenbaum

Efficient Thermal Analysis for Heat Dissipation in Three-Dimensional Chip-Stacking Packaging 103
Cheng-Yuan Lu, Chien-Min Lin, Ruey-Beei Wu

Design of an Ultra-High-Speed Digital Interface Based on a Coplanar Stripline 106
Mun-Ju Kim, Byung-Cheol Min, Hyun-Chul Choi, Kang-Wook Kim

Design and Analysis of High-Density Silicon Interposer Channel and Power Distribution Network 109
Haeyeon Kim, Joonsang Park, Hyunah Park, Keeyoung Son, Hyunsik Kim, Taeil Bae, Haekang Jung, Joungho Kim

Signal Integrity Analysis of PCIe Channel with Floating Board-To-Board Connectors in
Automotive Infotainment System 112
Junghyun Lee, Keeyoung Son, Junho Park, Joonsang Park, Keunwoo Kim, Hyunjun An, Seonguk Choi, Jihun Kim, Hyunah Park, Sumi Choi, Sanghyuk Son, Joungho Kim

Design and Analysis of Extended Scale Cache (ESC) Stacked-GPU-HBM Module Architecture
Considering Power Integrity (PI) 115
Hyunah Park, Seonguk Choi, Haeyeon Kim, Taein Shin, Keeyoung Son, Jiwon Yoon, Junghyun Lee, Haeseok Suh, Taesoo Kim, Jungmin Ahn, Hyunjun An, Joungho Kim

Explainable Reinforcement Learning(XRL)-Based Decap Placement Optimization for High
Bandwidth Memory (HBM) 118
Keunwoo Kim, Hyunwook Park, Keeyoung Son, Seonguk Choi, Taein Shin, Junghyun Lee, Jiwon Yoon, Hyunjun An, Haeyeon Kim, Wooshin Choi, Jung-Hwan Choi, Joungho Kim

Nonlinear Macromodeling of Voltage-Regulated Power Delivery Networks 121
Antonio Carlucci, Stefano Grivet-Talocia

Efficient Parametric Assessment of Worst-Case Voltage Droop in Power Delivery Networks 124
Tommaso Bradde, Antonio Carlucci, Riccardo Trinchero, Paolo Mandredi, Stefano Grivet-Talocia

Limit of the Impact of the Via Stub Length on the Via Impedance in Printed Circuit Boards 127
Katharina Scharff, Xiaomin Duan, Dierk Kaller

Reinforcement Learning Based Automatic Router for Power Delivery Network Prototypes 130
Felix Yuan, Abinash Roy

Improve CLK Phase Noise Performance by Mitigating Antiresonance Phenomenon of Power Net
with a π-Type Filtering Structure.. 133
Xinlin Tang, Shuxiang Li, Tao Fang, Yuan Fang

A Tunable Inductor Peaking Technique for Optical Communication Systems.................................... 136
Festim Iseini, Han-Ting Lin, Nicola Pelagalli, Andrea Malignaggi, Corrado Carta, Gerhard
Kahmen, Andreas Weisshaar

Tunable True-Time-Delay Unit Based on Bridged T-Coil.. 139
Han-Ting Lin, Festim Iseini, Andreas Weisshaar

An Efficient Machine Learning Approach for PSIJ Analysis in a Chain of CMOS Inverters 142
Ahsan Javaid, Ramachandra Achar, Jai Narayan Tripathi

Modeling Microwave S-Parameters Using Frequency-Scaled Rational Gaussian Process Kernels................. 145
Thijs Ullrick, Dirk Deschrijver, Wim Bogaerts, Tom Dhaene

PSIJ Based Optimal PDN Design for Cost-Effective SSD Using Reinforcement Learning 148
Taein Shin, Seonguk Choi, Jungmin Ahn, Junghyun Lee, Keunwoo Kim, Haeseok Suh, Hyunah
Park, Haeyeon Kim, Hyunjun An, Jinwook Song, Joungho Kim

A Study on How Capacitance of Power Filtering Circuit Influences the Antiresonance Frequency 151
Shuxiang Li, Xinlin Tang, Tao Fang, Yuan Fang, Greg Fu, Stephen Scearce

Simulation Method for Quasi-Static Solver to Effectively Model Parasitic Components Between
Package and PCB 154
Silvia Simone, Fabio Pareschi, Davide Lena, Gianluca Setti

Analysis of Echo and Crosstalk Cancellation in Simultaneous Bidirectional Transceivers for Dense
Die-To-Die Interconnects 157
Tong Liu, Taeyang Sim, Samuel Palermo

Transformer Based Channel Identification.. 160
Priyank Kashyap, Yeujiang Wen, Yongjin Choi, Chris Cheng, Paul D. Franzon

Design and Analysis of Ultra High Bandwidth (UHB) Interconnection-Based GPU-Ring for the AI
Superchip Module 163
Jungmin Ahn, Seonguk Choi, Taein Shin, Junghyun Lee, Jiwon Yoon, Keunwoo Kim,
Keeyoung Son, Haeseok Suh, Taesoo Kim, Hyunah Park, Hyunjun An, Jinwook Song,
Joungho Kim

Recent Advances in Signal Integrity Simulation and Analysis of Interposers................................ 166
Jonatan Aronsson, Feng Ling

Multiphysics-Informed ML-Assisted Chiplet Floorplanning for Heterogeneous Integration 169
Vinicius C. Do Nascimento, Seunghyun Hwang, Michael J. Smith, Qiang Qiu, Cheng-Kok
Koh, Ganesh Subbarayan, Dan Jiao

Accuracy Study of the Differential Surface Admittance Operator for Lossy Metal Characterization.............. 172
M. Huynen, V. Okhmatovksi, D. De Zutter, D. Vande Ginste

Compact Fiber Weave Model for Full Wave Solvers 175
Stefan De Araujo, Daniel De Araujo, Bhyrav Mutnury

Megtron 6 and 8 Characterization Methodology 178
Stefan De Araujo, Daniel De Araujo, Roger Delbue, Ryan Keegan

Full-Wave Analysis for Ground Via Placement with Layered Media Integral Equations 181
Alireza Niazi, Vladimir Okhmatovski

Tree-Based Boosting for Efficient Estimation of S-Parameters for Package Electrical Analysis.................... 184
Doganay Özese, Mustafa Gökçe Baydogan, Ahmet Cemal Durgun, Kemal Aygün

Causal RL Prediction of Fine-Pitch Interconnects Using Neural Networks... 187
Hasan Said Ünal, Ahmet Cemal Durgun

A Robust Optimization Approach for High Bandwidth Memory Interposer Using Machine
Learning 190
Anandajith Jinesh, Xuan Chen

Hand-Drawn Circuit Schematic Digitization and Netlisting Using Machine Learning with
Emphasis on Signal Integrity Applications... 193
Anuj Mathur, Ramachandra Achar

Analysis and Modeling of Controlled Silicon Substrate Roughness for Silver-Based Backside
Metallization in Power Electronics Packaging 196
Mohamed Lamine Faycal Bellaredj, Goran Miskovic, Lukas Vojkuvka

Worst-Case Voltage Droop Using Peak Distortion Analysis 199
Mohamed Sahouli, Isaac Ali, David Reinamendivil, Gerry Talbot

Equalization Techniques for Time Domain Signaling 202
Shakib Mahmood, Parneet Tethy, Richelle L. Smith, Carl W. Werner, Masum Hossain

A Highly-Scalable Parallel Boundary Element Method for the Full-Wave Electromagnetic Analysis
of Large Interconnect Networks and Entire Packages... 205
Damian Marek, Jasper Hatton, Yongzhong Li, Piero Triverio

Gradient-Based Method to Find Solution for Rational Polynomial Chaos Coefficients for
Uncertainty Quantification 208
Karanvir S. Sidhu, Roni Khazaka

On the Parallelization of the MultiAIM Algorithm for the Fast Electromagnetic Analysis of 3D ICs.............211
Yongzhong Li, Piero Triverio

Author Index

EPEPS
2024
Toronto, Canada

33rd IEEE Conference on Electrical Performance of Electronic Packaging and Systems
October 6-9, 2024

The template used to create this booklet originates from LaTeXTemplates.com and is based on the original version at:
https://github.com/maximelucas/AMCOS_booklet

Foreword

Welcome to EPEPS 2024!

Dear colleague,

we are delighted to welcome you to Toronto for the 33rd IEEE Conference on Electrical Performance of Electronic Packaging and Systems (EPEPS)! For thousands of years, Toronto has been the traditional land of several indigenous populations, including the Huron-Wendat, the Seneca, and the Mississaugas of the Credit. Now the largest city in Canada, Toronto is one of the most diverse and cosmopolitan cities in the world. Toronto's role as a hub for many cultures reminds us of the role that EPEPS has played for over thirty years for its scientific community. Throughout three decades, EPEPS has been the meeting point of leading experts in the area of electrical performance of electronic systems. EPEPS is now the premier international conference on advanced and emerging issues in the electrical modeling, analysis, and design of electronic interconnections, packages, and systems.

At EPEPS 2024, attendees will enjoy a top-quality technical program, featuring keynote speeches by Dr. Tony Chan-Carusone (Alphawave IP and University of Toronto) and Dr. Dean Gonzales (Advanced Micro Devices), as well as an invited talk by Prof. Zheng Zhang (University of California, Santa Barbara). On Sunday, the conference will begin with six tutorials by renowned experts in heterogeneous integration, chiplets, advanced packaging, SI/PI analysis, mmWave communication, and quantum computing. Overall, the technical program will offer 38 oral and 35 poster presentations, two sponsor demonstrations and two training sessions. Attendees will also find a small industrial exhibition in the main foyer adjacent to the conference rooms.

This year's social program will begin, on Monday evening, with a welcome reception at the conference venue, the brand-new Schwartz Reisman Innovation Campus. On Tuesday night, attendees will enjoy a spectacular conference gala at Toronto's most iconic landmark, the CN Tower. A cocktail reception on the main observation deck, with announcement of the conference awards, will open the gala. Afterwards, the conference dinner in the CN Tower's revolving restaurant at 351 meters (1,151 feet) above the ground will offer unique views of the Greater Toronto Area.

The EPEPS 2024 conference would not have been possible without the generous support of several institutions, sponsors and exhibitors. EPEPS 2024 is jointly organized and sponsored by the IEEE Electronics Packaging Society (EPS) and by the IEEE Microwave

Theory and Technology Society (MTT-S). Additionally, the conference enjoys the technical sponsorship of the IEEE Antennas & Propagation Society (AP-S). Thanks to the generous support of the EPS and MTT-S societies, and of the TC-16 "Microwave and Millimeter-Wave Packaging, Interconnect and Integration" committee of the MTT-S, we have been able to offer an unprecedented number of graduate student grants (12). These grants, assigned on a competitive basis, allow deserving graduate students to register for free to the conference. Graduate students travelling from outside Canada also receive a travel allowance. With support from the MTT-S TC-16 committee, we have also been able to offer 3 grants aimed at young professionals, and a tutorial in the area of mmWave communication. A strong participation by industry has been a distinguishing characteristic of EPEPS since its inception. We are pleased to present you the sponsors and exhibitors of EPEPS 2024:

- Ansys (platinum sponsor);
- Advanced Micro Devices, Amphenol, Keysight, Samsung, SK Hynix (gold sponsors);
- Nvidia, Qualcomm, Rohde & Schwarz (silver sponsors);
- the IEEE Antennas & Propagation Society, Texas Instruments (exhibitors).

The support of these organizations has been instrumental to offer a rich technical and social program, and is gratefully acknowledged.

Finally, we want to extend our gratitude to the University of Toronto, to the Schwartz Reisman Innovation Campus, and to the Edward S. Rogers Sr. Department of Electrical & Computer Engineering for technical and logistical support. In particular, we thank Prof. Deepa Kundur, Chris Balarajah, Megan Lindsay, Vanisa Dimitrova, Sherry Rezaeirad, Wendy Gamble, and Veronika Rimsha.

On behalf of the entire organizing committee, I wish you a pleasant and productive conference. Enjoy EPEPS 2024 and Toronto!

Piero Triverio, University of Toronto

General chair, EPEPS 2024

Committees

Organizing Committee

- Piero Triverio (*University of Toronto*), general chair
- Wendem Beyene (*Meta*), co-chair
- Vaishnav Srinivas (*Qualcomm*), finance chair
- Xu Chen (*University of Illinois, Urbana*), program and website chair
- Vladimir Okhmatovsky (*University of Manitoba*), program co-chair
- Kemal Aygun (*Intel*), awards chair
- Heidi Barnes (*Keysight*), awards co-chair
- Paolo Manfredi (*Politecnico di Torino*), technical program committee secretary

Technical Program Committee

- Ram Achar (*Carleton University*)
- Kemal Aygun (*Intel*)
- Heidi Barnes (*Keysight*)
- Wendem Beyene (*Meta*)
- Swagato Chakraborty (*Siemens*)
- Xu Chen (*University of Illinois, Urbana*)
- Chris Cheng (*Hewlett-Packard Enterprise*)
- Paul Franzon (*North Carolina State University*)
- Dipanjan Gope (*Indian Institute of Science*)
- Stefano Grivet-Talocia (*Politecnico di Torino*)
- Kevin Gu (*Metawave*)
- Jose Hejase (*Nvidia*)
- Roni Khazaka (*McGill University*)
- Joungho Kim (*KAIST*)
- Paolo Manfredi (*Politecnico di Torino*)
- Rajen Murugan (*Texas Instruments*)
- Vladimir Okhmatovsky (*University of Manitoba*)
- Zhen Peng (*University of Illinois, Urbana*)
- Tawfik Rahal-Arabi (*AMD*)
- Sourajeet Roy (*IIT, Roorkee*)
- Albert Ruehli (*Emeritus IBM, MST*)
- Rohit Sharma (*IIT, Ropar*)
- Vaishnav Srinivas (*Qualcomm*)
- Junyan Tang (*IBM*)
- Jai Narayan-Tripathi (*IIT, Jodhpur*)
- Piero Triverio (*University of Toronto*)
- Dries Vande-Ginste (*Ghent University*)
- Andreas Weisshaar (*Oregon State University*)
- Tzong-Lin Wu (*National Taiwan University*)
- Ali Yilmaz (*University of Texas, Austin*)

Paper Review Committee

- Giulio Antonini (*University of L'Aquila*)
- Nikhita Baladari (*NXP*)
- Fadime Bekmambetova (*Nanoacademic Technologies Inc.*)
- Xie Biancun (*Nvidia*)
- Tommaso Bradde (*Politecnico di Torino*)
- Xuan Chen (*AMD*)
- Chung-Kuan Cheng (*University of California at San Diego*)
- Paul Dahlen (*IBM*)
- Daniel DeAraujo (*Siemens*)
- Dirk Deschrijver (*Ghent University*)
- Matt Doyle (*IBM*)
- Xiaomin Duan (*IBM*)
- Ege Engin (*San Diego State University*)
- Francesco Ferranti (*Vrije Universiteit Brussel*)
- Fei Guo (*Marvell Technology*)
- Sunil Gupta (*Meta Platforms*)
- Anand Haridass (*Microsoft*)
- Shaowu Huang (*Marvell*)
- Rubaiyat Islam (*AMD*)
- Marco Kassis (*Cadence*)
- Jingook Kim (*UNIST*)
- Gokul Kumar (*Micron*)
- Naiguang Lei (*Synopsys*)
- Bowen Li (*Google*)
- Tianjian Lu (*Apple*)
- Giorgi Maghlakelidze (*Cisco*)
- Nikita Mahjabeen (*NXP*)
- Zhen Mu (*Cadence*)
- Bhyrav Mutnury (*Dell*)
- Behzad Nouri (*Carleton University*)
- Andrew Page (*IBM*)
- Pavel Roy Paladhi (*Nvidia*)
- Sung Joo Park (*Samsung*)
- Utkarsh Patel (*Cadence*)
- Zhiguo Qian (*Intel*)
- Shashwat Sharma (*Nvidia*)
- Sameer Shekhar (*Intel*)
- Eakhwan Song (*Kwangwoon University*)
- Jairam Sukumar (*Qualcomm*)
- Dani Tannir (*Lebanese American University*)
- Riccardo Trinchero (*Politecnico di Torino*)
- Tao Wang (*Qualcomm*)
- Boping Wu (*Huawei*)
- Zhuo Yan (*Apple*)
- Ming Yi (*Nvidia*)
- Huan Yu (*Apple*)
- Xinyue Zhang (*University College Dublin*)
- Biyao Zhao (*Nvidia*)
- Tingdong Zhou (*NXP*)
- Yaping Zhou (*Nvidia*)
- Zhen Zhou (*Intel*)

Institutional Sponsors

Sponsors

Technical Sponsors

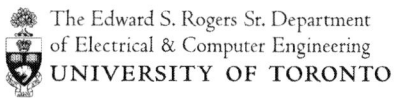

Sponsors and Exhibitors

Platinum Sponsor

Gold Sponsors

Silver Sponsors

Exhibitors

Schedule

Sunday, October 06

Time	Session	Room
08:00 - 09:00	Breakfast	
09:00 - 09:10	Welcoming Remarks	W280
09:10 - 11:10	Tutorials 1	W280
11:10 - 11:30	Coffee Break	
11:30 - 13:30	Tutorials 2	W280
13:30 - 14:30	Lunch Break	
14:30 - 16:30	Tutorials 3	W280

Monday, October 07

Time	Session	Room
08:00 - 09:00	Breakfast	
09:00 - 09:10	Opening Ceremony	W280
09:10 - 10:10	Keynote	W280
10:10 - 11:10	Heterogeneous Integration I	W280
11:10 - 11:30	Coffee Break	
11:30 - 12:30	Heterogeneous Integration II	W280
12:30 - 14:00	Lunch	
12:30 - 14:00	TPC (private meeting)	W240
14:00 - 15:20	Machine Learning I	W280
15:20 - 15:30	Sponsor Demo: Amphenol	W280
15:30 - 15:50	Coffee Break	
15:50 - 17:10	Device Modeling	W280
17:10 - 19:10	Welcome reception	
17:10 - 19:10	Poster Session I	Foyer

Tuesday, October 08

Time	Session	Room
08:00 - 09:00	Breakfast	
09:00 - 10:00	Keynote	W280
10:00 - 11:00	Macromodeling I	W280
11:00 - 11:20	Coffee Break	
11:20 - 12:40	Macromodeling II	W280
12:40 - 14:00	Lunch	
14:00 - 15:00	Computational Electromagnetics	W280
15:00 - 15:10	EPEPS 2025 Announcement	W280
15:10 - 15:30	Coffee Break	
15:30 - 17:10	Signal and Power Integrity I	W280
15:30 - 16:30	Ansys Training Program	W240
17:10 - 18:20	Poster Session II	Foyer
18:20 - 18:30	Bus pickup	Foyer
19:00 - 19:15	Check in at the CN Tower	CN Tower
19:15 - 20:30	Cocktail Reception and Awards Ceremony	CN Tower
20:30 - 22:00	Conference Dinner	CN Tower

Wednesday, October 09

Time	Session	Room
08:00 - 09:00	Breakfast	
09:00 - 10:00	Invited Talk	W280
10:00 - 11:00	Machine Learning II	W280
11:00 - 11:10	Sponsor Demo: Keysight	W280
11:10 - 11:30	Coffee Break	
11:30 - 12:10	Optics	W280
11:30 - 12:10	Keysight Training Program	W240
12:10 - 12:50	Computer Aided Design	W280
12:50 - 14:20	Lunch	
14:20 - 15:20	Signal and Power Integrity II	W280
15:20 - 15:40	Closing Ceremony	W280

Keynotes

Accelerating AI with Chiplet Technology

Speaker: Prof. Tony Chan Carusone, Alphawave & University of Toronto

Abstract: Chiplet technology is revolutionizing our digital infrastructure. The reductions in cost, time-to-market, and power consumption of chiplet-based solutions are compelling, particularly for AI hardware. Custom silicon for AI significantly benefits from the chiplet approach, which allows for the integration of dense logic, memory, and high-speed connectivity. Chiplets provide the flexibility to create systems-in-package that balance cost, power, and performance for specific workloads without reinventing the wheel for each new design. As chiplet adoption grows, it drives increased bandwidth requirements within packages and across die-to-die interfaces. Scaling AI performance requires low-latency inter-die communication and lots of high-speed optical connectivity.

Biography: Dr. Tony Chan Carusone has taught and researched integrated circuits and systems for high-speed connectivity in industry and academia for over 20 years. He has been the Chief Technology Officer of Alphawave Semi since 2022 and a faculty member at the University of Toronto since completing his Ph.D. there in 2002. He has well over 100 publications, including 11 award-winning best papers at leading conferences for work on chip-to-chip and optical communication, analog-to-digital conversion, and precise clock generation. He also co-authored the latest editions of the classic textbooks "Analog Integrated Circuit Design" and "Microelectronic Circuits," the best-selling engineering textbook of all time. Tony has also been a consultant to the semiconductor industry for over 20 years, working with both startups and some of the largest technology companies in the world. He is a Fellow of the IEEE.

Advanced Chiplet Package Signal Integrity for Future Data Center and AI

Speaker: Dr. Dean Gonzales, AMD

Abstract: Advanced Packaging Chiplet technologies have enabled large-scale deployment of heterogenous compute to address the disparate demands of the mega data center and the voraciously growing machine intelligence market. These advanced server CPU and GPU based systems require very high chip-to-chip IO connectivity and memory bandwidths with strict power, area, and latency optimization for scaling up and scaling out of rack-based systems. While dense integration brings together new capacity to solve the industry's most pressing problems, these Chiplets and high-power systems have abundant mechanical complexity and sensitivity to manufacturing variation. Even minor imperfections can have a profound impact on signal-to-noise margin at large scale system production. This discussion is an overview of the unique signal and power integrity design challenges for building these advanced Chiplet based products and systems with low defect rate requirements.

Biography: Dean Gonzales is an AMD Fellow with three decades of experience working on the design of silicon, advanced packaging technology, and computer systems. His signal integrity focus helps drive AMD pathfinding to improve power, area, and performance of silicon interconnects. Dean brings to this talk a passion for I/O electrical standardization and compliant channel design and modeling methodologies.

Invited Talks

Machine Learning for EDA, or EDA for Machine Learning?

Speaker: Prof. Zheng Zhang, University of California, Santa Barbara

Abstract: The rapid advancement of machine learning (especially deep learning) in the past decade has impacted, both positively and negatively, many research fields. Driven by the great success of machine learning in image and speech domains, there have been increasing interests in "Machine Learning for EDA". In the first part of the talk, I will explain the main challenge of data sparsity when applying existing machine learning techniques to EDA. Then I will show how some data-efficient scientific machine learning techniques, specifically uncertainty quantification and physics-constraint operator learning, can be utilized to build high-fidelity surrogate models for variability analysis and for 3D-IC thermal analysis, respectively. These techniques can greatly reduce the number of required device/circuit simulation data samples.

Another important but highly ignored direction is "EDA for Machine Learning". The five decades of EDA research has produced a huge body of solid theory and efficient algorithms for analyzing, modeling and optimizing complex electronic systems. Many of the white-box EDA ideas may be leveraged to solve black-box AI problems. In the second part of the talk, I will show how the self-healing idea and compact modeling idea from EDA can be utilized to improve the trustworthiness and sustainability of deep learning models (including large-language models).

Biography: Dr. Zheng Zhang is an Associate Professor of Electrical and Computer Engineering at University of California, Santa Barbara. He received his PhD degree in Electrical Engineering and Computer Science from MIT in 2015. His research is focused on uncertainty quantification and tensor computation for semiconductor chip design automation, and for responsible and sustainable AI systems. He is a recipient of NSF CAREER award, 3 best journal paper awards from IEEE Transactions in the EDA field, and two best dissertation awards from ACM SIGDA and MIT Microsystems Technology Labs. His work has been recognized by the IEEE CEDA Early Career Award and ACM SIGDA Outstanding New Faculty Award.

Tutorials

Signal and Power Integrity Analysis Using LIM – Recent Advances

Speaker: Prof. Jose Schutt-Aine, University of Illinois Urbana-Champaign

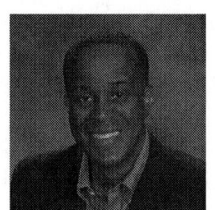

Abstract: With the increase in complexity and size of modern circuits, signal integrity has become an important aspect in the the study of the performance of electronic systems. Circuit designers are constantly in need of robust and efficient circuit simulation methods that can capture complex electromagnetic behaviors of networks and devices in fast turnaround time. As a result, there is a constant need for and push toward faster and more accurate circuit simulation techniques. The latency insertion method (LIM) has emerged as an approach of choice for performing fast simulations of very large circuits. By exploiting latencies in a circuit, LIM implements an algorithm that achieves linear numerical complexity. This results in a computationally efficient algorithm that is able to simulate large circuits significantly faster than traditional matrix inversion-based methods used in simulators such as SPICE.

In this tutorial we review the fundamentals of the latency insertion method and explore applications related to new technologies. Recent advances, several examples and case studies related to signal/power integrity and large circuits will be presented.

Biography: José E. Schutt-Ainé is with the faculty in the Electrical and Computer Engineering Department at the University of Illinois at Urbana-Champaign. His research interests are on signal integrity for high-speed digital and high-frequency applications. Dr. Schutt-Ainé is an IEEE Fellow, EPS Distinguished Lecturer, and served as Co-Editor-in-Chief of the IEEE Transactions on Components, Packaging and Manufacturing Technology (T-CPMT) from 2007 to 2018. He is currently chairing the Co-design chapter of the IEEE-EPS Co-Design Chapter of the Heterogeneous Integration Roadmap.

Practical aspects of FD-TDR co-modification for high-speed structures' matching and what-if simulations

Speaker: Pavel Vilner, Nvidia

Abstract: Time Domain Reflectometry (TDR) is a widely used technique in design of high-speed (HS) serial digital channels. It is frequently employed as a post-processing step for VNA measurements and Frequency Domain (FD) electromagnetic simulations' results to assess the localized channel impedance as the assumed propagating signal sees it. The TDR well-known connection to the channel's topology and design features makes it useful in identifying changes needed to achieve the desired system performance. Thus, when the TDR or FD results are not satisfactory, two questions often arise – (1) what the FD data will look like if some TDR feature is changed, and (2) what TDR feature needs to be changed to modify an undesired aspect of the FD data. Answering those questions requires co-modifying the TDR result and its respective FD representation.

While the Fourier theory underpinning FD-TDR co-modification is well known, this presentation deals with finer practical points needed to implement it. Extrapolation of the original FD data to achieve the desired bandwidth, application of the modification window functions, considerations in modifying the complex-valued FD data and influence of computational precision are discussed. Examples in MATLAB are given, and their usefulness in improving their respective HS structures' topology is showcased.

Biography: Pavel Vilner is a Senior Manager in the Nvidia Networking Signal Integrity Group (Toronto office), working on connectivity solutions for Nvidia NVLink, InfiniBand and Ethernet products. His interests include high-speed serial links' performance analysis, modeling and measurements of passive structures in high-speed serial interconnects and computational electromagnetics. He graduated from the Technion (Israel Institute of Technology) in 2019 with M.Sc. in Electrical Engineering and B.Sc. in Physics.

Superconducting Circuit Quantum Computers: Fundamental Concepts and Scaling Challenges

Speaker: Prof. Thomas Roth, Purdue University

Abstract: Superconducting circuit devices are one of the most mature hardware platforms for developing quantum computers, but significant engineering advances are needed for them to reach their revolutionary potential in practical applications. One primary challenge is to scale the number of quantum bits (qubits) in devices by orders of magnitude while continuing to improve key performance metrics. In this tutorial, we will introduce the fundamental concepts of how superconducting circuit quantum computers operate assuming no prior knowledge in quantum mechanics. We will discuss the key aspects of the primary qubit type currently being used, as well as how microwave fields and electronic biases are used to control and measure qubit states. We will then briefly discuss current hardware and computational modeling efforts underway that are working to address the scaling challenges of these systems. We will conclude with comments on directions for future work and the roles the electronics packaging community can play in making these revolutionary technologies a reality.

Biography: Thomas E. Roth is an Assistant Professor in the Elmore Family School of Electrical and Computer Engineering. He received all his degrees in electrical and computer engineering, with the B.S. degrees from Missouri University of Science and Technology and the M.S. and Ph.D degrees from the University of Illinois at Urbana-Champaign. Prior to joining Purdue, he was a Senior Member of the Technical Staff at Sandia National Laboratories in the Radar Electromagnetics & Sensor Technologies department where he was named a 2019 Up & Coming Innovator. He is the recipient of Young Scientist Awards at the 2023 Photonics & Electromagnetics Research Symposium and the URSI International Symposium on Electromagnetic Theory 2023 (1st place), a recipient of the 2023 IEEE Ulrich L. Rohde Innovative Conference Paper Award on Computational Techniques in Electromagnetics, as well as the 2023 Ruth and Joel Spira Outstanding Teacher Award at Purdue University. His research focuses on multiscale and multiphysics computational electromagnetics techniques, particularly for analyzing and designing quantum information processing devices.

Chiplet Design and Heterogeneous Integration Packaging

Speaker: Dr .John H. Lau, Unimicron Technology Corporation

Abstract: Chiplet is a chip design method and heterogeneous integration is a chip packaging method. Chiplet design and heterogeneous integration packaging have been generated lots of tractions lately. For the next few years, we will see more implementations of a higher level of chiplet designs and heterogeneous integration packaging, whether it is for cost, time-to-market, performance, form factor, or power consumption. In this lecture, the following topics will be covered.

- System-on-Chip (SoC)
- Why Chiplet Design?
- Chiplet Design and Heterogeneous Integration Packaging
  – Chip Partition and Chip Split
    - Chip partition and Heterogeneous Integration
    - Chip split and Heterogeneous Integration
    - Advantages and Disadvantages
- Communication between Chiplets (e.g., Bridges)
    - Bridge Embedded in Build-up Package Substrate
    - Bridge Embedded in Fan-Out EMC with RDLs
    - UCIe
    - Hybrid Bonding Bridge
- Chiplet Design and Heterogeneous Integration Packaging
  - Multiple System and Heterogeneous Integration
    - Multiple System and Heterogeneous Integration with Package Substrate (2D IC Integration)
    - Multiple System and Heterogeneous Integration with Thin Film layer on the Package Substrate (2.1D IC Integration)
    - Multiple System and Heterogeneous Integration with TSV-less (Organic) Interposer (2.3D IC Integration)
    - Multiple System and Heterogeneous Integration with Passive TSV-Interposer (2.5D IC Integration)
    - Multiple System and Heterogeneous Integration with Active TSV-Interposer (3D IC Integration)
- Advanced Packaging Driving by Artificial Intelligent
- Summary
- Potential R&D Topics in Chiplet Design and Heterogeneous Integration Packaging
- Trends in Chiplet Design and Heterogeneous Integration Packaging

Biography: John H Lau, with more than 40 years of R&D and manufacturing experience in semiconductor packaging has published more than 533 peer-reviewed papers (380 are the principal investigator), 50 issued and pending US patents (32 are the principal inventor), and 23 textbooks (all are the first author) such as Chiplet Design and Heterogeneous Integration Packaging (Springer, 2023) and Flip Chip, Hybrid Bonding, Fan-In, and fan-Out Technology (Springer, 2024). John is an elected IEEE fellow, IMAPS Fellow, and ASME Fellow and has been actively participating in industry/academy/society meetings/conferences to contribute, learn, and share.

Abstraction for Heterogeneous Integration

Speakers: Dr. Vaishnav Srinivas, Qualcomm & Prof. Madhavan Swaminathan, Pennsylvania State University

Abstract: Over the past few decades, the ASIC design methodology has evolved based on a sophisticated abstraction approach from RTL to GDS. This has enabled the design to transition through different areas of expertise seamlessly as the language of communication is transparent and easy to understand, owing to the abstraction. RTL gets synthesized, the netlist then gets floorplanned, placed and routed, signed off for timing and physical verification. Such an approach has enabled many positives, including (1) EDA focused on each step in the process to include the appropriate abstraction, e.g., a .lib to contain the timing information, or LEF/DEF to contain the physical information; (2) Ability to suitably model for each step as the quality of the collaterals have different level of detail, e.g., power estimation at RTL, gate, layout stages; and (3) Not least, the ability for humans, and now AI, to understand the whole flow and enable co-design at various levels, e.g., chip architects can assess crude area impact of their decisions very quickly.

This tutorial hopes to show that a similar approach for packaging and system design is the need of the hour during the heterogeneous integration revolution we are in today. The packaging electrical, thermal and mechanical analysis lacks a systematic abstraction approach and often gets tied up in details that make the estimation-design loops quite long. The proposed abstraction approach will also define key intercepts with the ASIC design methodology, so co-design between die/package/pcb can be realized for 2.5D/3D designs. These include (1) Chip partitioning; (2) IO standards and SIPI specification; (3) Floorplanning; and (4) System Signoff. The tutorial will also highlight benefit of abstraction for multiphysics signoff and AI/ML based methods.

Biography: Vaishnav Srinivas received the B.Tech. degree from IIT Madras, the M.S. degree from the University of California at Los Angeles, and the Ph.D. degree in electrical engineering from the University of California at San Diego. He is a Senior Director of Engineering with Qualcomm Technologies Inc., San Diego, CA, USA, where he leads a team working on electrical systems engineering for Qualcomm Technologies' chipsets, including circuit-system co-design for high-speed interfaces and mixed-signal circuits, signal and power integrity, PDN design, system validation and debug for interfaces and components. He leads a cross-functional circuit-system-technology exploration effort for next-generation interconnects, including memory, peripherals, inter-die, and intra-die interconnects. He has 35 US Patents and over 15 journal and conference publications.

Biography: Madhavan Swaminathan is the Department Head of Electrical Engineering and is the William E. Leonhard Endowed Chair at Penn State University. He also serves as the Director for the Center for Heterogeneous Integration of Micro Electronic Systems (CHIMES), an SRC JUMP 2.0 Center. Prior to joining Penn State University, he was the John Pippin Chair in Microsystems Packaging & Electromagnetics in the School of Electrical and Computer Engineering (ECE), Professor in ECE with a joint appointment in the School of Materials Science and Engineering (MSE), and Director of the 3D Systems Packaging Research Center (PRC), Georgia Tech (GT). Prior to GT, he was with IBM working on packaging for supercomputers. He is the author of 650+ refereed technical publications and holds 31 patents. He is the primary author and co-editor of 3 books and 5 book chapters, founder and co-founder of two start-up companies, and founder of the IEEE Conference on Electrical Design of Advanced Packaging and Systems (EDAPS), a premier conference sponsored by the IEEE Electronics Packaging Society (EPS). He is a Fellow of IEEE, Fellow of the National Academy of Inventors (NAI), Fellow of Asia-Pacific Artificial Intelligence Association (AAIA), and has served as the Distinguished Lecturer for the IEEE Electromagnetic Compatibility (EMC) society. He has been recognized through many awards with the most recent one being the 2024 IEEE Rao R. Tummala Electronics Packaging Award (highest award within the Electronics Packaging Society) for contributions to semiconductor packaging and system integration technologies that improve the performance, efficiency, and capabilities of electronic systems. He received his MS and PhD degrees in Electrical Engineering from Syracuse University in 1989 and 1991, respectively.

Packaging Technology for Next Generation mmWave Commmunications: Scalable Heterogeneous AiP Modules and the Future Role of Chiplets

Speaker: Dr. Atom Watanabe, IBM Research

Abstract: This talk cover the emerging packaging technology trends for mmWave communications with a focus on the design and integration methodologies for scalable phased array antenna modules. The presentation will first describe a heterogeneous integration strategy used to facilitate the effective integration of various active ICs, passive components, and decoupling capacitors into a substrate to form a 5G scalable phased array antenna module. The talk will also discuss the anticipated advantages of adopting a chiplet-based approach for digital baseband processing enabling the implementation of full end-to-end antennas-to-AI systems for the next generation of energy-efficient and adaptive mmWave networks.

Biography: Atom O. Watanabe is a Research Scientist who currently works at the IBM T. J. Watson Research Center as an IC packaging architect for advanced packaging and heterogeneous chiplet integrations. His current research interests and expertise include modeling and design, signal/power integrity analysis, and hardware characterization for millimeter-wave packages, co-packaged optics, high-performance computations, and quantum computers. Dr. Watanabe's contributions to the field include over 40 publications in renowned journals such as Applied Physics Letters, IEEE Transactions on Components, Packaging, and Manufacturing Technology, IEEE Transactions on Microwave Theory and Techniques, and IEEE Transactions Electromagnetic Compatibility, as well as numerous presentations at top-tier conferences like the Electronic Components and Technology Conference (ECTC) and Chiplet Summit. Furthermore, Dr. Watanabe has received multiple paper awards in recognition of his work. He has been serving on technical program committees for IEEE Microwave Theory and Technology Society and Chiplet Summit. He obtained a PhD in electrical and computer engineering from the Georgia Institute of Technology.

Program

Sunday, October 06

08:00 - 09:00 **Breakfast**

09:00 - 09:10 **Welcoming Remarks**   Room: W280
    Chairs: Piero Triverio

09:10 - 11:10 **Tutorials 1**   Room: W280
    Chairs: Stefano Grivet-Talocia, Bhyrav Mutnury

        09:10 - 10:10 **Signal and Power Integrity Analysis Using LIM – Recent Advances** (# s2)
        Author: Jose Schutt-Aine

        10:10 - 11:10 **Practical aspects of FD-TDR co-modification for high-speed structures' matching and what-if simulations** (# s6)
        Author: Pavel Vilner

11:10 - 11:30 **Coffee Break**

11:30 - 13:30 **Tutorials 2**   Room: W280
    Chairs: Stefano Grivet-Talocia, Bhyrav Mutnury

        11:30 - 12:30 **Superconducting Circuit Quantum Computers: Fundamental Concepts and Scaling Challenges** (# s5)
        Author: Thomas Roth

        12:30 - 13:30 **Chiplet Design and Heterogeneous Integration Packaging** (# s1)
        Author: John Lau

13:30 - 14:30 **Lunch Break**

14:30 - 16:30 **Tutorials 3**   Room: W280

Chairs: Stefano Grivet-Talocia, Bhyrav Mutnury

14:30 - 15:30 **Abstraction for Heterogeneous Integration** (# s4)
Authors: Vaishnav Srinivas, Madhavan Swaminathan

15:30 - 16:30 **Packaging Technology for Next Generation mmWave Commmu-nications: Scalable Heterogeneous AiP Modules and the Future Role of Chiplets** (# s3)
Author: Atom Watanabe

Monday, October 07

08:00 - 09:00 **Breakfast**

09:00 - 09:10 **Opening Ceremony**   Room: W280
    Chairs: Piero Triverio

09:10 - 10:10 **Keynote**   Room: W280
    Chairs: Piero Triverio

      09:10 - 10:10 **Accelerating AI with Chiplet Technology** (# s7)
      Author: Tony Chan Carusone

10:10 - 11:10 **Heterogeneous Integration I**   Room: W280
    Chairs: Jinwook Song, Jose Schutt-Aine

      10:10 - 10:30 **Multiphysics Simulation and Measurement Correlation of a Multichip Module IC Package Current Sensor** (# 19)
      Authors: Rajen Murugan, Jie Chen, Guangxu Li, Suzuki Yutaka, Sylvester Ankamah-Kusi

      10:30 - 10:50 **Yield-Aware Interposer Design for UCIe Interconnects** (# 42)
      Authors: Ram Krishna, Ashita Victor, Srujan Penta, Xu Chen, Muhannad Bakir, Nam Sung Kim, Elyse Rosenbaum

      10:50 - 11:10 **Efficient Thermal Analysis for Heat Dissipation in Three-Dimensional Chip-Stacking Packaging** (# 43)
      Authors: Cheng-Yuan Lu, Chien-Min Lin, Ruey-Beei Wu

11:10 - 11:30 **Coffee Break**

11:30 - 12:30 **Heterogeneous Integration II**   Room: W280
    Chairs: Jinwook Song, Jose Schutt-Aine

      11:30 - 11:50 **Eye-Diagram Edge Estimation (EEE) Network for Through Silicon Via Design in Next-Generation High Bandwidth Memory** (# 33)
      Authors: Hyunjun An, Junghyun Lee, Keeyoung Son, Seonguk Choi, Taein Shin, Keunwoo Kim, Jiwon Yoon, Taesoo Kim, Jungmin Ahn, Hyunah Park, Haeseok Suh, Joungho Kim

11:50 - 12:10 **Design and Analysis of High-Density Silicon Interposer Channel and Power Distribution Network** (# 46)
Authors: Haeyeon Kim, Joonsang Park, Hyunah Park, Keeyoung Son, Hyunsik Kim, Taeil Bae, Haekang Jung, Joungho Kim

12:10 - 12:30 **Optimization of TSV Array Based on Mathematical Model for HBM3** (# 11)
Authors: Yen-Tung Chen, Yu-Ying Cheng, Tzong-Lin Wu

12:30 - 14:00 **Lunch**

12:30 - 14:00 **TPC (private meeting)**   Room: W240

14:00 - 15:20 **Machine Learning I**   Room: W280
Chairs: Chris Cheng, Xu Chen

14:00 - 14:20 **A Hybrid Polynomial Chaos Expansion and Gaussian Process Regression Method for Forward Uncertainty Quantification of Integrated Circuits** (# 14)
Authors: Paolo Manfredi, Riccardo Trinchero

14:20 - 14:40 **Using Generative AI to Predict DC Electrical Performance** (# 28)
Author: Eric Bracken

14:40 - 15:00 **Transformer Based Channel Identification** (# 65)
Authors: Priyank Kashyap, Yeujiang Wen, Yongjin Choi, Chris Cheng, Paul Franzon

15:00 - 15:20 **Tree-Based Boosting for Efficient Estimation of S-Parameters for Package Electrical Analysis** (# 74)
Authors: Doganay Ozese, Mustafa Gökçe Baydoğan, Ahmet Durgun, Kemal Aygun

15:20 - 15:30 **Sponsor Demo: Amphenol**   Room: W280
Chairs: Chris Cheng, Xu Chen

15:30 - 15:50 **Coffee Break**

15:50 - 17:10 Device Modeling   Room: W280
Chairs: Daniel de Araujo, Joungho Kim

15:50 - 16:10 Tunable True-Time-Delay Unit Based on Bridged T-Coil (# 56)
Authors: Han-Ting Lin, Festim Iseini, Andreas Weisshaar

16:10 - 16:30 Modeling Multiplexed Qubit Readout with a Josephson Traveling-Wave Parametric Amplifier (# 16)
Authors: Samuel Elkin, Michael Haider, Thomas Roth

16:30 - 16:50 Latency Insertion Method for Fast Electro-Thermal Simulation of FinFET with Self-Heating Effect (# 32)
Authors: Yi Zhou, José Schutt-Ainé

16:50 - 17:10 Design and Analysis of L3 Cache Embedded-GPU-High Bandwidth Memory Architecture with Reduced Energy and Latency for AI Computing (# 22)
Authors: Haeseok Suh, Jiwon Yoon, Keeyoung Son, Seonguk Choi, Keunwoo Kim, Junghyun Lee, Taein Shin, Hyunjun An, Taesoo Kim, Jungmin Ahn, Hyunah Park, Hyunsik Kim, Taeil Bae, Haekang Jung, Joungho Kim

17:10 - 19:10 Welcome reception

17:10 - 19:10 Poster Session I   Room: Foyer
Chairs: Katharina Scharff, Andreas Weisshaar

Fan-out Region Crosstalk Optimization of High-Density PCIe 6.0 SMT Connectors (# 10)
Authors: Dan Liu, Yangfan Zhong, Minzheng Tian, Mengmeng Guo, Bing Wei, Weizhe Li, Jingbo Li, Tina Bao

A DDR5 Interposer De-embedding Method Based on Transfer Function (# 13)
Authors: Aobo Li, Jun Wang, Yan Xu, Kangkang Zhang, Xiuqin Chu

Parametric S-Parameter Prediction Using Deep Learning (# 15)
Authors: Lihong Feng, Vinayak Bansal, Valentin de la Rubia, Peter Benner

Comparative Evaluation of 100G-PAM4 Ethernet Link Performance in Air and Immersion Cooling Conditions (# 17)
Authors: Oluwafemi Akinwale, Dan Liu, Kai Wang, Yangfan Zhong, Cesar Mendez-Ruiz, Kusuma Matta

Cascading of 2D and 3D Simulations of ASIC Substrate Interconnect up to 100 GHz (# 18)
Authors: Zhekun Peng, Junyong Park, Sathvika Bandi, Santosh Pappu, Srinivas Venkataraman, Xu Wang, Granthana Rangaswamy, Donghyun Kim

Impact of Non-Functional Pads Location on Eye Diagram Performance (# 20)
Authors: Mehdi Mousavi, Kevin Cai, Junyong Park, Chaofeng Li, Reza Asadi, Shameem Ahmed, Bidyut Sen, Donghyun Kim, Manish K. Mathew

High-speed Interconnect Design of Silicon Interposer based Heterogeneous Integration for AI Computing (# 21)
Authors: Keeyoung Son, Seonguk Choi, Keunwoo Kim, Jiwon Yoon, Junghyun Lee, Haeseok Suh, Hyunjun An, Joungho Kim

Analysis of Nonlinear Phase Interactions of a Differential Line in the Presence of a Signal Skew (# 23)
Authors: Byung Cheol Min, Mun Ju Min, Hyun Chul Choi, Kang Wook Kim

A 155 MHz Low-Jitter PLL for Enhanced Signal Integrity in High-Speed Interconnects (# 25)
Authors: Mulat Ayinet Tiruye, Olani Baissa Gerba, T. Hui Teo

Single-Layer Wiring Design in UCIe to Realize Low-Cost Interposer Substrate (# 26)
Authors: Soshi Shimomura, Yutaka Uematsu, Katsuya Kikuchi, Haruo Shimamoto, Yuuki Araga, Shinichi Ouchi

PCIe Gen 6.0 SSD Receiver PAM 4 SI Analysis Based on End Port Time domain Measurements for Unknown System Channel (# 27)
Authors: Jinwook Song, Jinan Lee, Jonghee Jeong, Seokwoo Hong, Sungwoo Jin, Sungwon Roh, Taehyun Shim, Juneyoung Kim, Sunghoon Chun, Hyun-

woo Kim, Chorom Jang, Youngjun Ko, Dongho Choi, Kyungsuk Kim

Crosstalk Analysis in Add-In Card structure for High-Speed SerDes Channels with PCIe Gen6 (# 35)
Authors: Sungjin Yoon, Manho Lee, Kwangho Kim, Hyeongi Lee, Chulhee Cho, Youngjae Lee, Wooshin Choi, Young-Chul Cho, Jung-Hwan Choi, Young-Soo Sohn

Agile Analysis for Worst-Case Eye-Diagrams in Multi-Line Links of CoWoS Packaging (# 37)
Authors: Zhu-Chen Chang, Chien-Min Lin, Ruey-Beei Wu

A Signal Integrity Comparison of VIPPO Technology for PCIe 5.0 DC Blocking Capacitors (# 40)
Authors: Andrew Page, Matteo Cocchini

Analysis of Interconnects in Multilayer SIW Bandpass Filters Design (# 41)
Authors: Lu Qiu, Xiao-Wei Zhu, Xian-Long Yang

Design of an Ultra-High-Speed Digital Interface Based on a Coplanar Stripline (# 44)
Authors: Mun-Ju Kim, Byung-Cheol Min, Hyun-Chul Choi, Kang-Wook Kim

Signal Integrity Analysis of PCIe Channel with Floating Board-to-Board Connectors in Automotive Infotainment System (# 47)
Authors: Junghyun Lee, Keeyoung Son, Junho Park, Joonsang Park, Keunwoo Kim, Hyunjun An, Seonguk Choi, Jihun Kim, Hyunah Park, Sumi Choi, Sanghyuk Son, Joungho Kim

Tuesday, October 08

08:00 - 09:00 **Breakfast**

09:00 - 10:00 **Keynote**   Room: W280
    Chairs: Piero Triverio

        09:00 - 10:00 **Keynote 2 - Advanced Chiplet Package Signal Integrity for Future Data Center and AI** (# s8)
        Author: Dean Gonzales

10:00 - 11:00 **Macromodeling I**   Room: W280
    Chairs: Stefano Grivet-Talocia, Roni Khazaka

        10:00 - 10:20 **A Model Amongst us of the highest Order who can never be Replicated** (# s10)
        Author: Madhavan Swaminathan

        10:20 - 10:40 **Nonlinear macromodeling of voltage-regulated power delivery networks** (# 50)
        Authors: Antonio Carlucci, Stefano Grivet-Talocia

        10:40 - 11:00 **Operator Inference for Rigid-Flex Printed Circuit Boards Subject to Large Deformations** (# 39)
        Authors: Pascal den Boef, Wil Schilders, Joseph Maubach, Nathan van de Wouw, Diana Manvelyan

11:00 - 11:20 **Coffee Break**

11:20 - 12:40 **Macromodeling II**   Room: W280
    Chairs: Stefano Grivet-Talocia, Roni Khazaka

        11:20 - 11:40 **Gradient-based method to find solution for Rational Polynomial Chaos coefficients for Uncertainty Quantification** (# 83)
        Authors: Karanvir Singh Sidhu, Roni Khazaka

        11:40 - 12:00 **Automated Accurate Quadratic Formulation of Nonlinear Circuits** (# 34)
        Authors: Germin Ghaly, Emad Gad, Michel Nakhla

12:00 - 12:20 Electrothermal Co-Design Modeling and Analysis of an Ultra-Low On-Resistance Power Switch (# 6)
Authors: Sylvester Ankamah-Kusi, Blake Travis, Swathi Kamath, Rajen Murugan, Tom Kronenberg

12:20 - 12:40 Modeling Microwave S-parameters using Frequency-scaled Rational Gaussian Process Kernels (# 58)
Authors: Thijs Ullrick, Dirk Deschrijver, Wim Bogaerts, Tom Dhaene

12:40 - 14:00 **Lunch**

14:00 - 15:00 **Computational Electromagnetics**   Room: W280
Chairs: Shashwat Sharma, Vladimir Okhmatovski

14:00 - 14:20 On the Parallelization of the MultiAIM Algorithm for the Fast Electromagnetic Analysis of 3D ICs (# 84)
Authors: Yongzhong Li, Piero Triverio

14:20 - 14:40 Accuracy Study of the Differential Surface Admittance Operator for Lossy Metal Characterization (# 70)
Authors: Martijn Huynen, Vladimir Okhmatovski, Daniël De Zutter, Dries Vande Ginste

14:40 - 15:00 A Highly-Scalable Parallel Boundary Element Method for the Full-Wave Electromagnetic Analysis of Large Interconnect Networks and Entire Packages (# 82)
Authors: Damian Marek, Jasper Hatton, Yongzhong Li, Piero Triverio

15:00 - 15:10 **EPEPS 2025 Announcement**   Room: W280
Chairs: Wendem Beyene

15:10 - 15:30 **Coffee Break**

15:30 - 17:10 **Signal and Power Integrity I**   Room: W280
Chairs: Jose Hejase, Ram Achar

15:30 - 15:50 Analysis of Echo and Crosstalk Cancellation in Simultaneous Bidi-

rectional Transceivers for Dense Die-to-Die Interconnects (# 62)
Authors: Tong Liu, Taeyang Sim, Samuel Palermo

15:50 - 16:10 **Efficient parametric assessment of worst-case voltage droop in power delivery networks** (# 51)
Authors: Tommaso Bradde, Antonio Carlucci, Riccardo Trinchero, Paolo Manfredi, Stefano Grivet-Talocia

16:10 - 16:30 **PSIJ based Optimal PDN Design for Cost-Effective SSD using Reinforcement Learning** (# 59)
Authors: Taein Shin, Seonguk Choi, Jungmin Ahn, Keunwoo Kim, Junghyun Lee, Haeseok Suh, Hyunah Park, Haeyeon Kim, Hyunjun An, Jinwook Song, Joungho Kim

16:30 - 16:50 **PCIe Gen 6.0 SSD PSIJ Estimation Based on Early Design Stage Jitter Sensitivity Measurements** (# 29)
Authors: Youngjun Ko, Jinwook Song, Seokwoo Hong, Jinan Lee, Jonghee Jeong, Hyunwoo Kim, Chorom Jang, Sungwoo Jin, Sungwon Roh, Dongho Choi, Kyungsuk Kim, Sunghoon Chun

16:50 - 17:10 **Application of CAMM2 Connector on PCIe Gen 6.0 SSD Host Interface for Low Near-End Crosstalk** (# 30)
Authors: Sungwoo Jin, Jinwook Song, Seokwoo Hong, Youngjun Ko, Hyunwoo Kim, Sungwon Roh, Chorom Jang, Dongho Choi, Kyungsuk Kim, Sunghoon Chun

15:30 - 16:30 **Ansys Training Program**   Room: W240

15:30 - 16:00 **Electro-Thermal-Mechanical Design Workflow for Printed Circuit Boards and Electronic Packages** (# s12)
Author: Satyajeet Padhi

16:00 - 16:30 **Ensuring the Power and Signal Integrity of Your High-Performance PCBs and Packages Using Ansys SIwave** (# s13)
Author: Laila Salman

17:10 - 18:20 **Poster Session II**   Room: Foyer

Chairs: Damian Marek, Wendem Beyene

Design and Analysis of Extended Scale Cache (ESC) Stacked-GPU-HBM Module Architecture Considering Power Integrity (PI) (# 48)
Authors: Hyunah Park, Seonguk Choi, Haeyeon Kim, Taein Shin, Keeyoung Son, Jiwon Yoon, Junghyun Lee, Haeseok Suh, Taesoo Kim, Jungmin Ahn, Hyunjun An, Joungho Kim

Explainable Reinforcement Learning(XRL)-based Decap Placement Optimization for High-Bandwidth Memory (HBM) (# 49)
Authors: Keunwoo Kim, Hyunwook Park, Keeyoung Son, Seonguk Choi, Taein Shin, Junghyun Lee, Jiwon Yoon, Hyunjun An, Haeyeon Kim, Wooshin Choi, Jung-Hwan Choi, Joungho Kim

Limit of the Impact of the Via Stub Length on the Via Impedance in Printed Circuit Boards (# 52)
Authors: Katharina Scharff, Xiaomin Duan, Dierk Kaller

Improve CLK Phase Noise Performance by Mitigating Antiresonance Phenomenon of Power Net with a Pi-Type Filtering Structure (# 54)
Authors: Xinlin Tang, Shuxiang Li, Tao Fang, Yuan Fang

A Study on How Capacitance of Power Filtering Circuit Influences the Antiresonance Frequency (# 60)
Authors: Shuxiang Li, Xinlin Tang, Tao Fang, Yuan Fang, Greg Fu, Stephen Scearce

Simulation method for Quasi-static solver to effectively model parasitic components between Package and PCB (# 61)
Authors: Silvia Simone, Fabio Pareschi, Davide Lena, Gianluca Setti

Design and Analysis of Ultra High Bandwidth (UHB) Interconnection-based GPU-Ring for the AI Superchip Module (# 66)
Authors: Jungmin Ahn, Seonguk Choi, Taein Shin, Junghyun Lee, Jiwon Yoon, Keunwoo Kim, Keeyoung Son, Haeseok Suh, Taesoo Kim, Hyunah Park, Hyunjun An, Jinwook Song, Joungho Kim

Recent Advances in Signal Integrity Simulation and Analysis of Interposers (# 68)
Authors: Jonatan Aronsson, Feng Ling

Multiphysics-Informed ML-Assisted Chiplet Floorplanning for Heterogeneous Integration (# 69)
Authors: Vinicius C. Do Nascimento, Seunghyun Hwang, Michael Smith, Qiang Qiu, Cheng-Kok Koh, Ganesh Subbarayan, Dan Jiao

Compact Fiber Weave Model for Full Wave Solvers (# 71)
Authors: Stefan de Araujo, Daniel de Araujo, Bhyrav Mutnury

Megtron 6 and 8 Characterization Methodology (# 72)
Authors: Stefan de Araujo, Daniel de Araujo, Roger Delbue, Ryan Keegan

Full-Wave Analysis for Ground Via Placement with Layered Media Integral Equations (# 73)
Authors: Alireza Niazi, Vladimir Okhmatovski

Causal RL Prediction of Fine-Pitch Interconnects Using Neural Networks (# 75)
Authors: Hasan Said Unal, Ahmet Cemal Durgun

A robust optimization approach for High Bandwidth Memory interposer using Machine Learning (# 76)
Authors: Anandajith Jinesh, Xuan Chen

Analysis and Modeling of Controlled Silicon Substrate Roughness for Silver-Based Backside Metallization in Power Electronics Packaging (# 78)
Authors: Mohamed Bellaredj, Goran Miskovic, Luka Vojkuvka

Worst-Case Voltage Droop Using Peak Distortion Analysis (# 80)
Authors: Mohamed Sahouli, Isaac Ali, David Reinamendivil, Gerry Talbot

Equalization Techniques for Time Domain Signalling (# 81)
Authors: Shakib Mahmood, Parneet Tethy, Richelle L. Smith, Carl W. Werner,

Masum Hossain

18:20 - 18:30 **Bus pickup**   Room: Foyer

19:00 - 19:15 **Check in at the CN Tower**   Room: CN Tower

19:15 - 20:30 **Cocktail Reception and Awards Ceremony**   Room: CN Tower

20:30 - 22:00 **Conference Dinner**   Room: CN Tower

Wednesday, October 09

08:00 - 09:00 **Breakfast**

09:00 - 10:00 **Invited Talk**   Room: W280
    Chairs: Xu Chen

        09:00 - 10:00 **Machine Learning for EDA, or EDA for Machine Learning?** (# s9)
           Author: Zheng Zhang

10:00 - 11:00 **Machine Learning II**   Room: W280
    Chairs: Paolo Manfredi, Zheng Zhang

        10:00 - 10:20 **Reinforcement Learning Based Automatic Router for Power Delivery Network Prototypes** (# 53)
           Authors: Felix Yuan, Abinash Roy

        10:20 - 10:40 **An Efficient Machine Learning Approach for PSIJ Analysis in a Chain of CMOS Inverters** (# 57)
           Authors: Ahsan Javaid, Ramachandra Achar, Jai Tripathi

        10:40 - 11:00 **Hand-drawn Circuit Schematic Digitization and Netlisting using Machine Learning with Emphasis on Signal Integrity Applications** (# 77)
           Authors: Anuj Mathur, Ramachandra Achar

11:00 - 11:10 **Sponsor Demo: Keysight**   Room: W280
    Chairs: Paolo Manfredi, Zheng Zhang

11:10 - 11:30 **Coffee Break**

11:30 - 12:10 **Optics**   Room: W280
    Chairs: Andreas Weisshaar, Xuan Chen

        11:30 - 11:50 **A Tunable Inductor Peaking Technique for Optical Communication Systems** (# 55)
           Authors: Festim Iseini, Han-Ting Lin, Nicola Pelagalli, Andrea Malignaggi, Corrado Carta, Gerhard Kahmen, Andreas Weisshaar

        11:50 - 12:10 **Signal Integrity of Die-to-Die Interface with Advanced Packages**

for Co-Packaged Optics (# 8)
Authors: Jongchul Shin, Hamid Eslampour, Sangnam Jeong, Woopoung Kim, Seokbeom Yong, Sung-Oh Ahn, Eunkyeong Park, Sangsub Song

11:30 - 12:10 **Keysight Training Program**　　　　　　　　　　Room: W240
Chairs: Heidi Barnes

11:30 - 12:10 **Digital Twin PI Simulations for 2000 Amp AI, Cloud Compute, and Multi-Die Packages** (# s11)
Author: Heidi Barnes

12:10 - 12:50 **Computer Aided Design**　　　　　　　　　　Room: W280
Chairs: Ram Achar, Tommaso Bradde

12:10 - 12:30 **Delay Rational Macromodelling of Noisy Tabulated Frequency Responses** (# 9)
Authors: Alexander Kirchberger, Anestis Dounavis

12:30 - 12:50 **An Efficient SPICE-compatible Model for Fast Co-simulation of Signal and Power Integrity on Multilayer PCB with Arbitrary Shape** (# 31)
Authors: Hyunwoo Kim, Dongryul Park, Seunghun Ryu, Seonghi Lee, Sanguk Lee, Jinwook Lee, Dongkyun Kim, Seungyoung Ahn

12:50 - 14:20 **Lunch**

14:20 - 15:20 **Signal and Power Integrity II**　　　　　　　　　　Room: W280
Chairs: Tawfik Rahal-Arabi, Heidi Barnes

14:20 - 14:40 **Application of the Reverse Pulse Technique for Worst Case Transient Analysis in HPC PDN Design** (# 12)
Authors: Chad Smutzer, Jordan Keuseman, Alexander Hickman, Clifton Haider

14:40 - 15:00 **Optimizing Power and Power Delivery For Data Center GPUs** (# 3)
Authors: Tawfik Rahal-Arabi, Paul Van der Arend, Ashish Jain, Mehdi Saidi, Rashad Oreifej, Sriram Sundaram, Rajit Seahra

15:00 - 15:20 **8.4GTS LPDDR5x and 5.6GTS DDR5 Combo PHY** (# 7)

Authors: Nasirul Chowdhury, Jon Guerber, Yin-Huan Hwang, Takao Oshita, Anindya Saha, Allen Waters

15:20 - 15:40 **Closing Ceremony**   Room: W280
Chairs: Piero Triverio

Optimizing Power and Power Delivery For Data Center GPUs

Tawfik Rahal-Arabi, Paul Van der Arend, Ashish Jain, Mehdi Saidi, Rashad Oreifej, Sriram Sundaram, Rajit Seahra

Advanced Micro Devices

Abstract—**GPUs are used in products from ultra-low power mobile devices to high performance machine learning accelerators in data centers. Across the products, power and power delivery have become top limiters to performance and are key considerations in the early stages of product definition and design. In particular, the power and power delivery problem has been significantly exacerbated with the recent trends in the growth of AI workloads. In this paper, we present some of the data centers driven power optimizations used in latest generation of AMD GPUs including the recently announced AMD Instinct™ MI300 GPU. To this end, we cover power and power delivery optimization techniques spanning the product life cycle from architecture, physical design, validation, test, and manufacturing with an eye on the challenges in the road ahead.**

I. INTRODUCTION

The recent growth of large language models (LLMs) has put to question, once again, the ability of data centers to cope with electricity demand. The rise of AI has led to an increase in data center deployments with power emerging as the main bottleneck [1]. While there is a range of predictions on electricity demand for AI and potential efficiency improvements, [3] and [4] to name a few, there is little doubt that electricity consumption will continue to increase due to the performance demands of AI workloads. In order to meet the projected AI growth, it is important to keep GPU power in check, as it dominates AI compute power. In this paper, we present techniques used in design to save average power, peak power, and minimize $\frac{\delta i}{\delta t}$. We then show how we saved power leveraging AI algorithms with EDA tools. In post-silicon, we show optimizations that yielded significant savings and, finally, conclude the paper by highlighting future challenges and opportunities.

II. ASSISTING POWER DELIVERY WITH ANALOG CIRCUITS

GPUs used for machine learning applications have a significantly higher power budget than other components in the data center. Along with the large power supply and associated large current requirements, given that CMOS systems in the latest technology node operate at sub 1V voltage levels, there is the additional burden of dealing with supply noise associated with power delivery. Without supply noise mitigation, systems would have to include large voltage margins to ensure correct operation. The solutions to power supply noise generally

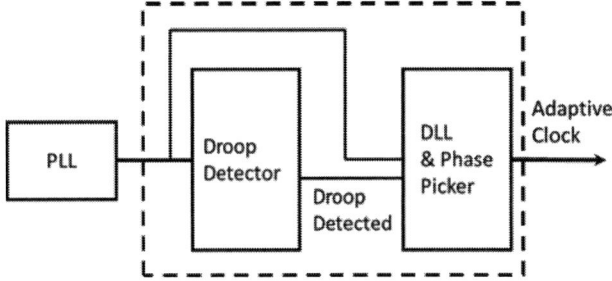

Fig. 1. Adaptive clocking concept

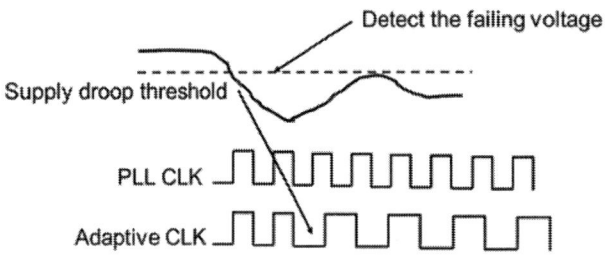

Fig. 2. Adaptive clocking resonse to droop events.

fall in two categories: *suppression* and *mitigation*. The suppression category relies on a using capacitance to dampen the current swings in the system. As shown in [5], these include a balanced decoupling network with metal-insulator-metal (MIM) capacitors or oxide gate capacitors at the die level, high frequency decoupling capacitors at the package level, and bulk capacitors at the board level. The relationships between each stage of the decoupling have been highlighted in several papers, [2] and [5] to mention a few. As outlined in [2] and [5] it is not physically possible to eliminate all voltage droops. Hence, several mitigation techniques such as the one in [6], were developed as early as 2005 to deal with high $\frac{\delta i}{\delta t}$. AMD GPUs use a form of adaptive clocking that modulates clock frequency in response to voltage droop events such that device timing is met at the lowest voltage observed during these droop events. This mitigates the need to add voltage margin to account for these voltage droop events, resulting in a more efficient voltage-frequency (V-F) curve and improved performance/watt. Figure 1 and Figure 2 show an implementation of adaptive clocking. Because large droop events are rare, the performance loss is, on average, tolerable.

979-8-3503-5124-8/24 $31.00 © 2024 IEEE

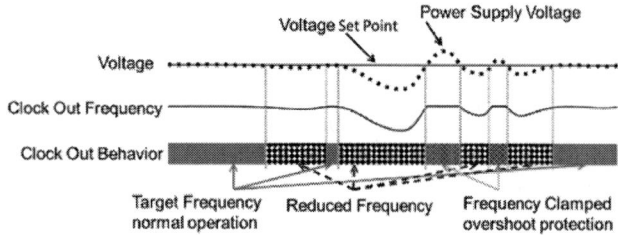

Fig. 3. Clock stretching proportionally to voltage droops.

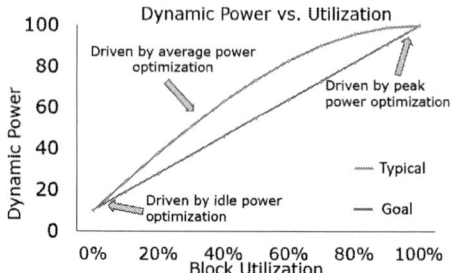

Fig. 4. Dynamic capacitance (CAC) versus utilization.

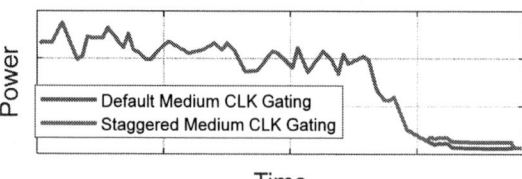

Fig. 5. Staggering clock gating to improve overshoot and undershoot of the power delivery network (PDN).

III. DYNAMIC POWER OPTIMIZATIONS AND POWER DELIVERY TRADE-OFFS

The foundation of low-power design is in the design of the RTL to reduce dynamic switching capacitance. This takes the form of clocks and/or data gating when the data is not being sampled. The relationship between a GPU block utilization, which is a measure of the number of gates switching in that block, and the dynamic switching capacitance (CAC) should be, ideally, linear as shown in Figure 4. The lowest point on the curve is when no work is done (idle power) and the switching capacitance should be zero. Minimizing power at the 100% utilization reduces peak power and $\frac{\delta i}{\delta t}$. Our goal in the RTL stage is to minimize the distance between the ideal curve and actual curve for the workloads of interest across a range of utilizations. To achieve the RTL goal, gating is done at fine grain (flop level), medium grain (several flops), or coarse grain (an entire unit). Gating clocks, however, may have the adverse effect of increasing power supply noise, so a balanced approach must be taken to trade off average power savings with clock gating and worsening $\frac{\delta i}{\delta t}$. To highlight this trade-off, we show two examples. In the first example, we staggered the medium clock gating such that the gating is spread over dozens of clocks instead of a single clock. As shown in Figure 5, the power reaches idle state a little later than it does with default clock gating and raises the average idle power by a small amount. On the other hand, Figure 6 shows the impact on voltage overshoot and undershoot in the event of a large $\frac{\delta i}{\delta t}$ with and without the clock gating staggering. In the case of the non-staggered clock gating, since we do not know when these large $\frac{\delta i}{\delta t}$ events happen, we would need to compensate for the voltage undershoot by raising the voltage to the GPU by the difference between the non-staggered and staggered curves of Figure 6 and, consequently, increase the idle, average, and peak power far more than the increase in power due to the staggering of the clock gating. It is clear from this analysis that staggering the clock gating has a good return on investment (ROI). In the second example, we disable the fine-grain clock gating. This has a large impact on average power due to the pervasiveness of the fine-grain gating. Surprisingly, the undershoot for this particular workload improved with the fine grain clock gating, so this is a win-win situation where there is no trade-off between average power and $\frac{\delta i}{\delta t}$. This is because fine-grained clock gating is spatially and temporally well distributed versus medium-grained clock gating, where

the gating is confined to a small spatial and temporal space. These are just two examples of trade-offs between average power and $\frac{\delta i}{\delta t}$, but this trade-off is considered in the analysis of every important feature in the product.

IV. LOW POWER AUTOMATION

In addition to RTL and design techniques, dynamic power is optimized through a tight focus on EDA tool and internal CAD flows. We have measured power improvements over the past two years, using four fixed designs with a constant technology environment, and show normalized results in Figure 7, with 10% savings in dynamic power. These savings are achieved through a holistic examination of the synthesis, placement, clocking, and routing phases of the flow. In synthesis, we have focused on improved logic structuring, ranging from early-stage logic optimization of large structures through to gate level Boolean re-mapping, with the primary goal of minimizing switching on high activity nodes. Power is lowered in placement through minimizing the capacitance of high activity logic by reducing wire length. Forcing high activity logic to have lower wire length can increase wire length for other circuits, potentially causing timing and congestion issues. We have minimized these side effects, allowing us to increase the effort spent to optimize power critical logic while meeting other goals. For clocking, power is minimized by building a clock distribution network that has the highest global skew and transition requirements necessary to achieve timing. By relaxing these constraints, we can build a more

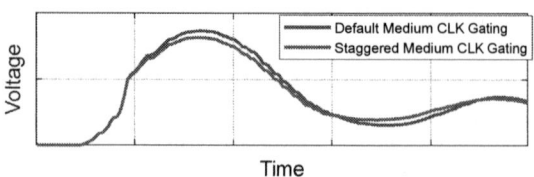

Fig. 6. Overshoot and undershoot with and without the clock staggering.

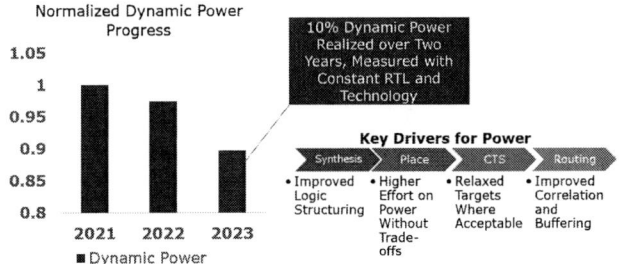

Fig. 7. Reduction in dynamic power with CAD tool improvements.

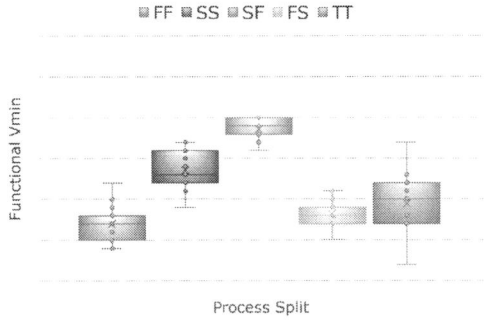

Fig. 8. Functional Vmin distribution across process.

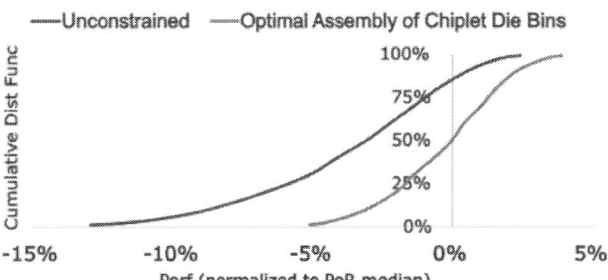

Fig. 9. Performance optimization and variation tightening via novel chiplet binning and assembly in large multi-chiplet GPU.

efficient distribution network, measured by both transistor count and wire length. Finally, routing of the flow is optimized through a focus on correlation and buffering techniques.

V. POST SILICON OPTIMIZATIONS

A. Per Part V_{min}

Large dynamic range of operating voltage is very desirable for GPU design to provide efficient performance across wide range of workloads. For example, in a typical training workload, all GPUs tend to be thermal power (TDP) limited and operate at the lowest possible functional voltage (V_{min}), while in an inference workloads, some GPUs may utilize more power than other GPUs, and consequently operate at significantly higher voltage than V_{min}. To enable such wide dynamic voltage range, physical design of the graphics core tiles is typically optimized for multi-voltage corners. This strategy allows for optimal trade-offs with threshold voltage (V_t), leakage, and graceful frequency scaling with operating voltage. V_{min}, however, varies across parts as shown in Figure 8. Instead of assuming a worst case V_{min} over the full distribution, parts are individually programmed to operate at their specific V_{min} for optimal performance in their operating conditions.

B. Chiplet Optimizations

Chiplet and advanced packaging technologies have enabled AMD to build large and complex data center GPU accelerators at high yields. A component of parallel processing, typical of GPUs, is referred to as barrier synchronization, where GPUs must wait until all other GPUs have also reached the barrier. Accordingly, a system that includes multiple chiplets/GPUs

working together will be performance constrained by the performance of the slowest chiplet/GPU. To get around this issue, AMD GPUs use power management firmware to non-uniformly distribute the GPU power budget amongst the different chiplets to minimize the performance difference between the chiplets. For example, a chiplet which intrinsically consumes lower power at given frequency is given less power budget as compared to a chiplet that intrinsically requires higher power at the same frequency. In this fashion, for the same total package power budget, the intrinsic performance variation between chiplets is minimized. In addition, product test and binning are further enhanced with assembly instruction to optimally choose, which chiplets, based on their power and performance characteristics, go on a given package, to achieve the maximum performance for the package. Indeed, Figure 9 shows that by using the chiplet binning to tighten the variations across the manufacturing process, significant improvement in performance is achieved.

VI. SUMMARY AND CONCLUSIONS

In this paper, we covered design optimizations to lower power and mitigate the impact of power supply noise on performance for our modern GPUs, including the recently announced AMD Instinct™ MI300 GPU. In post-silicon, we showed how we leveraged our advanced packaging to further improve power and performance of our GPUs.

REFERENCES

[1] I. E. Agency. (2022) Global data centre energy demand by data centre type. [Online]. Available: https://www.iea.org/data-and-statistics/charts/global-data-centre-energy-demand-by-data-centre-type

[2] G. Ji, T. Arabi, and G. Taylor, "Design and validation of a power supply noise reduction technique," *IEEE Transactions on Advanced Packaging*, vol. 28, no. 3, pp. 445–448, 2005.

[3] J. Koomey and S. Naffziger, "Energy efficiency of computing: What's next?" *Electronic Design*, 2016. [Online]. Available: http://electronicdesign.com/microprocessors/energy-efficiency-computing-what-s-next

[4] E. Masanet, A. Shehabi, and J. Koomey, "Characteristics of low-carbon data centres," *Nature Clim Change 3*, pp. 627–630, 2013. [Online]. Available: https://doi.org/10.1038/nclimate1786

[5] A. Waizman and C.-Y. Chung, "Extended adaptive voltage positioning (eavp) [power delivery network design]," in *IEEE 9th Topical Meeting on Electrical Performance of Electronic Packaging (Cat. No.00TH8524)*, 2000, pp. 65–68.

[6] K. Wong, T. Rahal-Arabi, M. Ma, and G. Taylor, "Enhancing microprocessor immunity to power supply noise with clock-data compensation," *IEEE Journal of Solid-State Circuits*, vol. 41, no. 4, pp. 749–758, 2006.

Electrothermal Co-Design Modeling and Analysis of an Ultra-Low On-Resistance Power Switch

Sylvester Ankamah-Kusi, Blake Travis, Swathi Kamath, Rajen Murugan, Tom Kronenberg
Texas Instruments, Inc.
Dallas, TX, USA
s-ankamah-kusi@ti.com

Abstract—A metal-oxide-semiconductor field-effect transistor (MOSFET) power switch is an electronic device that acts as a voltage-controlled current source. One critical aspect of the power switch's performance is its ability to dissipate heat into the environment efficiently. However, optimal dissipation is influenced by the combined effect of the device's electrical and thermal behavior. While standard methods such as increasing die and package sizes, using heat sinks, and employing top-side and immersion cooling techniques have improved heat dissipation, they mainly focus on the device's thermal behavior with little consideration for the electrical aspect. Additionally, these approaches come with added costs, making them less feasible for high-volume applications. This paper details a coupled electrothermal modeling, analysis, and silicon-package physical co-design optimization of an ultra-low on-resistance MOSFET power switch integrated circuit. Achieving optimal performance at the lowest cost is demonstrated successfully here by implementing the coupled modeling methodology early in the design phase of the power switch.

Keywords—Electrothermal, heat sink, immersion cooling, integrated circuit, package footprint

I. INTRODUCTION

Efficient heat dissipation in semiconductor devices refers to the ability of the device to swiftly dissipate heat into the surrounding environment, thereby reducing the maximum junction temperature. The junction temperature of a device is given by equation (1) below:

$$T_j = T_a + \theta_{ja} * I^2 * R_{dson} \qquad (1)$$

where T_j is the junction temperature, T_a is the ambient temperature, θ_{ja} is the thermal resistance, I is the operating current, R_{dson} is the electrical on-resistance. Equation (1) underscores the interplay between thermal resistance and electrical resistance in determining the junction temperature. Consequently, an electrothermal approach becomes imperative for optimizing the overall thermal efficiency of the device.

The active circuitry (i.e. power switch) is the main heat generation component of the semiconductor device, with the bulk silicon, package, and the printed circuit board (PCB) being the main heat dissipation components as shown in Fig. 1 below.

Several research efforts have focused on enhancing the thermal properties of packaging materials. Wei et al. introduced a high thermal conductivity die-attach epoxy that achieved a thermal conductivity of 110 W/m-K [1]. Shibuya et al. achieved a thermal conductivity of 8.7 W/m-K, significantly higher than 1W/m-K for traditional silica-filled epoxy mold compound by integrating graphene rather than silica fillers in the mold compound [2].

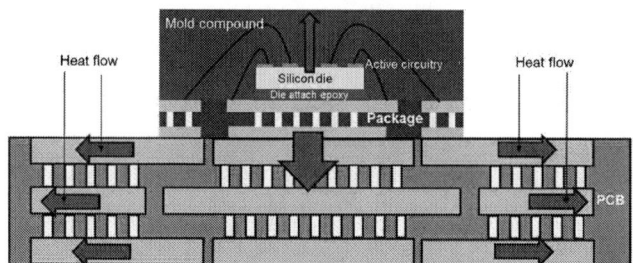

Fig. 1. Heat dissipation within a semiconductor device.

Beyond integrating advanced thermal materials, alternative techniques, such as exposed die with top-side and immersion cooling, have been explored to address heat dissipation challenges. However, implementing these methods incurs additional costs, rendering them less desirable, especially when striving to achieve a low-cost target.

This paper develops and implements a coupled multiphysics and system co-design modeling and analysis methodology to design an ultra-low MOSFET power switch device package. Section II provides key functional details of the power switch device. Section III details the coupled electrothermal and co-design modeling flow. Section IV discusses the coupled optimization of modeling and physical co-design, which provides an alternative to the high cost of current solutions.

II. POWER SWITCH DEVICE DETAILS

The device under test (DUT) being considered here is an ultra-low on-resistance power switch integrated into an eFuse, designed to regulate current flow to a designated load. Fig. 2 below shows a simplified functional block diagram. Operational control is initiated when both the power supply (VDD) and input signal (IN) pins register as high. At this point, the device monitors the Enable (EN) pin; upon detection of a high signal on EN, the switch begins conduction, controlled by the gate driver. This enables the controlled flow of current from the input to the output. Conversely, a low signal on EN prompts the switch to deactivate, interrupting the current flow. Crucially, a built-in over-temperature protection mechanism intervenes when the

979-8-3503-5124-8/24 $31.00 © 2024 IEEE

junction temperature surpasses predefined limits, leading to an automatic shutdown of the switch.

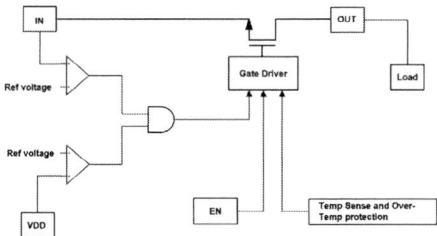

Fig. 2. Simplified function block diagram of the power switch.

Due to the high load currents, a key objective of this design is to optimize both the electrical and thermal resistances of the power switch, to maintain a low junction temperature for the device, with a focus on the electromagnetic interaction between the silicon and package. Fig. 3 below shows the package design for the power switch. The device is packaged in a 5mm x 5mm substrate-based packaging technology.

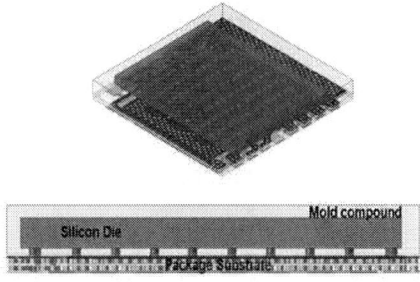

Fig. 3. Power Switch package drawing. Top and cross-sectional views.

Here we focus our efforts on the design of the input and output nets of the power switch, since these are the major current carrying pins. Fig. 4. shows the input and output pin assignment on the package top trace and footprint.

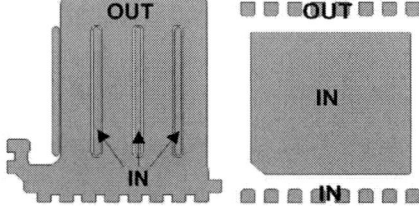

Fig. 4. Package top trace and footprint design.

III. ELECTROTHERMAL CO-DESIGN MODELING FLOW

A predictive electrothermal co-design modeling methodology was developed to assess the overall performance of the device (silicon + package) using a combination of a 3D resistance extraction solver and a CFD thermal solver. The methodology was validated successfully on multiple devices. More details on the implementations of the methodology are discussed in references [3-5]. The general steps involved in the developed methodology are shown in the modeling flow chart on Fig. 5.

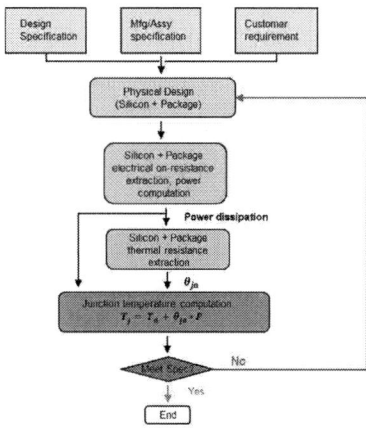

Fig. 5. Die-Package electrothermal co-design modeling methodology.

The flow is initiated from the initial physical design of the die and package. The electrical on-resistance of the design is extracted and the corresponding power dissipation is computed. The power is fed as an input to the CFD thermal solver to model and extract the thermal resistance of the device. Having obtained the dissipated power and the thermal resistance, the junction temperature of the device is computed using equation (1). This flow is repeated iteratively, with silicon and package re-designs until the junction temperature does not exceed the target temperature, at which point the flow is exited.

In the next section, we discuss the design optimization techniques used to improve the electrothermal performance of the power switch.

IV. ELECTRO-THERMAL CO-DESIGN OPTIMIZATION

To begin with, we discuss the challenges associated with the baseline design. The footprint of the package consists of a thermal pad, and 2 sets of signal pins, as shown in Fig. 4 (left picture). As shown, the input net of the power switch is connected to both the thermal pad and the bottom row of signal pins, while the output net is connected to the top row of signal pins. As such, the thermal pad serves as a huge current sink for the input net, providing enough area for the current to flow through. Consequently, all the current from the input net flows through the thermal pad, while very little to no current flows through the input net signal pins at the bottom of the package. This results in vertical and uniform flow of current. However, in the output net, current takes a lateral path from silicon, through the package layers to the output net signal pins. The current density distribution in both nets is shown in Fig. 6 below.

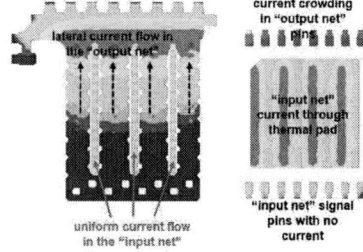

Fig. 6. Simulation of current density distribution within the package.

979-8-3503-5124-8/24 $31.00 © 2024 IEEE

While the vertical and uniform current flow within the input net results in low electrical on-resistance, the lateral flow of current in the output net leads to current crowding within the package, and consequently an increase in the electrical on-resistance. The simulated electrical on-resistance, power dissipation, thermal resistance, and junction temperature at ambient temperature is shown in Table 1.

To reduce the electrical on-resistance, an "inter-leaved" package design scheme is created. In this design, the thermal pad of the package is split up between the input net and the output net in an alternating fashion, as shown in Fig. 7(a).

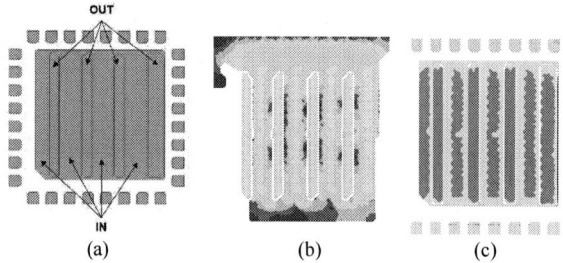

Fig. 7. (a) Inter-leaved package design, (b) input and (c) output current density distribution.

With the inter-leaved design, the current path in both the input and output nets is now vertical and uniform, with no lateral flow of current, as shown in Fig. 7(b) and 7(c) respectively. With the elimination of current crowding, a 28% reduction in both on-resistance and power dissipation is observed (see Table 1). However, an increase in thermal resistance (27% from baseline) is observed due to the reduction in the thermal conductive area from implementing the interleaving scheme.

To achieve a more optimal solution, an innovative re-assignment of the signal pins of the input net to the output net, as shown in Fig. 8 (a) was implemented.

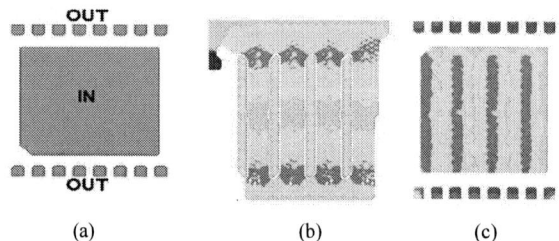

Fig. 8. (a) Package design with pin re-assignment, (b) input, and (c) output current density distribution.

As already highlighted, in the baseline design, since the thermal pad is assigned to the input net, most of the current from the input net flows through the thermal pad, while the signal pins assigned to the input net are left unutilized. By re-assigning these signal pins to the output net, we create a parallel path for current flow in the output net, where half the current flows laterally towards the top row of output net signal pins, while the other half flows laterally towards the bottom row of output net signal pins, as shown in Fig. 8(b) and 8(c) respectively. This reduces the amount of current crowding, and consequently the on-resistance, since all the current is not drifting in one

direction. With the pin re-assignment scheme, a 14% lower on-resistance and power dissipation are achieved as compared to the baseline design.

Table 1 below provides a comparison of all 3 designs. We can observe that while the "inter-leaved" design has the best on-resistance, it does not show good thermal resistance. Similarly, while the baseline design has the best thermal resistance, it has the worst on-resistance.

TABLE1: ELECTROTHERMAL COMPARISON OF THE OPTIMIZATION SCHEMES.

Parameter	Baseline	Inter-leaved	Pins re-assigned
On-res (mΩ)	0.9	0.65	0.77
Power (W)	2.72	1.97	2.33
θ_{ja} (°C/W)	23.9	29.2	23.0
Tj (°C)	87.54	82.44	80.57

Using equation (1), we are able to determine that the design with pins re-assigned yields the best electro-thermal performance, optimizing the on-resistance and thermal resistance to attain the lowest junction temperature.

V. CONCLUSION

The efficient dissipation of heat in a semiconductor device relies on its electrical and thermal performance. An electrothermal approach is necessary to optimize semiconductor devices for effective heat dissipation. This paper outlines an electrothermal co-design methodology utilized to improve the overall performance of semiconductor devices. Through modeling, we have showcased how these design techniques offer a cost-effective solution for enhancing the electrothermal performance of an ultra-low on-resistance power switch.

REFERENCES

[1] Dr. Wei Yao, Raj Peddi, Kily Wu, Hoseung Yoo, "High Thermal Conductive Semi-Sintering Die Attach Paste", 2018 IEEE 68th Electronic Components and Technology Conference

[2] M. Shibuya, L. Nguyen, "High Thermal Conductivity Mold Compounds for Advanced Packaging Applications", 2017 IEEE 67th Electronic Components and Technology Conference.

[3] Rajen Murugan, Nathan Ai, and C.T. Kao, "System-Level Electro-Thermal Analysis of RDS(ON) for Power MOSFET", IEEE SEMI-THERM33, March 13-17, 2017.

[4] Jie Chen, Rajen Murugan, Steven Kummerl, Usman Chaudhry, Edwin Lim, Tatsuhiro Shimizu, Thatcher Klumpp, and Jack Grantham (2018) System ElectroThermal Co-Design of a Zero-Drift Current-Shunt Monitor with Precision Integrated Shunt Resistor. International Symposium on Microelectronics: Fall 2018, Vol. 2018, No. 1, pp. 000193-000197.

[5] Rajen Murugan, CT Kao, Nathan Ai, Jie Chen, and Todd Harrison, "System ElectroThermal Transient Analysis of a High Current (40A) Synchronous Step-Down Converter", INTERPACK, International Technical Conference on Packaging and Integration of Electronic and Photonic Microsystems Conference and Exhibition Hilton Anaheim, Anaheim, CA, October 7-9, 2019.

Signal Integrity of Die-to-Die Interface with Advanced Packages for Co-Packaged Optics

Jongchul Shin
Advanced Packaging Group
Samsung Semiconductor Inc.
San Jose, USA
jc.shin@samsung.com

Hamid Eslampour
Advanced Packaging Group
Samsung Semiconductor Inc.
San Jose, USA
h.eslampour @samsung.com

Sangnam Jeong
Advanced Packaging Group
Samsung Semiconductor Inc.
San Jose, USA
snam. jeong @samsung.com

Woopoung Kim
Advanced Packaging Group
Samsung Semiconductor Inc.
San Jose, USA
woopoung.kim@samsung.com

Seokbeom Yong
AVP Product Development 2
Samsung Electronics Co., Ltd.
Hwaseong-si, South Korea
ss.yong@samsung.com

Sung-Oh Ahn
AVP Product Development 2
Samsung Electronics Co., Ltd.
Hwaseong-si, South Korea
star.ahn@samsung.com

Eunkyeong Park
AVP Product Development 2
Samsung Electronics Co., Ltd.
Hwaseong-si, South Korea
ekyeong.park@samsung.com

Sangsub Song
AVP Product Development 2
Samsung Electronics Co., Ltd.
Hwaseong-si, South Korea
sangsub.song@samsung.com

Abstract— **To meet the high demands in Artificial Intelligence (AI) and Machine Learning (ML) applications, the conventional electrical interconnect has limitations on bandwidth, latency, and power consumptions. Co-Packaged Optics (CPO), and advanced heterogenous integration of optics and silicon on a single package, has attracted much attention to achieve higher bandwidth and power efficiency. Samsung has dedicated to develop advanced packaging technologies, such as 2.5D I-CubeS with a silicon interposer, 2.3D I-CubeE with an embedded silicon-bridge and organic redistribution layer (RDL), 2.3D I-CubeR with an organic RDL, and 3D X-Cube with a fine bump pitch Thermo-compression Bonding (TCB) and hybrid Cu bonding. Using I-CubeS or I-CubeE technologies, the feasible CPO architecture can be built by optoelectronic engine arrangement, stacking of photonic integrated circuit (PIC) and electronic integrated circuit (EIC), and PIC coupling type. This paper presents the signal integrity of die-to-die (D2D) interface for the logic of XPU to the logic of EIC. The signal integrity under simplified I/O models for a transmitter and receiver, W-element interconnect of Si-bridge, and through-silicon via (TSV) in the PIC was compared and verified from Universal Chiplet Interconnect express (UCIe) specification.**

Keywords—Co-packaged Optics, heterogeneous integration, advanced packaging, signal integrity, UCIe

I. INTRODUCTION

As the workload demands in Artificial Intelligence (AI) and Machine Learning (ML) applications are rapidly growing, requirements of higher bandwidth, lower latency, and efficient power consumptions are also increasing. However, the conventional electrical I/O with copper connectivity has limitations to support large clusters of XPU architectures due to propagation losses, which result in limited bandwidth and high-power consumptions. Co-Packaged Optics (CPO), an advanced heterogeneous integration of optoelectronic transceiver and various silicon dies in a single package, is an emerging technology that addresses both the bandwidth and power consumption. The current status and many challenges of CPO have been presented in [1], and one of the main challenges is the integration of the photonics integrated circuit (PIC) and electronic integrated circuit (EIC) into the packaging solution. Advanced packaging technologies that enable heterogenous integration of various Chiplet such as high bandwidth memory (HBM), XPU, PIC, and EIC have received more attention as they not only minimize the die-to-die (D2D) channel length, but also the photonic to electrical domain conversion is done close to the point of computation.

There has been a wide range of development activities in defining the optimum optoelectronic transceiver technology both on the system and package level such as the deployment of external laser source or integrated laser, and the selection of PIC coupler design from various available options including edge, grating, evanescent coupling [2]. Depending on how these design features are selected, the feasible packaging solution can be significantly impacted. In addition, the thermal management, power delivery network (PDN), and power integrity are all significantly impacted. However, study on the signal integrity for various CPO architectures has not been implemented much.

In this paper, the signal integrity of D2D interface in a 3.5D packaging that is emerging as an ideal architecture of CPO is introduced. Section II introduces the 2.5D and 3D advanced packaging technologies provided by Samsung. Section III describes the CPO architecture, which integrates the 3D stacked PIC and EIC in an organic redistribution layer (RDL) interposer with Si-bridge known as I-Cube-E, resulting in a 3.5D packaging solution. Section IV presents the signal integrity analysis of D2D interface based on Universal Chiplet Interconnect express (UCIe). Lastly, Section V gives concluding remarks on the simulation results.

II. ADVANCED PACKAGING FOR CPO

Samsung's advanced packaging technologies include I-CubeR with organic RDL interposer and fine line width and space (L/S) which enables multi-reticle (> 4x) interposer, I-CubeS with through-silicon via (TSV) silicon interposer, I-CubeE with an embedded silicon-bridge and organic RDL interposer, X-Cube with finer bump pitch, and bump-less X-

979-8-3503-5124-8/24 $31.00 © 2024 IEEE

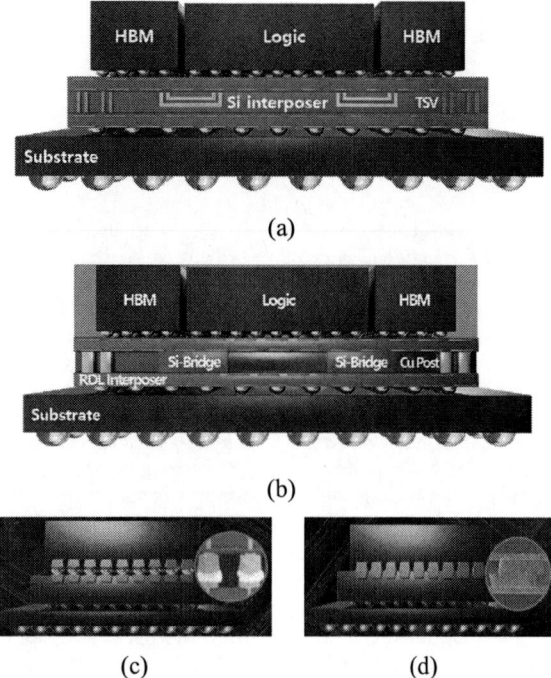

(a)

(b)

(c)　　　　　(d)

Fig. 1. Samsung's advanced packaging solutions (a) I-CubeS, (b) I-CubeE, (c) X-Cube with μ-bump, (d) bumpless X-Cube with hybrid Cu bonding

Cube with hybrid Cu bonding [3]-[4]. Fig. 1 represents 2.5D I-CubeS, 2.3D I-CubeE, and 3D X-Cube technologies from Samsung. Although I-Cube-E technology provides equivalent signal integrity and power integrity performance to I-Cube-S, it provides benefits on cost efficiency, scalability, and productivity, while meeting the high-density D2D interconnect with submicron L/S. Similar to I-CubeS, I-CubeE allows the embedding of integrated silicon capacitor (ISC) in the interposer for the PDN improvement.

III. CPO PLATFORM FOR SIMULATION

The feasible package configuration is determined by the optoelectronic engine arrangement, whether a monolithic design or a discrete PIC, EIC design, and the PIC coupling type such as grating or edge couplers. Driven by the bandwidth increase, latency reduction, and power consumption reduction (pJ/bit), the stacking of EIC and PIC is emerging as the ideal design for the optoelectronic engine. On the other hand, thermal coupling to the PIC, unwanted temperature gradients on the PIC, and thermal power dissipation (TDP) are some of the challenges raised by this stacking configuration. Fig. 2 depicts the CPO configuration that is compatible with both I-Cube-E and I-Cube-S design using the vertical coupling type PIC.

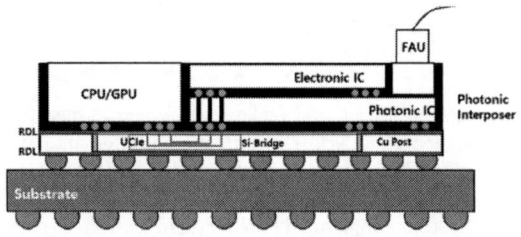

Fig. 2. The CPO platform with the I-CubeE advanced packaging technology

IV. SIMULATION RESULTS

In this chapter, the signal integrity simulation for the aforementioned CPO is presented. A scope of the simulation includes only D2D interconnection from logic of the XPU to logic of the EIC based on the UCIe. In this configuration, the XPU is connected to the EIC through the frontside RDL, the Si-bridge, TSV in the PIC interposer, and with bump pitches of the XPU, TSV, and EIC, all assumed to be identical at 50um.

To analyze logic to logic interface, the simulation setup described in the UCIe specification [5] is adopted. In this simulation, the interconnect model mainly includes Si-bridge interconnect, via models in the front side RDL, and TSV in the PIC interposer. The interconnect in Si-bridge is modeled as the W-element with 8 microstrip lines with a 2mm channel length. The via with 50um height in the front RDL over the bridge and the TSV with 100um height in the PIC interposer are is modeled with 3D EM extraction using ANSYS HFSS.

For the transmitter side, the internal impedance of the transmitter and the output pad capacitance are respectively set as 25 Ω and 0.25 pF, and the transmitting sources are generated with pseudo random bit sequence (PRBS). For the receiver model, the pad capacitance is assumed to be 0.2 pF and the receiver is set to be unterminated. The data rate per a pin to be analyzed is 24 Gbps and 32 Gbps and the IO voltage is set to be 0.4 V.

To determine feasibility, the insertion loss and crosstalk using Voltage Transfer Function (VTF) defined in the UCIe specification are first observed. It is noted that the VTF crosstalk is defined as the power sum of the ratios of the aggressors to the

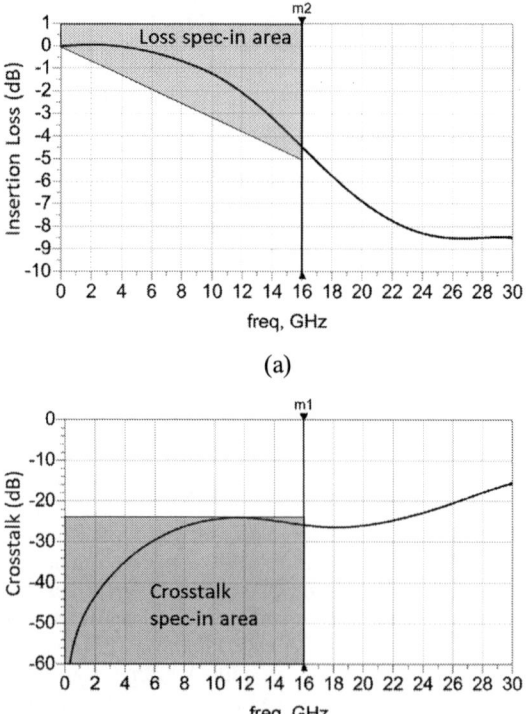

(a)

(b)

Fig. 3. Simulated VTF of (a) insertion loss, (b) crosstalk

979-8-3503-5124-8/24 $31.00 © 2024 IEEE

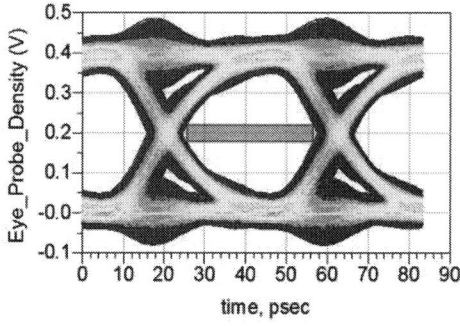

(a)

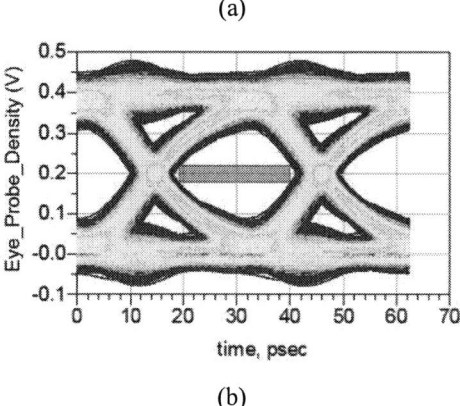

(b)

Fig. 4. Simulated eye diagrams with 2mm channel length at data rate (a) 24 Gbps, (b) 32 Gbps

victim source. A mask is respectively defined by a linear mask for the insertion loss from DC to −5 dB and a flat mask for the crosstalk −24 dB at Nyquist frequency, 12 GHz and 16GHz respect to 24 Gbps and 32 Gbps. As shown in Fig. 3 (a) and (b) for the insertion loss and crosstalk, they are positioned inside the spec-in area, −2.08 dB at 12GHz and −4.47 dB at 16GHz for the loss and −24.1 dB at 11.6GHz for the crosstalk.

Next, the simulated eye diagrams at 24 Gbps and 32 Gbps data rates are shown in Fig. 4(a) and (b) respectively. The eye mask to determine feasibility is defined to be 65% of unit interval (UI) in the eye width and ± 40mV at the reference voltage in the eye height. It is noted that this defined mask is assumed that the equalizations for both transmitter and receiver are enabled and the crosstalk among the signals is included. As shown in Fig. 4(a) and (b), the measured eye openings at 24Gbps

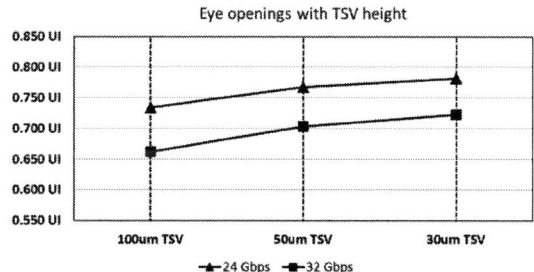

Fig. 5. Eye openings at 24 Gbps and 32 Gbps with the TSV height

and 32 Gbps are respectively 0.734 UI and 0.662 UI. It is shown that the eyes at both 24 Gbps and 32 Gbps can meet the mask without equalizations.

Furthermore, the effect on the TSV height in the PIC interposer was observed. Fig. 5 illustrates the value of eye openings depending on the TSV height in the PIC interposer while maintaining other conditions such as the 2 mm channel length and I/O characteristics. It is shown that minimizing TSV height from 100 um to 30 um improves 10 % of the signal integrity, from 0.662 UI to 0.723 UI at 32 Gbps.

Lastly, to check feasibility at the higher data rate, 40 Gbps, the effect on the load capacitance in the receiver was investigated and the simulated eye openings are shown in Fig. 6. As shown in Fig. 6, the measured eye at 40 Gbps is 0.668 UI, and this shows that controlling the load capacitance up to 0.1 pF may enable the D2D interface to work at 40 Gbps.

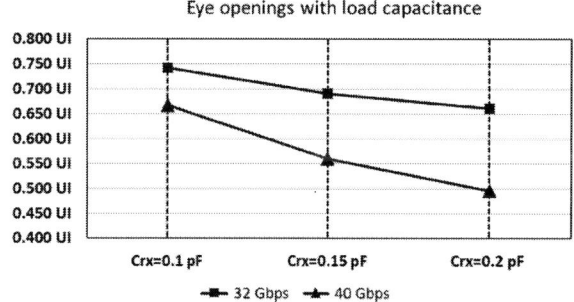

Fig. 6. Eye openings at 32 Gbps and 40 Gbps with the load capacitance

V. CONCLUSIONS

The signal integrity of D2D for the innovative CPO architecture that can be built on Samsung's I-CubeE or I-CubeS advanced packaging has been presented. Although the signal integrity performance for the I-CubeS and I-CubeE are equivalent, the I-CubeE can support more effectively the CPO because of cost efficiency, scalability, and productivity. The simulation demonstrated that the D2D interface of the CPO architecture with I-CubeE can support 24 Gbps and 32 Gbps within 2 mm channel length without equalizations. Further investigations showed that the data rate 40 Gbps, higher than 32 Gbps, can be supported if the receiver I/O can be designed up to 0.1 pF loading capacitance.

REFERENCES

[1] M. Tan et al. "Co-packaged optics (CPO): status, challenges, and solutions," *Frontiers of Optoelectronics*, 16(1), Mar 20, 2023.

[2] Y. Yang et al. "Silicon Photonics Chip I/O for Ultra High-Bandwidth and Energy-Efficient Die-to-Die Connectivity," in *IEEE Custom Integrated Circuits Conference (CICC)*, April, 2024.

[3] M. Kang, "Heterogeneous Integration Platform for Next Generation Computing," in *Sixth Annual Symposium on Heterogeneous Integration*, February, 2023.

[4] Y. Li and W. Kim, "Heterogeneous Packaging Technologies for Chiplet and Memory Integration," *Advancing Microelectronics*, vol. 51, no. 1, pp. 12–15, 2024.

[5] Universal Chiplet Interconnect Express (UCIe) 1.1 specification.

Delay Rational Macromodelling of Noisy Tabulated Frequency Responses

Alexander Kirchberger and Anestis Dounavis

Department of Electrical and Computer Engineering, Western University, London, ON, Canada

akirchbe@uwo.ca, adounavis@eng.uwo.ca

Abstract—**This paper presents a scheme to fit delay rational macromodels of electrically long electric networks from noisy frequency domain tabulated data. Delay regions of the tabulated data are estimated by evaluating the energy of the frequency responses over time. Then, the frequency responses of each delay term are obtained as a rational approximation using a vector fitting-instrumental variable technique. The combination of rational approximations of each delay region provides a delay rational macromodel for the entire network.**

Index Terms—**Index Terms—Delay Extraction, Vector Fitting, Delay Rational Macromodels, Noisy Frequency Data.**

I. INTRODUCTION

In recent years, complexity, circuit densities, and the higher operating frequencies of electric circuits have made signal integrity a challenging task for the case where there are non-uniformities, complex geometries, or process variations. Under these conditions, the behaviour of interconnects as well as other electromagnetic modules such as connectors, vias and packages are often characterized by tabulated data obtained either from electromagnetic simulations or from experimental measurements using a Vector Network Analyser (VNA). As a result, the development of efficient macromodelling methodologies have become an important issue to evaluate distributed systems characterized by tabulated data.

Among rational fitting techniques, Vector Fitting (VF) has emerged as a popular macromodelling technique [1]–[3]. However, when the frequency response data is described by noisy data, such as measurements obtained by a VNA, VF may have difficulties in estimating rational Transfer Functions (TFs) since the least squares pole relocation solution is biased by noise [4], [5]. Another issue when developing macromodels is that for electrically long electromagnetic components, delay rational approximations are often more efficient when compared to purely rational TFs. This efficiency is derived from extracting the propagation delays from tabulated data, resulting in lower order TFs, leading to more compact macromodels with fewer poles when compared to using only rational approximation.

In [4], [5] a Vector Fitting-Instrumental Variable (VF-IV) algorithm was proposed to obtain efficient rational TFs of networks characterized by noisy frequency domain data. However, the VF-IV has not been applied to electrically long distributed systems modelled as delay rational TFs. In this paper, the VF-IV approach is extended to model noisy frequency domain data corresponding to electrically long distributed electromagnetic systems modeled as delay rational approximations. A numerical example of a one port network is provided to illustrate the validity of the proposed approach.

II. PROPOSED ALGORITHM

In order to describe the proposed algorithm consider a distributed system to be modeled as a delay rational approximation in the following form:

$$H_l(s) = \sum_{m=1}^{M} H_l^{(m)}(s) e^{-s\tau_l^{(m)}} \tag{1}$$

where $\tau_l^{(m)}$ and $H_l^{(m)}$ are the mth propagation delay and mth rational approximation, respectively and l corresponds to the lth TF. The proposed process to find the delay rational approximation is broken up into two steps, finding each delay term, and finding a rational approximation associated with each delay term.

A. Delay Extraction

The conventional method to obtain delay regions and estimates of each delay is by performing the Gabor Transform to obtain the time frequency representation of the response as described in [6]–[8]. The time frequency transform that relates the data $\hat{H}_l(s)$ to $F_l(f, \tau)$ is:

$$F_l(f, \tau) = \int_{\infty}^{-\infty} \hat{H}_l(\zeta) W(\zeta - \omega) e^{j2\pi\zeta\tau} d\zeta \tag{2}$$

where $W(\zeta - \omega)$ is a window centered at $\zeta = \omega$. Equation (2) can be thought of as an Inverse Fourier Transform (IFT), retaining only the frequency components under the window. Once the time frequency representation is obtained, a time energy function can be obtained as:

$$n_l(\tau) = \int_{\infty}^{-\infty} |F_l(f, \tau)|^2 df \tag{3}$$

Delay regions can be found by finding local minima under a threshold local maxima of the energy time function over a threshold. Estimates for the delay of each region can be taken by the max value of each region. If there is no satisfactory split point between two delay regions, the two delays regions can be estimated as one.

979-8-3503-5124-8/24 $31.00 © 2024 IEEE

From here the inverse time frequency transform of each delay region is found to obtain each tabulated frequency response given by:

$$\hat{H}_l^{(m)}(\zeta) = \frac{1}{2\pi} \iint_{\Omega_l^{(m)}} F_l(\omega, \tau) W(\zeta - \omega) e^{-j\omega\tau} d\omega d\tau \quad (4)$$

The delays of each region is estimated as the largest point of the time energy function over that region. Then to optimize the delay values a methodology similar to [9] is used. However, in the proposed methodology Relaxed Vector Fitting Instrumental Variable (RVF-IV) is used in each delay estimate iteration whereas [9] uses VF.

B. Rational Approximation of Frequency Responses

Once the optimized delay terms are found a rational approximation is performed for each region. Using RVF, the system of equations for $H_l^{(m)}$ is expressed as:

$$\begin{bmatrix} \mathbf{X}^{(m)} & -\tilde{\mathbf{H}}_l^{(m)}\mathbf{X}^{(m)} \end{bmatrix} \begin{bmatrix} \mathbf{C}_l^{(m)} \\ \mathbf{C}_{pl}^{(m)} \end{bmatrix} = 0 \quad (5)$$

where $\mathbf{C}_l^{(m)}$ is a vector that contains the unknown residues of $(\sigma H)_l^{(m)}(s)$ and $\mathbf{C}_{pl}^{(m)}$ contains the unknown residues of $\sigma(s)$. The data $\tilde{H}_l^{(m)}(s)$ used for creating $\tilde{\mathbf{H}}_l^{(m)}(s)$ is given by:

$$\hat{H}_l^{(m)}(s) = \tilde{H}_l^{(m)}(s)e^{-s\tau_l^{(m)}} \quad (6)$$

A detailed description of $\mathbf{X}^{(m)}$, $\tilde{\mathbf{H}}_l^{(m)}$, $(\sigma H)_l^{(m)}(s)$ and $\sigma(s)$ is described in [1], [3], [10]. Next, a QR decomposition is used, to decouple $\mathbf{C}_l^{(m)}$ from $\mathbf{C}_{pl}^{(m)}$, given by:

$$\begin{bmatrix} \mathbf{X}^{(m)} & -\tilde{\mathbf{H}}_l^{(m)}\mathbf{X}^{(m)} \end{bmatrix} = \tilde{\mathbf{Q}}_l^{(m)} \begin{bmatrix} \tilde{\mathbf{R}}_{l11}^{(m)} & \tilde{\mathbf{R}}_{l12}^{(m)} \\ 0 & \tilde{\mathbf{R}}_{l22}^{(m)} \end{bmatrix} \quad (7)$$

The system can be now be represented as:

$$\begin{bmatrix} \tilde{\mathbf{R}}_{122}^{(m)} \\ \cdots \\ \tilde{\mathbf{R}}_{L22}^{(m)} \\ \mathbf{r}^T \end{bmatrix} \mathbf{C}_{pl}^{(m)} = \begin{bmatrix} 0 \\ \cdots \\ 0 \\ K \end{bmatrix} \quad (8)$$

where $\mathbf{r}$ and K corresponds to the relaxation constraint and can be found in [2], [10]. Since equation (8) is an overdetermined system, when fitting a noisy frequency response the least squares solution for $\mathbf{C}_{pl}^{(m)}$ can be biased by the noise as illustrated in [4], [5]. In order reduce the biasing of the least squares solution, the instrumental variable method [11] formulates the least square solution as:

$$\left(\sum_{l=1}^{L} [(\tilde{\mathbf{R}}_{IVl22}^{(m)})^T \tilde{\mathbf{R}}_{l22}^{(m)}] + \mathbf{r}\mathbf{r}^T \right) \mathbf{C}_{pl}^{(m)} = \mathbf{r}K \quad (9)$$

where $\tilde{\mathbf{R}}_{IVl22}^{(m)}$ and $\tilde{\mathbf{Q}}_{IVl}^{(m)}$ are from a QR decomposition of $\mathbf{X} - \mathbf{\Psi}\mathbf{X}$, and $\mathbf{\Psi}$ is data obtained from the previous rational approximation. Since the errors of the original noisy data and the previous rational approximation are less correlated, this leads to less biasing and results in a more accurate rational approximation for each delay region. This will be illustrated in the numerical examples section.

TABLE I
POLES, RESIDUES AND DELAYS OF TF FOR NUMERICAL EXAMPLE

Delay 1 (0 ns)	
Poles (GHz)	Residues (GHz)
$-0.6132 \pm j3.4551$	$-0.7877 \pm j0.0809$
$-0.3940 \pm j7.3758$	$-0.2067 \pm j0.0131$
$-0.1680 \pm j14.3024$	$-0.2682 \pm j0.0145$
$-0.4097 \pm j17.7864$	$-0.0652 \pm j0.0146$
$-0.2991 \pm j28.4622$	$-0.2426 \pm j0.0145$
$-0.6447 \pm j35.2669$	$-0.4043 \pm j0.0297$
$-1.0135 \pm j37.9655$	$-0.6787 \pm j0.1465$
$-0.5711 \pm j57.4748$	$-0.2726 \pm j0.1037$
d = 0.61	
Delay 2 (15 ns)	
$-0.3940 \pm j7.3758$	$-0.2067 \pm j0.0131$
$-0.3097 \pm j13.7864$	$-0.1278 \pm j0.0147$
$-0.2367 \pm j16.1892$	$-0.0384 \pm j0.0192$
$-0.1938 \pm j32.5637$	$-0.0757 \pm j0.0125$
$-0.4032 \pm j47.5864$	$-0.0202 \pm j0.0076$
d = 0.07	

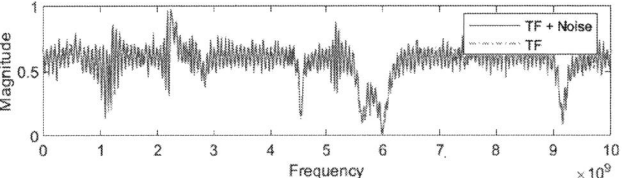

Fig. 1. Data created by TF plus 20 dB white noise

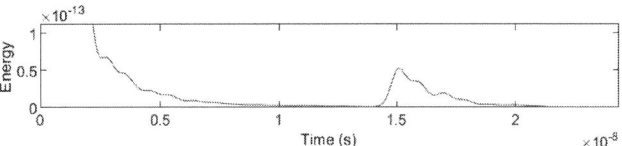

Fig. 2. Time energy function obtained using data with 20 dB white noise

III. NUMERICAL EXAMPLE

A synthetic TF of a delay rational function is described in Table I. A Signal-to-Noise Ratio (SNR) of 20 dB is analysed. The TF coefficients in Table I and SNR of 20 dB were chosen since this resulted in a very challenging TF to model using traditional means. Fig. 1 shows the response of the original TF and the noisy frequency response overlayed using 1000 data points spaced evenly between 0-10 GHz. Using (2) and (3) the time energy function is given in Fig. 2 where each delay region can be easily seen. The delay estimates taken as the local maxima of the time energy function are [0 ns and 15.65 ns] for each delay region. Using [9] with RVF-IV the optimized delay terms found were [0 ns and 14.999 ns]. With a knowledge of the delay terms Fig. 3 shows the magnitudes of the rational approximation of each delay region using RVF-IV and RVF for the 10th iteration. For the proposed algorithm the solution of (9) was applied starting at the fourth iteration to ensure that the $\tilde{\mathbf{R}}_{IVl22}^{(m)}$ estimates were reasonably accurate. For this example RVF failed to capture all the poles of the first and second delay TFs. For $\tau_1 = 0ns$ it failed to capture

979-8-3503-5124-8/24 $31.00 © 2024 IEEE

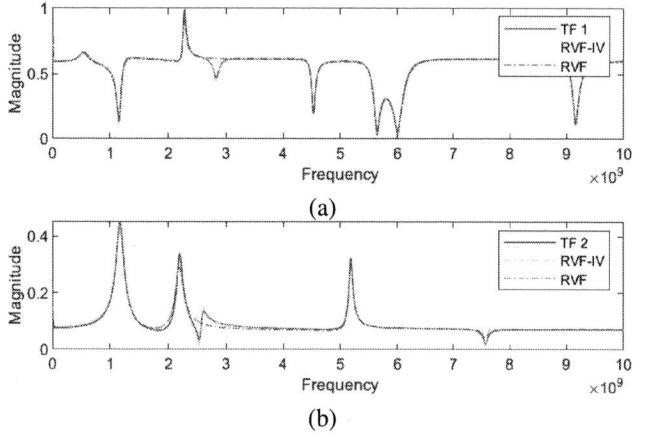

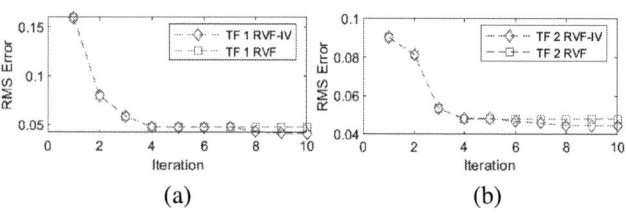

Fig. 3. TF rational approximation of magnitude (a) For $\tau_1 = 0ns$, (b) For $\tau_2 = 14.999ns$

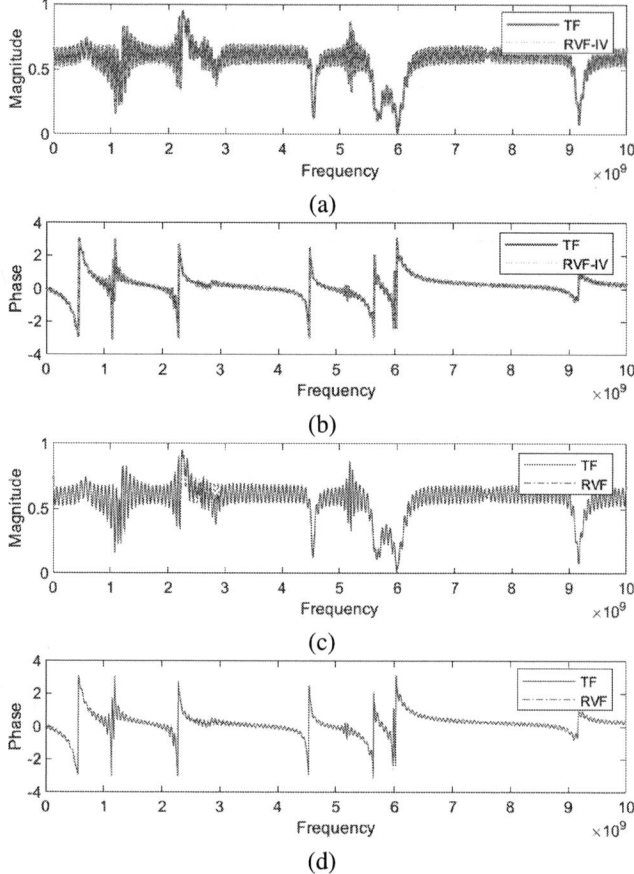

Fig. 5. Rational approximation of overall TF (a) Magnitude plot using RVF-IV, (b) Phase plot using RVF-IV, (c) Magnitude plot using RVF, (d) Phase plot using RVF

Fig. 4. RMS error vs iteration (a) For $\tau_1 = 0ns$, (b) For $\tau_2 = 14.999ns$

$-0.4097 \pm j17.7864$, and for $\tau_2 = 14.999ns$ it failed to capture $-0.2367 \pm j16.1892$, as shown from the plots of Fig. 3. Fig 4. shows the Root Mean Squares (RMS) error of each TF with respect to the noisy data after 10 iterations. For the first and second TF the RMS error with respect to the noisy data using RVF-IV was 0.0411 and 0.0466 respectively, while for RVF it was 0.0476 and 0.0482 respectively. Fig. 5 shows the magnitude and phase of each TF obtained by RVF-IV and RVF from the 20 dB noisy data. The overall RMS error when compared to the original noiseless data is 0.0129 for the RVF-IV approach and 0.0318 for the RVF approach.

IV. CONCLUSION

In this paper a scheme to obtain delay rational macromodels from noisy data based on RVF-IV is presented. The numerical examples illustrate that the proposed RVF-IV algorithm is able to significantly improve the convergence properties when compared to RVF when obtaining delay rational approximations from noisy data. This leads to improved delay rational approximations with less RMS error.

REFERENCES

[1] B. Gustavsen and A. Semlyen, "Rational approximation of frequency domain responses by vector fitting," *IEEE Transactions on Power Delivery*, vol. 14, no. 3, pp. 1052–1061, 1999.

[2] B. Gustavsen, "Improving the pole relocating properties of vector fitting," *IEEE Transactions on Power Delivery*, vol. 21, no. 3, pp. 1587–1592, 2006.

[3] D. Deschrijver, M. Mrozowski, T. Dhaene, and D. De Zutter, "Macromodeling of multiport systems using a fast implementation of the vector fitting method," *IEEE Microwave and Wireless Components Letters*, vol. 18, no. 6, pp. 383–385, 2008.

[4] A. Beygi and A. Dounavis, "An instrumental variable vector-fitting approach for noisy frequency responses," *IEEE Transactions on Microwave Theory and Techniques*, vol. 60, no. 9, pp. 2702–2712, 2012.

[5] M. Sahouli and A. Dounavis, "An instrumental-variable qr decomposition vector-fitting method for modeling multiport networks characterized by noisy frequency data," *IEEE Microwave and Wireless Components Letters*, vol. 26, no. 9, pp. 645–647, 2016.

[6] S. Grivet-Talocia, "Delay-based macromodels for long interconnects via time-frequency decompositions," in *2006 IEEE Electrical Performane of Electronic Packaging*, 2006, pp. 199–202.

[7] M. Sahouli and A. Dounavis, "Delay extraction-based modeling using loewner matrix framework," *IEEE Transactions on Components, Packaging and Manufacturing Technology*, vol. 7, no. 3, pp. 424–433, 2017.

[8] R. Stockwell, L. Mansinha, and R. Lowe, "Localisation of the complex spectrum: The s transform," *IEEE Transaction on Digital Signal Processing*, vol. 44, pp. 99–114, 01 1996.

[9] B. Gustavsen, "Time delay identification for transmission line modeling," in *Proceedings. 8th IEEE Workshop on Signal Propagation on Interconnects*, 2004, pp. 103–106.

[10] A. Chinea and S. Grivet-Talocia, "On the parallelization of vector fitting algorithms," *IEEE Transactions on Components, Packaging and Manufacturing Technology*, vol. 1, no. 11, pp. 1761–1773, 2011.

[11] L. Ljung, *System Identification: Theory for the User*, ser. Prentice Hall information and system sciences series. Prentice Hall PTR, 1999.

979-8-3503-5124-8/24 $31.00 © 2024 IEEE

Fan-out Region Crosstalk Optimization of High-Density PCIe 6.0 SMT Connectors

Dan Liu
Alibaba Cloud
Alibaba Group
Shenzhen, China
ld259920@alibaba-inc.com

Yangfan Zhong
Alibaba Cloud
Alibaba Group
Shenzhen, China
yangfan.zyf@alibaba-inc.com

Minzheng Tian
Hardware Department
IEIT SYSTEMS Co.,Ltd.
Jinan, China
tianminzheng@ieisystem.com

Mengmeng Guo
Hardware Department
IEIT SYSTEMS Co.,Ltd.
Jinan, China
Guomengmeng@ieisystem.com

Bing Wei
Hardware Department
IEIT SYSTEMS Co.,Ltd.
Jinan, China
weibing@ieisystem.com

Weizhe Li
Data Center and AI Group
Intel Corporation
Shanghai, China
weizhe.li@intel.com

Jingbo Li
Data Center and AI Group
Intel Corporation
Hillsboro, USA
jingbo.li@intel.com

Tina Bao
Data Center and AI Group
Intel Corporation
Hillsboro, USA
tina.bao@intel.com

Abstract—This study compares various fan-out design approaches for high-density SMT connectors in modular systems. Both simulations and measurements show significant crosstalk differences, which can be instrumental in layout optimization for PCIe 6.0 signal integrity design.

Keywords—Layout optimization, Crosstalk, PCIe 6.0

I. INTRODUCTION

Systems built on modular hardware are gaining popularity in data centers. This is because the best total cost of ownership (TCO) model prefers maximizing component sharing to be cost-effective for various server constructions and configurations. Modularization drives compact PCB board designs, shortens high-speed signal PCB traces and utilizes high-speed cable assemblies to connect various modules. Consequently, high-density surface mount technology (SMT) connectors are broadly implemented on those compact boards. However, in high-speed channel designs, the limited routing area around SMT connectors does not allow enough spacing or shielding for crosstalk optimization. Using the hot swap back plane (HSBP) design in Fig. 1 as an example, the high-density SMT connectors are soldered on the top and bottom of a compact board. The pitch between a pair of SMT connectors is only 300 mils, which is very limited for routing optimization. In contrast, PCIe 6.0 signaling employs PAM4 modulation, which is significantly more sensitive to noise and interference compared to PCIe 5.0 with NRZ signaling [1] [2]. Consequently, a high-density PCIe 6.0 PAM4 design presents greater challenges for crosstalk optimization.

In order to better understand the crosstalk effects of different fan-out from high-density PCIe 6.0 SMT connectors, six fan-out cases are meticulously designed with the consideration of various design constraints from system requirements. The following section describes the details of these study cases. Section III shows the crosstalk simulation results from the six cases under study while section IV presents the results of performed measurements. In addition, the measurement on a PCIe 6.0 SMT connector customized for immersion-cooling server design is also conducted, which has a footprint and form factor compatible with traditional air-cooling connector design. Finally, section V concludes the findings and discusses additional enablers for further crosstalk optimization.

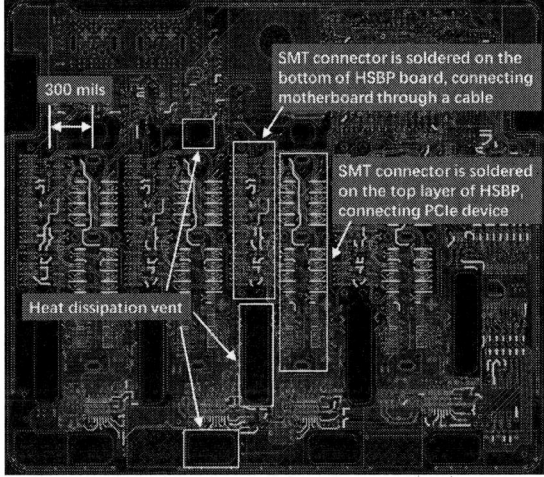

Fig. 1. A compact board in real HSBP design.

II. FAN-OUT STUDY CASES DESCRIPTION

In this study, the geometry of the fan-out from a high-density PCIe 6.0 SMT connector includes the microstrip on the PCB top layer, and the differential plated-through-hole (PTH) vias for traces switching to inner PCB layers. The primary crosstalk source of the fan-out is attributed to both of horizontal coupling from the microstrip routing, and the vertical coupling from long via barrels.

Fig. 2 shows six fan-out study cases for a PCIe 6.0 SMT connector. A 22-layer PCB stack-up in [3] with enhanced ultra-low loss material is being used. The board thickness is about 120 mils. The pitch of the differential PTH via is set to 35 mils. The PCIe 6.0 SMT connector has 0.6mm pitch and is compliant with PCIe 6.0 connector specification [4]. For each case, two fan-out scenarios are designed with different layer transitions: (1) from the top layer to layer 3 where the differential-via barrel length is ~12 mils after back-drill, and (2) from the top layer to layer 20 where the differential-via barrel length is ~110 mils. Detailed descriptions of the six cases can be found in Table 1.

979-8-3503-5124-8/24 $31.00 © 2024 IEEE

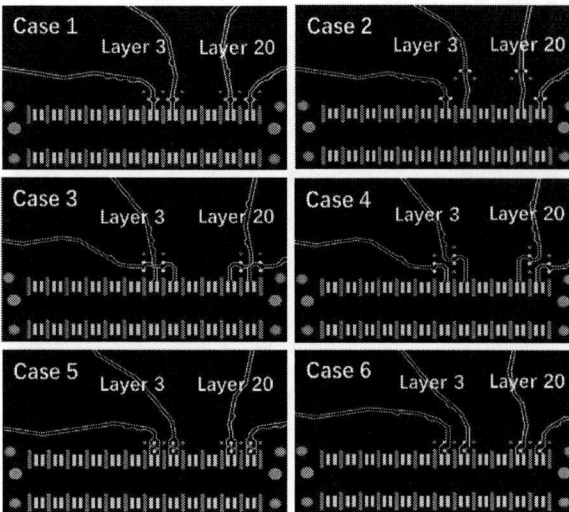

Fig. 2. Six cases of fan-out from high-density PCIe 6.0 SMT connector.

TABLE I.   CASE STUDY DESCRIPTION

Cases	Description
Case 1	Shortest microstrip routing. Symmetric fan-out routing. Differential vias are laid out parallel to the connector body.
Case 2	Leveraging from case 1, the pair-to-pair spacing of differential vias are increased from 18 mils to 105 mils.
Case 3	Asymmetric fan-out routing. Differential vias and ground vias are laid out vertically to the connector body.
Case 4	Leveraging from case 3, the differential vias are staggered by 35 mils. One additional ground via is implemented.
Case 5	Leveraging from case 3, the arrangement of the differential vias and ground vias are slightly tuned for space saving.
Case 6	Leveraging from case 5, microstrip length are mismatched.

III. SIMULATION RESULTS

Comprehensive simulation is performed for six fan-out study cases. Fig. 3 shows the simulated crosstalk of the fan-out from the top layer to layer 3. Fig. 4 displays the crosstalk simulation results of the fan-out from the top layer to layer 20. The far-end crosstalk (FEXT) data at 16 GHz are summarized in Table 2. The key observations are shown below:

- The vertical coupling from long via barrels is dominant in the fan-out crosstalk sources. At 16 GHz, there is about 15-20 dB reduction when the fan-out is from the top layer to layer 3, compared to the fan-out from the top layer to layer 20.

- According to the simulation, case 1 has the strongest FEXT noise among all the cases. Case 2 and 4 have the least FEXT noise. This is due to the larger pair-to-pair spacing and better via arrangement.

- Given the FEXT expectation of -45 dB or less for PCIe 6.0 components, the fan-out with shorter via barrels can easily meet the FEXT requirement. However, for the fan-out with longer via barrels, only case 2 and case 4 can meet the requirement.

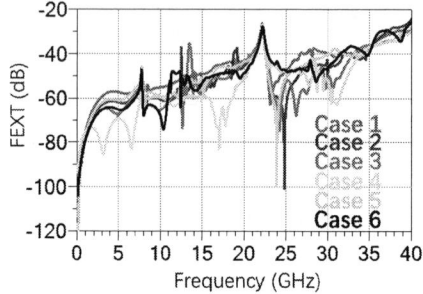

Fig. 3. FEXT simulation results of fan-out from top layer to layer 3.

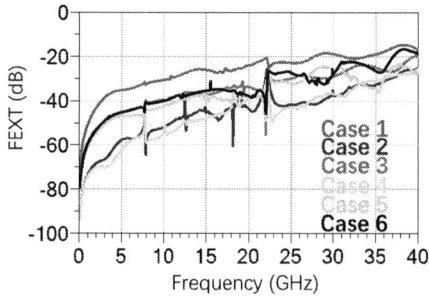

Fig. 4. FEXT simulation results of fan-out from top layer to layer 20.

TABLE II.   FEXT SIMULATION RESULTS SUMMARY

FEXT (dB) @ 16GHz	Top Layer to Layer 3	Top Layer to Layer 20
Case 1	-47.0	-27.1
Case 2	-54.3	-46.5
Case 3	-51.8	-36.6
Case 4	-64.3	-46.9
Case 5	-50.4	-35.8
Case 6	-50.4	-35.0

IV. MEASUREMENT RESULTS

The six fan-out scenarios are also implemented on a testing board for laboratory measurements. Fig. 5 and 6 show the measured FEXT levels, which are consistent with simulation results in Fig. 3 and 4, respectively. The good correlation validates that the simulation methodology is capable of predicting the FEXT trends in real designs.

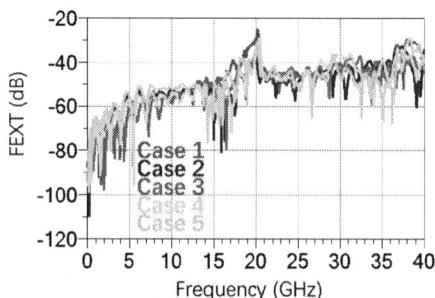

Fig. 5. FEXT measurement results of fan-out from top layer to layer 3.

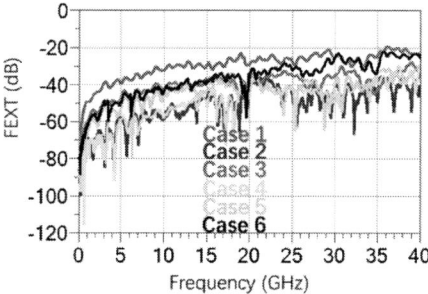

Fig. 6.   FEXT measurement results of fan-out from top layer to layer 20.

TABLE III.   FEXT MEASUREMENT RESULTS SUMMARY

FEXT (dB) @ 16GHz	Top Layer to Layer 3	Top Layer to Layer 20
Case 1	-47.0	-29.7
Case 2	-60.0	-50.9
Case 3	-56.0	-35.4
Case 4	-64.0	-48.6
Case 5	-54.0	-37.8
Case 6	N/A	-34.6

Furthermore, according to the previous studies [5] [6], when high-speed interconnects designed for an air environment are deployed in immersion-cooling data centers, significant degradation in SI performance of PCIe 6.0 PAM4 is observed due to the different electrical properties of the coolant. Therefore, a novel PCIe 6.0 high-speed SMT connector, customized for immersion-cooling designs and compatible in footprint and form factor with traditional air connector designs, has been developed. This novel immersion SMT connector still meets the requirements of PCIe 6.0 connector specification. The measurements are performed on fan-out case 1 and case 2 test boards, which are respectively soldered with air connectors and immersion connectors.

Fig. 7 shows the measured impedance between air and immersion designs under corresponding environments. Fig. 8 shows comparison of the measured FEXT. There is no obvious difference between the fan-out case 1 with the air connector and the immersion connector – the trend of the curves matches very well. The same is true for the fan-out case 2.

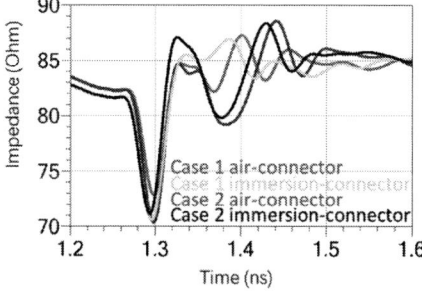

Fig. 7.   TDR measurement results of fan-out from top layer to layer 20 in air and in immersion liquid.

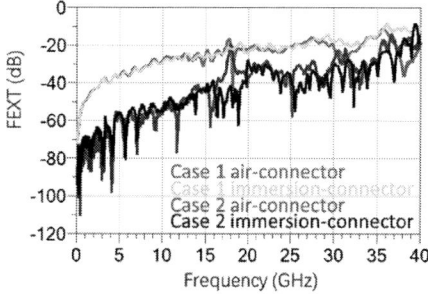

Fig. 8.   FEXT measurement results of fan-out from top layer to layer 20 in air and in immersion liquid.

V. CONCLUSIONS

As shown in both simulation and measurement, not all the crosstalk effects of different fan-out designs from high-density PCIe 6.0 SMT connectors can fully meet the design expectations for PCIe 6.0. This is especially true for thick PCB boards and limited routing areas. Therefore, it is critical to optimize the FEXT in the fan-out region for practical PCIe 6.0 channel designs.

Other than the placement of differential vias, further reducing the pitch size of the differential PTH vias also can reduce FEXT. With a smaller pitch, tighter coupling within the differential vias improves the noise immunity, thereby improving the overall FEXT performance for the fan-out design. Another feasible enabler is to use more advanced PCB materials in thick PCB designs to reduce the thickness of the board. Consequently, the length of the differential PTH vias will be reduced, thus achieving the goal of decreasing the vertical coupling from the long via barrels. However, it should be noted that advanced PCB materials will lead to significant PCB cost increases.

In addition, it is worth mentioning that no obvious SI performance difference is observed between the air connector and the immersion connector on the same fan-out board. This indicates that there are opportunities to share the same PCIe 6.0 PCB board designs in both air and immersion environments. When the hardware system design is intended for deployment in immersion cooling, it is only required to implement an immersion connector that complies with the CEM specification on the PCB board.

REFERENCES

[1] Fabio A. Ruiz-Molina et al., "Crosstalk Analysis for PCIe 6.0 (PAM 4) Under Different Transmitter Conditions", IEEE 31st Conference on Electrical Performance of Electronic Packaging and Systems, 2022.

[2] Intel, "PAM4 Signaling Fundamentals", Application Note, 2019.

[3] Intel RDC# 788196, Oak Stream PDG

[4] PCIE Exspress Card Electromechanical Specification Revision 6.0.0, December 20, 2023.

[5] D. Liu, J. Li, K. Wang et al., "PCIe 6.0 (PAM4) Signal Integrity Challenges in Immersion-Cooling Data Centers", DesignCon 2023.

[6] J. Li, K. Wang, D. Liu et al., "Immersion-Cooling Impact on PCIe 5.0 (NRZ) and PCIe 6.0 (PAM4) Link Performance from Measurements", DesignCon 2024.

979-8-3503-5124-8/24 $31.00 © 2024 IEEE

Optimization of TSV Array Based on Mathematical Model for HBM3

Yen-Tung Chen
Graduate Institute of Communication
Engineering
National Taiwan university
Taipei, Taiwan
r11942194@gmail.com

Yu-Ying Cheng
Graduate Institute of Communication
Engineering
National Taiwan University
Taipei, Taiwan
f09942033@ntu.edu.tw

Tzong-Lin Wu
Graduate Institute of Communication
Engineering
National Taiwan University
Taipei, Taiwan
tlwu@ntu.edu.tw

Abstract— **In this paper, we present a novel approach to optimize the through-silicon vias (TSV) structure by the multi-conductor transmission line (MTL) theory and a genetic algorithm (GA) for high bandwidth memory (HBM). A comprehensive mathematical model is developed for the TSV array based on MTL theory. The S-parameter estimated by the model is consistent with the full-wave simulated results. Moreover, GA, which iteratively adjusts the TSV parameters, is employed to minimize the insertion loss. The results demonstrate significant improvements in TSV performance, including reduced signal degradation and enhanced overall efficiency.**

Keywords— *muti-conductor transmission line (MTL), insertion loss (IL), genetic algorithm (GA), through-silicon via (TSV), high bandwidth memory (HBM)*

I. INTRODUCTION

With the increasing density and stacking of through-silicon vias (TSVs) in high-bandwidth memories (HBMs), the complexity of noise sources becomes a significant threat to signal integrity (SI). Previous studies have addressed the SI problem by improving TSV design and enhanced materials. Although these approaches are effective, they are often overly reliant on expensive electronic design automation (EDA) tools.

To address the high costs of EDA tools and improve simulation efficiency, an approximate mathematical model of the TSV array is necessary. Previous studies have provided different mathematical models to analyze the multi-signal and multi-ground TSV array structures based on multiconductor transmission line theory (MTL) [1]-[3]. The mathematical model presented in this paper incorporates inter-metal dielectric (IMD) layers and hybrid bonding to approach the realistic structure. Furthermore, a well-established optimization algorithm called genetic algorithm (GA) [4] is applied to capture the complex interaction between structural design and electrical properties, and the TSV array is optimized to minimize the insertion loss.

The rest of this article is segmented into subsequent sections. Section II details the development of the mathematical model for TSV analysis, Section III describes

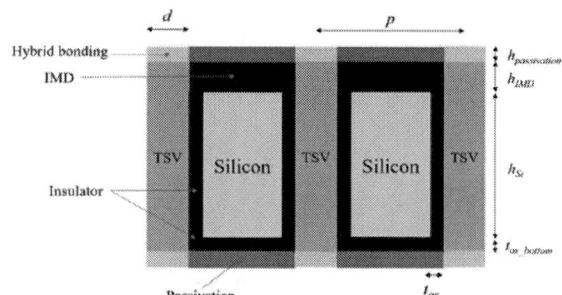

Fig. 1. The sectional view of TSV structure.

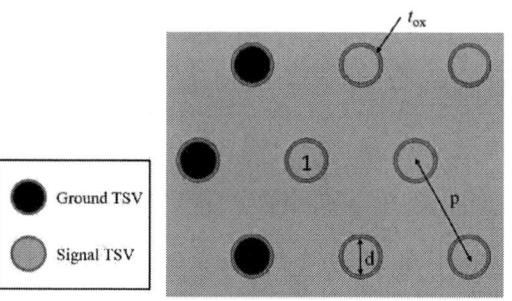

Fig. 2. The top view of 3×3 TSV array.

the optimization of TSV structures using GA, and Section IV provides a summary and conclusions of our findings.

II. MATHEMATICAL MODEL FOR TSV ARRAY

Fig. 1 shows the detailed TSV structure with hybrid bonding, where the inter-metal dielectric (IMD) and insulator are silicon dioxide (SiO2). Fig. 2 illustrates a 3×3 TSV array containing 6 signal TSVs and 3 ground TSVs. The signal TSV in the center is set as target #1. d is the diameter of the TSV, p is the pitch between TSVs, and t_{ox} is the oxide thickness. Based on the MTL theory, the mathematical model of the TSV matrix can be developed as follows.

First, the inductance matrix **L** for $N \times N$ TSV array is defined as

979-8-3503-5124-8/24 $31.00 © 2024 IEEE

$$\mathbf{L} = \begin{bmatrix} L_{11} & L_{12} & \cdots & L_{1N} \\ L_{21} & L_{22} & \cdots & L_{2N} \\ \vdots & \vdots & \ddots & \vdots \\ L_{N1} & L_{N2} & \cdots & L_{NN} \end{bmatrix} \qquad (1)$$

According to [1]-[3], the self-loop inductance, L_{ii}, is calculated by

$$L_{ii} = \frac{\mu_0 h_{si}}{\pi} \ln\left(\frac{P_{i0}}{d/2}\right) \qquad (2)$$

where P_{i0} is the pitch between TSV #i and the reference TSV, and μ_0 is the vacuum permeability and h_{si} is the height of the silicon substate. The mutual inductance between TSV #i and #j, L_{ij}, is estimated by.

$$L_{ij} = \frac{\mu_0 h_{si}}{2\pi} \ln\left(\frac{P_{i0}P_{j0}}{P_{ij}(d/2)}\right) \qquad (3)$$

where P_{ij} is the pitch between TSV #i and #j.

Since the silicon substrate is homogeneous, the capacitance between TSV #i and #j, $\mathbf{C}_{si}$ can be calculated as

$$\mathbf{L}_{si}\mathbf{C}_{si} = \mu_0 \varepsilon_{si} h_{si}^2 \mathbf{I} \qquad (4)$$

Where $\mathbf{L}_{si}$ is the silicon substrate inductance matrix estimated by (2) and (3) using the equivalent diameter ($d + 2\,t_{ox}$). ε_{si} is the permittivity of silicon, and $\mathbf{I}$ is the identity matrix. Moreover, from the principle of duality, the conductance matrix $\mathbf{G}_{si}$ can be obtained by

$$\frac{\mathbf{G}_{si}}{\mathbf{C}_{si}} = \frac{\sigma_{si}}{\varepsilon_{si}} \qquad (5)$$

Moreover, the internal impedance matrix $\mathbf{Z}_{int}$ can be represented as

$$\mathbf{Z}_{int} = \begin{bmatrix} Z_{int,1} & 0 & 0 & 0 \\ 0 & Z_{int,2} & 0 & 0 \\ 0 & 0 & \ddots & 0 \\ 0 & 0 & 0 & Z_{int,N} \end{bmatrix} + \begin{bmatrix} Z_{int,11} & Z_{int,12} & \cdots & Z_{int,1N} \\ Z_{int,21} & Z_{int,22} & \cdots & Z_{int,2N} \\ \vdots & \vdots & \ddots & \vdots \\ Z_{int,N1} & Z_{int,N2} & \cdots & Z_{int,NN} \end{bmatrix}$$

(6)

where the internal impedance of TSV [3][4], $Z_{int,ij}$, is

$$Z_{int,i} = \frac{h_{si}\sqrt{j\varpi\mu_0\rho}}{2\pi(d/2)} \frac{I_0(\frac{d}{2}\sqrt{j\varpi\mu_0/\rho})}{I_1(\frac{d}{2}\sqrt{j\varpi\mu_0/\rho})} \qquad (7)$$

where I_0 and I_1 are the zero and first-order modified Bessel function of first kind, and the resistivity of TSV is represented as ρ. In addition, a random ground TSV is selected as a reference. $Z_{int,ij}$ equals the reference TSV impedance, since the reference conductor is finite size [1].

Then, for the multi-ground problem [3], $\mathbf{Z}_{int}$ is decomposed as

$$\mathbf{Z}_{int} = \begin{bmatrix} (\mathbf{Z}_1)_{s\times s} & (\mathbf{Z}_2)_{s\times(g-1)} \\ (\mathbf{Z}_3)_{(g-1)\times s} & (\mathbf{Z}_4)_{(g-1)\times(g-1)} \end{bmatrix} \qquad (9)$$

where s is the number of signal TSV, and g represents the number of ground TSV. Therefore, $\mathbf{Z}_{int}$ can be reduced as

$$\mathbf{Z}_r = \mathbf{Z}_1 - \mathbf{Z}_2\mathbf{Z}_4^{-1}\mathbf{Z}_3 \qquad (10)$$

The impedance of oxide is

$$Z_{ox} = \frac{1}{j\varpi} \frac{\ln(\frac{d/2+t_{ox}}{d/2})}{2\pi\varepsilon_{ox}h_{si}} \qquad (11)$$

where ε_{ox} is the permittivity of silicon dioxide. The reduced impedance matrix of oxide $\mathbf{Z}_{ox,r}$ [3][5] can be calculated by the above-reduced matrix method.

After the reduced $RLGC$ matrix is defined, $\mathbf{Z}$ and $\mathbf{Y}$ are the impedance and conductance matrices, respectively, which can be represented as

$$\begin{cases} \mathbf{Z} = \mathbf{Z}_r + j\varpi\mathbf{L}_r \\ \mathbf{Y} = ((\mathbf{G}_{si,r} + j\varpi\mathbf{C}_{si,r})^{-1} + \mathbf{Z}_{ox,r})^{-1} \end{cases} \qquad (12)$$

Then, the Z-parameter is derived by

$$\begin{cases} \mathbf{Z}_{11} = \mathbf{Z}_{22} = \mathbf{Z}\left[\sqrt{\mathbf{YZ}}\right]^{-1}\left[\sinh\left(\sqrt{\mathbf{YZ}}\right)\right]^{-1}\left[\cosh\left(\sqrt{\mathbf{YZ}}\right)\right] \\ \mathbf{Z}_{12} = \mathbf{Z}_{21} = \mathbf{Z}\left[\sqrt{\mathbf{YZ}}\right]^{-1}\left[\sinh\left(\sqrt{\mathbf{YZ}}\right)\right]^{-1} \end{cases}$$

(13)

Finally, the Z-parameter can be converted to the S-parameter based on [1] for the non-homogeneous structure shown in Fig. 1.

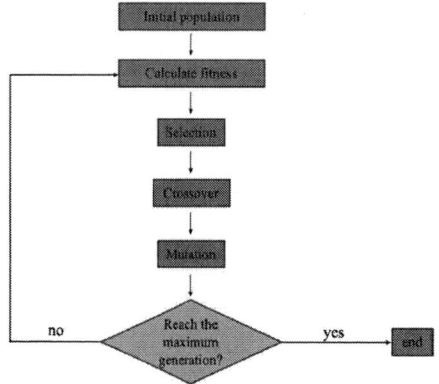

Fig. 3. The standard flow chart of the genetic algorithm.

III. OPTIMIZATION METHOD AND RESULTS

A. Genetic algorithm

Genetic algorithm (GA) is a powerful technique inspired by the process of natural selection and evolution to find the structure of the optimal solution [6]. Fig. 3 is the standard flow chart of GA. The initial population is generated, where each chromosome is composed of 3 parameters [p, d, h_{si}]. p ranges from 5 μm to 100 μm, d ranges from 3 μm to 100 μm,

979-8-3503-5124-8/24 $31.00 © 2024 IEEE

and h_{si} ranges from 30 μm to 100 μm. The fitness of each chromosome, represented by $|S_{21}|$, is calculated based on the proposed mathematical model.

In the process of selection, we sort the chromosomes according to their fitness. If the chromosome exhibits better fitness, it will be sorted towards the front of the matrix. During the crossover step, two adjacent chromosomes are selected, and the crossover point is chosen randomly. The genes of the two chromosomes on the crossover point are exchanged to generate two new chromosomes. In the mutation step, the one-point mutation method is used to randomly change the gene on the chromosome with a defined mutation probability. This iteration loop continues until the maximum number of generations is reached.

TABLE I.   INITIAL STRUCTURAL PARAMETERS

p	10 μm
d	30 μm
h_{si}	30 μm
t_{ox}	1 μm
$h_{passivation}$	0.17 μm
h_{IMD}	5 μm
$t_{ox,bottom}$	1 μm

TABLE II.   OPTIMIZED PARAMETERS FOR DIFFERENT OXIDE THICKNESS

t_{ox} (μm)	Parameter after optimizing		
	p (μm)	d (μm)	h_{si} (μm)
0.5	5	3	30
1	6	3	30
1.5	8	4	30

B. Simulation result

Table I shows the initial TSV structural parameters [7]. Table II presents the optimized TSV structures at different t_{ox}. Fig. 4(a) illustrates the comparison results before and after optimization. At 5 GHz, $|S_{21}|$ is improved 0.06 dB after the optimization. Fig. 4(b) presents the $|S_{21}|$ of TSV#1 with different t_{ox}, demonstrating that the smaller t_{ox} has the worse $|S_{21}|$. Because t_{ox} increases, the oxide capacitance decreases, and then the capacitance between the TSVs decreases, which improves $|S_{21}|$.

IV. CONCLUSION

This paper introduces a method to optimize TSV structures in HBM using MTL theory and GA. The recent mathematical model accurately predicts the insertion loss of TSV, aligning well with full-wave simulations. GA improves insertion loss by optimizing TSV parameters, suggesting a design rule with optimal pitch-to-diameter ratio near 2 and shorter TSV heights to improve S-parameters. Integrating MTL theory combined with GA offers an alternative tool for optimizing TSV arrays, improving HBM performance and guiding future TSV design.

ACKNOWLEDGMENT

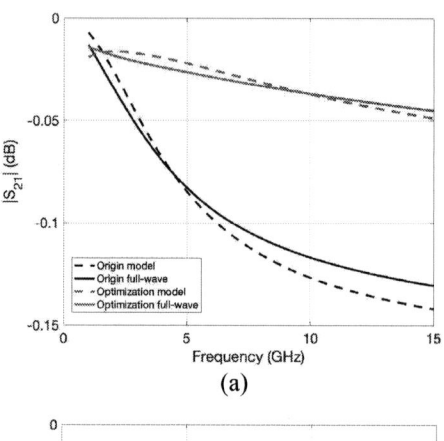

(a)

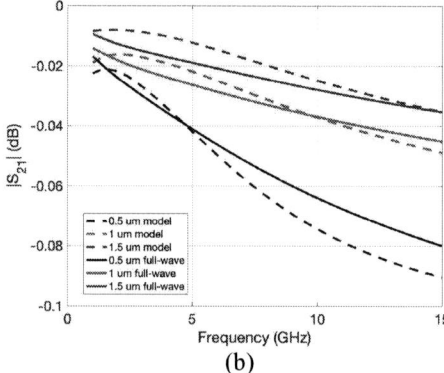

(b)

Fig. 4. $|S_{21}|$ of the signal TSV #1 (a) before and after optimization, and (b) for the different t_{ox}.

The authors would like to thank Powerchip Semiconductor Manufacturing Corporation (PSMC) for the funding support.

REFERENCES

[1] C. R. Paul, Analysis of Multiconductor Transmission Lines. New York: Wiley, 1994.

[2] Y. -J. Chang et al., "Novel crosstalk modeling for multiple through-silicon-vias (TSV) on 3-D IC: Experimental validation and application to Faraday cage design," *2012 IEEE 21st Conference on Electrical Performance of Electronic Packaging and Systems,* Tempe, AZ, USA, 2012.

[3] C. Qu, R. Ding, X. Liu and Z. Zhu, "Modeling and Optimization of Multiground TSVs for Signals Shield in 3-D ICs," in *IEEE Transactions on Electromagnetic Compatibility*, vol. 59, no. 2, pp. 461-467, April 2017.

[4] S. Vujević, D. Lovrić and V. Boras, "High-Accurate Numerical Computation of Internal Impedance of Cylindrical Conductors for Complex Arguments of Arbitrary Magnitude," in *IEEE Transactions on Electromagnetic Compatibility*, vol. 56, no. 6, pp. 1431-1438, Dec. 2014.

[5] A. E. Engin, "Passive Multiport RC Model Extraction for Through Silicon Via Interconnects in 3-D ICs," in *IEEE Transactions on Electromagnetic Compatibility*, vol. 56, no. 3, pp. 646-652, June 2014.

[6] K. F. Man, K. S. Tang and S. Kwong, "Genetic algorithms: concepts and applications [in engineering design]," in *IEEE Transactions on Industrial Electronics*, vol. 43, no. 5, pp. 519-534, Oct. 1996.

[7] L. -H. Huang, Y. -Y. Cheng and T. -L. Wu, "HBM3 PPA Performance Evaluation by TSV Model with Micro-Bump and Hybrid Bonding," *2023 IEEE 32nd Conference on Electrical Performance of Electronic Packaging and Systems (EPEPS)*, Milpitas, CA, USA, 2023, pp. 1-3

Application of the Reverse Pulse Technique for Worst Case Transient Analysis in HPC PDN Design

Chad M. Smutzer, Jordan R. Keuseman, Alexander P. Hickman and Clifton R. Haider
Mayo Clinic Special Purpose Processor Development Group (SPPDG)
Rochester, MN, USA
Email: smutzer.chad@mayo.edu, keuseman.jordan@mayo.edu, alexander.hickman0@gmail.com, haider.clifton@mayo.edu

Abstract—**The reverse pulse technique (RPT) predicts peak voltage deviation in the presence of worst case transient (WCT) load current. While RPT has been vetted through simulation, this paper addresses the practical application in lab hardware.**

Keywords—Worst Case Transient, Reverse Pulse Technique

I. INTRODUCTION

Modern high-performance compute (HPC) systems include processing devices consuming several hundred Watts of power. Many IC operating voltages have decreased to well below 1V leading to very narrow voltage noise tolerance margins. A traditional solution to reducing voltage noise has been to lower the power delivery network (PDN) impedance using the frequency-domain target impedance method (FDTIM) [1]. However, this approach has practical limits [2] that prevent a holistic error-prevention solution.

Processor load activity will vary as operational functions are exercised. The reverse pulse technique (RPT) is a methodology borrowed from oceanic "rogue wave" analysis [3] and applied in electronics design for bounding the pathological maximum voltage deviation in the presence of a worst-case transient (WCT) load current event [4],[5].

The theoretical aspects of RPT are well established and have been exemplified numerous times through simulation and modeling [4],[5],[6]. In this paper, we demonstrate the technique in the lab using test instrumentation and an exemplar resonant PDN. Two separate options for acquiring the voltage step response are described; one via direct measurement and the other through a synthetization process. The WCT current pulse train is computed from the voltage step response. This current load is subsequently applied to the physical PDN with a programmable arbitrary waveform generator (AWG) and current injection hardware. The measured voltage response is compared to prediction.

II. BASICS OF RPT PROCESS FLOW

The process for evaluating maximum expected PDN voltage variation includes a combination of hardware measurement and post-processing mathematics. A very basic flow diagram is illustrated in Fig. 1. Two unique methods of acquiring the voltage step response V(t) are represented as Path A and Path B in the diagram. Future references to voltage and current measurements or calculations are denoted with A and B depending upon which path was exercised. Path A is the direct time-domain measurement option whereby the voltage response to a current step load I(t)_step is obtained with an oscilloscope at the current injection point. Path B synthesizes the voltage response from the measured linear time invariant (LTI) PDN network Z(f) and a time-varying current step load stimulus.

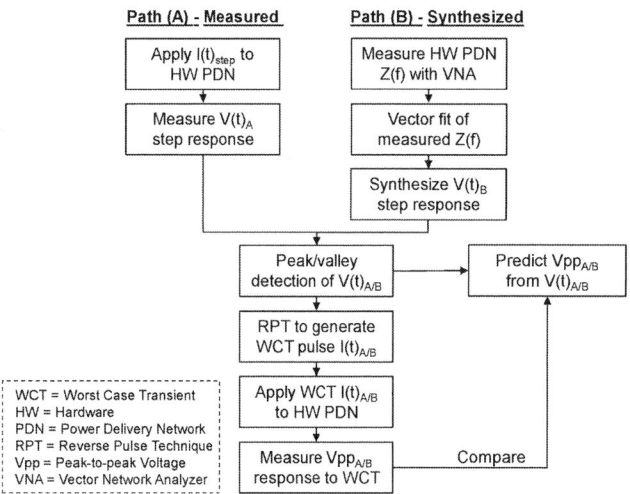

Fig. 1 – RPT Measurement and Calculation Process Flow Diagram

Each RPT implementation path offers unique advantages and disadvantages. The direct measurement of Path A is the most straightforward, but test instrumentation noise can pollute the results and falsify the current pulse train creation. In addition, this path requires assembled production hardware and high-bandwidth current injection capability. The synthesized approach in Path B allows for better noise rejection through VNA measurement or pre-fabrication electro-magnetic (EM) simulation models but is also susceptible to mathematical errors during post-processing computation.

Once the voltage step response V(t) is obtained, the process to create the WCT pulse train and acquire the predicted and measured maximal ripple voltage Vpp_max is identical. Using a filtered peak detection algorithm, the predicted Vpp_max is calculated from the summation of positive peaks and negative valleys in V(t). The WCT pulse train I(t)_WCT is constructed from the timestamp interval data in the V(t) response. The time-domain V(t) data is encoded with the resonant and Q-based decay information necessary to generate a square pulse train that will pathologically stimulate the PDN. A programmable AWG is used to drive load cells with I(t)_WCT at the current injection point of the PDN. The corresponding peak-to-peak ripple voltage is measured with an oscilloscope and compared to the

979-8-3503-5124-8/24 $31.00 © 2024 IEEE

predicted value. In the next sections, the RPT process will be demonstrated in lab measurements on exemplar hardware.

III. ACQUIRING THE VOLTAGE RESPONSE

For this example, the RPT process was intentionally demonstrated using a PDN with well-defined and exaggerated resonant peaks in the impedance profile Z(f) that readily produced obvious voltage variations. A simple printed circuit board (PCB) was designed with passive R/L/C elements to generate PDN resonances at 70 kHz, 600 kHz and 4 MHz.

For Path A, the measurement setup for capturing the step response V(t)A is depicted in Fig. 2 (a). A fixed rise-time voltage step was programmed into the arbitrary waveform generator (AWG) and converted to a 0A→1A current step load with commercial current injection hardware. The measured PDN voltage response to the load step is plotted as V(t)A in Fig. 3 (b) and the y-scale is normalized to Volts per Amp. For completeness, a 1A→0A load release response is also plotted.

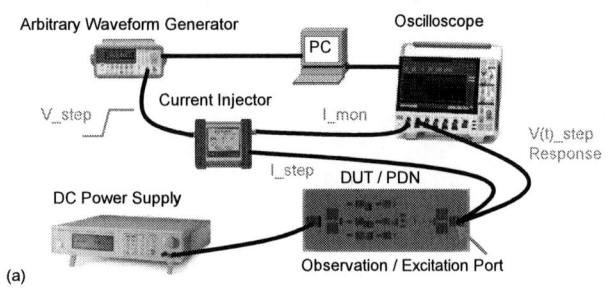

Step Response V(t) Measurement

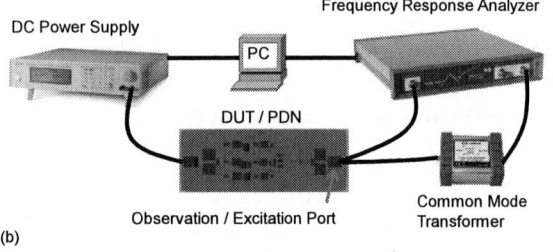

2-port Shunt-Thru Z(f) Measurement

Fig. 2 – Measurement Connectivity Diagrams: V(t)_step and Z(f)

As an alternative to directly measuring the time-domain voltage response V(t), Path B starts with a 2-port VNA measurement of the PDN impedance Z(f) as depicted in Fig. 2 (b). The DC power supply is necessary for capturing the low-frequency response in the complete PDN network. Additionally, a common-mode transformer is used to isolate leakage current in the test equipment return path that can cause measurement error. With this hardware setup, the 2-port shunt through methodology [7] is used for capturing a broadband S-parameter measurement in Touchstone format and subsequently converted to the frequency-based impedance Z(f).

A vector-fit algorithm was applied to Z(f) to produce an equivalent pole/residue model of the PDN network that is required for time-domain analysis mathematics. The fit correlates well with the measurement as illustrated in Fig. 3 (a).

A synthesized voltage response, V(t)B plotted in Fig. 3 (c), was then derived using the Matlab timeresp function with the rational fit Z(f) PDN model and a 0A→1A time varying load step stimulus. The step signal has a finite risetime approximately matching the current injection hardware edge rate capability.

The voltage responses attained from Path A and B are roughly equivalent. Minor variations are likely attributed to differences in model and test equipment bandwidth that will be discussed later. A prediction for maximum peak-to-peak voltage ripple was calculated through a summation of the peaks (additive) and valleys (subtractive) in these step response curves. Here, the predicted Vpp_max of 826 mV/A and 755 mV/A was calculated from V(t)A and V(t)B; respectively.

From this point forward, the RPT process is identical for operations on both V(t)A and V(t)B.

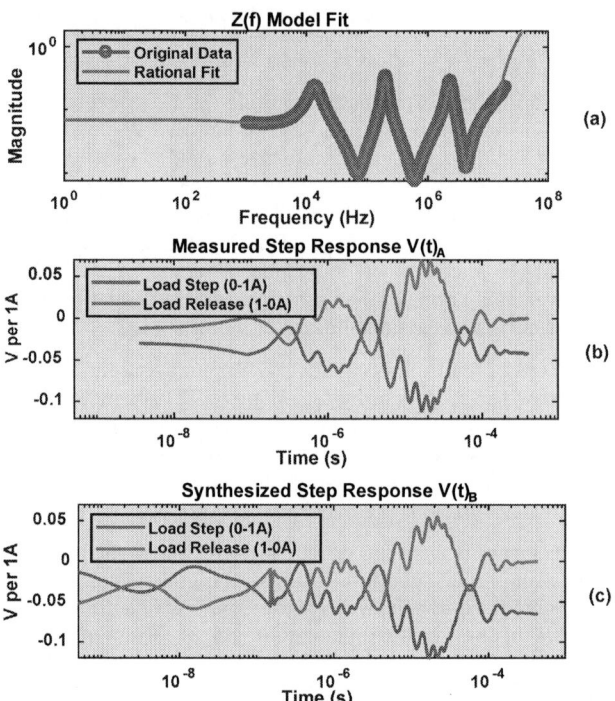

Fig. 3 – Voltage Step Response Plots V(t)A and V(t)B

IV. WCT PULSE TRAIN AND RPT VOLTAGE RESPONSE

Ringing visible in the voltage step responses illustrates the peaks, valleys and periodicity that contain coded information about the PDN resonances. A reverse decomposition of the voltage step responses [4] leads to a singular current pulse train with spectral content that will pathologically stimulate the PDN to produce the worst-case peak-to-peak voltage variation. For this PDN DUT hardware, a WCT pulse train was calculated from each step response V(t)A and V(t)B. As expected, the pulse trains derived from the similarly behaving voltage step responses are also very comparable.

Using the same current injection hardware with a programmable AWG, the WCT waveforms plotted as I(t)A and I(t)B in Fig. 4 (a) were applied to the PCB excitation port. The hardware configuration, as depicted in Fig. 2 (a), is identical to

979-8-3503-5124-8/24 $31.00 © 2024 IEEE

the setup for the step response except the applied voltage waveform from the AWG is the calculated WCT for the given PDN.

The voltage responses to each of the WCT current loads were captured with an oscilloscope and are plotted in Fig. 4 (b). Since the pulse train current load stimuli were very similar, it was expected that the final WCT voltage responses would also align as shown. For clarity, the inverse pulse train waveforms and voltage responses are not depicted but were also collected and are necessary to calculate the total maximum peak-to-peak voltage variation.

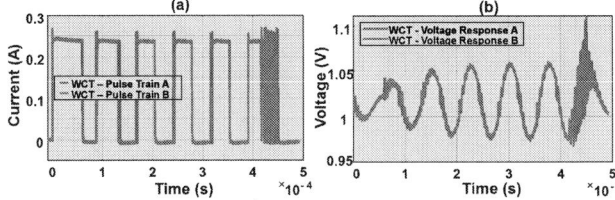

Fig. 4 – WCT Current Pulse Train and Network Voltage Response

In this example, a WCT current pulse train amplitude of 250 mA was selected. Using the predicted maximum ripple voltage scale factors of 826 mV/A and 755 mV/A previously acquired from the voltage step responses and normalized to the 250 mA applied load amplitude, the maximum predicted peak-to-peak ripple voltages of 206.5 mV and 188.8 mV were calculated for Path A and B; respectively. The voltage responses in Fig. 4 (b) represents the maximum voltage excursion in the presence of the applied current pulse trains. This measurement, along with the minimum voltage response from the applied inverse pulse train (not shown), resulted in total measured peak-to-peak ripple of 201.3 mV and 189.5 mV. These values correlate to the predictions within -2.5% and 0.4% for Path A and B; respectively.

V. A Discussion of Results and Practical Applications

Although the RPT process was demonstrated on simplified PDN hardware, the same principles would be applied in production designs. Ideally, the PCB would be populated with the VR, capacitors, and a die-less package for the intended load IC, e.g., ASIC or FPGA. In this case, the current step, and subsequent voltage measurements, would be applied at the exposed IC power pads. However, acquiring die-less packages is often impractical. With additional programming and calibration efforts, a populated production ASIC or FPGA could provide the high-bandwidth step current and WCT pulse but may not be as precise or reliable as dedicated current injection test instrumentation. In any case, capturing the non-linear behavior of the application VR is important.

These physical implementation challenges are partially responsible for the exploration of Path B. With trustworthy linear network models of the PDN path from VR to die, the peak-to-peak ripple voltage can still be predicted through the RPT process without fabricated hardware or packaged devices. Hence, Path B can provide post-layout and pre-fabrication insight into power delivery performance and boundary conditions. It is noted that the non-linear VR behavior and its intrinsic noise is ignored in Path B.

The difference in the results between Path A and Path B occurred during the V(t) acquisition stage. The ideal pulse generated in software from V(t) and programmed into the AWG has idealized edge transitions and noiseless steady-state current values. However, the applied load, as measured at the output of the current injection hardware, is incapable of matching the ideal conditions produced from software algorithms [8]. These signal imperfections are likely the cause of minor errors between predicted and measured worst case ripple values for both paths.

VI. Summary

The reverse pulse technique was demonstrated with test instrumentation on a resonant power delivery network. Two options for acquiring the voltage step response were presented with reasonable equivalence. It was also shown that worst case ripple voltage for a given PDN can be predicted prior to hardware fabrication and assembly. Less than 5% difference was observed between measured and predicted maximum peak-to-peak ripple voltage.

Only pathological current load events, stimulating the primary PDN resonances, could create these worst-case voltage ripple excursions. The intent of the RPT is to bound the PDN performance rather than to predict the practical operating conditions. If operational current load conditions are conveniently known, then PDN design quality can be readily evaluated using more traditional analysis techniques for design checkout.

Application of the RPT in an ultra-low PDN impedance (< 500 µΩ) product is a natural extension to the work presented in this paper. Overcoming challenges with programmable load cell bandwidth and measurement observability in a complex structure are additional considerations. It is also anticipated that identifying and resolving less obvious peaks/valleys in a flatter voltage response will add complexity to generating a true WCT pulse train.

References

[1] Smith, L. et al., "Power Distribution System Design Methodology and Capacitor Selection for Modern CMOS Technology," IEEE Transactions on Advanced Packaging, Vol. 22, No. 3, August 1999.

[2] Smutzer, C. M., Gilbert, B. K., and Daniel, E. S., "Practical limitations of state-of-the-art passive printed circuit board power delivery networks for high performance compute systems," 2013 17th IEEE Workshop on Signal and Power Integrity, Paris, France, 2013, pp. 1-4.

[3] Drabkin, et al, "Aperiodic Resonant Excitation of Microprocessor Power Distribution Systems and the Reverse Pulse Technique," Proceedings of EPEP 2002, p. 175.

[4] Novak, I., "Systematic Estimation of Worst-Case PDN Noise: Target Impedance and Rogue Waves," PCB Design 007 QuietPower, November 2015.

[5] Sandler, S., "Target Impedance Limitations and Rogue Wave Assessments on PDN Performance," DesignCon, 2015.

[6] Nagy, I., "Rogue Wave Estimation in PDNs Using the Multi-Tone Technique," Printed Circuit Design and Fab, March 2022.

[7] Novak, I., "Measuring MilliOhms and PicoHenry's in Power Distribution Networks," DesignCon 2000, February 1-4.

[8] Keuseman, J. R., Daun-Lindberg, T., White, C. K., Haider, C. R., and Gilbert, B. K., "Enhancing Ultra-low Impedance Measurement Techniques for High-performance Power Delivery Networks," 2020 IEEE International Symposium on Electromagnetic Compatibility & Signal/Power Integrity (EMCSI), Reno, NV, USA, 2020, pp. 1-5

A DDR5 Interposer De-embedding Method Based on Transfer Function

Aobo Li
Key Laboratory of High-Speed Circuit Design and EMC Ministry of Education
Xidian University
Xi'an, China
liaobo@stu.xidian.edu.cn

Jun Wang
Key Laboratory of High-Speed Circuit Design and EMC Ministry of Education
Xidian University
Xi'an, China
wangjun313@xidian.edu.cn

Yan Xu
Key Laboratory of High-Speed Circuit Design and EMC Ministry of Education
Xidian University
Xi'an, China
22021211249@stu.xidian.edu.cn

Kangkang Zhang
Key Laboratory of High-Speed Circuit Design and EMC Ministry of Education
Xidian University
Xi'an, China
22021211668@stu.xidian.edu.cn

Xiuqin Chu
Key Laboratory of High-Speed Circuit Design and EMC Ministry of Education
Xidian University
Xi'an, China
xqchu@mail.xidian.edu.cn

Abstract—**A novel method for DDR interposer de-embedding is introduced in this article, utilizing transfer function and microwave network theory. The de-embedding transfer function is computed and then convolved with the waveform measured at the probe tip of the interposer to restore the waveform at the DRAM ball. The interposer is characterized as a 3-port network, and its parasitic effects can be entirely eliminated using the proposed technique. The accuracy of this method is assessed through validation in the time domain simulations.**

Keywords—de-embedding, signal integrity, DDR5 interposer

I. INTRODUCTION

Signal measurement plays a vital role in validating DDR5 memory interfaces. The signal test point is specified at the DRAM ball, but direct access is hindered by advanced packaging technologies like Ball Grid Array (BGA). In some cases, signals can be accessed through vias on the PCB backside, impacting signal integrity significantly due to high-speed signal characteristics in DDR5. To address this, DDR interposers are increasingly used for memory interface validation [1]. Acting as small PCBs, interposers route signals to their edges, enabling signal pins breakout at the DRAM ball for oscilloscope probing. Channel discontinuities and parasitic effects introduced by the interposer necessitate de-embedding to obtain accurate waveforms. Previous methods have modeled the interposer as a 2-port or 3-port network, with the latter better capturing interposer characteristics [2]. A novel de-embedding method presented in this study computes a transfer function based on microwave network theory to remove parasitic effects completely and recover accurate waveforms at the DRAM ball.

Several methods have been utilized in the past for interposer de-embedding. In [3], the interposer is characterized as a 2-port network linking the probe tip and DRAM ball. However, the 2-port network approach only views the interposer as an extension to the probe, thereby de-embedding solely the impact of the stub while neglecting a part of the interposer's via. On the other hand, in [3], the interposer is depicted as a 3-port network. Unlike the 2-port network, the 3-port network can comprehensively capture the performance characteristics of the interposer. Nevertheless, this method still includes the insertion loss caused by the interposer and does not completely de-embed the interposer.

This study introduces a novel approach to interposer de-embedding. By leveraging transfer function principles and microwave network theory, a de-embedding transfer function is computed [5]. This function is then convolved with the measured waveform at the probe tip to reconstruct the waveform at the DRAM ball. Modeling the interposer as a 3-port network allows for the complete removal of its parasitic effects.

II. DE-EMBEDDING MODEL AND THEORY DERIVATION

A. De-embedding Structure

A standard DDR5 memory interface consists of a DDR controller, SOC package, PCB, DRAM package, and ODT termination resistor. The interposer is positioned between the memory chip and the PCB. The DDR5 memory interface structure with interposer is shown in Fig. 1.

Fig. 1 DDR5 memory interface structure with interposer

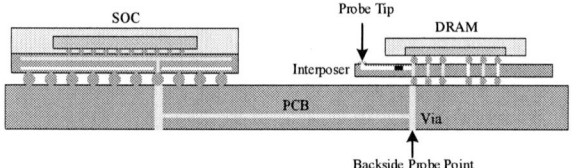

(a)

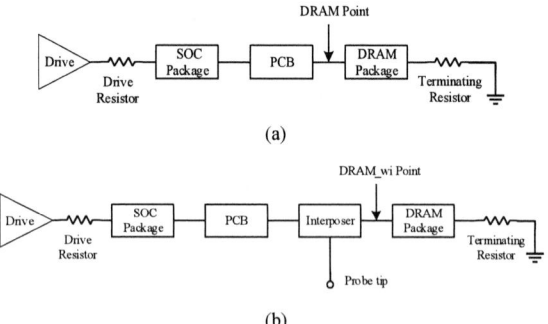

(b)

Fig. 2 (a) Block diagram of memory interface, and (b) Block diagram of memory interface with interposer

S-parameters are employed to capture the performance characteristics of the interconnects. The SOC package, PCB, and DRAM package are represented by 2-port S-parameters. The memory interface's block diagram is illustrated in Fig. 2(a). On the other hand, the interposer is depicted using a 3-port S-parameter, as shown in Fig. 2(b). Waveform

---

This work was supported in part by the National Natural Science Foundation of China under Grant 62171338, 61871453 and 61901331, and in part by the Natural Science Research Program of Shaanxi under grant 2021TD-07.

measurements are conducted by placing the oscilloscope probe at the probe tip. Through the proposed method, precise waveforms at the DRAM point can be derived from the waveform at the probe tip.

B. Mathematics Derivation

The transfer function from the probe tip to the DRAM's DRAM_wi point is initially calculated. The circuit diagram depicting the 3-port Z-parameter matrix of the interposer can be seen in Fig. 3.

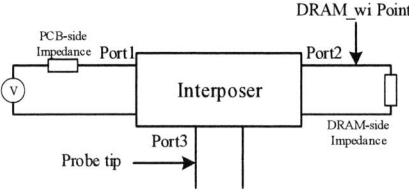

Fig. 3 Circuit diagram of Z-parameter of the interposer

The U_i and I_i represent the current and voltage of port i, respectively. The probe tip corresponds to port 3, while port 3 is situated at the DRAM-side. The impedance at the DRAM-side is denoted as Z_{in}. Following the definition of the Z-parameter of the interposer, the relationship between the voltage at the probe tip and the DRAM_wi point is illustrated in (1):

$$U_{dram\_wi} = U_2 = z_{21}I_1 + z_{22}I_2 + z_{23}I_3$$
$$U_{probe} = U_3 = z_{31}I_1 + z_{32}I_2 + z_{33}I_3 \quad (1)$$

where z_{ij} are entries of Z-parameter matrix of the interposer; U_{probe} is the voltage at the probe tip; U_{dram_wi} is the voltage at the DRAM_wi point and I_2 is the current into port 2. The connection between U_{dram_wi} and is detailed as follows:

$$U_{dram\_wi} = -I_2 Z_{in} \quad (2)$$

The DRAM-side impedance is determined in (3) utilizing the Z-parameters matrix of the DRAM package and the termination resistor:

$$Z_{in} = \frac{z_{11}^{DRAM} z_{22}^{DRAM} - z_{12}^{DRAM} z_{21}^{DRAM} + z_{11}^{DRAM} Z_L}{z_{22}^{DRAM} + Z_L} \quad (3)$$

where Z_L is the value of the termination resistor and z_{ij}^{DRAM} are the entries of Z-parameter matrices of the DRAM package.

Given the high input impedance of the oscilloscope probe, the current at the probe tip I_3 is considered negligible. By substituting (2) into (1), the transfer function from the probe tip to the DRAM_wi point can be expressed and denoted as TF_1:

$$TF_1 = \frac{U_{DRAM\_wi}}{U_{probe}} = \frac{z_{21} Z_{in}}{z_{31} Z_{in} - z_{21} z_{23} + z_{31} z_{22}} \quad (4)$$

Subsequently, the transfer function from the DRAM_wi point to the DRAM point is calculated. Both interconnects, with and without the interposer, are linked to the same voltage source. The frequency response from the source to the DRAM_wi point is labeled as H_1, while the frequency response from the source to the DRAM point is denoted as H_2, as shown in (5). The transfer function from the DRAM_wi point to the DRAM point, represented as TF_2, equals the ratio of H_2 to H_1:

$$H_1 = \frac{U_{DRAM\_wi}}{U_s}; H_2 = \frac{U_{DRAM}}{U_s} \quad (5)$$

$$TF_2 = \frac{U_{DRAM}}{U_{DRAM\_wi}} = \frac{H_2}{H_1} \quad (6)$$

The determination of H_1 and H_2 can be achieved by cascading the S-parameter and simplifying interconnects to 2-port S-parameter networks with a source and load impedance.

Due to the high input impedance of the oscilloscope probe at the probe tip, the 3-port S-parameter of the interposer can be simplified to a 2-port network with port 3 terminated with an open circuit:

$$s'_{ij} = s_{ij} - \frac{s_{i3} s_{3j}}{1 + s_{33}} (i, j = 1, 2) \quad (7)$$

where s_{ij} are entries of S-parameter matrices of the interposer and s'_{ij} are entries of S-parameter matrices of simplified 2-port interposer.

To acquire the S-parameter of the whole structure, the S-parameters of the SOC package, PCB, and 2-port interposer are cascaded through T-parameter [6], denoted as network U, as indicated in Fig. 4(a). The network D is obtained by cascading the S-parameter of the SOC package and PCB, as depicted in Fig. 4(b). The source impedance and the load impedance at the DRAM-side of both interconnects are denoted as Z_s and Z_{in}, respectively.

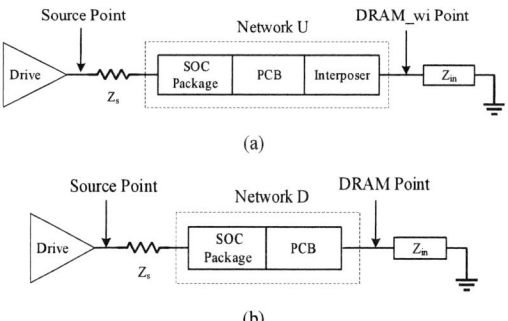

Fig. 4 Simplified network of memory interfaces (a) network U, and (b) network D

H_1 and H_2 are determined using the simplified 2-port S-parameter networks displayed in Fig. 4:

$$H_1 = \frac{S_{21}^U (1 + \Gamma_l)(1 - \Gamma_s)}{2(1 - S_{22}^U \Gamma_l)(1 - \Gamma_{in}^U \Gamma_s)} \quad (8)$$

$$H_2 = \frac{S_{21}^D (1 + \Gamma_l)(1 - \Gamma_s)}{2(1 - S_{22}^D \Gamma_l)(1 - \Gamma_{in}^D \Gamma_s)} \quad (9)$$

in which $\Gamma_l = \frac{Z_{in} - Z_0}{Z_{in} + Z_0}; \Gamma_s = \frac{Z_s - Z_0}{Z_s + Z_0} \quad (10)$

$$\Gamma_{in}^U = S_{11}^U + (S_{12}^U S_{21}^U \frac{\Gamma_l}{1 - S_{22}^U \Gamma_l}); \Gamma_{in}^D = S_{11}^D + (S_{12}^D S_{21}^D \frac{\Gamma_l}{1 - S_{22}^D \Gamma_l})$$

$$(11)$$

where S_{ij}^D are entries of S-parameter matrices of network D; S_{ij}^U are entries of S-parameter matrices of network U.

The transfer function TF_2 is the ratio of H_2 to H_1:

$$TF_2 = \frac{U_{DRAM}}{U_{DRAM\_wi}} = \frac{H_2}{H_1} = \frac{S_{21}^D(1-S_{22}^U\Gamma_l)(1-\Gamma_{in}^U\Gamma_s)}{S_{21}^U(1-S_{22}^D\Gamma_l)(1-\Gamma_{in}^D\Gamma_s)} \quad (12)$$

Subsequently, the de-embedding transfer function, represented as $TF_{de\text{-}embed}$, is obtained by combining TF_1 and TF_2:

$$TF_{de-embed} = TF_1 \times TF_2 \quad (13)$$

This de-embedding transfer function in the frequency domain is converted to the impulse response in the time domain through Inverse Fast Fourier Transform (IFFT). The de-embedded waveform at the DRAM point is computed by convolving the impulse response with the measured waveform at the probe tip:

$$u_{de-embed}(t) = IFFT[TF_{de-embed}(f)] \otimes u_{probe}(t) \quad (14)$$

III. VALIDATIONS

The validation of the proposed de-embedding method is conducted by comparing the de-embedded results with the simulated results in the circuit simulator. The simulation diagram in the circuit simulator is shown in Fig. 5.

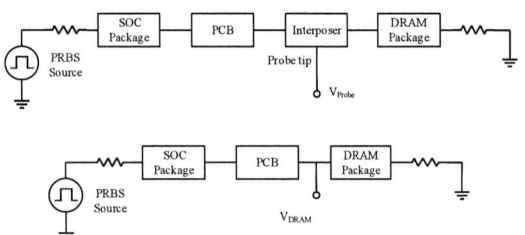

Fig. 5 Simulation diagram of memory interface with and without interposer

Waveforms at the DRAM point V_{DRAM} and probe tip V_{probe} are initially simulated using same PRBS sequence at a bitrate of 6.4Gbps to stimulate the memory interface with interposer and without interposer, respectively. The de-embedded waveform $V_{de\text{-}embed}$ at the DRAM point is computed by proposed de-embedding method with the simulated waveform V_{probe}. The comparison between the simulated waveform V_{DRAM} and de-embedded waveform $V_{de\text{-}embed}$ is presented in Fig. 6, displaying a good alignment between the proposed method and the simulated result. Furthermore, there is a noticeable difference between the waveform at the probe tip and the waveform at the DRAM. The simulated and computed eye diagrams are depicted in Fig. 7, demonstrating minimal errors and further validating the accuracy of the method.

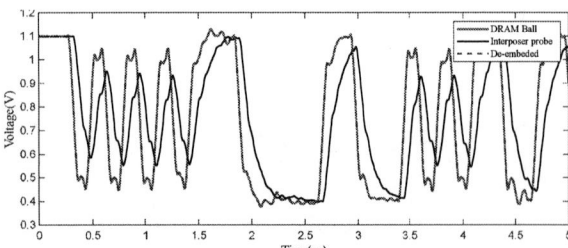

Fig. 6 Waveform comparison between V_{DRAM}, V_{probe} and $V_{de\text{-}embed}$

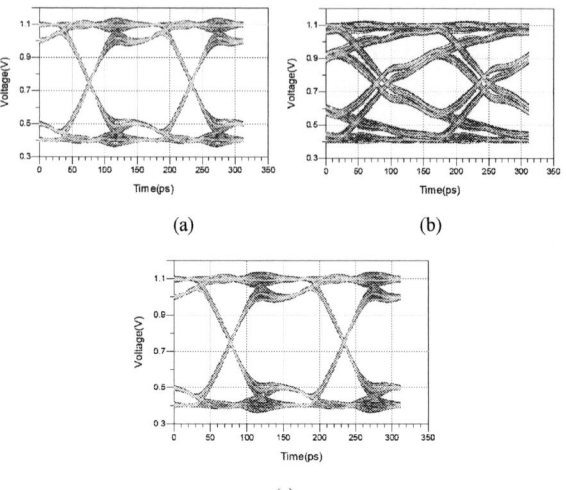

Fig. 7 (a) Eye diagram of simulation waveform at DRAM, and (b) Eye diagram of simulation waveform at probe tip, and (c) Eye diagram of de-embedded waveform at DRAM

IV. CONCLUSION

In this paper, a technique for interposer de-embedding is explored to accurately retrieve the waveform at the DRAM ball from the waveform measured at the probe tip of the interposer. The interposer is characterized as a 3-port network, allowing for full elimination of the parasitic effects introduced by the interposer. The effectiveness of the method is assessed through validation in simulations conducted in the time domains. Moreover, the method can be easily implemented using programming languages.

REFERENCES

[1] Keysight Technology, "W5642A DDR5 78-ball BGA Interposer," 2020. [Online]. Available: https://www.keysight.com/us/en/assets/7120-1071/data-sheets/W5643A-DDR5-78-ball-BGA-Interposer.pdf.

[2] J. Socha, J. Dandy, P, Pun and P. Thota, "Designing High Performance Interposers with 3-port and 6-port S-parameters," *DesignCon 2015, Proceddings*, Santa Clara, CA, 2015.

[3] W. Yuan, S. Kim, W. H. Ryu, S. Moon and S. Lee, "Robust PoP probing solutions for high-performance application processor developments," *2013 IEEE 22nd Conference on Electrical Performance of Electronic Packaging and Systems*, San Jose, CA, USA, 2013, pp. 159-162.

[4] J. J. Kim et al., "Signal Integrity Design and Analysis of a Multilayer Test Interposer for LPDDR4 Memory Test With Silicone Rubber-Based Sheet Contact," in *IEEE Transactions on Electromagnetic Compatibility*, vol. 59, no. 4, pp. 1239-1251, Aug. 2017.

[5] M. K. Sampath and N. Atout, "Signal integrity validation of de-embedding techniques using accurate transfer functions," *2012 IEEE 16th International Symposium on Consumer Electronics*, Harrisburg, PA, USA, 2012, pp. 1-4.

[6] David M.Pozar, Microwave Engineering, 3rd ed., Wiley,

A Hybrid Polynomial Chaos Expansion and Gaussian Process Regression Method for Forward Uncertainty Quantification of Integrated Circuits

Paolo Manfredi and Riccardo Trinchero

EMC Group, Department of Electronics and Telecommunications, Politecnico di Torino
Corso Duca degli Abruzzi 24, 10129 Torino, Italy
E-mail: {paolo.manfredi, riccardo.trinchero}@polito.it

Abstract—This paper introduces a novel kernel-based formulation for the efficient uncertainty quantification of integrated circuits. The method combines the polynomial chaos expansion (PCE) and Gaussian process regression (GPR) frameworks, the former to provide closed-form statistical information and the latter for an efficient training. In essence, the PCE coefficients are computed using a suitable Bayesian formulation that involves the definition of a special implicit kernel based on an infinite sequence of Hermite polynomials. The proposed method is illustrated based on a network with microstrip lines.

Index Terms—Gaussian process regression, kernel method, least-square support-vector machine, machine learning, polynomial chaos, surrogate modeling, uncertainty quantification.

I. INTRODUCTION

In recent years, polynomial chaos expansion (PCE) emerged as a primary tool for uncertainty quantification (UQ) in signal and power integrity [1], [2]. The main advantage of PCE lies in the closed-form statistical and sensitivity information, which is analytically derived from the model coefficients, and in its optimal convergence rate, making it more efficient than Monte Carlo (MC) for a moderate number of uncertain parameters.

Nevertheless, the burgeoning field of artificial intelligence led researches and engineers to start exploring alternative approaches based on machine learning methods due to their inherent data-driven nature. In particular, kernel-based methods such as least-square support-vector machine (LSSVM) and Gaussian process regression (GPR) exhibit attractive features due to their relatively simple form and flexible nature, which allows them to naturally scale well to a large number of uncertain parameters [1], [2], [3].

In this paper, we introduce a hybrid PCE-GPR approach that combines the advantages of both methods. With a suitable formulation that leverages a special kernel based on an infinite sequence of Hermite polynomials, the PCE coefficients are analytically derived from the GPR model via a simple and quick post-processing. Moreover, thanks to the inherent Bayesian setting of GPR, a prediction confidence is associated to the PCE, which accounts for the limited accuracy due to the finite number of training samples that is used.

II. REVIEW OF BASELINE SURROGATE MODELS

Let us consider a generic simulation problem expressed as

$$y = \mathcal{M}(\boldsymbol{x}), \qquad (1)$$

where $\mathcal{M}$ generically denotes the algorithm or simulator that allows obtaining the output of interest y for given a configuration of the input parameters $\boldsymbol{x}$. The goal of forward UQ is to statistically characterize the output y resulting from uncertainty on the inputs $\boldsymbol{x}$. We assume y to be a scalar and the inputs $\boldsymbol{x}$ to be independent and Gaussian-distributed.

A. PCE

The PCE seeks to approximate (1) using an expansion of orthogonal Hermite polynomials, i.e.,

$$y \approx \mathcal{M}_{\mathrm{PCE}}(\boldsymbol{x}) = c_0 + \sum_{k=1}^{K} c_k H_k(\boldsymbol{x}) \qquad (2)$$

where H_k are multivariate orthonormal polynomials that are built as the product of univariate Hermite polynomials.

A peculiarity of the PCE is that the first two statistical moments, the mean and the variance, are readily obtained from the coefficients as $\mu_y \approx c_0$ and $\sigma_y^2 \approx \sum_{k=1}^{K} c_k^2$, respectively. Several approaches exist for the calculation of the PCE coefficients, ranging from the accurate yet intrusive stochastic Galerkin method to non-intrusive techniques based on projection, interpolation, or regression [4] (Ch. 2).

B. LSSVM

The model can be expressed using two equivalent formulations, i.e.,

$$y \approx \mathcal{M}(\boldsymbol{x}) = b + \sum_{k=1}^{K} w_k \varphi_k(\boldsymbol{x}) = b + \sum_{l=1}^{L} \alpha_l k(\boldsymbol{x}, \boldsymbol{x}_l), \quad (3)$$

where φ_k are generic "feature-space functions" in the *primal space*, whereas $k(\cdot, \cdot)$ is a kernel function in the *dual space*. Moreover, $\{\boldsymbol{x}_l\}_{l=1}^{L}$ are a set of training points at which observations $y_l = \mathcal{M}(\boldsymbol{x}_l)$ of the actual system are collected. The two formulations are readily shown to be equivalent if the kernel function is defined explicitly as

$$k(\boldsymbol{x}, \boldsymbol{x}') = \sum_{k=1}^{K} \varphi_k(\boldsymbol{x}) \varphi_k(\boldsymbol{x}'). \qquad (4)$$

The bias term b and the dual-space coefficients $\boldsymbol{\alpha} = (\alpha_1, \ldots, \alpha_L)^{\mathsf{T}}$ are obtained by solving the linear system

$$\begin{pmatrix} \boldsymbol{K} + \gamma^{-1}\boldsymbol{I} & \boldsymbol{1} \\ \boldsymbol{1}^{\mathsf{T}} & 0 \end{pmatrix} \begin{pmatrix} \boldsymbol{\alpha} \\ b \end{pmatrix} = \begin{pmatrix} \boldsymbol{y} \\ 0 \end{pmatrix}, \qquad (5)$$

979-8-3503-5124-8/24 $31.00 © 2024 IEEE

where $\boldsymbol{K}$ is the $L \times L$ kernel matrix of the training points, with entries $K_{lm} = k(\boldsymbol{x}_l, \boldsymbol{x}_m)$ for $l, m = 1, \ldots, L$, $\boldsymbol{I}$ is the identity matrix of the same size, $\mathbf{1}$ is a column vector of ones, $\boldsymbol{y} = (y_1, \ldots, y_L)^\mathsf{T}$ is the vector of the training observations, and γ is a regularization hyperparameter.

The advantage of the LSSVM formulation is that the coefficients are computed by means of L observations only, regardless of the dimensionality of $\boldsymbol{x}$ and the number of primal-space functions K. Furthermore, the dual-space formulation is also valid for implicit (e.g., radial basis function or squared-exponential) kernels, which are more efficient to evaluate even in high-dimensional settings. However, in that case, an explicit primal-space formulation is lost due to the unavailability of the corresponding feature-space functions φ_k.

It was recently shown that the PCE model (2) can be cast as a LSSVM problem (3) if Mehler kernel is used [5]. The univariate kernel reads

$$k(x, x') = \sum_{k=0}^{\infty} \rho^k H_k(x) H_k(x') = \frac{e^{\frac{-\rho^2(x^2 + x'^2) - 2xx'}{2(1-\rho^2)}}}{\sqrt{1-\rho^2}} \quad (6)$$

and corresponds to the sum of an (infinite!) sequence of (univariate) Hermite polynomials. Comparing with (4) shows that the Hermite polynomials, rescaled by a factor $\sqrt{\rho^k}$, form the feature-space functions for this kernel.

The multivariate kernel is build as the product of univariate kernels and is further equipped with a variance σ^2, i.e.,

$$k(\boldsymbol{x}, \boldsymbol{x}') = \sigma^2 \prod_{j=1}^{d} k(x_j, x_j'). \quad (7)$$

Apart from trivial rescaling factors, this makes the corresponding feature-space functions generalize to the multivariate Hermite polynomials as in (2). Since the feature-space functions for this kernel are known to be the Hermite polynomials, the primal-space LSSVM coefficients w_k are readily obtained from the dual-space coefficients α_l and correspond to the PCE coefficients c_k (virtually, up to an infinite order!), whereas the bias term b is equivalent to c_0.

C. GPR

The model assumes that (1) be one specific realization of a prior Gaussian process $\mathcal{GP}(\mu(\boldsymbol{x}), k(\boldsymbol{x}, \boldsymbol{x}'))$, where the function $k(\cdot, \cdot)$ now plays the role of prior covariance.

Starting from a collection of training observations $\{(\boldsymbol{x}_l, y_l)\}_{l=1}^{L}$, similarly as for the LSSVM, posterior predictions are computed as

$$y \approx \mathcal{M}(\boldsymbol{x}) = \mu(\boldsymbol{x}) + \sum_{l,m=1}^{L} B_{lm}(y_m - \mu(\boldsymbol{x}_m)) k(\boldsymbol{x}, \boldsymbol{x}_l) \quad (8)$$

where B_{lm} are the elements of $\boldsymbol{B} = \boldsymbol{K}^{-1}$, the inverse of the kernel matrix with the same definition as in (5).

Notably, a (co)variance

$$c(\boldsymbol{x}, \boldsymbol{x}') = k(\boldsymbol{x}, \boldsymbol{x}') - \sum_{l,m=1}^{K} k(\boldsymbol{x}, \boldsymbol{x}_l) B_{lm} k(\boldsymbol{x}, \boldsymbol{x}_m), \quad (9)$$

is also associated to the model predictions, which reflects their uncertainty due to the limited amount of observations. This information can be used to obtain confidence bounds on pertinent statistical information when the GPR model is used as a surrogate in a MC analysis [3]. However, the calculation becomes prohibitive if the number of MC trials is large, due to the large size of the resulting covariance matrix.

III. PROPOSED HYBRID PCE-GPR METHOD

It is readily shown that the GPR and the dual-space LSSVM formulations are in fact equivalent if a constant prior mean function $\mu(\boldsymbol{x}) = b$ is assumed and the same LSSVM kernel is taken as the GPR prior covariance function. With the above settings, the proof is found by comparing (8) and the dual-space formulation in (3), and noting that the coefficients

$$\alpha_l = \sum_{m=1}^{L} B_{lm}(y_m - b) \quad (10)$$

are equivalent to the solution of (5) if the term $\gamma^{-1}\boldsymbol{I}$ is embedded in the kernel matrix $\boldsymbol{K}$. In the GPR settings, this is equivalent to assuming a Gaussian noise on the training data with standard deviation $\sigma_n = \sqrt{\gamma^{-1}}$ [3]. For the sake of simplicity, in the following we assume a noise-free interpolation, which is a standard scenario if no noise is expected on the observations and is equivalent to letting $\gamma \to \infty$ in (5).

In summary, based on the above considerations, a GPR model with constant trend that uses the Mehler kernel (6) as prior covariance is equivalent to a dual-space LSSVM formulation, whose primal-space formulation is in turn equivalent to a Hermite PCE. This Bayesian interpretation allows associating a predictive covariance in the form of (9), from which the covariance matrix of the PCE coefficients is analytically derived, expressing the estimation uncertainty resulting from using a limited amount of observations. Owing to the lack of space, the theoretical details are deferred to a future report.

At present, an analogous kernel is available for Legendre polynomials and uniform variability [5]. As in standard PCE, dealing with Gaussian correlation is straightforward and is readily achieved by preliminarily decorrelating the input variables. Suitable kernels should be instead identified for other distributions, including non-Gaussian correlated ones.

IV. APPLICATION EXAMPLE AND NUMERICAL RESULTS

The proposed hybrid PCE-GPR method is applied to the UQ of the transmitted voltage in the network of Fig. 1, which was one of the test cases in [4] (Ch. 3). The uncertainty is in the thickness, relative permittivity, and loss tangent of the substrate. These parameters are Gaussian-distributed with nominal values of $\varepsilon_r = 4.1$, $\tan\delta = 0.02$, and $h = 100\,\mu\text{m}$ and a standard deviation of 10%. We refer to [4] for additional details. For the sake of simplicity, we restrict the analysis to the transmitted voltage at a specific time point, i.e., $t = 1.46$ ns, corresponding to the location of the maximum overshoot.

We compare the advocated method against the following state-of-the-art techniques [4]: a) stochastic Galerkin method

979-8-3503-5124-8/24 $31.00 © 2024 IEEE

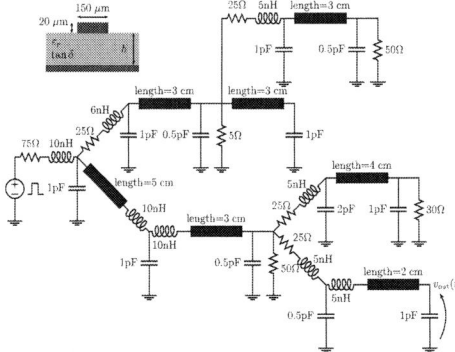

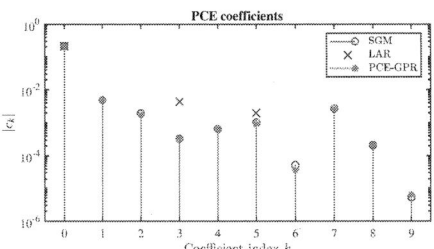

Fig. 1. Schematic of the considered test case.

Fig. 2. PCE coefficients of the transmitted voltage at 1.46 ns (log scale). Circles (∘): SGM; crosses (×): LAR; stars (∗): proposed PCE-GPR.

TABLE I
MEAN, VARIANCE, AND RMSE OBTAINED WITH THE VARIOUS METHODS.

	Mean (V)	Variance ($V^2 \times 10^5$)	RMSE (V)
MC	0.2120	3.5628	–
SGM	0.2120	3.5234	4.4809×10^{-4}
ST	0.2120	3.5546	4.7095×10^{-4}
LAR	0.2122	2.3751	2.4443×10^{-3}
Plain GPR	0.2122	2.7771	1.7566×10^{-3}
2-sigma interval	$[\overline{0.2120}, \overline{0.2124}]$	$[\overline{2.2139}, \overline{3.3402}]$	
Hybrid PCE-GPR	0.2120	3.5377	4.7845×10^{-4}
2-sigma interval	$[\overline{0.2120}, \overline{0.2120}]$	$[\overline{3.2843}, \overline{3.8181}]$	–

(SGM): requires a single deterministic simulation of a $(K+1)$-times augmented counterpart of the original stochastic circuit; b) stochastic testing (ST): solves the stochastic circuit at $K + 1$ specific configurations of the uncertain parameters and interpolates the resulting responses; c) least-angle regression (LAR): collects observations for some configurations of the uncertain parameters and performs a sparse regression, here implemented using UQLab [6]; d) plain GPR: baseline formulation considering a squared-exponential kernel. Furthermore, reference results are generated based on a MC simulation with 10'000 runs. For the PCE-based techniques, including the proposed hybrid method, we consider a second-order expansion as in [4], which features $K + 1 = 10$ terms. However, it is important to note that with PCE-GPR the expansion order does not need to be predetermined, as opposed to classical PCE-based methods. The 10 simulations of the ST method are also used to train the data-driven methods, i.e., LAR and the two GPR-based techniques.

Figure 2 compares the magnitude of the PCE coefficients obtained with the SGM method against the coefficients estimated using LAR and the proposed PCE-GPR method. Despite using exactly the same training data, LAR is able to identify only three coefficients (i.e., c_0, c_3, and c_6), two of which with rather low accuracy. Conversely, the proposed method obtains a very accurate estimate.

Table I collects the relevant figures related to the mean and the variance, as well as the root-mean-square error (RMSE) obtained with the various methods. For the GPR-based techniques, the moments are expressed in terms of expected value and 2-sigma confidence interval. First of all, it is observed that the SGM and ST provide very accurate results, whereas the LAR exhibits a larger error. The performance of the standard GPR is similar to LAR, whereas the proposed hybrid method achieves a better accuracy that is comparable to the one of the SGM and ST. However, PCE-GPR is expected to scale better to high-dimensional problems. Indeed, GPR methods were demonstrated to achieve accurate results with a moderate number of training samples (generally, in the order of a few hundreds) even in high-dimensional settings [3], which also makes the required factorization of matrix $\boldsymbol{K}$ computationally tractable. The confidence intervals of PCE-GPR are also narrower and enclose the reference MC result.

V. CONCLUSIONS

This paper introduced a hybrid approach combining the advantages of both PCE and kernel methods. The model uses a GPR formulation and a special kernel corresponding to an infinite sequence of Hermite polynomials. Thanks to the inherent Bayesian formulation, the PCE coefficients are analytically calculated with the inclusion of confidence bounds, which allow assigning a prediction confidence to statistical metrics. This information could be used to drive the acquisition of additional training samples in an active learning scenario. The method was illustrated based on the simulation of a transmission-line network, for which a very good accuracy, comparable to the state-of-the-art SGM, was obtained.

REFERENCES

[1] R. Trinchero et al., "Machine learning and uncertainty quantification for surrogate models of integrated devices with a large number of parameters," *IEEE Access*, vol. 7, pp. 4056–4066, 2018.

[2] T. Nguyen et al., "Comparative study of surrogate modeling methods for signal integrity and microwave circuit applications," *IEEE Trans. Compon. Packag. Manuf. Technol.*, vol. 11, no. 9, pp. 1369–1379, Sep. 2021.

[3] P. Manfredi and R. Trinchero, "A probabilistic machine learning approach for the uncertainty quantification of electronic circuits based on Gaussian process regression," *IEEE Trans. Comput.-Aided Design Integr. Circuits Syst.*, vol. 41, no. 8, pp. 2638–2651, Aug. 2021.

[4] S. Roy, *Uncertainty Quantification of Electromagnetic Devices, Circuits, and Systems*. Stevenage, United Kingdom: IET, 2021.

[5] P. Manfredi and R. Trinchero, "Nonparametric formulation of polynomial chaos expansion based on least-square support-vector machines," *Eng. Appl. Artif. Intell.*, vol. 133, p. 108182, 2024.

[6] S. Marelli and B. Sudret, "UQLab: A framework for uncertainty quantification in matlab," in *Vulnerability, uncertainty, and risk: quantification, mitigation, and management*, 2014, pp. 2554–2563.

Parametric S-Parameter Prediction Using Deep Learning

Vinayak Bansal
Department of Chemical Engineering
Indian Institute of Technology
Delhi, India
vinayak.bansal314@gmail.com

Lihong Feng
Computational Methods in Systems and Control Theory
Max Planck Institute for Dynamics of Complex Technical Systems
Magdeburg, Germany
feng@mpi-magdeburg.mpg.de

Valentin de la Rubia
Departamento de Matemática Aplicada a las TIC
Universidad Politécnica de Madrid
Madrid, Spain
valentin.delarubia@upm.es

Peter Benner
Computational Methods in Systems and Control Theory
Max Planck Institute for Dynamics of Complex Technical Systems
Magdeburg, Germany
benner@mpi-magdeburg.mpg.de

Abstract—We construct a neural network model of S-parameters, from which the S-parameters can be quickly predicted. Numerical tests on a filter model show that the proposed method accurately predicts the S-parameters with multiple sharp resonances.

Index Terms—S-parameter, convolutional autoencoder, feedforward neural network

I. INTRODUCTION

Transfer function analysis is a widely used technique for testing the electromagnetic performance of designed devices [1]–[7]. As a standard, (parameter-dependent) transfer functions, e.g., the S-parameters, are computed by solving a large-scale system obtained from the numerical discretization of Maxwells's equations. This is computationally expensive, as the large-scale system needs to be repeatedly solved during a frequency sweep. This becomes much more challenging when the system is parameter dependent and the design of the devices needs to be optimized by parameter optimization. In such cases, the large-scale system must be simulated not only at many frequency samples but also at many parameter samples, so as to extract the parameter-frequency behavior of the S-parameters.

To overcome the computational bottlenecks of direct simulation, model order reduction (MOR) is proposed to build a small-scale surrogate of the original large-scale system [8], [9]. This surrogate then replaces the original system during simulation tasks. In this situation, the original system is often called the full-order model (FOM) and the surrogate is called the reduced-order model (ROM). Many MOR methods are intrusive, i.e., they require access to the system matrices of the FOM. The ROM is derived by linear subspace projection operated on the large system matrices. Data-driven MOR [2], [7], [15] was proposed for problems whose system matrices are unavailable, but their input-output data can be obtained from black-box simulations of the FOM or from measurements.

Recently, machine learning, in particular, deep learning has drawn more and more attention from computational scientists because the trained neural network (NN) can be used as a surrogate model for very fast predictions. Deep-learning based surrogate modelling can be considered as a data-driven MOR technique, as only data are needed to train a neural network model. Different deep learning methods have been proposed for S-parameter prediction [4], [5], [7], [10]. Typically, the neural transfer function based methods [4], [5] use NNs to learn the poles/residues or the coefficients of polynomials in the assumed rational representation of the transfer functions, where the issue of pole-residue mismatching might occur. Additional processes need to be done to mitigate the pole discontinuity [11] to some extent. In [7], [10], the real and imaginary parts of the transfer function are directly learned by NNs, avoiding the issue of pole/residue mismatching. A deep feedforward NN (DFNN) is used in [7] to learn the parametric behavior of S-parameters over the frequency band of interest, where the parameter and frequency are treated equally as inputs to the DFNN. This limits the efficiency of DFNN, as a large amount of data of S-parameters corresponding to parameter samples and frequency samples must be used as training data, making the training process relatively long. Moreover, the DFNN is only accurate for smooth transfer functions with almost no sharp resonances in the frequency range. For transfer functions with multiple sharp resonances, DFNN quickly looses accuracy. A spectral transposed convolutional NN (STCNN) is proposed in [10], where S-parameters with multiple resonances can be learned with good accuracy.

We propose a new deep learning model for predicting S-parameters. This deep learning model combines a convolutional autoencoder (CAE) with a DFNN. This combination previously appeared in [12], [13] for spatio-temporal fluid dynamics prediction. We apply it to frequency-domain anal-

979-8-3503-5124-8/24 $31.00 © 2024 IEEE

ysis for S-parameter prediction. Due to the totally different problems considered, the application is nevertheless nontrivial. The main difference is that the numerically discretized data of the state vectors in space are compressed by CAE in [12], [13], whereas we use CAE to compress the frequency-domain data. The input of the DFNN in [12], [13] includes both time and the vector of parameters, while parameters are the only inputs of DFNN in this work. Compared with the existing deep learning methods [4], [5], [7], [10] for S-parameter prediction, we have several advantages and new contributions: 1) We use an autoencoder to compress the frequency-domain data, so that online prediction can be quickly implemented by combining a DFNN and the decoder of the CAE. Compared with using only DFNN [7], the efficiency of training and prediction is improved a lot. 2) The CAE-DFNN gives more accurate predictions than the STCNN [10]. 3) The CAE-DFNN directly predicts the real and imaginary parts of S-parameters. We don't use rational functions [4], [5] to approximate the possibly non-rational S-parameters, e.g. those of a time-delay system.

In the next section, we introduce the problem we aim to solve. Then the structure of CAE-DFNN in the training and prediction phase is described, respectively. Section III presents implementation details for an inline dielectric resonator filter model along with the prediction results of both the proposed CAE-DFNN and the STCNN [10].

II. PROBLEM SETTING AND STRUCTURE OF CAE-DFNN

A. Problem setting

Evaluation of S-parameters can be done by numerical simulation of the following finite-element-discretized FOM,

$$
\begin{aligned}
M(s,\mu)x(s,\mu) &= B(s,\mu), \\
y(\mu) &= C(s,\mu)x(s,\mu),
\end{aligned} \tag{1}
$$

where $s = 2\pi f_\jmath$ is the Laplace variable, with $f \in \mathcal{B}$ being the ordinary frequency and $\mathcal{B} = [f_{\min}, f_{\max}]$ being the frequency band of interest. $\mu \in \mathcal{P} \subset \mathbb{R}^p$ is the vector of physical or geometrical parameters. The system matrix is $M \in \mathbb{R}^{N \times N}, \forall s, \mu$. The state vector is $x \in \mathbb{R}^N, \forall s, \mu$, and the input, output matrices are $B \in \mathbb{R}^{N \times n_I}, C \in \mathbb{R}^{n_o \times N}, \forall s, \mu$, respectively. The output is $y \in \mathbb{R}^{n_o}$. In this work, we aim to develop a deep learning model for learning S-parameters in both the frequency and the parameter domain.

B. Structure of CAE-DFNN

We propose to use a combination of a CAE and a DFNN for fast S-parameter prediction. A CAE consists of two sub-networks, an encoder and a decoder. In the training phase, we first use the encoder of a CAE to generate embeddings of S-parameters. The decoder then returns the embedded latent space variable to those in the original S-parameter space. Let $X \in \mathbb{R}^{n_f \times 2n_I n_o}$ denote an instance of the input of this network. In the case of S-parameter prediction, X consists of the values of S-parameters evaluated at a specific physical parameter sample and at all frequency samples in the frequency band of interest. The encoder of a CAE generates an embedding of the input: $z \in \mathbb{R}^r, r \ll n_f \cdot 2n_I n_o$. The

decoder decodes this embedding to recover X. A DFNN is used to learn the embedding z from the geometrical parameter, μ. The learned embedding z is fed into the decoder to get an approximation of X. In the prediction phase, the encoder is not used anymore. Given any new (testing) parameter sample μ^*, the embedding z is learned by the DFNN from μ. Then z is fed into the decoder network to produce $\tilde{X}$, an approximation of the desired S-parameters at μ^* and all frequency samples in the frequency band of interest. We illustrate the CAE-DFNN structure in the left part of Fig. 1. For the online prediction phase, we only use DFNN-decoder, which is presented in the right part of Fig. 1.

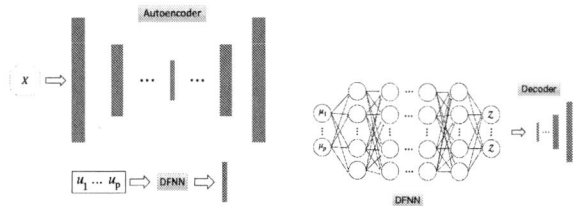

Fig. 1. CAE-DFNN. Left: training phase. Right: prediction phase.

III. IMPLEMENTATION DETAILS AND NUMERICAL RESULTS

We show the results of the proposed method for an inline dielectric resonator filter model [14]. We also compare our method with the STCNN method proposed in [10]. It is shown that the proposed CAE-DFNN is more accurate than the STCNN. All the numerical tests are done with the NVIDIA Tesla K80 GPU on Google Colab.

For this model, we have discretized $\mathcal{B}$ into 128 evenly distributed points. As a result of reciprocity, $S_{12} = S_{21}$, such that we only need to learn S_{11}, S_{12} and S_{22}. The parameters of the model are two dimensionless relative dielectric permittivities $\mu_1, \mu_2 \in \mathcal{P} = [76.5, 77.5]$. We have used Latin hypercube sampling to generate 500 sets of random parameter samples. Corresponding to these, we have generated 500 sets of S-parameters. Each set includes S-parameters at the given 128 frequency points. These S-parameters are complex in nature, hence, we have split the 3 complex quantities into 6 real ones. Finally, the i-th row of the input matrix X to the CAE is $X_i = (\mathrm{Re}(S_{11}(f_i)), \mathrm{Im}(S_{11}(f_i)), \mathrm{Re}(S_{12}(f_i)), \mathrm{Im}(S_{12}(f_i)), \mathrm{Re}(S_{22}(f_i)), \mathrm{Im}(S_{22}(f_i))), i = 1, \ldots, n_f$.

The data was split into 400 training samples, 50 testing samples and 50 validation samples. The CAE was trained for 2000 epochs. The DFNN was also trained on the same metrics for 1000 epochs. Training CAE-DFNN used 650s. It took 3.2s for predicting S-parameters at all the 50 testing samples. From our observation, the training time is upper bounded by O(n), where n is the size of the data set, while the testing time remains constant w.r.t. the testing data size. The STCNN was trained for 10,000 epochs. Training STCNN was finished in 100 seconds. STCNN predicted S-parameters at 50 testing samples in 2ms. However, the prediction time increases linearly with the testing data size.

We present the results of the STCNN and those of the CAE+DFNN at an unseen testing parameter sample in Fig. 2 for comparison. Based on the results at all the testing pa-

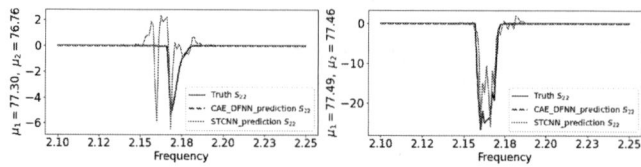

Fig. 3. Instability and inaccuracy of STCNN for magnitudes of s_{22}

existing and closely relevant deep-learning model. One future work might be on a NN framework not only with accurate prediction but also preserving stability.

REFERENCES

[1] H. Wang, L. Sun, J. Liu, H. Zou, Z. Yu, and J. Gao, "Transfer function analysis and broadband scalable model for on-chip spiral inductors," *IEEE Trans. Microw. Theory Techn.*, vol. 59, no. 7, pp. 1696–1708, 2011.

[2] ——, "Toward fully automated high-dimensional parameterized macro-modeling," *IEEE Transactions on Components, Packaging and Manufacturing Technology*, vol. 11, no. 9, pp. 1402–1416, 2021.

[3] F. Ferranti, D. Romano, L. Lombardi, G. Antonini, Y. Tao, and M. Nakhla, "Efficient frequency-domain uncertainty quantification using parameterized model order reduction," in *Proc. of the 2022 International Symposium on Electromagnetic Compatibility (EMC Europe 2022)*. IEEE, 2022, pp. 868–872.

[4] J. Zhang, J. Chen, W. L. Q. Guo, F. Feng, and Q.-J. Zhang, "Parameterized modeling incorporating MOR-based rational transfer functions with neural networks for microwave components," *IEEE Microwave and Wireless Components Letters*, vol. 32, no. 5, pp. 379–382, 2022.

[5] Y. Zhuo, F. Feng, J. Zhang, and Q.-J. Zhang, "Parametric modeling incorporating joint polynomial-transfer function with neural networks for microwave filters," *IEEE Transactions on Microwave Theory and Techniques*, vol. 70, no. 11, pp. 4652–4665, 2022.

[6] V. de la Rubia, "Physics-based greedy algorithm for reliable fast frequency sweep in electromagnetics via the reduced-basis method," *IEEE Transactions on Antennas and Propagation*, vol. 70, no. 11, pp. 10 724–10 735, 2022.

[7] E. Mattucci, L. Feng, P. Benner, D. Romano, and G. Antonini, "Fast frequency-domain analysis for parametric electromagnetic models using deep learning," in *IEEE 32nd Conference on Electrical Performance of Electronic Packaging and Systems (EPEPS)*. IEEE, 2023.

[8] A. C. Antoulas, C. A. Beattie, and S. Gugercin, *Interpolatory Methods for Model Reduction*, ser. Computational Science & Engineering. Philadelphia, PA: Society for Industrial and Applied Mathematics, 2020.

[9] P. Benner, S. Grivet-Talocia, A. Quarteroni, G. Rozza, W. H. A. Schilders, and L. M. Silveira, Eds., *Model Order Reduction. Volume 1: System- and Data-Driven Methods and Algorithms.* Berlin: De Gruyter, 2021.

[10] H. M. Torun et al, "A spectral convolutional net for co-optimization of integrated voltage regulators and embedded inductors," in *In Proc. International Conference on Computer-Aided Design.* IEEE, 2019, pp. 1–8.

[11] J. Feng, Q. Li, F. Feng, L. Zhu, and Q. Zhang, "Systematic pole-zero sorting method for neuro-TF modeling of electromagnetic response," *Opt. Express*, vol. 32, 2024.

[12] S. Nikolopoulos, I. Kalogeris, and V. Papadopoulos, "Non-intrusive surrogate modeling for parametrized time-dependent partial differential equations using convolutional autoencoders," *Engineering Applications of Artificial Intelligence*, vol. 109, p. 104652, 2022.

[13] S. Fresca, L. Dedé, and A. Manzoni, "A comprehensive deep learning-based approach to reduced order modeling of nonlinear time-dependent parametrized PDEs," *J. Sci. Comput.*, vol. 87, no. 61, 2021.

[14] S. Chellappa, L. Feng, V. de la Rubia, and P. Benner, "Inf-sup-constant-free state error estimator for model order reduction of parametric systems in electromagnetics," *IEEE Trans. Microw. Theory Techn.*, vol. 71, no. 11, pp. 4762–4777, 2023.

[15] P. Manfredi and S. Grivet-Talocia, "Fast Stochastic Surrogate Modeling via Rational Polynomial Chaos Expansions and Principal Component Analysis," *IEEE Access*, vol. 9, pp. 102732–102745, 2021.

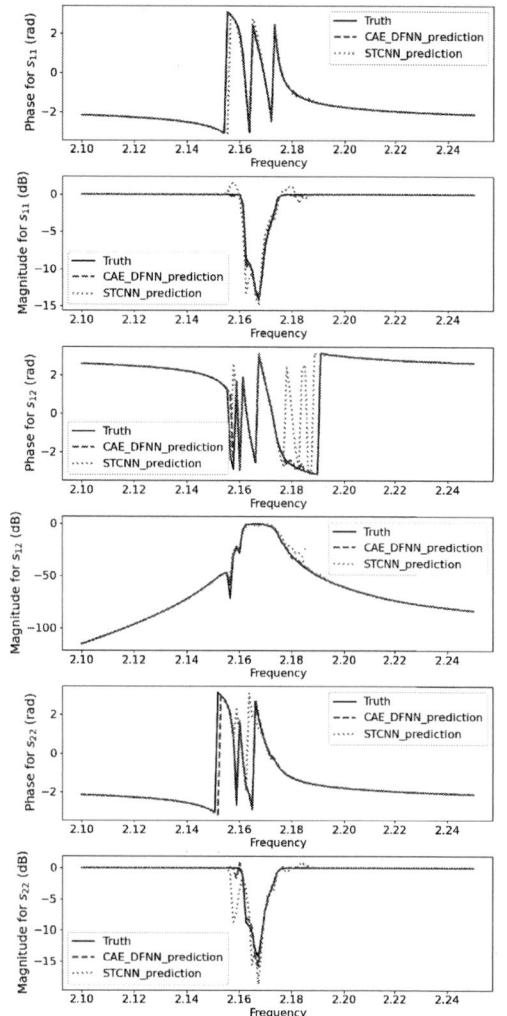

Fig. 2. S-parameters responses at $\mu^* = (77.15, 77.41)$

rameter samples, we observed that STCNN couldn't capture the locations of the spikes at some sets of the testing parameter samples. In addition, the STCNN is found to give locally oscillating results at some sample sets, where the truth should be smooth. For example, plots in Fig. 3 show that at some testing parameter samples, the STCNN-predicted values oscillate around the true values of the input and are inefficient in capturing the spikes in the magnitude. The CAE+DFNN on the other hand was able to predict the spikes properly.

IV. CONCLUSIONS

We propose a deep learning technique for learning parametric transfer functions using a convolutional autoencoder and a feedforward neural network. Numerical results on a filter model show that the overall accuracy of the proposed deep-learning surrogate is acceptable, which is higher than an

979-8-3503-5124-8/24 $31.00 © 2024 IEEE

Modeling Multiplexed Qubit Readout with a Josephson Traveling-Wave Parametric Amplifier

Samuel T. Elkin
Elmore Family School of Electrical and
Computer Engineering
Purdue University
West Lafayette, IN 47906, USA
Email: selkin@purdue.edu

Michael Haider
TUM School of Computation,
Information and Technology
Technical University of Munich
85748 Garching, Germany
Email: michael.haider@tum.de

Thomas E. Roth
Elmore Family School of Electrical and
Computer Engineering
Purdue University
West Lafayette, IN 47906, USA
Email: rothte@purdue.edu

Abstract—**To model multiplexed readout, a key ingredient for scaling quantum computers, we develop a numerical method for co-simulation of qubits and a Josephson traveling-wave parametric amplifier. The integrated characterization reveals behavior absent from independent analyses.**

Index Terms—Multiphysics modeling, Josephson traveling-wave parametric amplifier, multiplexed readout.

I. Introduction

A key challenge for scaling quantum computers using superconducting transmon qubits is in multiplexing readout of the qubit states. Typically, a "dispersive readout" procedure is performed where the qubit is coupled to a microwave resonator detuned from the qubit transition frequency. The resonator frequency is shifted based on the qubit state, allowing the state to be measured by probing the resonator rather than directly driving the qubit [1]. In multi-qubit processors, each qubit is coupled to a microwave resonator with a distinct resonator frequency to enable independent readout.

To avoid inducing errors due to the dispersive readout pulses, an extremely weak drive must be used. Successfully measuring the readout pulse then requires an initial amplifier with an ultra-low noise figure to maintain a usable signal-to-noise ratio at the output. The preeminent option for this is a parametric amplifier, which provides gain through mixing with a strong pump tone [2]. However, typical parametric amplifiers are very narrowband, leading to each qubit needing its own amplifier. A more promising alternative for scaling system sizes is to use Josephson traveling-wave parametric amplifiers (JTWPAs) that achieve high gain with ultra-low noise over a bandwidth of several GHz. This wide bandwidth is sufficient for frequency multiplexing the simultaneous readout of many qubits with a single amplifier, as desired for scalability.

In current implementations of multiplexed readout, the qubits and JTWPA are designed and characterized independently [3]. Ideally, both subsystems could be combined on a single chip to reduce system complexity and impede errors in their integration. However, a lack of tools to model their interaction makes this difficult. To address this, we propose a numerical method for simulating multiplexed readout in

such devices by combining the methods of [4] and [5] to co-simulate a transmission line network that incorporates qubits and a JTWPA. The method discretizes transmission lines using finite element time domain (FETD), which are coupled into quantum descriptions of transmon qubits and classical models of Josephson junctions in the JTWPA using leap-frog time-marching. We present results highlighting how this integrated characterization predicts new, non-trivial behavior that is important in the design of these integrated systems.

II. Formulation

This work combines the Maxwell-Schrödinger method of [4], which describes interactions between qubits and transmission lines, with the numerical method of [5] for modeling a JTWPA. In Sections II-A and II-B, we review the basic principles of these techniques. Because both methods take the same approach to discretize transmission lines, they can easily be integrated, as will be demonstrated in Section III.

A. Maxwell-Schrödinger Modeling

Here, we briefly review the formulation of [4] to explain the fundamental principles of the semiclassical Maxwell-Schrödinger method that treats transmission lines classically and qubit dynamics quantum mechanically. The derivation uses Hamiltonian mechanics to derive self-consistent equations of motion (EoMs) for a transmon qubit capacitively coupled to a transmission line, which is shown in Fig. 1 of [4].

By deriving a Hamiltonian for this system and applying Hamilton's equations, we obtain the EoMs

$$\phi''(z,t) - LC\ddot{\phi}(z,t) = -\delta(z-z_0)2e\beta L \langle \dot{n}(t) \rangle, \quad (1)$$

$$\left[4E_{\mathrm{C}} \left(-i\frac{\partial}{\partial\varphi} - n_{\mathrm{g}} \right)^2 - E_{\mathrm{J}}\cos\varphi \right] \psi(\varphi,t)$$
$$- i2e\beta\dot{\phi}(z_0,t)\frac{\partial}{\partial\varphi}\psi(\varphi,t) = i\hbar\dot{\psi}(\varphi,t), \quad (2)$$

where

$$\langle \dot{n}(t) \rangle = \frac{\partial}{\partial t} \int_{-\pi}^{\pi} \psi^*(\varphi,t) \left(-i\frac{\partial}{\partial\varphi}\psi(\varphi,t) \right) \mathrm{d}\varphi. \quad (3)$$

In these expressions, $\phi(z,t)$ is the node flux at position z along the transmission line, which has per-unit-length (PUL) inductance L and capacitance C. The node flux is related to voltage and current via $V(z,t) = \dot{\phi}(z,t)$ and $I(z,t) = -L^{-1}\phi'(z,t)$, where $\phi'(z,t) \equiv \frac{\partial}{\partial z}\phi(z,t)$ and $\dot{\phi}(z,t) \equiv \frac{\partial}{\partial t}\phi(z,t)$.

The qubit is characterized by charging energy E_C, Josephson energy E_J, and offset charge n_g. It is coupled to the transmission line at point z_0 by capacitance C_g, which forms a voltage divider of ratio $\beta = C_g/(C_g + C_q)$ with the transmon ground capacitance C_q. Further, the wavefunction $\psi(\varphi,t)$ is expressed as a continuous function of the phase φ, which is treated as a position variable, and e is the elementary charge.

To discretize $\phi(z,t)$ in space, FETD with first-order (triangular) basis functions is used. While $\psi(\varphi,t)$ could be discretized with respect to φ by the same means, that approach is relatively inefficient. Instead, we evaluate the eigenstates and eigenenergies of the bare qubit Hamiltonian, and approximate $\psi(\varphi,t)$ in (2) using its expansion in terms of a subset of eigenstates. In this work, 10 eigenstates are included for each qubit. Temporal discretization of both dynamical variables is performed using central differencing. Leap-frog time-marching is used to evaluate (1) and (2) in tandem, meaning $\phi(z,t)$ and $\psi(\varphi,t)$ are updated at alternating half-timesteps.

B. JTWPA Modeling

This section reviews the formulation of [5]. To develop this method, Lagrangian mechanics are used to derive EoMs for a single unit cell of a JTWPA of length a, which consists of a Josephson junction inserted between two transmission lines, as depicted in Fig. 1 of [5]. This procedure leads to the EoMs

$$\phi''(z,t) - L_1 C_1 \ddot{\phi}(z,t) = 0, \quad 0 < z < z_1, \quad (4)$$

$$\phi''(z,t) - L_2 C_2 \ddot{\phi}(z,t) = 0, \quad z_2 < z < a, \quad (5)$$

$$L_1^{-1}\phi'(z_1,t) = -C_J\ddot{\phi}_J(t) - I_c \sin\left(\frac{2e}{\hbar}\phi_J(t)\right), \quad (6)$$

$$L_2^{-1}\phi'(z_2,t) = -C_J\ddot{\phi}_J(t) - I_c \sin\left(\frac{2e}{\hbar}\phi_J(t)\right), \quad (7)$$

where C_i and L_i are the PUL capacitance and inductance of transmission line i, and the junction capacitance C_J and critical current I_c characterize the Josephson junction. The Josephson junction has junction flux $\phi_J(t) \equiv \phi(z_1,t) - \phi(z_2,t)$, and couples to each transmission line at points z_1 and z_2. By summing (6) and (7), an EoM for $\phi_J(t)$ is obtained:

$$C_J\ddot{\phi}_J(t) + I_c \sin\left(\frac{2e}{\hbar}\phi_J(t)\right)$$
$$= -\frac{1}{2L_1}\phi'(z_1,t) - \frac{1}{2L_2}\phi'(z_2,t). \quad (8)$$

Spatial discretization of $\phi(z,t)$ is once again performed using FETD with first-order basis functions, and temporal discretization of both dynamical variables is again implemented using central differencing with leap-frog time-marching. The benefit to leap-frogging here is that it allows $\phi_J(t)$ to be treated as a distinct dynamical variable whose time-marching equation contains all the nonlinearity, allowing the matrix equation for

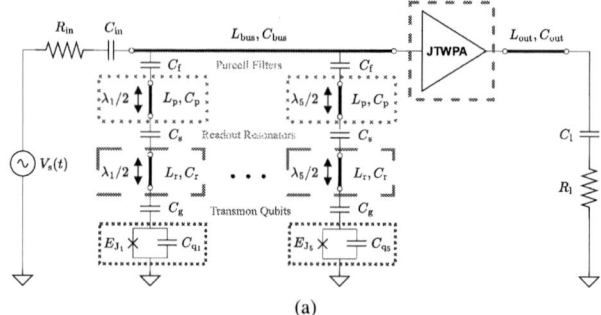

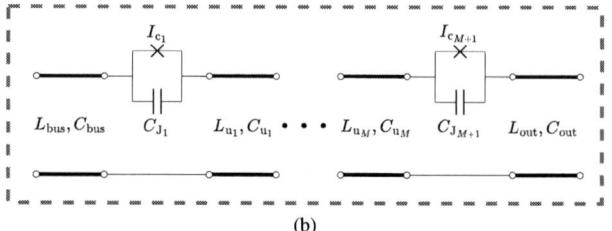

Fig. 1. (a) Schematic of the five-qubit topology used to simulate multiplexed readout. Each transmission line has PUL inductance $L_i = 110.4\,\mu\text{H/m}$ and $C_i = 3.9\,\text{nF/m}$, and $C_g = C_s = C_f = 10.3\,\text{fF}$. Further, $R_\text{in} = R_l = 53\,\Omega$. The remaining parameters are defined in the text. (b) Schematic of the JTWPA with M unit cells, which has its parameters listed in Table 1 of [5].

$\phi(z,t)$ to be evaluated linearly. Further, central differencing causes the nonlinear term in the time-marching equation for $\phi_J(t)$ to only be dependent on the junction flux at prior timesteps, allowing that equation to be evaluated linearly as well. Unit cells modeled in this manner can be combined to form a full JTWPA, as described in [5].

C. Integrated Modeling

In both methods summarized above, the dynamics of transmission lines are described by $\phi(z,t)$, the same discretization strategy is employed, and pointwise coupling is used to enable interaction between subsystems. As a result, the two methods can easily be combined by incorporating both coupling techniques within a single model, as is done in this work. While multiple qubits can be included using this process, quantum crosstalk between the qubits is neglected. The method will be updated to incorporate these effects in future work.

III. RESULTS

In this section, we describe how multiplexed readout is simulated in a five-qubit topology using the integrated method. The model is used to measure dispersive shifts χ_k for the kth qubit, which are validated against analytical estimates. We also demonstrate how the simulation can be used to analyze changes induced by the coupling between the subsystems.

The full schematic for this model is provided in Fig. 1. For the kth qubit, E_{C_k} is tuned by varying C_{q_k} with a fixed ratio $E_{J_k}/E_{C_k} = 87$ to target the desired operating frequency. Here, the transition frequency for Qubit 1 is set to $5.7\,\text{GHz}$, and reduced by $200\,\text{MHz}$ for each successive qubit. Each qubit is connected to a resonator sized for $1.5\,\text{GHz}$ detuning from

979-8-3503-5124-8/24 $31.00 © 2024 IEEE

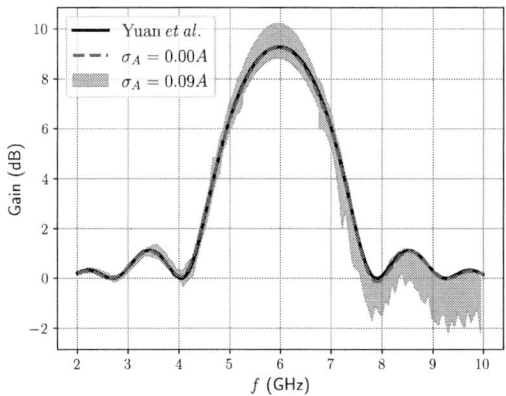

Fig. 2. JTWPA gain compared to an analytical result [2]. For $\sigma_A = 0.09$, the shaded area contains the gain from 50 simulations using data from [5].

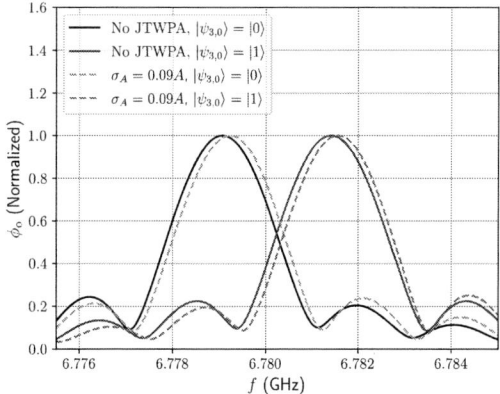

Fig. 3. Spectrum of transmitted flux with and without the non-ideal JTWPA. For $\sigma_A = 0.09A$, results are taken from a single instance of parameters. Curves are independently normalized so the changes can be observed.

the qubit transition frequency. The resonators are connected to the bus through a "Purcell filter" (another resonator), which is included to limit the decay rate of each qubit and inhibit interaction between resonators [3]. The input capacitance C_{in} and load capacitance C_l are included to enable a correction method for late-time instability, as explained in [5], and are sized at $100\,\text{pF}$ to have a negligible effect on normal operation. The voltage source $V_s(t)$ supplies narrowband readout pulses at each resonator frequency and generates the pump for the JTWPA. While this is unlike typical experimental setups, no significant variations in the behavior of the qubits were observed with the introduction of the pump.

To validate the qubit subsystem independently from the JTWPA, the schematic of Fig. 1 was simulated with the bus directly terminated in the load. The dispersive shift χ_k for the kth qubit was calculated by measuring the distance between peaks in the spectrum of the transmitted flux $\phi_o(f)$ for initial states $|\psi_{k,0}\rangle = |0\rangle$ and $|\psi_{k,0}\rangle = |1\rangle$. Each χ_k was estimated analytically using expressions from [6] and found to agree well with the numerical results, as shown in Table I. The JTWPA was also independently validated by calculating gain using the procedure detailed in [5], and comparing to the analytical result from [2]. Additionally, the area A of each junction in the JTWPA, where I_J, $C_J \propto 1/A$, is varied on a normal distribution with standard deviation σ_A to study the impact on gain. Strong agreement between numerical and analytical results was found for $\sigma_A = 0$, as shown in Fig. 2.

In Fig. 3, we show the spectra of $\phi_0(f)$ near Resonator 3.

When the non-ideal JTWPA with $\sigma_A = 0.09A$ is introduced, the resonant peaks are shifted and their shapes change, which can impact the overall fidelity in estimating the qubit state from measurements. While this behavior is challenging to model analytically, our method can easily capture it, allowing a better estimate of overall readout fidelity at the design stage.

IV. CONCLUSION

In this work, numerical methods for modeling transmon qubits and JTWPAs were combined to simulate multiplexed readout. Strong agreement was found between the dispersive shifts estimated analytically and those measured using this model. Further, JTWPA-induced changes in the resonator spectra were shown, validating the need for this approach.

To expand on this method in the future, we plan to incorporate the collapse of the qubit wavefunctions, which will allow the impact of the JTWPA on the qubit states to be analyzed further. Support for quantum coupling between qubits will also be added to properly describe their collective state. The model will also be updated to incorporate more modern JTWPA topologies with superior performance to determine if additional non-ideal behaviors become apparent.

REFERENCES

[1] T. E. Roth, R. Ma, and W. C. Chew, "The transmon qubit for electromagnetics engineers: An introduction," *IEEE Antennas and Propagation Magazine*, vol. 65, no. 2, pp. 8–20, 2023.

[2] Y. Yuan, M. Haider, J. A. Russer *et al.*, "Circuit quantum electrodynamic model of dissipative-dispersive Josephson traveling-wave parametric amplifiers," *Phys. Rev. A*, vol. 107, p. 022612, Feb 2023.

[3] J. Heinsoo, C. K. Andersen, A. Remm, S. Krinner, T. Walter, Y. Salathé, S. Gasparinetti *et al.*, "Rapid high-fidelity multiplexed readout of superconducting qubits," *Phys. Rev. Appl.*, vol. 10, p. 034040, Sep 2018.

[4] T. E. Roth and S. T. Elkin, "Maxwell-Schrödinger modeling of a superconducting qubit coupled to a transmission line network," *IEEE Jour. on Multsc. and Multphys. Comp. Tech.*, vol. 9, pp. 61–74, Jan 2024.

[5] S. T. Elkin, M. Haider, and T. E. Roth, "Multiphysics numerical method for modeling josephson traveling-wave parametric amplifiers," *IEEE Jour. on Multsc. and Multphys. Comp. Tech.*, vol. 9, pp. 247–257, 2024.

[6] F. Swiadek, R. Shillito, P. Magnard, A. Remm *et al.*, "Enhancing dispersive readout of superconducting qubits through dynamic control of the dispersive shift: Experiment and theory," *arXiv:2307.07765*, 2023.

TABLE I
DISPERSIVE SHIFTS χ_k FOR EACH QUBIT k

	Analytical (MHz)	**Numerical (MHz)**	**Relative Error**
χ_1	3.157	3.146	-0.35%
χ_2	2.697	2.715	0.68%
χ_3	2.322	2.318	-0.17%
χ_4	1.998	1.987	-0.56%
χ_5	1.678	1.656	-1.33%

979-8-3503-5124-8/24 $31.00 © 2024 IEEE

Comparative Evaluation of 100G-PAM4 Ethernet Link Performance in Air and Immersion Cooling Conditions

Dan Liu
Alibaba Cloud
Alibaba Group
ShenZhen, China
ld259920@alibaba-inc.com

Oluwafemi Akinwale
Datacenter and AI Group
Intel Corporation
Santa Clara, California, USA
oluwafemi.akinwale@intel.com

Yangfan Zhong
Alibaba Cloud
Alibaba Group
ShenZhen, China
yangfan.zyf@alibaba-inc.com

Kai Wang
Datacenter and AI Group
Intel Corporation
Portland, Oregon, USA
Kai.a.wang@intel.com

Cesar Mendez-ruiz
Datacenter and AI Group
Intel Corporation
Zapopan, Mexico
cesar.mendez-ruiz@intel.com

Kusuma Matta
Datacenter and AI Group
Intel Corporation
Santa Clara, California, USA
kusuma.matta@intel.com

Abstract—This paper comprehensively evaluates signal integrity (SI) performance in 100G-PAM4 Ethernet links under two different cooling methodologies: air cooling and immersion cooling. It focuses on how these environments affect the BER in high-speed data transmissions.

Keywords—100G Ethernet, QSFP connector, immersion cooling

I. INTRODUCTION

As next-generation server processors escalate power requirements, cloud service providers encounter significant challenges in managing power constraints within data centers. The power usage effectiveness (PUE) is defined as the total power consumption of a data center divided by the power used to run the IT equipment within it. It is widely used in the industry to evaluate the energy efficiency of data centers. A PUE of 1.0 is ideal [1]. The power consumption of AI data centers around the world has increased several-fold, with the global average PUE of data centers at 1.55. However, it is worth mentioning that by implementing immersion-cooling technology in super large-scale deployment, PUE and optimized total cost of ownership (TCO) will yield in energy savings of more than 30%. Immersion cooling technology is emerging as an effective solution for the energy efficiency in mission-critical data centers. Fig. 1 illustrates a data center where systems are fully immersed in liquid.

But now, the immersion cooling introduces unique technical considerations. This paper builds on our previous research [2] and explores the technical challenges of transitioning QSFP100 connectors from air to liquid environments. Notable issues include obvious impedance discontinuities, increased insertion losses, and deviations in resonance frequency. It primarily due to the higher dielectric constants (Er) of popular coolants, typically ranging between 1.8 and 2.2. These issues impact the performance of high-speed links, especially in the noise-sensitive 100G-PAM4 Ethernet configurations.

Aiming for cost-effectiveness and accelerated design cycles for high-speed channel configurations across varied data center environments, this study comprehensively evaluates 100G-PAM4 Ethernet link performance, contrasting conventional air-cooling methods with immersion-cooling techniques. In the following section, by leveraging channel operating margin (COM) and IBIS-AMI simulations, we dissect the end-to-end link dynamics, emphasizing the ramifications of variances in PCB trace lengths and cable lengths from established benchmarks. Section III shows the bit error rates (BER) measurement results from 5 cases under study. Finally, section IV concludes the findings and discusses optimization strategy to compensate the immersion impact on signal integrity (SI).

Fig. 1. Datacenter immersion systems in a liquid tank.

II. SIMULATION RESULTS

Previous publications have discussed DDR5, PCIe 5.0, and PCIe 6.0 immersion SI data [1] [3]. This study extends those findings by comparing 100G-PAM4 Ethernet and artificial intelligence (AI) link performance under air-cooling and immersion-cooling conditions.

Fig. 2 depicts how QSFP connectors, typically used in Ethernet and AI Serdes, perform when immersed in liquids. Despite a significant impedance discontinuity observed in TDR simulations in air and in Er=2 liquid—dropping from 90 ohms to 70 ohms, as shown in Fig. 3. However, at 100 Gbps PAM4, a difference in COM simulation results was noted [2].

Fig. 2. Immersed QSFP connector.

979-8-3503-5124-8/24 $31.00 © 2024 IEEE

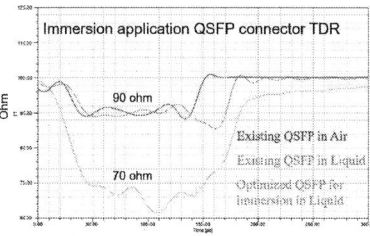

Fig. 3. QSFP connector simulated TDR in air and in Er=2 liquid. The impedance-optimized QSFP is designed for ~110 ohms in air and ~90 ohms in liquid.

Using the Ethernet high-speed channel of the network interface card (NIC) to the switch as an example, comprehensive COM and IBIS-AMI simulations are performed for typical topologies, considering various end-to-end loss by varying the routing trace length on NIC and switch boards and the cable length across different configurations in real server system designs. In addition, three scenarios of QSFP connectors in Fig. 3 are implemented: (1) the existing QSFP connector in air, (2) the existing QSFP connector in liquid, and (3) the impedance-optimized QSFP connector for immersion in liquid. The typical Ethernet topology under simulation is shown in Fig. 4.

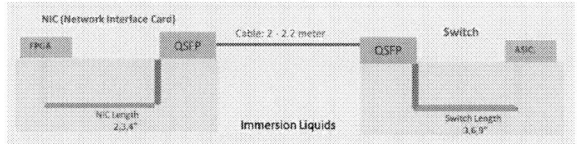

Fig. 4. Ethernet topology analysis.

Fig. 5 presents three scenarios of simulation data:

- The red data points represent the COM margin by using the existing QSFP connector in an air-cooling environment. All red data points pass 3 dB COM channel margin requirement.

- The blue data points depict the same Ethernet topology but simulated with an immersion dielectric liquid (Er=2.0). The blue data using the existing connector show a significant drop in COM dB margin, approximately 2 dB, although they remain above the 3 dB COM channel margin requirement. However, long channel COM data already show very marginal.

- The pink data point shows an impedance-optimized QSFP connector with the same assumption as the blue data point. The COM dB margin closely matches that observed in the air simulation of the same topology.

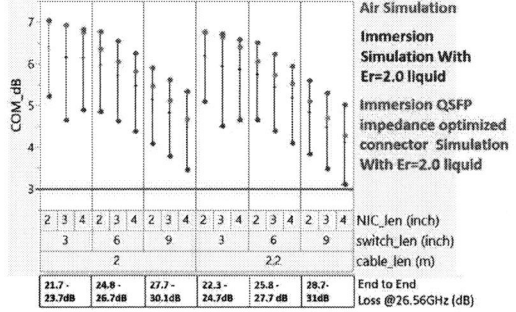

Fig. 5. 100G COM simulation data with optimized QSFP connector, target above 3 dB.

The primary differences observed in the COM simulation can be attributed to how the impedance of the QSFP connector changes between air and immersion environments as depicted in Fig. 3. The optimized QSFP connector for immersion effectively improve more than 1 dB SI margin in liquid.

The same topologies used in the COM simulation were analyzed with a 100G/Lane IBIS-AMI model to simulate BER before receiver Forward Error Correction (FEC). In Fig. 6, similar simulation results can be observed using the IBIS-AMI simulation method. The air-cooling simulations show a lower BER, while immersed in dielectric liquids results in a higher BER. However, the immersion BER remains below the 1e-4 threshold set by IEEE [4] and OIF Ethernet LR requirements [5].

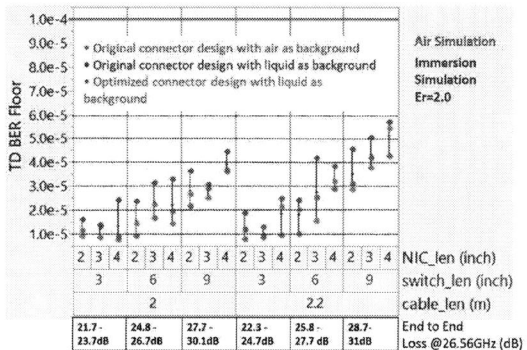

Fig. 6. IBIS-AMI time domain BER, target below 1e-4.

III. MEASUREMENT RESULTS

A 100G-PAM4 IP evaluation kit loopback channel involves two module compliance boards (MCB) and a QSFP cable, links the IP kit's transmitter to its receiver. This setup was then tested for BER with the chip silicon in place. The QSFP connectors and cables were submerged in liquids with Er of 1.9 and 2.1, respectively. Comprehensive measurement is conducted for five study cases of Ethernet high-speed channel configurations across varied data center environments.

Table 1 describes the details of 5 cases. The loss impact of different direct attach copper (DAC) cable lengths used in each case is assessed. IEEE 802.3ck specification [4] dictates a pin-to-pin loss of 28.5 dB at 26.56 GHz, translating to a total pad-to-pad loss of 38 dB to 40 dB when including the loss from two packages. The pad-to-pad loss is calculated for each case according to the above.

TABLE I.   CASE STUDY DESCRIPTION

Cases	Description
Case 1	31.8 dB pad-to-pad loss at 26.56 GHz. A 2.2-meter-long AWG26 DAC cable. Liquid Er=1.9.
Case 2	27.1 dB pad-to-pad loss at 26.56 GHz. A 1.2-meter-long AWG28 DAC cable. Liquid Er=1.9.
Case 3	34 dB pad-to-pad loss at 26.56 GHz. A 2.5-meter-long AWG26 DAC cable. Liquid Er=2.1.
Case 4	27.6 dB pad-to-pad loss at 26.56 GHz. A 1.5-meter-long AWG26 DAC cable. Liquid Er=2.1.
Case 5	27.1 dB pad-to-pad loss at 26.56 GHz. A 1.2-meter-long AWG28 DAC cable. Liquid Er=2.1.

The BER standard for 100G-PAM4 is set to 1e-4 before applying FEC. In Fig. 7, case 3 shows a pad-to-pad loss of 34 dB, where the measured 100G-PAM4 BER deteriorates from 1e-7 in air to 3.1e-5 in immersion. However, the BER measurement results in immersion show about 100-fold degradation, compared to that in air.

Similar to what was observed in the simulation results, the transition of the deployment environment from air to liquid causes significant impedance discontinuities and reflections, which in turn resulted in a higher BER at data rates of 100 Gbps, illustrated in Fig. 7.

The comparison of air BER data (in blue) and immersion BER data (in red) within the same case demonstrates a clear reduction in BER because of the dielectric properties of the liquid (Er). Due to the repeatability and uncertainty of BER measurements, especially when BER is very low (<1e-7), it is not advisable to compare BER across different cases.

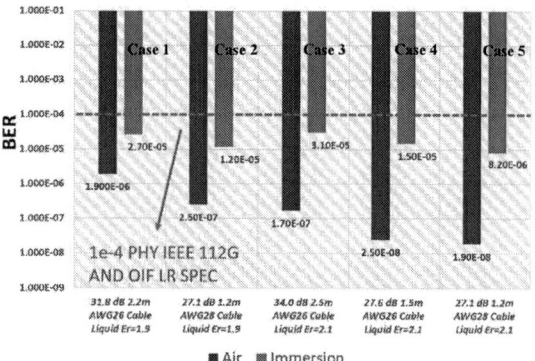

Fig. 7. Measured BER at 100G-PAM4 in air and in immersion.

IV. CONCLUSIONS

Our study evaluates 100G-PAM4 Ethernet link performance in air and immersion cooling and provides an analysis of 100G-PAM4 Ethernet and AI link SI performance under various deployment conditions.

Simulations of typical cases indicate that the existing QSFP connector, when used in immersion cooling, has a ~2dB reduction in COM margin compared to air cooling, which stays consistent across channel loss. Although these margins meet the 3 dB requirement, for a 28dB channel, the margin is critically close to the failure threshold. This presents a risk for high-volume manufacturing (HVM) due to potential variances in real-world hardware designs. To mitigate this risk, a design

of experiments (DOE) methodology in simulations is recommended. This approach will cover corner cases and provide a more robust risk assessment.

The IBIS-AMI simulations indicate that while the existing QSFP connector shows the worst BER performance in liquid, it offers better margin results for a 28 dB channel than COM simulations suggest. This improved margin is likely due to the IBIS-AMI model's ability to more effectively simulate the adaptability and resilience of current receiver designs in liquid environments. Data also suggest that an immersion-optimized connector can outperform a standard connector in air for specific scenarios. This points to potential design advantages in optimizing connectors for immersion cooling to leverage these performance gains.

Experimental measurements confirm that 100G-PAM4 link BER performance degrades approximately 100-fold in immersion cooling compared to air cooling. This substantial degradation underscores the significant impact of immersion cooling on signal integrity. Developing an immersion-optimized QSFP connector can effectively compensate for the dielectric changes introduced by immersion cooling.

The insights from our study highlight the critical need for precise design considerations for high-speed Ethernet channels under varying cooling scenarios. Combining an optimized QSFP connector for immersion cooling and a DOE methodology for simulations will significantly enhance the robustness and reliability of Ethernet high-speed channels in diverse deployment conditions. These findings are particularly relevant for data centers that support AI-driven applications, where reliable, high-speed data transmission is crucial.

REFERENCES

[1] D. Liu, J. Li, K. Wang et al., "PCIe 6.0 (PAM4) Signal Integrity Challenges in Immersion-Cooling Data Centers", DesignCon 2023.

[2] D. Liu, A. Nowak, X. Jiang et al., "112G-PAM4-QSFP Interconnect: A Study in Air Cooling and Immersion Cooling," DesignCon 2022

[3] D. Liu, K. Wang, S. Li et al., "Risks and Enablers of Server Platform Design in Immersion Cooling," DesignCon 2022.

[4] "IEEE Standard for Ethernet Amendment 4: Physical Layer Specifications and Management Parameters for 100 Gb/s, 200 Gb/s, and 400 Gb/s Electrical Interfaces Based on 100 Gb/s Signaling," in IEEE Std 802.3ck-2022 (Amendment to IEEE Std 802.3-2022 as amended by IEEE Std 802.3dd-2022, IEEE Std 802.3cs-2022, and IEEE Std 802.3db-2022) , vol., no., pp.1-316, 28 Dec. 2022, doi: 10.1109/IEEESTD.2022.9999414

[5] OIF CEI-112G Standards. [Online]. Available: https://www.oiforum.com/technical-work/hot-topics/common-electrical-interface-cei-112g-2

Cascading of 2D and 3D Simulations of ASIC Substrate Interconnect up to 100 GHz

Zhekun Peng
EMC Laboratory
Missouri University of Science and Technology
Rolla, MO, USA
pengzhe@mst.edu

Junyong Park
EMC Laboratory
Missouri University of Science and Technology
Rolla, MO, USA
junyongpark@mst.edu

Sathvika Bandi
EMC Laboratory
Missouri University of Science and Technology
Rolla, MO, USA
sbgnk@mst.edu

Santosh Pappu
Meta Platforms Inc.
Menlo Park, CA, USA
skpappu@meta.com

Srinivas Venkataraman
Meta Platforms Inc.
Menlo Park, CA, USA
srinivasv@meta.com

Xu Wang
Meta Platforms Inc.
Menlo Park, CA, USA
xuwang@meta.com

Granthana Rangaswamy
Meta Platforms Inc.
Menlo Park, CA, USA
granthana@meta.com

DongHyun Kim
EMC Laboratory
Missouri University of Science and Technology
Rolla, MO, USA
dkim@mst.edu

Abstract—**A method of cascading 3D models and 2D models to model the full channel of ASIC package substrate interconnects is proposed, showing good match to full-wave results in S-parameter and TDR up to 100 GHz.**

Keywords— *ASIC package substrate interconnects, cascaded S-parameters, 2D cross section analysis.*

I. Introduction

The generation of Serializer-Deserializer (SerDes) has undergone a series of evolutionary changes, progressing from 56 G to 112 G, 224 G, and 448 G in the near future. The signal integrity requirements for packaging, such as the bandwidth limit and the insertion loss, increase with the evolution of the technology [1, 2]. To evaluate the entire channel, the general method is to break down the package structure, as shown in Fig. 1, for separate evaluation. As the Nyquist frequency increases to higher levels (56 GHz for 224 G/lane PAM4, and 74.67 GHz for 448 G/lane PAM8), the simulation time can still cost longer time. During the design phase, the objective is to identify an optimal solution that minimizes insertion loss and crosstalk while ensuring an acceptable balance between time and resources. Concurrently, the optimization stage is unable to generate extensive real models for full-wave simulation. A methodology for rapid assessment of different channel lengths is required.

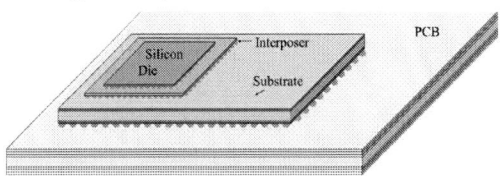

Fig. 1. Stacked structure of packaging.

Cascading S-parameters is a widely employed method for modeling the interconnects on the packaging, such as PCB routing, via modeling and 3D stacked ICs [3, 4, 5]. This approach involves breaking down the model into subparts, which reduces its complexity while also ignoring many interactions between subparts. Moreover, the use of appropriate cutting planes enables the simplification of some subparts through 2D modeling, thereby facilitating the generation of models with varying lengths and a more expedient optimization process. The approach represents a promising strategy for streamlining the optimization

---

This work was supported in part by the National Science Foundation (NSF) under Grant IIP-1916535.

procedure, although it has not been extensively employed in the package substrate simulation.

The paper proposed a method of combining partial 3D models, which is unavoidable for essential 3D full-wave simulation, and 2D cross-sectional analysis, which serves as a substitute for evenly distributed sections, to estimate the signal integrity performance of a high-speed channel. The channel is situated within the scenario of an Application Specified Integrated Circuit (ASIC) based substrate interconnect channel simulation, designed for 224 G/lane PAM4 application, with the objective of supporting the simulation up to 100 GHz. The simulation results are compared through frequency domain (S-parameters) and time domain (Time Domain Reflectometry, TDR), respectively. Several lengths of the interconnects are generated to observe the impact of the cascading methodology. The advantages and disadvantages of the repetitive 3D models and 2D generation method are compared.

II. 3D/2D Model Cascading

Due to the limitation of 2D analysis on inability to analyze the complex structure, some subparts still requires full-wave simulation. The simulation requires dedicated segmentation to ensure the model's suitability for cascading and 2D analysis.

A. Full-wave whole package substrate channel simulation

The channel on the package substrate is depicted in Fig. 2. The substrate is connected to the interposer and silicon die through C4 (controlled collapse chip connection) bump, routing in one of the build-up layers, going through the core and other build-up layers by vias, and connected to the PCB through the solder ball. High-speed channels utilize the differential pair to minimize the common-mode noise. For high-frequency and high-accuracy simulation, the model requires to be accurately set to the wave port to match the input impedance of the real bump/solder ball mechanism [6]. Otherwise, there will be significant mismatch on the port side. A large metal plane, aligned with the top/bottom edge of the C4 bump/solder ball, is used (not shown in Fig. 2) to create a reference plane for ports. A large number of shielding vias are placed around the entire channel, which increases the complexity of the simulation. Furthermore, for high-speed channel simulation, the dielectric material, conductor type and surface roughness must be accurately modeled to match the correct insertion loss along the channel. In this paper, the simulation uses Ansys HFSS, 3D Layout and 2D Extractor [7] to set the model setting and perform the simulation up to 100-GHz.

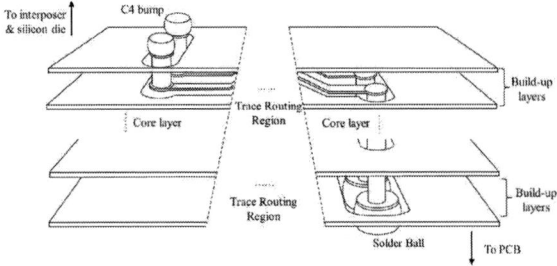

Fig. 2. Interconnects on the package substrate.

B. 3D model cutting and cascading

Performing 3D model cutting is a preliminary step of the cascading and duplication process, High-speed channel typically perform intra-pair de-skew, which minimizes mode-conversion and radiation [8], and neckdown for trace to breakout. Additionally, the via-transition can also introduce unexpected differences in the result. Involving such discontinuity structures in the repetitive middle part will introduce additional uncontrollable factors to the result. Therefore, a more optimal solution for cutting the model is to select a segment devoid of any discontinuity, as illustrated in Fig. 3. The middle part of the model can be represented as a simple stripline and subjected to 2D cross-section analysis, which will be discussed in the next section.

The cascading is initiated by simulating the C4 bump part, the middle transmission line part and the solder ball part separately. These three S-parameters are then cascaded through SPICE tool as the connections in Fig. 3. The cascading requires the wave to be continuous at the cut plane [3]:

$$\bar{a}_{out}^{nk} = \bar{b}_{in}^{(n+1)k} \tag{1}$$

$$\bar{b}_{out}^{nk} = \bar{a}_{in}^{(n+1)k} \tag{2}$$

where $\bar{a}_{out}^{nk}$ and $\bar{b}_{out}^{nk}$ denotes the incident and reflected wave at k^{th} output port of nth network, $\bar{a}_{in}^{(n+1)k}$ and $\bar{b}_{in}^{(n+1)k}$ denotes the incident and reflected wave at k^{th} output port of $(n+1)^{th}$ network.

The cascading methodology demonstrates a good match in the frequency domain for both S_{DD11} and S_{DD21} up to 100 GHz, as illustrated in Fig. 4. A minor discrepancy at specific frequencies may be attributed to different mesh settings.

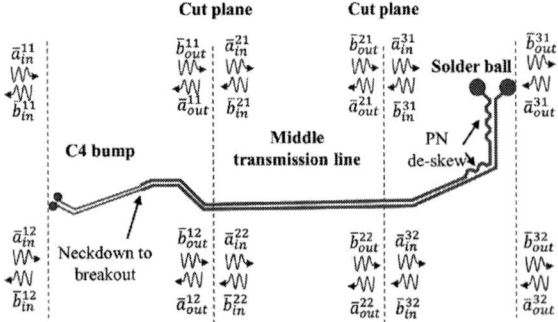

Fig. 3. Top view of routing details from C4 bump to solder ball. Cut planes are placed to avoid involving discontinuity in the middle part.

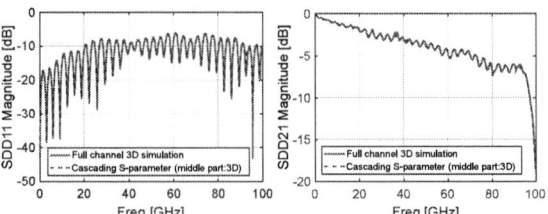

Fig. 4. S_{DD11} and S_{DD21} result comparison of cascading 3D model with full channel 3D simulation.

C. 2D analysis replacing the middle part

The mechanism of the 3D model cutting allows to simplify the middle part as a stripline model. The differential pair, together with the reference plane, is uniform along the length direction, making it the possible to be characterized through 2D analysis. The 2D cross-section analysis is performed in Ansys 2D Extractor, as shown in Fig. 5 (a). The insertion loss is influenced by the conductor type and surface roughness, which are compared in Fig. 5 (b). The full-wave simulation result for the entire channel is compared with the cascading method by replacing the middle part with 2D analysis result. The S-parameter result and TDR comparison are shown in Fig. 6 and Fig. 7. Minor discrepancy mainly comes from simplified 2D model (e.g. shielding vias are ignored). Both frequency domain and time domain results demonstrate a high degree of correlation with the full-wave simulation. Table I shows the time and memory usage of both methods, showing the advantages of 2D analysis over 3D full-wave simulation.

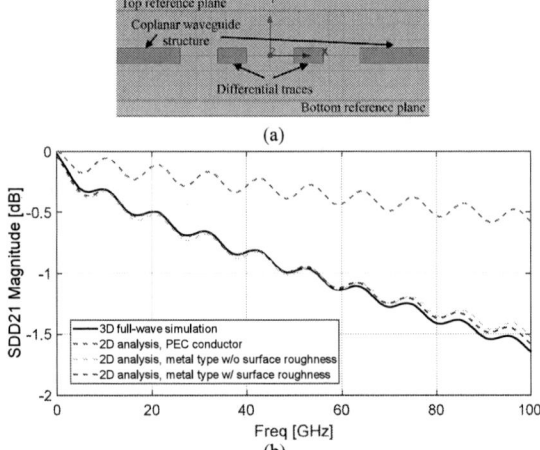

Fig. 5. 2D analysis of (a) cross-section geometry, (b) comparison of results for different 2D analysis settings.

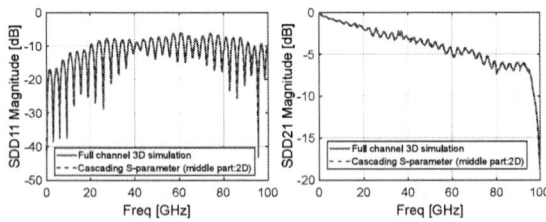

Fig. 6. S_{DD11} and S_{DD21} result comparison of replacing middle part with 2D model in cascading method with full channel 3D simulation.

TABLE I. TIME AND MEMORY USAGE FOR BOTH METHODS

Method	Time	Avg. Memory Usage
3D full-wave simulation	15 h 44 min	48 GB
2D cross-section anaylsis	6 min	60 MB

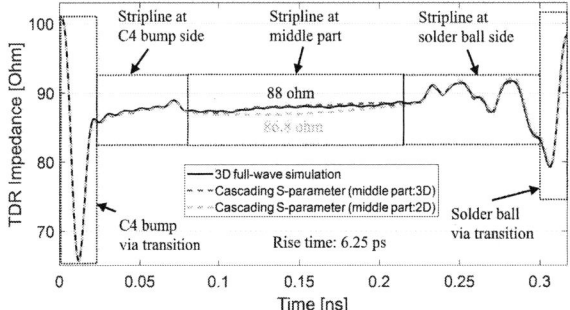

Fig. 7. TDR results comparison. Rise time is set to be 6.25 ps.

III. COMPARISON OF 3D MODEL DUPLICATION AND 2D GENERATION RESULT

Two methods exist for generating different lengths of the channel: duplication of the 3D model with S-parameter chain concatenation or generation of the 2D model. Both methods are effective, though each has advantages and disadvantages.

3D model duplication indicates the addition of a fixed number of lengths to the existing channel. In contrast, the generation of a 2D model is more flexible in terms of the desired length. Ansys 2D extractor provides the necessary flexibility for the generation of S-parameters based on the targeted length. Consequently, a number of cases with different lengths are generated by duplicating the middle stripline length by 2x, 3x and 4x using both methods. The S_{DD21} results are shown in Fig. 8. There are some minor discrepancies between 3D model duplication and 2D model generation. A comparison of unit insertion loss for both cases at the Nyquist frequency (56 GHz) is presented in Table II. The trend is not monotonic due to the oscillation caused by impedance mismatch, while the value provides a reasonable estimation of the channel performance. The corresponding TDR results are shown in Fig. 9. As the length of channel is increased, the TDR impedance grows slowly due to the channel loss. The 2D generation result is found to be slightly different from the 3D duplication, which represents an disadvantage of the 2D generation method.

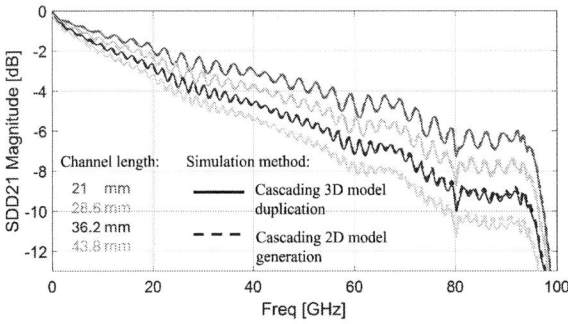

Fig. 8. S_{DD21} results when extending the whole channel to four different lengths using 3D model duplication and 2D model generation.

TABLE II. UNIT INSERTION LOSS FOR MODELED CHANNELS WITH DIFFERENT LENGTHS AT 56 GHz

Full channel length	3D Duplication	2D generation
21 mm	-3.97 dB (0.189 dB/mm)	-3.97 dB (0.189 dB/mm)
28.6 mm	-5.52 dB (0.193 dB/mm)	-5.52 dB (0.193 dB/mm)
36.2 mm	-6.46 dB (0.178 dB/mm)	-6.39 dB (0.176 dB/mm)
43.8 mm	-7.10 dB (0.162 dB/mm)	-7.02 dB (0.160 dB/mm)

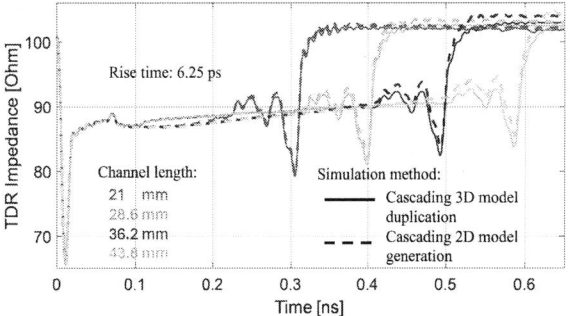

Fig. 9. TDR when extending the whole channel to four different lengths using 3D model duplication and 2D model generation. Rise time is 6.25 ps.

IV. CONCLUSION

The paper proposed a method of cascading 2D and 3D models for simulation of different lengths of high-speed channels on the package substrate designed for 224G/lane PAM4 application. The method was first validated by cutting the existing model into three parts through a detailed analysis of the discontinuity of the channel. The results demonstrated a satisfactory match between the cascaded S-parameters of each separated 3D model and the replacement of the middle stripline with 2D analysis for both insertion loss and TDR. A number of different lengths were generated to facilitate a comparative analysis of both methods. The results indicates that 3D model duplication yielded a closer match to the exact full channel, while 2D generation exhibit greater flexibility in the length generation. The cascading method offers a convenient approach for estimating the performance of various lengths of high-speed channels without the necessity of generating extensive models for full-wave simulation.

REFERENCES

[1] H. Liu, Q. Ding and J. Jiang, "112G PAM4/56G NRZ Interconnect Design for High Channel Count Packages," 2018 IEEE 27th Conference on Electrical Performance of Electronic Packaging and Systems (EPEPS), San Jose, CA, USA, 2018, pp. 237-239.

[2] Mike Li, Jenny Xiaohong Jiang, Yee Lun Ong, et al. "224G Package and PCB Investigations and COM Reference Model" [Online]. Available: https://www.ieee802.org/3/df/public/22_03/mli_3df_01a_220316.pdf.

[3] F. de Paulis, Y. -J. Zhang and J. Fan, "Signal/Power Integrity Analysis for Multilayer Printed Circuit Boards Using Cascaded S-Parameters," in IEEE Transactions on Electromagnetic Compatibility, vol. 52, no. 4, pp. 1008-1018, Nov. 2010.

[4] Z. Z. Oo, E. -X. Liu, X. C. Wei, Y. Zhang and E. -P. Li, "Cascaded Microwave Network Approach for Power and Signal Integrity Analysis of Multilayer Electronic Packages," in IEEE Transactions on Components, Packaging and Manufacturing Technology, vol. 1, no. 9, pp. 1428-1437, Sept. 2011.

[5] S. Piersanti, F. de Paulis, A. C. Scogna, M. Swaminathan and A. Orlandi, "Electromagnetic simulation of 3D stacked ICs: Full model vs. S-parameter cascaded based model," 2014 IEEE International Symposium on Electromagnetic Compatibility (EMC), Raleigh, NC, USA, 2014, pp. 57-62.

[6] J. Sun, Z. Qian, C. S. Geyik and K. Aygün, "Accurate BGA Package Solder Joint Modeling for High Speed SerDes Interfaces," 2020 IEEE 29th Conference on Electrical Performance of Electronic Packaging and Systems (EPEPS), San Jose, CA, USA, 2020, pp. 1-3.

[7] *Ansys® Electromagnetics Suite, Release 2024 R1, HFSS, HFSS 3D layout Design and 2D Extractor.*

[8] J. Li and J. Fan, "Radiation Physics and Design Guidelines of High-Speed Connectors," in IEEE Transactions on Electromagnetic Compatibility, vol. 58, no. 4, pp. 1331-1338, Aug. 2016.

979-8-3503-5124-8/24 $31.00 © 2024 IEEE

Multiphysics Simulation and Measurement Correlation of a Multichip Module IC Package Current Sensor

Rajen Murugan, Jie Chen, Guangxu Li, Yutaka Suzuki, Sylvester Ankamah-Kusi
Texas Instruments, Inc.
Dallas, TX, USA
r-murugan@ti.com

Abstract—**As the semiconductor market shifts its focus to 3D integration, lowest cost, and high performance, the need for advanced analog packaging technologies becomes paramount. Integrating multichip and passives onto cost-effective package technologies create opportunities for highly competitive semiconductor products. However, this aggressive integration also leads to complex multiphysics and multiscale interactions, especially for small-form-factor modules. This work describes the multiphysics (electrical-thermal-mechanical) co-design modeling methodology that resulted in the industry's first highly accurate, voltage-output multichip module current-sense integrated circuit. The reliability of the multiphysics modeling methodology was confirmed by directly comparing it to measurements taken on a 36V, bi-directional, precision current sense amplifier with an integrated shunt resistor evaluation module (EVM). The study demonstrates a strong correlation (within +/- 0.3-2.5% difference) between simulation and laboratory measurements, underscoring the practical relevance and impact of our research.**

Keywords—*Advanced packaging, Current sensor, Multichip module, Multiphysics modeling, Correlation*

I. INTRODUCTION

Current measurements in electronic systems are commonly utilized for providing feedback, ensuring operations are within acceptable margins, and detecting potential fault conditions. However, measuring the current signal directly can be quite challenging. There are multiple techniques available to measure current indirectly. For instance, current passing through a wire produces a magnetic field that can be detected by magnetic sensors such as hall-effect and fluxgate. An accurate, yet cost-effective method involves measuring the voltage developed across a resistor as current passes through it. This type of resistor is called a current sensing, or shunt, resistor and is based on the concept of Ohm's law [1]. The shunt resistor is typically preferred as it provides for a physically smaller, more accurate and temperature stable measurement compared to a magnetic-based solution. For the system's current information to be evaluated and analyzed, it must be digitized and sent to the system controller. There are many methods for measuring and converting the signal developed across the shunt resistor. The most common approach involves using an analog front-end (AFE) to convert the current sensing resistor's differential signal to a single-ended signal. This single-ended signal is then connected to an analog-to-digital converter (ADC) that is connected to a microcontroller as shown in Fig. 1.

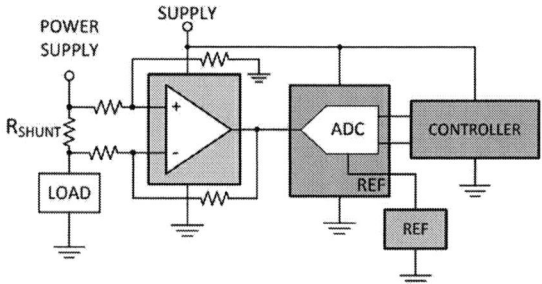

Fig. 1. A typical current sensing signal chain.

In order to optimize the current sensing signal chain, it is essential to select the appropriate shunt resistor value and amplifier gain for the current range and full-scale input range of the ADC. However, there is a trade-off between power loss (voltage drop across the resistor) and the amplifier's input offset voltage. An ideal product should be able to handle a relatively large current with minimal impact from electrothermal (i.e., Joule heating) and thermomechanical reliability (i.e., thermomechanical). Predicting the multiphysics couplings is crucial for achieving the device's optimal performance.

The electrothermal simulations and silicon measurements for the 36V, bi-directional, precision integrated shunt resistor amplifier was published in [2]. Good correlations were observed between the simulations and measurements for the minimum and maximum device temperature within (+/- 2.2%). This work extends the predictive modeling to include the thermomechanical component. In the process, we demonstrate the complete development and implementation of the electrical-thermal-mechanical predictive modeling. Section II provides key functional details of the current sensor device under test. Section III details the components of the predictive coupled co-design and multiphysics modeling flow. Simulation to measurement comparisons are presented in Section IV.

II. CURRENT SENSOR DEVICE DETAILS

The device features a 2-mΩ precision current-sensing resistor and a 36-V common-mode, zero-drift topology precision current-sensing amplifier integrated into a single TSSOP (Thin Shrink Small Outline) package. High-precision measurements are enabled through the matching of the shunt resistor value and the current-sensing amplifier gain, providing a highly accurate, system-calibrated solution (see Fig. 2).

979-8-3503-5124-8/24 $31.00 © 2024 IEEE

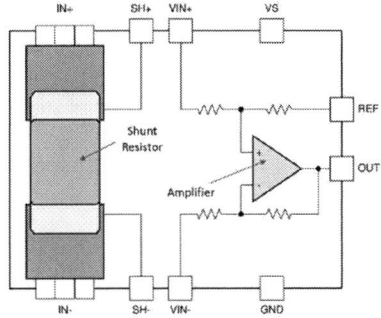

Fig. 2. The functional block diagram of the current sensing device.

The low-drift current-sensing resistor is capable of high-precision measurements within the temperature range of -40°C to 125°C. Its integrated design ensures measurement stability across temperatures. The onboard current-sensing resistor is designed as a 4-wire (or Kelvin) connected resistor, enabling accurate measurements through a force-sense connection (see Fig. 3). Connecting the amplifier input pins (VIN– and VIN+) to the sense pins of the shunt resistor (SH– and SH+) helps reduce parasitic impedances commonly found in typical very low sensing-resistor level measurements. Under specific conditions, such as no airflow, a maximum ambient temperature of 85°C, and 1-oz. copper input power planes, the device can handle continuous current levels up to 15A.

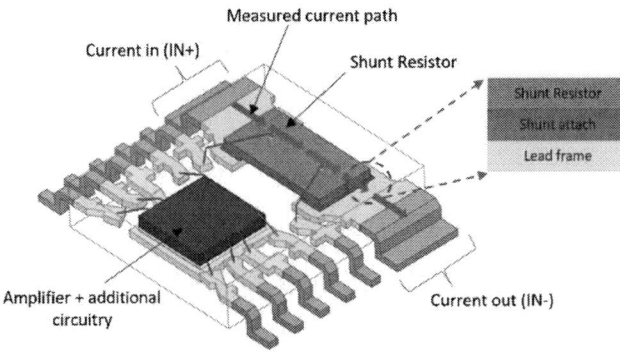

Fig. 3. MCM package with integrated amplifier and shunt resistor.

The device is packaged in a 16-pin TSSOP package of 5.0x4.4x1.0mm in size. The current to be measured enters the package through the fused pins (IN+) lead frame, through the shunt where the differential wirebond senses the voltage drop, and out the package through fused pins (IN-). The current path includes the conductive shunt attach epoxy as shown in the zoom cross-section in Fig. 3. What is the maximum current (i.e., beyond 15A nominal) that can be supported before the onset of a reliability failure? The multiphysics modeling methodology developed in the next section is used to quantify the maximum current at multiple junctions of the package.

III. MULTIPHYSICS MODELING FLOW

The multiphysics modeling flow developed here is based on the weak formulation theory [3]. Fig. 4 shows the stepwise flowchart for the multiphysics analysis. The device is ready for modeling once the MCM package physical design is optimized

using robust signal and power integrity design guidelines. In the modeling phase, each physics model is solved separately, and data is transferred from each model while maintaining the domain coupling. The electrical analysis is formulated via the electrical potential 3D current continuity equation and the point form of Ohm's law (which relates the current density to the electric field). The boundary conditions are Dirichlet and impedance boundary conditions. The impedance boundary condition is useful at boundaries where the electromagnetic field penetrates only a short distance outside the boundary).

The dissipated power calculated in the electrical analysis serves as the heat source for the thermal analysis. The thermal analysis is formulated via the coupled conservation of energy and 3D Fourier heat conduction equations. The coupled electrothermal thermal analysis considers the nonlinear temperature-dependent of the materials properties. The boundary conditions are Dirichlet and air convection.

The derived localized temperature distribution gradient from the electrothermal analysis is then used sequentially to perform the thermomechanical analysis in order to assess the stress distributions of the package system. The mechanical formulation is based on linear thermoelasticity theory – these include the infinitesimal strain relations, the constitutive equations, and the conservation of momentum.

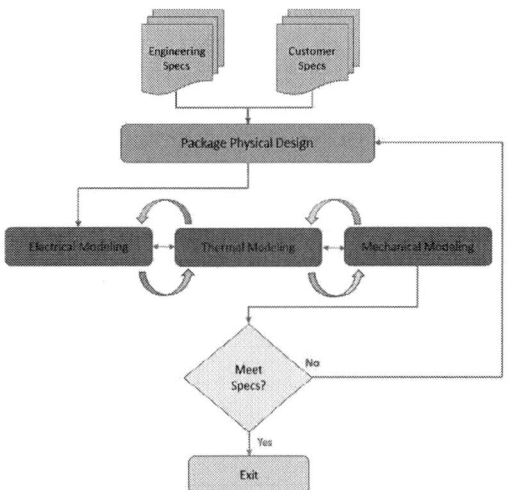

Fig. 4. Multiphysics modeling to assess device maximum current tolerance before the onset of reliability issue.

IV. SIMULATION VS MEASUREMENT CORRELATION

The electrothermal simulation compares was employed to predict the temperature of the package system. Table 1 below summarizes the simulated temperature at a few critical locations in the system – the shunt, the mold top, and the conductive shunt attach epoxy for current sweep of 15A to 50A. For more details of the simulated and measurement setup see [2]. Here, we describe the thermomechanical simulation to measurements correlation. The temperature distribution from the electrothermal simulation for a given current input into the package was applied as temperature loading for the mechanical model.

979-8-3503-5124-8/24 $31.00 © 2024 IEEE

TABLE I. TEMPERATURE OF SHUNT, MOLD TOP, AND SHUNT EPOXY ATTACH FROM CURRENT SWEEP (15A TO 50A).

I (A)	Power (W)	Max Temp (°C)		
		Shunt	Mold top	Shunt attach epoxy
15	1.1	100.8	99.2	87.8
20	2.0	167.5	164.6	144.1
30	4.8	356.2	349.5	**323.8**
40	8.7	615.2	603.2	522.0
50	13.6	947.3	928.4	801.6

The package physical design was meshed with hexahedral elements (3D element with 20 nodes) near the shunt die regions to achieve good accuracy while other regions have tetrahedral mesh to speed up messing process. Elastic properties were used for all materials. Molding temperature around 175°C was set to stress free condition in the model. The temperature gradient combined with the coefficient of thermal expansion (CTE) mismatch among different package materials causes package deformation and stress build up.

The interested region here is the shunt attach material which physically connects lead frame and shunt resistor. Figure 5 shows maximum values for 4 different stress components at the conductive die attach material for different current values. As expected the increase current generated more heat and caused higher temperature at shunt region and larger stress values at shunt epoxy attach interface as well. For a given material, crack or fracture can occur once the stress is beyond material strength. On the other hand, interfacial adhesion strength and material fracture strength will reduce at high temperature for polymer materials. Voids or microcracks increase due to shunt attach material decomposition at higher temperature. The combination of higher stress and deteriorated properties makes die attach material much easier to crack or delamination at higher temperature.

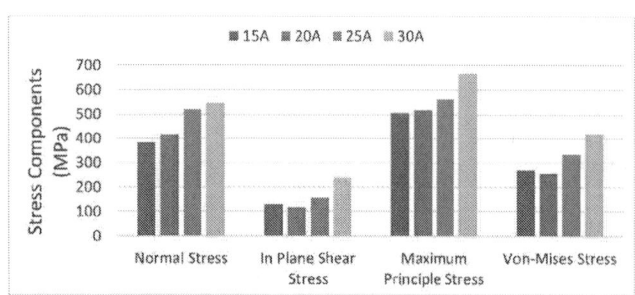

Fig. 5. Maximum values for different stress components at shunt attach material under shunt for different current values.

Based on the electrothermal modeling the temperature of the conductive shunt epoxy attach at 30A is approximately 323.8°C (see Table 1). Higher temperature (which corresponds to higher current) yields the highest stress value for the four stress components as per thermomechanical modeling findings. For the current values simulated, 30A yields the highest stress. It is unclear if the stress values at 30A will cause reliability issues. To assess the safe operating area of the device, we performed

dynamic mechanical analysis (DMA) characterization of the conductive shunt epoxy material. The thermomechanical properties of organic material representative of die-attach material and mold compound are sensitive to temperature and dynamic response. DMA is a widely adopted method that effectively determines polymers' mechanical behaviors and transitions under dynamic response [4].

As shown in Fig. 6 below, the decomposition of the shunt attach epoxy material, based on DMA testing, is between 325°C and 332°C (as indicated by the red circle). It should be noted that the temperature corresponding to 30A of current (which also yielded a specific stress value) corresponds to the temperature at which the decomposition of the material begins. Hence, we conclude that the 30A current value for this device is outside of the devices safe operating area because it leads to a temperature and stress distribution across the device, resulting in the die attach's decomposition.

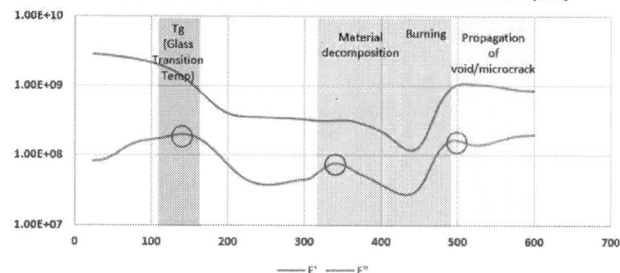

Fig. 6. DMA characterization of the conductive shunt attach material. E'(ω) is the dynamic storage elastic modulus, and E'' (ω) is the dynamic loss elastic modulus. X-axis is temperature and Y-axis is stress/strain.

V. CONCLUSION

Advanced analog MCM packaging is prone to high-level multiphysics interactions due to the aggressive market needs for integration, miniaturization, and cost-effective solutions. A predictive multiphysics (electrical-thermal-mechanical) modeling methodology was developed, implemented, and validated with measurements on an MCM IC package current sensor device. A good correlation was observed between simulation and measurement (i.e., within +/- 0.3-2.5% difference), reinforcing the reliability of our research.

REFERENCES

[1] Dr. S. Hill, "Integrating the Current Sensing Resistor", TI Tech Notes, October 2016, http://www.ti.com/lit/an/sboa170a/sboa170a.pdf.

[2] Jie Chen, Rajen Murugan, Steven Kummerl, Usman Chaudhry, Edwin Lim, Tatsuhiro Shimizu, Thatcher Klumpp, and Jack Grantham (2018) System ElectroThermal Co-Design of a Zero-Drift Current-Shunt Monitor with Precision Integrated Shunt Resistor. International Symposium on Microelectronics: Fall 2018, Vol. 2018, No. 1, pp. 000193-000197.

[3] Zhang Q. and Cen S., "Multiphysics Modeling: Numerical Methods and Engineering Applications", Tsinghua University Press Computational Mechanics Series, 1st Ed., Dec 15, 2015.

[4] S. R. Kumbhar, S. Maji and B. Kumar, "Dynamic mechanical analysis of Magnetorheological Elastomer," 2013 International Conference on Energy Efficient Technologies for Sustainability, Nagercoil, India, 2013, pp. 870-873, doi: 10.1109/ICEETS.2013.6533500.

Impact of Non-Functional Pads Location on Eye Diagram Performance

Mehdi Mousavi
EMC Laboratory
Missouri University of
Science and Technology
Rolla, USA
smousavi@mst.edu

Kevin Cai
Unified Computing
Systems
Cisco Systems, Inc
San Jose, CA, USA
kecai@cisco.com

Junyong Park
EMC Laboratory
Missouri University of
Science and Technology
Rolla, USA
junyongpark@mst.edu

Chaofeng Li
EMC Laboratory
Missouri University of
Science and Technology
Rolla, USA
clf83@mst.edu

Manish K. Mathew
EMC Laboratory
Missouri University of
Science and Technology
Rolla, USA
mkmbzm@mst.edu

Reza Asadi
EMC Laboratory
Missouri University of
Science and Technology
Rolla, USA
reza.asadi@mst.edu

Shameem Ahmed
Unified Computing
Systems
Cisco Systems, Inc
San Jose, CA, USA
shameem@cisco.com

Bidyut Sen
Unified Computing Systems
Cisco Systems, Inc
San Jose, CA, USA
bisen@cisco.com

DongHyun (Bill) Kim
EMC Laboratory
Missouri University of
Science and Technology
Rolla, USA
dkim@mst.edu

Abstract— **This study shows that placing four NFPs in selective PCB layers significantly improves signal integrity, reducing minimum jitter from 18.28 ps to 12.81 ps and maximum jitter from 27.66 ps to 22.03 ps.**

Keywords— Signal Integrity, Non-functional Pads (NFP), Eye Diagram.

I. INTRODUCTION

The design of printed circuit boards (PCBs) for high-speed applications requires careful attention to signal integrity, particularly the impedance of signal vias. Recent developments have introduced models capa ble of analyzing via impedance up to 100 GHz [1-3]. This study focuses on the impact of non-functional pad (NFP) placements on signal via impedance, building on previous work that highlights the importance of strategic NFP placement for signal integrity [4]. By examining various configurations of NFPs, we aim to provide insights into optimizing PCB designs for high-speed channels, referencing foundational analyses on differential vias and signal integrity [5].

II. METHODOLOGY

A. Simulation Setup

In our research, we used a 3D full-wave electromagnetic (EM) simulation tool to analyze the frequency dependent electrical performance of signal vias in PCBs, focusing on the influence of NFPs. The via diameter was 8 mils, and it was placed 40 mils apart from adjacent ground vias, with an anti-pad diameter of 30 mils. 0.6 mils thick and 16.1 mils wide microstrip lines were connected to the via on top and bottom. These traces have an impedance of 48.84 ohms.

Two configurations were analyzed in the simulations: one with two NFPs and another with four per signal via. This was to examine the influence of NFP quantity on via impedance and signal integrity.

B. Configuration of non-functional pads and location choices

The objective of our analysis was to optimize PCB signal integrity by analyzing the placement of NFPs in two distinct configurations. The first configuration involved two NFPs, leveraging the PCB's 11 internal layers to explore 55 unique placement combinations (noted as M and N for the first and second pad's layer positions, respectively). The second setup employed four NFPs, expanding the analysis to 330 possible combinations (denoted as M, N, P, and Q for the layers of each pad), to assess the broader impact of NFP arrangement on signal integrity. These configurations were visualized using a model structure depicted in Fig. 1, facilitating a comprehensive understanding of the electromagnetic interactions within the PCB and the potential for optimized NFP placement to enhance signal performance.

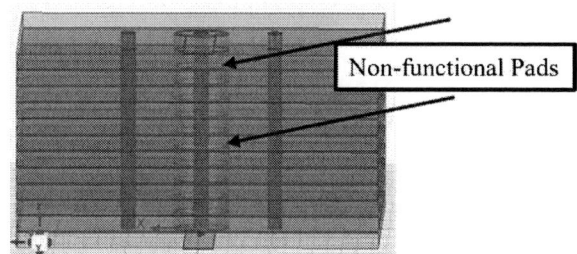

Fig. 1. HFSS Structure consisting of 12 layers, two ground vias, microstrip lines on top and bottom connected to signal via.

III. EXPERIMENTAL PROCEDURE

A. Simulation Parameters

Our experimental simulations utilized full-wave simulation tools and Advanced Design System (ADS) software to assess the impact of NFPs on the impedance and signal integrity of high-speed PCBs. The simulations focused on two configurations, one with two NFPs which is case 1 and another with four which is case 2, both involving trace lengths of 5 inches + via + 5 inches. Key simulation parameters included using a PRBS31 pattern at 32 Gbps for signal testing, setting rise and fall times at 5 ps, and targeting eye height and jitter in eye diagram analysis to gauge signal integrity. The aim was to identify configurations that yield the most open-eye diagram, a marker of optimal signal performance, with all tests assuming a source and load impedance of 50 ohms

B. Data Collection Method

Data collection in the study comprised two main steps: extracting S-parameters using full-wave simulations to evaluate impedance and signal integrity for different PCB configurations and trace lengths and analyzing eye diagrams with ADS using a PRBS31 signal to measure eye height and jitter. This dual approach allowed for a comprehensive analysis of how NFP configurations and trace lengths impact signal integrity.

IV. RESULTS

The study revealed that in high-speed PCBs, altering NFP configurations from two to four NFPs significantly improves signal integrity. Specifically, in a configuration with a 5-inch trace length plus via, switching to four NFPs increased the minimum eye height from 0.011V to 0.051V (a 78.43% improvement) and the maximum eye height from 0.052V to 0.071V (26.76 % improvement). Additionally, jitter reductions were observed, with minimum jitter decreasing from 18.28 ps to 12.81 ps (29.92 %) and maximum jitter from 27.66 ps to 22.03 ps (20.35 %). These results indicate a marked enhancement in signal integrity, consistently shown by more open-eye diagrams across varying trace lengths.

Fig. 2 shows the histogram of eye jitter variations for a PCB configuration with two NFPs and a 5-inch trace length plus via (case 1). The significant jitter observed at positions #10 and #11 (27.66 ps) highlights suboptimal NFP placement, causing increased signal degradation due to electromagnetic interference and reflection.

Fig. 3 (case 1 with two NFPs) presents the histogram of eye height variations for the same configuration. The worst-case scenario, indicated by the lowest eye height of 0.011 V, demonstrates poor signal integrity due to inadequate impedance matching.

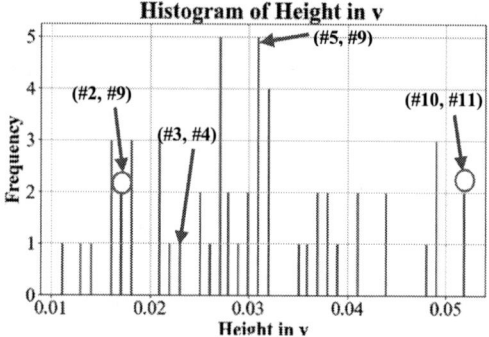

Fig. 3. Histogram of eye diagram height (55 cases) for two NFPs. Red arrows with circles indicate best and worst locations based on jitter; plain red arrows show maximum and minimum heights. The histogram is not smooth due to discrete and varied NFP placements, causing specific height measurements to cluster.

Figs. 4 and 5 illustrate the improvements in eye jitter and eye height when increasing NFPs to four. The optimal configuration (#3, #7, #9, and #10) achieves the lowest jitter (12.81 ps) and highest eye height (0.071 V), suggesting that this arrangement minimizes signal reflections and enhances impedance consistency, resulting in a clearer and more stable signal.

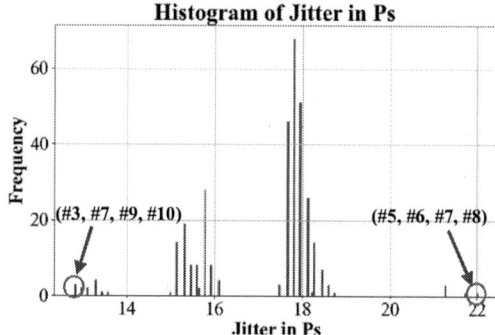

Fig. 4. Histogram of jitter in ps (330 cases) for four NFPs with the same trace length. Red arrows with circles indicate best and worst locations based on jitter.

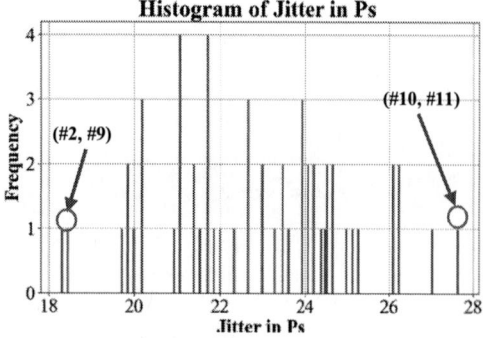

Fig. 2. Histogram of jitter in ps (55 cases) for two NFPs. Red arrows with circles indicate best and worst locations based on jitter.

979-8-3503-5124-8/24 $31.00 © 2024 IEEE

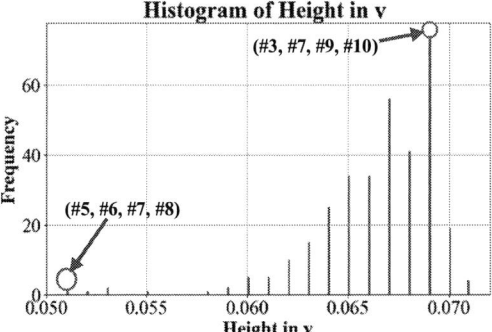

Fig. 5. Histogram of eye diagram height in V (330 cases) for four NFPs with the same trace length. Red arrows with circles indicate best and worst locations based on jitter.

The best and worst locations are determined primarily by jitter, with eye height as a secondary criterion. For instance, the optimal placement in case 1 is at locations #2 and #9 (18.28 ps jitter), and the least effective is at locations #10 and #11 (27.66 ps jitter). In contrast, with four NFPs, the best result is achieved at locations #3, #7, #9, and #10 (12.81 ps jitter), with the worst at locations #5, #6, #7, and #8 (22.03 ps).

The analysis reveals that trace length directly impacts signal quality, with more NFPs generally leading to better signal integrity. However, the in-depth analysis for configurations with four NFPs required considerably more time than those with two. These comprehensive findings guide optimal NFP placement for enhanced signal performance in high-speed PCB designs.

TABLE I.   BEST AND WORST COMBINATINO OF LOCATIONS FOR DIFFERENT SETS OF COMBINATION OF NON-FUNCTIONAL PADS

Cases	Case 1. Two non-functional pads	Case 2. Four non-functional pads
Best	(#2, #9)	(#3, #7, #9, #10)
Worst	(#10, #11)	(#5, #6, #7, #8)

V. Conclusion

Our findings highlight that in high-speed PCBs, the strategic placement and quantity of NFPs are vital for signal integrity. The study underscores the relationship between data rate, PCB stack-up, and NFP layer location, which needs to be considered when designing high-speed digital channels in PCB. Optimal NFP placement significantly improves signal performance by minimizing signal degradation and enhancing impedance consistency across the PCB. Fig 2 through 5 clearly demonstrate that configurations with four NFPs provide superior signal integrity, as evidenced by improved eye height and reduced jitter. Compared to the case without NFPs, which resulted in a closed-eye diagram, the configurations with NFPs show significant enhancements.

We conclude that the strategic placement of NFPs, particularly in specific layers, is crucial for improving signal integrity. For the given design, placing four NFPs in selective layers significantly enhances signal performance. However, increasing the number of NFPs requires more time and effort to investigate all possible configurations, presenting a trade-off between improved signal integrity and increased design complexity. Future work will focus on developing more efficient methods to optimize NFP placement, balancing performance gains with the time required for thorough investigation. This guidance should help designers in making informed decisions about NFP usage in high-speed PCB designs.

References

[1] C. Li et al., "High-Speed Differential Via Optimization using a High-Accuracy and High-Bandwidth Via Model," 2023 IEEE Symposium on Electromagnetic Compatibility & Signal/Power Integrity (EMC+SIPI), Grand Rapids, MI, USA, 2023, pp. 280-285, doi: 10.1109/EMCSIPI50001.2023.10241408.

[2] C. Li et al., "Mode-decomposition-based Equivalent Via (MEV) Model and MEV Model Application Range Analysis," 2023 Joint Asia-Pacific International Symposium on Electromagnetic Compatibility and International Conference on ElectroMagnetic Interference & Compatibility (APEMC/INCEMIC), Bengaluru, India, 2023, pp. 1-4, doi: 10.1109/APEMC57782.2023.10217672.

[3] C. Li, K. Cai, M. Ouyang, Q. Gao, B. Sen and D. Kim, "Mode-Decomposition-Based Equivalent Model of High-Speed Vias up to 100 GHz," in IEEE Transactions on Signal and Power Integrity, vol. 2, pp. 74-83, 2023, doi: 10.1109/TSIPI.2023.3268255.

[4] M. Ouyang et al., "Optimizing the Placement of Non-Functional Pads on Signal Vias Using Multiple Reflection Analysis," 2022 IEEE Int. Symp. on Electromagnetic Compatibility & Signal/Power Integrity (EMCSI), Spokane, WA, USA, 2022, pp. 169-174, doi: 10.1109/EMCSI39492.2022.9889339.

[5] C. Cho et al., "Differential Via Optimization for PCIe Gen5 Channel based on Particle Swarm Optimization Algorithm," 2022 IEEE 31st Conference on Electrical Performance of Electronic Packaging and Systems (EPEPS), San Jose, CA, USA, 2022, pp. 1-3, doi: 10.1109/EPEPS53828.2022.9947193.

[6] Y. Deng, Z. Li, Y. Yu, B. Li, X. Wang and Z. Wu, "S Parameters Optimization of High-Speed Differential Vias Model on A Multilayer PCB," 2022 23rd International Conference on Electronic Packaging Technology (ICEPT), Dalian, China, 2022, pp. 1-4, doi: 10.1109/ICEPT56209.2022.9873518.

High-speed Interconnect Design of Silicon Interposer based Heterogeneous Integration for AI Computing

Keeyoung Son, Seonguk Choi, Keunwoo Kim, Jiwon Yoon, Junghyun Lee, Haeseok Suh, Hyunjun An and Joungho Kim
Korea Advanced Institute of Science and Technology (KAIST), School of Electrical Engineering, Daejeon, Republic of Korea
keeyoung@kaist.ac.kr

Abstract— In this paper, signal integrity (SI) design and analysis of high-speed interconnect of silicon interposer based heterogeneous integration for artificial intelligence (AI) computing was carried out. With the increasing popularity of AI services, there has been a significant surge in the demand for AI computing capabilities. As a result, high-performance, high-density AI computing modules based on multi-GPU architectures, rather than single-GPU configurations, have emerged as a prominent solution. For high-density GPU integration, the silicon interposer based heterogeneous integration of multi-GPU architecture is arisen. In this research, we design multi-GPU integration on silicon interposer and high-speed GPU links considering routability. Furthermore, we analyzed the designed high-speed GPU link considering SI, but also fabrication cost. Consequently, multiple GPUs are integrated on a silicon interposer, but it has been determined that adequate SI for GPU links cannot be secured on the interposer. Instead, utilizing a package for these connections ensures the necessary SI and offers significant advantages in fabrication cost with reducing metal layers of silicon interposer.

Keywords—Heterogeneous integration; Interconnect; Multi-GPU; Signal integrity; Silicon interposer;

I. INTRODUCTION

The growing popularity of artificial intelligence (AI) services has led to a substantial increase in the demand for advanced AI computing performance. Consequently, multi-GPU architectures have become a prominent solution, offering high-performance and high-density AI computing modules in place of traditional single-GPU configurations. To maximize AI computing performance by efficiently utilizing multiple GPUs, it is essential to ensure sufficient bandwidth for the interconnects between GPUs, known as GPU links [1, 2]. Therefore, to achieve higher bandwidth and integrate more GPUs within a limited space, the GPUs need to be placed in closer proximity, since reducing interconnect length enhances signal bandwidth. Accordingly, silicon interposer-based heterogeneous integration of multi-GPU architectures, which allows GPUs to be placed in closer rather than conventional multi-GPU architecture which integrated on PCB, has emerged as a key approach.

Fig. 1. shows the development trend of multi-GPU on silicon interposer-based heterogeneous integration. Initially, AI computing modules were designed with 1 GPU and 4 high bandwidth memory (HBM) integrated on a silicon interposer. As the demand for AI computing increased, the number of HBMs and GPUs on the interposer increasing, leading to the current configuration of 2 GPUs and 8 HBMs on a single interposer. Given this trend, it is expected that in the future, four or more GPUs will be integrated on a single interposer. When

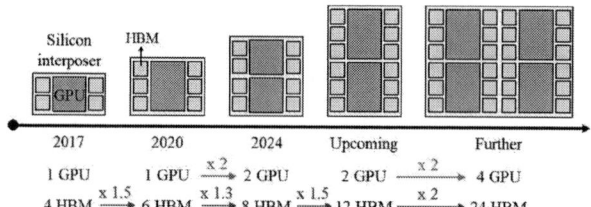

Fig. 1. Development trend of multi-GPU on silicon interposer based heterogeneous integration for AI computing.

there were 2 GPUs, communication between the GPUs on the interposer was achieved via short GPU links within interposer, while communication with GPUs outside the interposer was done through interconnects with a GPU link switch. However, with 4 or more GPUs, the increased distance for communication between GPUs on the interposer can pose signal integrity (SI) issues.

In this paper, SI design and analysis of high-speed interconnect of multi-GPU on silicon interposer based heterogeneous integration was carried out. We designed multi-GPU on silicon interposer with GPU link interface floorplan for high AI computing performance. Also we designed the high-speed GPU links via silicon interposer and package (PKG) considering routability. Then, we analyzed the SI of GPU link in frequency-domain and time-domain, respectively. As a results, although multiple GPUs are integrated on a silicon interposer, but it has been determined that guaranteed SI for GPU links cannot be secured on the interposer. Instead, utilizing a PKG for these interconnects ensures SI, and also offers significant advantages in fabrication cost with reducing metal layers of silicon interposer redistribution layer (RDL). This research shows future direction of high-speed interconnect design for silicon interposer based heterogeneous integration architecture for high AI computing performance.

II. DESIGN OF MULTI-GPU ARCHITECTURE ON SILICON INTERPOSER BASED HETEROGENEOUS INTEGRATION

Fig. 2. (a). shows the designed AI computing module configured by 4 GPU integration on silicon interposer. Designed AI computing module consists 4 GPUs, 24 HBMs, and 3 GPU link switches. Also, each GPU is interconnected with others by GPU link as shown as fig. 2. (b). In terms of GPUs above silicon interposer, it is directly connected by GPU link. In terms of GPUs on other AI computing module, it connected through GPU link switches. For GPU links required to connect GPUs integrated on a silicon interposer, there will inevitably be crossing trace between GPUs. Therefore, when designing GPU links for multi-GPU architecture integration on

---

979-8-3503-5124-8/24 $31.00 © 2024 IEEE   46

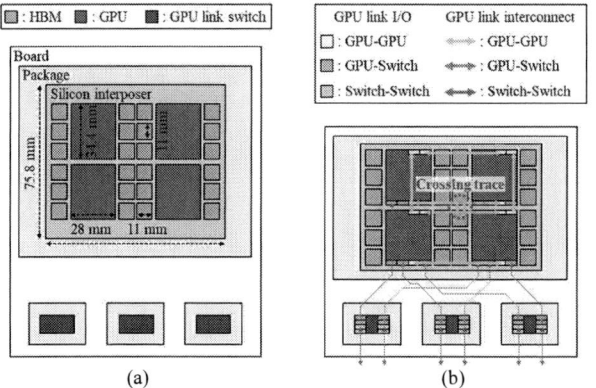

Fig. 2. (a) Designed AI computing module configured by multi-GPU integration on silicon interposer. (b) GPU link interface floorplan of designed AI computing module and GPU link routing.

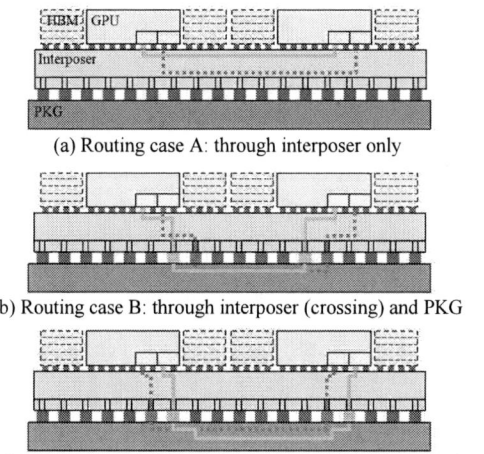

(a) Routing case A: through interposer only

(b) Routing case B: through interposer (crossing) and PKG

(c) Routing case C: through interposer and PKG (crossing)

Fig. 3. Interconnect routing cases of GPU link for designed multi-GPU integration on silicon interposer.

TABLE I
COMPONENT-WISE INTERCONNECT LENGTH OF GPU LINK

Routing	Interposer Channel	Interposer TSV	PKG Channel	PKG Via	Full Channel
Case A	55.5 mm (M1/M5)	X	X	X	55.5 mm
Case B	11.1 mm (M1/M5)	0.2 mm	44.4 mm (M1)	X	55.7 mm
Case C	2.2 mm (M1)	0.2 mm	2 mm (M1) 51.3 mm (M3)	0.27 mm	55.97 mm
			53.3 mm (M1)	X	55.7 mm

silicon interposer, it is essential to consider crossing trace points with layer change of signal trace.

Fig. 3. shows the 3 feasible GPU link interconnect routing cases for designed multi-GPU architecture integration on silicon interposer considering trace crossing points. Case A used only interposer with microstrip and strip differential channel. Case B used both silicon interposer and PKG, the trace crossing is at silicon interposer. Case C also used both interposer and PKG, however the trace crossing is at PKG. The component-wise interconnect length of GPU link with worst SI performance is summarized at Table I. The length of interposer channel is determined considering escape routing and trace crossing. Also, the routing case C has longest channel length.

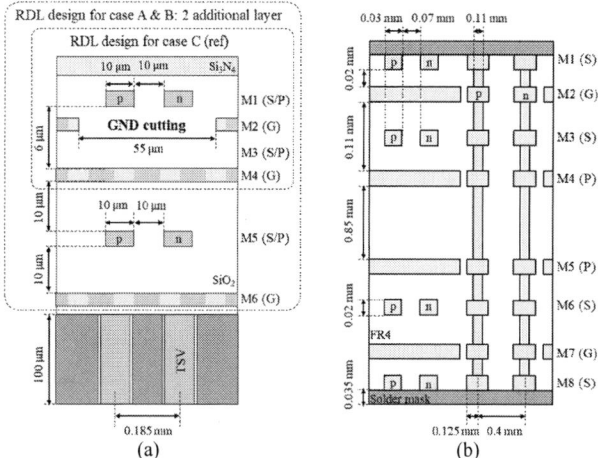

Fig. 4. Stack-up design of (a) silicon interposer and (b) package for GPU link routing.

Every routing case used silicon interposer and PKG, therefore we designed the stack-up of silicon interposer and PKG [3, 4, 5]. Fig. 4. shows the designed stack-up with physical dimensions and materials. The channel and via of silicon interposer and PKG are designed considering 85 Ω matching, which is the standard impedance of high-speed interconnect. Furthermore, the PKG via is assumed as back-drilled condition for secured SI. In terms of silicon interposer microstrip differential channel, the M2 ground is cut for impedance matching. This is because the dielectric height between M1 and M2 cannot be handled due to the memory channel design which interconnect GPU and HBM. Although the M2 ground is cut, the M4 and M6 ground is original because of there are no memory channel routing on M5. Each component was designed to have similar loss characteristics for microstrip and strip channels. To analyze the worst case of routed interconnects, the Table I was authorized by using microstrip instead of strip, as strip is less sensitive to crosstalk.

III. SIGNAL INTEGRITY ANALYSIS OF GPU LINK AS HIGH-SPEED INTERCONNECT OF MULTI-GPU ARCHITECTURE

To determine the proper routing schematic of GPU link as high-speed interconnect, we conducted frequency-domain and time-domain SI analysis of each routing case. We assumed that the datarate of GPU link as 50 Gbps considered the datarate of other high-speed I/O such as PCIe, NVLink, infinity fabric and others. The fig. 5. (a). shows the simulated differential insertion loss (DIL) of GPU link depends on routing cases. The case A which is consist of only interposer has worst DIL, because of the high DC resistance of interposer channel rather than PKG channel. At the Nyquist frequency, the routing case C has improved DIL than case B. This is because that the case C has lower proportion of interposer channel than case B. However, beyond Nyquist frequency, due to the case C includes both interposer TSV and PKG via which caused impedance mismatch within high-speed interconnect, additional loss and other SI drawbacks, it has worse DIL than case B. Additionally, above 75 GHz, due to the PKG trace has higher dielectric loss than interposer channel, the routing case A has better DIL than other cases.

979-8-3503-5124-8/24 $31.00 © 2024 IEEE   47

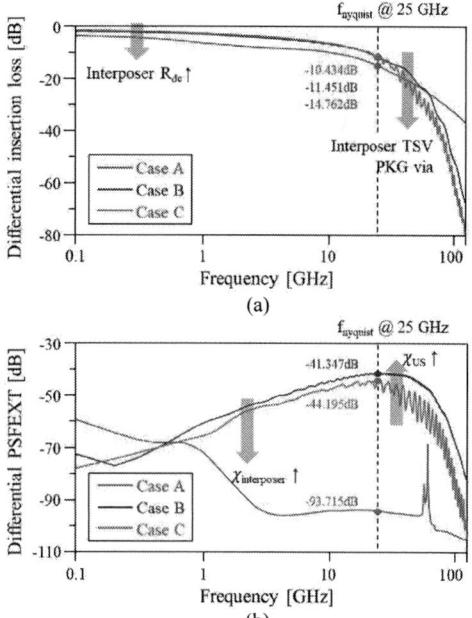

(a)

(b)

Fig. 5. Simulated (a) differential insertion loss and (b) differential power-sum FEXT (PSFEXT) of GPU link depends on routing cases.

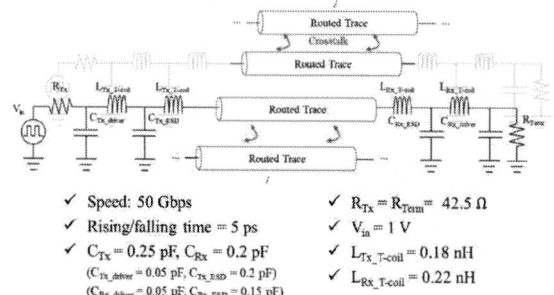

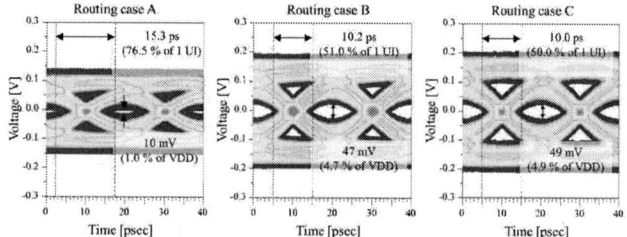

✓ Speed: 50 Gbps
✓ Rising/falling time = 5 ps
✓ C_{Tx} = 0.25 pF, C_{Rx} = 0.2 pF
  (C_{Tx_driver} = 0.05 pF, C_{Tx_ESD} = 0.2 pF)
  (C_{Rx_driver} = 0.05 pF, C_{Rx_ESD} = 0.15 pF)
✓ R_{Tx} = R_{Term} = 42.5 Ω
✓ V_{in} = 1 V
✓ L_{Tx_T-coil} = 0.18 nH
✓ L_{Rx_T-coil} = 0.22 nH

Fig. 6. Eye-diagram set-up for GPU link.

Fig. 7. Simulated eye-diagram of GPU link depends on routing cases.

The simulated differential power-sum FEXT (PSFEXT) of GPU link depends on routing cases are shown in fig. 5. (b). In case A, the DIL was the worst. However, case A exhibited the best differential PSFEXT. This is because that the interposer channel offers superior routability compared to the PKG, allowing for larger spacing between lanes. Conversely, case B has the worst differential PSFEXT, primarily due to the PKG channel being predominantly composed of microstrip, which are more vulnerable to crosstalk.

For time-domain analysis, we conducted eye-diagram simulation of designed GPU link depends on routing cases. Fig. 6. shows the eye-diagram set-up for GPU link without equalizer for low power consumption based on short interconnect length [5, 6]. The simulated eye-diagrams of GPU link are shown in Fig. 7. It shows that the routing case A has worst eye opening and cases B and C have similar eye openings which have larger eye height and eye width rather than them of case A, about 5 times and 2 times respectively. This is because that DIL of cases B and C have better value at Nyquist frequency than case A. Although the case A has lowest PSFEXT, every routing case have low PSFEXT value which affect quite small to eye opening.

This result show that multi-GPU on silicon interposer based heterogeneous integration requires interconnect routing on PKG, not only silicon interposer. Additionally, case C not only has the largest eye opening but also uses only 4 metal layers of silicon interposer RDL, unlike cases A and B using 6 metal layers. To sum up, routing case C utilizes a package for these interconnects, providing guaranteed signal integrity (SI), and offers significant advantages in fabrication costs by reducing the number of metal layers in the silicon interposer.

IV. CONCLUSION

In this paper, SI design and analysis of high-speed interconnect of multi-GPU on silicon interposer based heterogeneous integration was carried out. We designed multi-

GPU on silicon interposer. Also we designed GPU links routing via silicon interposer and PKG considering routability. Then, we analyzed the SI of GPU links in frequency-domain and time-domain, respectively. As a result, although multiple GPUs are integrated on silicon interposer, it has been verified that routing high-speed GPU links on the interposer cannot guarantee SI. Instead, using PKG for these interconnects ensures SI and reduces fabrication costs by reducing metal layers of interposer RDL. This research shows the future direction of high-speed interconnect design for silicon interposer based heterogeneous integration architecture for high AI computing performance.

ACKNOWLEDGMENT

We would like to acknowledge the technical support from ANSYS Korea.

REFERENCES

[1] Arunkumar, Akhil, et al. "MCM-GPU: Multi-chip-module GPUs for continued performance scalability." ACM SIGARCH Computer Architecture News 45.2 (2017): 320-332.

[2] Rajbhandari, Samyam, et al. "Zero-infinity: Breaking the gpu memory wall for extreme scale deep learning." Proceedings of the international conference for high performance computing, networking, storage and analysis. 2021.

[3] Son, Keeyoung, et al. "Signal integrity analysis of high speed channel considering thermal distribution." 2021 IEEE 30th Conference on Electrical Performance of Electronic Packaging and Systems (EPEPS). IEEE, 2021.

[4] Cho, Kyungjun, et al. "Signal integrity design and analysis of differential high-speed serial links in silicon interposer with through-silicon via." IEEE Transactions on Components, Packaging and Manufacturing Technology 9.1 (2018): 107-121.

[5] Kim, Seongguk, et al. "Signal Integrity Analysis of High-speed PCIe Channel with Board-to-Board Interconnect for High-Performance Server." 2023 IEEE Electrical Design of Advanced Packaging and Systems (EDAPS). IEEE, 2023.

[6] Poulton, John W., et al. "A 1.17-pJ/b, 25-Gb/s/pin ground-referenced single-ended serial link for off-and on-package communication using a process-and temperature-adaptive voltage regulator." IEEE Journal of Solid-State Circuits 54.1 (2018):

979-8-3503-5124-8/24 $31.00 © 2024 IEEE

Design and Analysis of L3 Cache Embedded-GPU-High Bandwidth Memory Architecture with Reduced Energy and Latency for AI Computing

Haeseok Suh, Jiwon Yoon, Keeyoung Son, Seonguk Choi, Keunwoo Kim, Junghyun Lee, Taein Shin, Hyunjun An, Taesoo Kim, Jungmin Ahn, Hyunah Park, Hyunsik Kim*, Taeil Bae*, Haekang Jung* and Joungho Kim

School of Electrical Engineering (EE), Korea Advanced Institute of Science and Technology (KAIST)

SK hynix Inc.

haeseoksuh@kaist.ac.kr

Abstract—For the first time, this paper proposes a L3 cache embedded-GPU-High bandwidth memory (L3E-GPU-HBM) for reduced latency and enhanced energy efficiency of large scale memory intensive AI computing. Accessing HBM in conventional GPU-HBM architecture involves significant latency and requires high data movement energy. To address the challenge, we propose L3E-GPU-HBM in which L3 cache is embedded in interposer between GPU and HBM. To implement the proposed architecture, embedded SRAM interconnect (ESI) chip is employed, which consists of local silicon interconnect (LSI) die and L3 cache die, merged by hybrid bonding. Then, using Chip-on-Wafer-on-Substrate with Local interconnect (CoWoS-L) method, ESI chip is placed inside the reconstituted interposer (RI). The ESI chip functions as both interconnect and L3 cache between L2 cache of GPU and HBM. For verification of the proposed architecture, the circuit model of driver and channel is utilized to obtain the wire latency and energy. The result showed that the proposed L3E-GPU-HBM architecture reduced the wire latency and energy compared to conventional GPU-HBM architecture by 17% and 33%, respectively.

Index Terms—Embedded SRAM Interconnect, GPU, High Bandwidth Memory, L3 Cache, Energy, Wire Latency

I. INTRODUCTION

In recent years, the demand for high performance AI accelerators is increasing due to emergence of immense generative AI models. The model parameters are increasing 750 times per year, creating massive energy consumption on data movement [1]. To overcome this challenge, the GPU-High memory bandwidth (HBM) architecture on silicon interposer has become a prominent solution. However, with the ever increasing amount of data movement, number of HBM access substantially increased for large scale memory intensive AI computing. In conventional GPU-HBM architecture, memory hierarchy consists of register, L1 cache, L2 cache, and HBM. With register being the fastest and smallest memory, if a hit occurs, GPU obtains the data from the register. On the other hand, if a miss occurs, GPU accesses the next memory in the hierarchy. So, HBM is accessed when L2 cache miss occurs. HBM access incurs 5 times the latency and consumes 10 times the energy compared to L2 cache access, due to longer channel length [2]–[4]. Consequently, maximizing cache hit to minimize HBM access is imperative in decreasing latency and energy. The intuitive solution is increasing L2 cache capacity inside GPU. However, increasing L2 cache inside

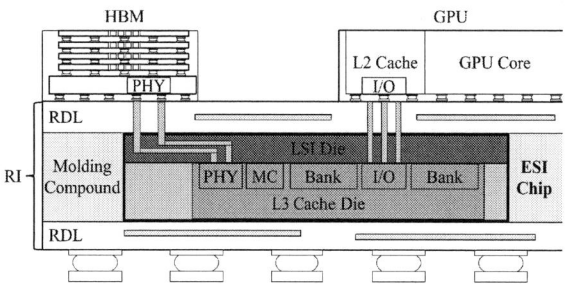

Fig. 1. Conceptual view of the proposed L3 cache embedded-GPU-High bandwidth memory (L3E-GPU-HBM) architecture.

GPU has restriction due to limited reticle size and end of transistor scaling. A architecture placing additional L3 cache on interposer between GPU and HBM has been previously done to reduce HBM access [5]. However, this approach yet provokes large latency and energy due to interposer connection to L3 cache and extended interconnection length to HBM.

Hence, we propose an L3 cache embedded (L3E)-GPU-HBM architecture to minimize HBM access by providing L3 cache in proximity of GPU. The embedded SRAM interconnect (ESI) chip is placed inside the reconstituted interposer (RI) between GPU and HBM to operate as both L3 cache and interconnect. This leads to decreased latency and energy, due to increased form factor. To verify the proposed architecture, driver and channel are designed to analyze the costs of L3 cache access and HBM access. The result verified that L3E-GPU-HBM architecture performs at lower wire latency and higher energy efficiency compared to conventional GPU-HBM architecture.

II. PROPOSAL OF L3E-GPU-HBM ARCHITECTURE

Fig. 1 presents the proposed L3E-GPU-HBM architecture for reducing HBM access. The ESI chip is placed below GPU and HBM, using Chip-on-Wafer-on-Substrate-Local Interconnect (CoWoS-L) technology [6]. Conventional CoWoS-L technology utilizes RI, which is composed of redistribution layer (RDL) and local silicon interconnect (LSI) with molding compound surrounded. In RI, LSI is a small interposer channel for die to die communication. For the proposed architecture, instead of LSI, ESI chip is embedded to function as both L3 cache and LSI. The ESI chip provides additional level of cache

979-8-3503-5124-8/24 $31.00 © 2024 IEEE

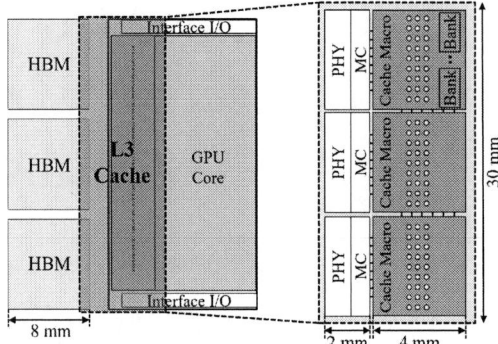

Fig. 2. Floorplan of L3E-GPU-HBM architecture with physical dimension of L3 cache die embedded in the interposer.

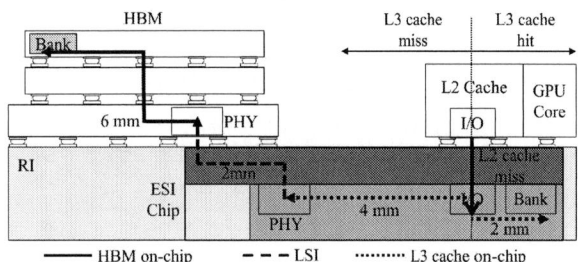

Fig. 3. Interconnect path from L2 cache to L3 cache bank and HBM bank. L3 cache hit leads to L3 cache bank, whereas L3 cache miss leads to HBM bank.

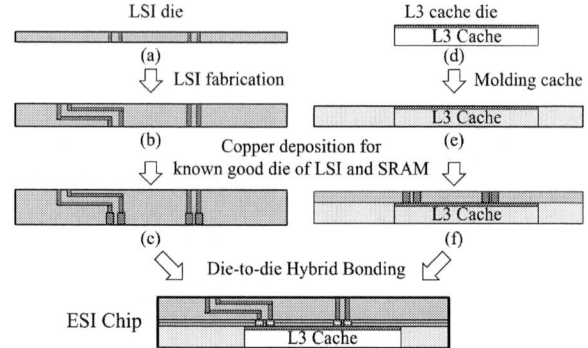

Fig. 4. Fabrication process of ESI chip using die-to-die hybrid bonding.

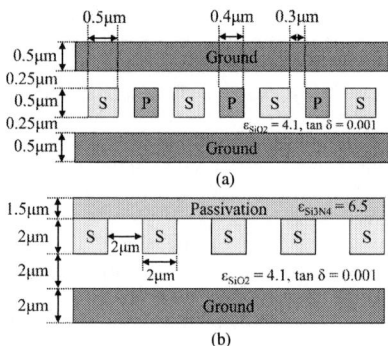

Fig. 5. The designed channel dimensions for (a) on-chip channel of L3 cache and HBM, (b) off-chip channel of LSI.

below GPU, which leads to improvement in wire latency and energy.

Fig. 2 shows the floorplan of the proposed architecture. The physical layer (PHY) and memory controller (MC), which were originally located in GPU, is relocated to L3 cache since L3 cache provides the connection to HBM. The area that was originally allocated for PHY and MC in GPU is replaced with via I/O pad between L2 cache and L3 cache. The total area of L3 cache is 180 mm^2, which consists of three PHY, MC, and cache macro. PHY and MC occupy one-third of the area, and facilitate communication with HBM. To allocate more area to cache macros, the area of PHY and MC is halved with signal I/O halved. To keep the HBM bandwidth identical to conventional architecture, the data rate is doubled. Each cache macros are capable of 40 MB L3 cache capacity including all the peripheral circuits [7]. 240 MB of total L3 cache capacity is provided in the proposed architecture, which decreases the HBM access by 50% for MLPerf inference [5].

Based on the physical dimensions, the path and total length of interconnection in case of L2 cache miss is shown in Fig. 3. In case of L3 cache hit, the length of on-chip channel from L3 cache I/O to L3 cache bank is 2 mm, reducing wire latency and energy. For L3 cache miss, the data path is from L3 cache I/O to HBM bank. The path is composed of 4 mm of L3 cache on-chip channel, 2 mm of LSI, and 6 mm of HBM on-chip channel.

To fabricate the ESI chip, the LSI die and L3 cache die is integrated using die to die hybrid bonding [8]. Fig. 4 illustrates the fabrication process of ESI chip. Fig. 4 (a) to (b) shows LSI die fabrication with RDL process. Fig. 4 (d) to (e) depicts the molding of L3 cache die to match the size with LSI die. Fig. 4 (c) and (f) demonstrates the copper deposition process for hybrid bonding. The two dies merged with hybrid bonding to form ESI chip, as shown in Fig. 4 (g). Finally, the ESI chip is placed in the RI using CoWoS-L technology.

Fig. 5 shows the designed stack-up of channel, based on previous study [9]. Fig. 5 (a) depicts L3 cache and HBM on-chip channel. On-chip channel is identical for both L3 cache and HBM. Repeaters are used per 1.67 mm of on-chip channel to compensate for the high frequency attenuation. For LSI shown in Fig. 5 (b), microstrip structure is exploited to reduce channel capacitance, which leads to wire latency and energy reduction. One signal layer is sufficient since the amount of signal channel is halved. The I/O driver operates at 4 Gb/s with 0.7 V for L3 cache on-chip channel, and 1.1 V for HBM on-chip channel. For LSI, the driver operates at 0.4V and 12.8 Gb/s, which is twice the bandwidth of HBM3. [10].

III. WIRE LATENCY AND ENERGY ANALYSIS OF L3E-GPU-HBM ARCHITECTURE

To verify the wire latency and energy improvement of the proposed architecture, the analysis is performed by two stages. First, wire latency and energy are analyzed by circuit simulation of driver and channel models. Then, the overall system latency and energy are calculated considering the hit ratio of L3 cache.

TABLE I. Equivalent RLC Circuit Values for On-chip and LSI Drivers and Channels.

	On-chip	LSI	Interposer
Data rate	4 Gb/s	12.8 Gb/s	6.4 Gb/s
R_{Tx}	85 Ω	24 Ω	16 Ω
$C_{Tx, C_{Load}}$	0.1 pF	0.4 pF	
$R_{channel}$	78 Ω/mm	7.6 Ω/mm	
$L_{channel}$	0.645 nH/mm	0.253 nH/mm	
$C_{channel}$	0.365 pF/mm	0.241 pF/mm	

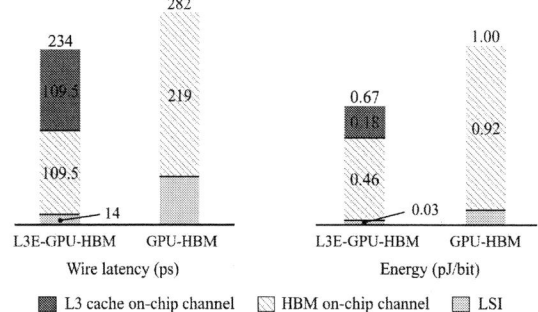

Fig. 7. Comparison of wire latency and energy per bit between L3E-GPU-HBM architecture and conventional GPU-HBM architecture

HBM architecture's wire latency and energy are decreased by 17% and 33%, respectively.

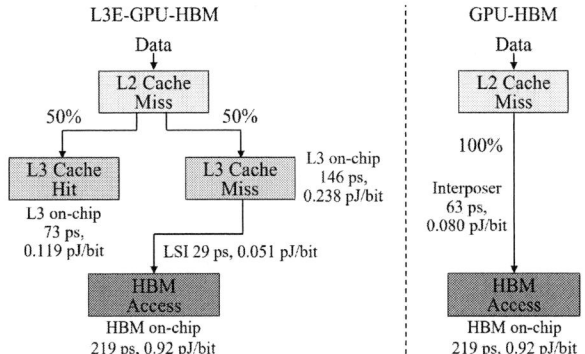

Fig. 6. Wire latency and energy values of L2 cache miss for L3E-GPU-HBM and GPU-HBM.

The extracted circuit values of driver and channel are shown in Table I. Based on the designed parameters of on-chip channel and LSI, the drivers are designed to match the required rise time. The RC models of driver are calculated based on JEDEC standard [10]. The RLC model is extracted from electromagnetic simulation. Using the models, the wire latency is obtained with time-domain circuit simulation, by time delay between half of the input voltage and half of the output voltage. The energy is acquired by dividing dynamic power $\alpha V_{dd}^2 C f$ with data rate in which α is toggle rate set as 0.5, C is total capacitance, and f is Nyquist frequency.

Fig. 6 shows the data path for L2 cache miss scenarios along with L3 cache hit and miss rates. For L3E-GPU-HBM architecture, L2 cache miss leads to either L3 cache hit or L3 cache miss. L3 cache hit merely results in 73 ps of wire latency and consumes 0.119 pJ/bit; however, L3 cache miss requires 394 ps of latency and 1.21 pJ/bit. The wire latency and energy values differ depending on whether L2 cache miss led to L3 cache hit or miss. In contrast, L2 cache miss in conventional GPU-HBM architecture always requires 282 ps and 1.00 pJ/bit, due to absence of L3 cache.

To achieve the overall latency and energy efficiency of the proposed L3E-GPU-HBM architecture, the decreased amount of HBM access due to L3 cache needs to be taken into account. With total of 240 MB L3 cache, 50% of HBM access leads to L3 cache hit, and the remaining 50% results in HBM access. Hence, the values for L3 cache hits and L3 cache misses are each reduced to half of their original values, resulting in average wire latency of 234 ps and energy of 0.67 pJ/bit for L2 cache miss. Fig. 7 illustrates that compared to conventional GPU-HBM architecture, the proposed L3E-GPU-

IV. Conclusion

This paper, for the first time, proposes a L3E-GPU-HBM architecture for large scale memory intensive AI computing. Addition of L3 cache in the GPU memory hierarchy lowers wire latency and energy consumption resulting from decreased HBM access. The proposed architecture decreased wire latency by 17% and energy by 33%, relative to conventional architecture. Therefore, with the escalating demands of data movement in AI computing, L3E-GPU-HBM architecture is an outstanding solution to decrease latency and energy. Additionally, the proposed interposer-embedded structure with functionality can be utilized in various other architectures that employs multiple dies on interposer with heterogeneous integration.

Acknowledgment

This research was funded by SK hynix Inc. We would also like to acknowledge the technical support of ANSYS Korea.

References

[1] X. Amatriain et al., "Transformer models: an introduction and catalog," 2023, arXiv:2302.07730 [cs.CL].

[2] Chatterjee et al., "Architecting an Energy-Efficient DRAM System For GPUs," in IEEE Symposium on High-Performance Computer Architecture, Austin, TX, 2017.

[3] J. Shalf, "The future of computing beyond Moore's Law," Philosophical Transactions A, 2019.

[4] Z. Jia et al., "Dissecting NVIDIA Volta GPU architecture via microbenchmarking," 2018, arXiv:1804.06826v1 [cs.DC].

[5] Y. Fu et al., "GPU Domain Specialization via Composable On-Package Architecture," 2021, arXiv:2104.02188 [cs.AR].

[6] Y.-C. Hu et al., "CoWoS Architecture Evolution for Next Generation HPC on 2.5D System in Package," in Electronic Components and Technology Conference (ECTC), Orlando, FL, 2023.

[7] J. Chang et al., "A 5nm 135Mb SRAM in EUV and High-Mobility-Channel FinFET Technology with Metal Coupling and Charge-Sharing Write-Assist Circuitry Schemes for High-Density and Low-VMIN Applications," in International SoC Design Conference (ISOCC), San Francisco, CA, 2020.

[8] A. Elsherbini et al., "Hybrid bonding interconnect for advanced heterogeneously integrated processors," in IEEE Electronic Components and Technology Conference (ECTC), San Diego, CA, 2021.

[9] S. Kim et al., "Processing-in-memory in High Bandwidth Memory (PIM-HBM) Architecture with Energy-efficient and Low Latency Channels for High Bandwidth System," in IEEE Electrical Performance of Electronic Packaging and Systems (EPEPS), Montreal, Canada, 2019.

[10] "JEDEC Standard: High Bandwidth Memory DRAM (HBM3)," 2022.

979-8-3503-5124-8/24 $31.00 © 2024 IEEE

Analysis of Nonlinear Phase Interactions of a Differential Line in the Presence of a Signal Skew

Byung Cheol Min
Kyungpook National University
Electric and electronic
engineering department
Daegu, Korea
minbc4658@knu.ac.kr

Mun Ju Kim
Kyungpook National University
Electric and electronic
engineering department
Daegu, Korea
dranswn@knu.ac.kr

Hyun Chul Choi
Kyungpook National University
Electric and electronic
engineering department
Daegu, Korea
hcchoi@ee.knu.ac.kr

Kang Wook Kim
Kyungpook National University
Electric and electronic
engineering department
Daegu, Korea
kang_kim@ee.knu.ac.kr

Abstract—**Nonlinear phase interactions between two signal lines of a differential line are analyzed with various signal skew levels and compensation distances. This nonlinear phenomenon may limit the operation of DL-based digital lines above ~10 GHz.**

Keywords—signal integrity, differential line, EM coupling, phase difference, signal skew, high-speed digital signal

I. INTRODUCTION

For high-speed transmission of digital signals, these days, a differential line (DL) is typically used in digital PCBs. Recently, as the digital data speed reaches well up to microwave frequencies, digital signals, consisting of their harmonics, can be considered as very-wideband microwave signals [1], and the digital signal transmission lines should be designed and analyzed with microwave design principles.

A DL consists of two microstrip lines with opposite polarities, placed in parallel with a gap on a substrate. The DL is more resistant to the external noise as compared with a single-ended line. With high-speed digital signals, however, DLs have to deal with multiple unfavorable effects such as signal skew, unbalanced EMI, fiber-weave, etc. [2], [3]. Also, the issue of the optimum gap width, i.e., tight coupling (narrow gap) or loose coupling (wide gap), is still debated, but a number of researchers have recommended that a tight-coupled DL line performs better [4].

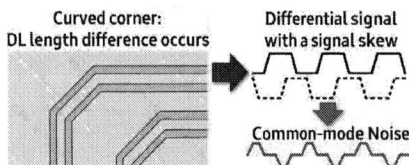

Fig. 1. A differential line (DL) with a curved corner, generating a signal skew.

At the signal launch, the two digital signals on a DL line are oppositely polarized and balanced. In a practical digital circuit board, which contains crowded signal lines, multiple chips, and interfaces, the signal lines are often bent or connected to vias or interfaces, and a length difference between two lines in the DL occurs, and then a signal skew is generated as shown in Fig. 1. The signal skew possesses common-mode components, which tend to radiate over the other parts of the PCB and contribute to the EM interference. Therefore, the digital circuit design

guideline recommends that the line difference should be compensated within 15 mm [5]. However, no detailed analysis has yet been reported on the characteristics and limits of this DL length compensation over the frequencies.

Phase interactions between two EM-coupled lines, especially for the balanced lines, have been discussed and analyzed in [6]. In [6], the balanced lines were shown to have the capability of autonomous phase balancing in the presence of phase imbalance. On the other hand, the DL is not a real balanced line, where the balanced (or differential) signals are maintained only when the lengths of the two DL signal lines are kept the same. As soon as the line length difference in a DL happens, a signal skew occurs. If the two DL signal lines are completely independent (e.g., very loose-coupled), the skew can be completely compensated when a length-matched section is applied. However, there exists some amount of EM coupling between the two DL lines, and therefore, in the presence of the signal skew, the amplitudes and phases of the two DL lines are interacted and thus changed to some directions. The amount and direction of interactions will depend on multiple parameters, such as the gap, line length, frequency, dielectric constant, etc., and cannot be easily estimated for a variety of line configurations.

These interactions will be greater with tight-coupled DL lines (narrow gap), i.e., the amplitude and phase will be significantly modified while propagating on the unbalanced DL portion before applying the length-matched section. As a result, the tight-coupled DL may not be a favorable option with respect to the signal skew phenomena. On the other hand, the tight-coupled DL line will be preferable with respect to the external unbalanced EMI [7], which means that the loose-coupled DL would be unattractive. Therefore, for the higher speed transmission of digital data required for the 5G/6G communications, the design guideline for the DL has to be re-estimated, considering the EM interaction between the DL lines.

This paper analyzes the phenomenon of nonlinear phase interactions, due to EM coupling, between the two DL lines in the presence of a signal skew. With examples of the tight- and loose-coupled DL lines, the phase interactions are analyzed with different signal skew levels and compensation distances. The phase balances between the two DL lines are monitored and verified using EM simulations and measurements.

979-8-3503-5124-8/24 $31.00 © 2024 IEEE

II. PROPERTIES OF A DIFFERENTIAL LINE

A differential line (DL) consists of two microstrip lines with opposite polarities, placed in parallel with a gap on a substrate. As shown in Fig. 2, electric field lines of the DL are formed mainly between each signal line and the ground plane, and some portion of field lines are tight- or loose-coupled between the signal lines depending on the gap width. Electric field distributions for tight-coupled and loose-coupled DL lines are illustrated in Figs. 2b and 2c.

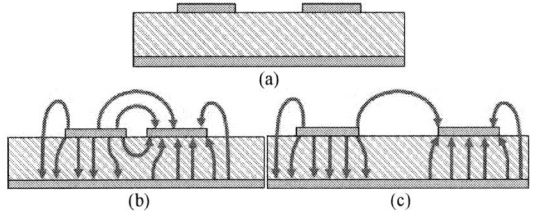

Fig. 2. Electric field distributions of a differential line (DL): (a) a DL, (b) a tight-coupled DL, (c) a loose-coupled DL.

The DL is typically designed to have a line impedance of 100 Ω. The dimension parameters (linewidth, gap width, and substrate height) of the DL determine the line impedance and the magnitude of the EM coupling. The amount of EM coupling between the two DL lines can be adjusted mainly by a gap width, but the proper coupling level is still debated [8], [9]. In this paper, the dimensions are chosen to obtain the characteristic line impedance of 100 Ω with the 6-mil FR-4 substrate (relative permittivity $\varepsilon_r = 4.3$), as listed in Table 1.

TABLE I. DIMENSIONS OF THE DLs WITH 100 Ω LINE IMPEDANCE.

DLs	Dimension Parameters		
	Linewidth w	Gap width g	Height h
Tight-coupled DL	7.78 mil (0.2 mm)	5 mil (0.13 mm)	6 mil (0.15 mm)
Loose-coupled DL	10.05 mil (0.26 mm)	20 mil (0.5 mm)	6 mil (0.15 mm)

III. DESIGN OF DL COMPENSATION STRUCTURES

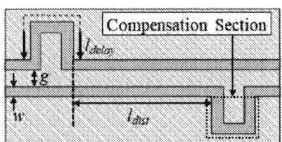

Fig. 3. Top view of a length-matched DL with a compensation section.

The conventional solution to cope with the signal skew is to compensate for the length difference by connecting an extra delay line to the shorter path of the DL. The structure of the length-matched DL is depicted in Fig. 3. The path length difference of the DL is modelled as a delay line (l_{delay}). The length-mismatched DL stays unbalanced before reaching the length compensation section placed with the distance l_{dist}. The amounts of phase interaction between the two DL lines are evaluated for a tight-coupled DL (gap of 5 mil) and a loose-coupled DL (gap of 20 mil). For each DL type, four different cases, i.e., two delay lengths l_{delay} (30 and 60 mil) and two distances up to the length compensation section l_{dist} (500 and 1000 mil), are considered, as listed in Table 2.

TABLE II. DIMENSIONS OF THE 4 CASES OF THE DLs.

Parameters	4 Cases of the DLs			
	Case 1	Case 2	Case 3	Case 4
Delay length l_{delay}	30 mil (0.76 mm)	60 mil (1.52 mm)	30 mil (0.76 mm)	60 mil (1.52 mm)
Distance l_{dist}	500 mil (12.7 mm)	500 mil (12.7 mm)	1000 mil (25.4 mm)	1000 mil (25.4 mm)

IV. FABRICATION AND MEASUREMENT

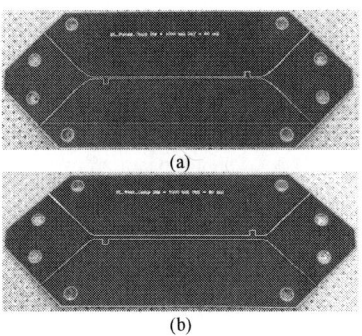

Fig. 4. Pictures of the fabricated length-matched DLs for Case 4: (a) Tight-coupled DL, (b) Loose-coupled DL.

To verify the phase balancing over the frequency through measurements, the length-matched DLs are fabricated with the FR-4 6-mil substrate, as shown in Fig. 4. The two lines of the DL at both transmit and receive ends are smoothly converted to two microstrip lines (using the DL-to-MSL transitions), respectively, and a 4-port VNA (Rohde & Schwarz ZNB40) is used to measure the 4-port S-parameters. The transition is designed to maintain a 100 Ω differential line impedance. Four end-launch connectors are installed at the receive/transmit ends.

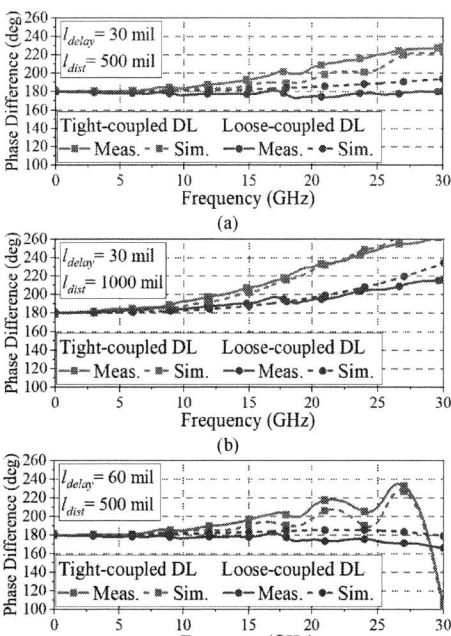

979-8-3503-5124-8/24 $31.00 © 2024 IEEE   53

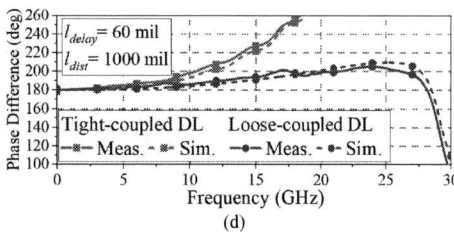

Fig. 5. Measured and EM-simulated phase differences of the length-matched DLs with FR-4 6-mil substrate: (a) Case 1, (b) Case 2, (c) Case 3, (d) Case 4.

In [10], a skew budget for a high-speed digital signal is recommended as 0.14 UI (phase difference of 25°), which means that for high-speed digital data transmission, the phase deviation from 180° should be less than 25°. Fig. 5 depicts the measured and EM-simulated phase differences of the length-matched DLs. A commercial 3D EM simulator, CST EM Suite 2023, has been used for the EM simulations. With Case 1, as shown in Fig. 5(a), the tight-coupled DL exceeds the phase margin of 25° after 19.8 GHz. Fig. 5(b) shows the DLs in Case 2, and the tight-coupled DL exceeds the phase margin at above 19.6 GHz, and the phase difference drops dramatically to 106.6° at 30 GHz. In the DLs in Case 3, in Fig. 5(c), the phase margin is exceeded after 14.3 GHz and 24.6 GHz, respectively. In Case 4, in Fig. 5(d), the tight-coupled DL exceeds the phase margin after 11.3 GHz. Deviations of the measured phase values as compared with the EM simulated results may have been caused by connectors and fiber-weave effects of FR4 substrates [2].

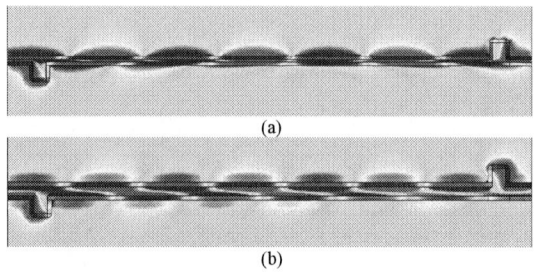

Fig. 6. Electric field distributions of the length-compensated DLs for Case 4: (a) Tight-coupled DL, (b) Loose-coupled DL.

Fig. 6 shows the electric field distributions of the length-matched DLs of Case 4 at 20 GHz. The positive and negative polarities of the electric fields are represented by red and blue colors, respectively. A signal is launched from the left side of the structure. In Fig. 6(a), the electric field distribution of the tight-coupled DL with the length compensation is shown. The phase balance of the tight-coupled DL changes nonlinearly as the signal travels along the DL, and the phase polarity turns inversely at the end of the structure. It seems that the signal on one line tries to pull the other signal to recover the phase balance through the EM coupling, but the recovery action is rather limited and unpredictable. On the other hand, with the loose-coupled DL, as shown in Fig. 6(b), the interactions are less obvious.

As can be seen in Figs. 5 and 6, nonlinear phase interactions over the frequency are clearly shown, especially with the tight-coupled DLs, and a proper phase compensation using the conventional length compensation cannot be achieved for frequencies over ~12.8 GHz. Loose-coupled DLs are better with nonlinear phase interactions, but are known to be susceptible to unbalanced EM interferences.

V. CONCLUSION

In this paper, nonlinear phase interactions between two lines of a DL in the presence of a signal skew are analyzed. The phase of the skewed signal can be modified nonlinearly before reaching the compensation section due to the EM coupling between the two DL lines. To verify these nonlinear phase interactions, tight- and loose-coupled DLs with various delay lengths and distances to the compensation section are EM-simulated and measured with the fabricated samples. As a result, the tight-coupled DL in Case 4, with the delay length of 60 mil and the distance-to-compensation of 1000 mil, exceeds the phase margin of 25° after 11.3 GHz. The loose-coupled DLs performed better in compensating for phase differences up to 20 GHz. Many design guides recommend the advantages of using tight-coupled DLs over loose-coupled DLs. However, it is proven that the tight-coupled DLs are more susceptible to nonlinear phase interactions. Therefore, the design guides for DLs should be re-evaluated, especially for over 10 GHz, which is required for the 5G/6G communications.

ACKNOWLEDGMENT

This research was supported in part by the National Research and Development Program through the National Research Foundation of Korea (NRF) funded by the Ministry of Education (No. NRF-2022R1I1A3064460), and in part by the BK21 FOUR project funded by the Ministry of Education (No. 4199990113966).

REFERENCES

[1] E. Bogatin, Signal and Power Integrity-Simplified, 3rd ed., New Jersey, USA: Prentice Hall, 2018.

[2] S. Singh; T. Kukal, "Timing skew enabler induced by fiber weave effect in high speed HDMI channel by angle routing technique in 3DFEM," in 2015 IEEE 24th Electrical Performance of Electronic Packaging and Systems (EPEPS), San Jose, CA, USA, 2015, pp. 163-166.

[3] T. -L. Wu, F. Buesink and F. Canavero, "Overview of Signal Integrity and EMC Design Technologies on PCB: Fundamentals and Latest Progress," in IEEE Transactions on Electromagnetic Compatibility, vol. 55, no. 4, pp. 624-638, Aug. 2013.

[4] "Should You Use Tight vs. Loose Differential Pair Spacing and Coupling?", 2021. [Online]. Available: https://resources.altium.com/p/s hould-you-use-tight-vs-loose- differential-pair-spacing-and-coupling

[5] "11 Best High-Speed PCB Routing Practices," 2020. [Online]. Available:

[6] Lee, J.S.; Min, B.C.; Kumar, S.; Choi, H.C.; Kim, K.W. "On Autonomous Phase Balancing of the Coplanar Stripline as a Feedline for a Quasi-Yagi Antenna," Electronics, vol. 12, no. 19, 4168, 2023.

[7] "Ultra-Fine Line Differential Pair Design with No Return Plane," 2023. [Online]. Available: https://www.signalintegrityjournal.com/articles/285 9-ultra-fine-line-differential-pair-design-with-no-return-plane

[8] M. Abu Khater, "High-Speed Printed Circuit Boards: A Tutorial," in IEEE Circuits and Systems Magazine, vol. 20, no. 3, pp. 34-45, 2020.

[9] H. Johnson, "High-Speed Digital Design," in IEEE Microwave Magazine, vol. 12, no. 5, pp. 42-50, Aug. 2011.

[10] "Beware of the Skew Budget: How Fiber Weave Effect can Affect Your High-Speed Design," 2022. [Online]. Available: https://www.signalinteg rityjournal.com/articles/2459-beware-of-the-skew-budget-how-fiber-weave-effect-can-affect-your-high-speed-design

979-8-3503-5124-8/24 $31.00 © 2024 IEEE

A 155 MHz Low-Jitter PLL for Enhanced Signal Integrity in High-Speed Interconnects

Mulat Ayinet, Tiruye
Dept. of Information Engineering
Computer Engineering
University of Pisa
Pisa, Italy
m.tiruye@studenti.unipi.it

Olani Baissa, Gerba
Dept. of Information Engineering
Computer Engineering
University of Pisa
Pisa, Italy
o.gerba@studenti.unipi.it

T. Hui, Teo
Engineering Product Development
Science, Mathematics and Technology
Singapore University of Technology and Design
Singapore
tthui@sutd.edu.sg

Abstract—**Designing high-speed interconnects faces challenges from factors like jitter, noise, and skew, which degrade performance. This study introduces a 155 MHz low-jitter PLL clock generator using CMOS 180 nm model files. Operating with a 10 MHz reference signal input, it achieves 2.832 ps average jitter, crucial for signal integrity in high-speed interconnects.**

Index Terms—**Low-jitter, High-speed interconnects, Signal integrity.**

I. INTRODUCTION

High-speed interconnect rely on precise clock generation to ensure data integrity and synchronization. Accurate and reliable clock signals are paramount in high-speed interconnects. Clock signals are critical for synchronous systems, ensuring signals meet setup and hold time requirements. However, real-world challenges such as clock skew, illustrated in Figure 1, and jitter from noise and environmental factors, as shown in Figure 2, pose significant obstacles in high-performance designs. These uncertainties impact sequential circuit performance by introducing timing variations [1], emphasizing the critical need for precise management to uphold system reliability and integrity [2]. The total timing uncertainty can be described in Equation 1.

$$t_{uncertainty} = t_{skew(\delta)} + t_{jitter} \tag{1}$$

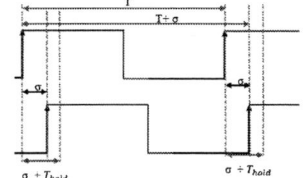

Fig. 1: Skew definition.

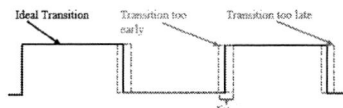

Fig. 2: Jitter definition.

Various methods, such as balanced paths like H-tree networks and matched RC trees [3], address these issues. However, the Phase-Locked Loop emerges as the preferred solution, providing low jitter [2] and high signal integrity for robust data transfer in high-speed interfaces.

This paper presents a comprehensive study on designing and optimizing a 155 MHz low-jitter PLL for enhanced signal integrity in high-speed interconnects. The proposed PLL operates with an input reference frequency of 10 MHz and generates an output frequency of 155 MHz, achieving an average jitter of 2.832 ps, critical for high-speed link designs.

II. BLOCK DIAGRAM OF PLL SYSTEM

A phase-locked loop (PLL) is a feedback system that synchronizes an oscillator's phase and frequency with a reference signal [4]. The basic PLL structure is illustrated in the block diagram in Figure 3 [5], comprising essential components like the phase-frequency detector, charge pump, loop filter, voltage-controlled oscillator, and frequency divider [6].

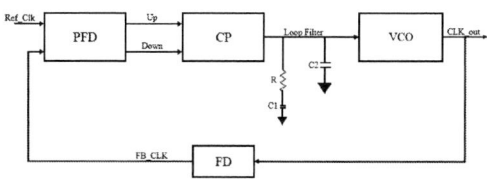

Fig. 3: PLL block diagram.

III. DESIGN PROCEDURE

The PLL is designed to meet specific criteria essential for high-speed interconnect systems, providing stable and precise clock signals crucial for maintaining signal integrity in fast data transmission environments, as shown in Table I.

TABLE I: Design Parameters and Values.

Design Parameter	Value
Output Voltage	> 1V
Input Frequency	10 MHz
Output Clock Frequency	>150 MHz
Phase Error	< 5 degrees
Jitter	< 10 ps

979-8-3503-5124-8/24 $31.00 © 2024 IEEE

A. Phase Frequency Detector(PFD)

The phase frequency detector ensures accurate phase and frequency locking in a PLL, preventing false lock conditions with harmonics of the input signal [7]. The gate-level realization in Figure 4 minimizes phase error, essential for achieving reliable 155 MHz data transfer speeds. Figure 5 illustrates the PFD design, detecting phase and frequency differences to generate error signals for the charge pump.

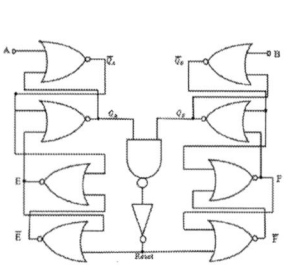 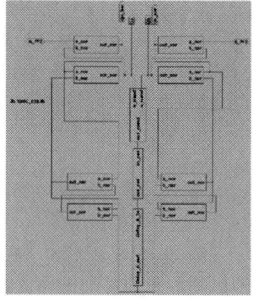

Fig. 4: Implementation of PFD.   **Fig. 5:** PFD design schematics.

B. Charge Pump (CP)

The charge pump converts PFD output into stable control voltage critical for the 155 MHz PLL design, minimizing phase error [8]. Figure 13 demonstrates its role in reducing jitter and improving signal integrity by maintaining consistent current. This stabilizes voltage input to the Loop Filter (LF), enhancing PLL stability and performance, depicted in Figure 6.

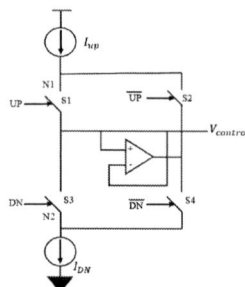 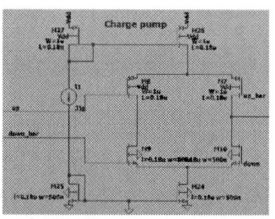

Fig. 6: Charge pump implementation.   **Fig. 7:** Charge pump design schematics.

C. Loop Filter (LF)

The RC loop filter smooths the control voltage from the CP by filtering out high-frequency noise [9], using a low-pass RC filter with specific component values (R = 43 kΩ , C1 = 28.2 pF, C2 = 2 pF). This design minimizes jitter and stabilizes the VCO control voltage, crucial for maintaining a stable 155 MHz output frequency with low noise.

D. Voltage Control Oscillator (VCO)

The VCO is integral to the PLL design, responsible for generating the output frequency. Implemented with a current-starved architecture and 5-stage design, it minimizes sensitivity to VDD variations, reducing jitter [10]. By converting the

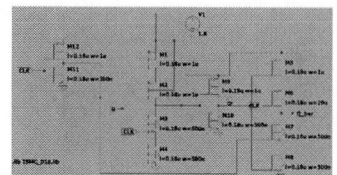 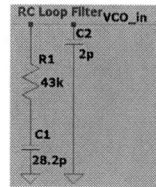

Fig. 8: Divider by 16 D-FF design schematics.   **Fig. 9:** Loop filter design schematics.

LF's stable control voltage, the VCO produces a precise 155 MHz oscillation, crucial for high-frequency stability and low phase noise in high-speed data transmission, as depicted in Figure 11.

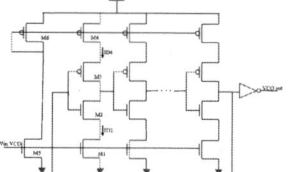 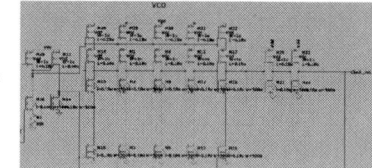

Fig. 10: VCO implementation.   **Fig. 11:** VCO design sechematics.

E. Frequency Divider (FBDIV)

The frequency divider, employing a divider-by-16 configuration with D flip-flops as illustrated in Figure 8, ensures precise PLL locking and maintains signal integrity by aligning the VCO output with the reference frequency, crucial for stable high-speed operation with low jitter [11]. The integrated PLL design, depicted in Figure 12, combines key components to achieve a stable 155 MHz output with minimal jitter.

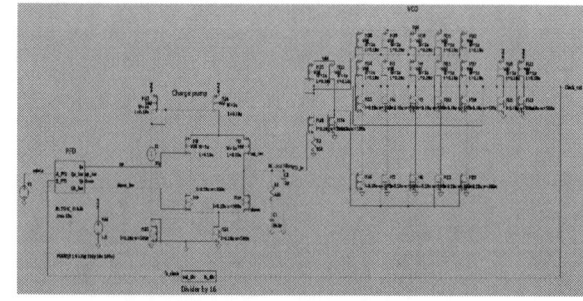

Fig. 12: The full PLL design schematics.

IV. RESULTS

The output of charge pump is shown in Figure 13 illustrates the charge pump's ability to maintain a steady current. This steady current is crucial for the stable operation of the PLL, as it ensures that the control voltage fed to the VCO is consistent. Any fluctuations in this current could lead to variations in the VCO output frequency, thereby increasing jitter.

Figure 14 displays the overall PLL output. The PLL consistently operates at a stable 155 MHz frequency for over 10 microseconds, highlighting its robustness and reliability

979-8-3503-5124-8/24 $31.00 © 2024 IEEE   56

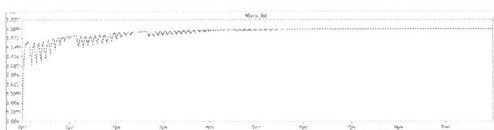

Fig. 13: Charge pump output.

in maintaining consistent output essential for high-speed data transfer applications. The zoomed view of the PLL output

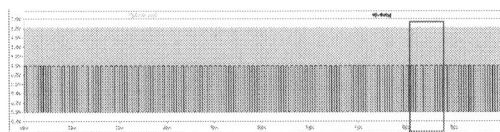

Fig. 14: PLL output.

in Figure 15 provides a detailed look at the PLL's stable frequency output, demonstrating effective suppression of phase noise and jitter over time.

Fig. 15: PLL output with zoom out.

Jitter Analysis: Jitter is critical in high-speed link design as it affects the timing precision of data transfers, essential for maintaining high signal integrity. Figure 16 illustrates the

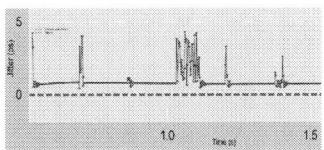

Fig. 16: Practical jitter analysis with noise model.

jitter analysis with a noise model. Instantaneous jitter exhibits higher standard deviation and peak-to-peak values, reflecting its sensitivity to high-frequency variations, as shown in Table II. Practical jitter, influenced by added noise, shows a higher mean but lower standard deviation and peak-to-peak values, suggesting effective smoothing of high-frequency variations by the noise model. Table III shows the proposed PLL's mean

TABLE II: Jitter Analysis Metrics.

Metric	Instantaneous Jitter	Practical Jitter
Mean (ps)	0.3204	2.832
Standard Deviation (ps)	7.192	3.711
Peak-to-Peak (ps)	5.711	2.910

jitter of 2.832 ps, outperforming other designs at 27.1 ps, 10.0 ps, and 3.0 ps.

V. CONCLUSION

In this study, we meticulously simulated a PLL circuit using CMOS 180 nm model files in LTspice and a standardized

TABLE III: Jitter Performance Comparison.

Design	Jitter (ps)
[12]	27.1
[13]	10.0
[14]	3.0
Proposed Design	**2.832**

1.8V voltage supply (VDD). The PLL excelled with an output frequency range of 0 MHz to 155 MHz, optimized through stability analysis of the loop filter using a 43 kΩ resistor and capacitors C1 (28.2 pF) and C2 (2 pF). It reliably produced the targeted 155 MHz output, validating our precise design approach. With a 10 MHz input, the PLL demonstrated a practical mean jitter of 2.832 ps, crucial for enhancing signal integrity in high-speed interconnects. This performance underscores its robustness in maintaining dependable data transfer rates and high signal fidelity across challenging conditions, effectively mitigating issues like jitter, noise, and skew in high-speed interconnect designs.

REFERENCES

[1] X. Gao, "Low jitter and low power plltowards the utopia," in *2019 International SoC Design Conference (ISOCC)*, pp. 38–39, 2019.

[2] K. Bidaj, *PLL Phase Noise & Jitter Modeling, for High Speed Serial Links*. Theses, Université de Bordeaux, Nov. 2016.

[3] A. Mandal, S. P. Khatri, and R. N. Mahapatra, *Clock Distribution for Fast Networks-on-Chip*, pp. 15–66. New York, NY: Springer New York, 2014.

[4] A. Yadlapati and H. K. Kakarla, "Low-power design-for-test implementation on phase-locked loop design," *Measurement and Control*, vol. 52, no. 7-8, pp. 995–1001, 2019.

[5] C. Shekhar and S. Qureshi, "Design of 50 mhz pll using indigenous scl 180 nm cmos technology," in *2021 IEEE International Symposium on Smart Electronic Systems (iSES)*, pp. 12–17, 2021.

[6] F. Noruzpur, S. Mahdavi, M. Poreh, and S. T. Ghasemi, "A new semi-digital low power low jitter and fast pll in 0.18m technology," in *2018 25th International Conference "Mixed Design of Integrated Circuits and System" (MIXDES)*, pp. 109–115, 2018.

[7] A. Fahim, *Clock Generators for SOC Processors: Circuits and Architectures*. Springer, 2005.

[8] A. Homayoun and B. Razavi, "On the stability of charge-pump phase-locked loops," *IEEE Transactions on Circuits and Systems I: Regular Papers*, vol. 63, no. 6, pp. 741–750, 2016.

[9] H. Wu, W. Wang, L. Ye, K. Song, and X. Sun, "A method to reduce jitter due to power noise by optimizing loop filter in pll based clock source," in *2022 Asia-Pacific International Symposium on Electromagnetic Compatibility (APEMC)*, pp. 616–619, 2022.

[10] M. Sivasakthi and P. Radhika, "Design and analysis of pvt tolerant hybrid current starved ring vco with bulk driven keeper technique at 45 nm cmos technology for the pll application," *AEU - International Journal of Electronics and Communications*, vol. 173, p. 154987, 2024.

[11] A. L. Makarevich, D. V. Garaga, S. M. Sokovnich, N. S. Kostyukevich, and S. A. Karapetyan, "Circuit design of a frequency synthesizer device for high-speed data transmission systems," in *2022 Systems of Signals Generating and Processing in the Field of on Board Communications*, pp. 1–4, 2022.

[12] S. Aditya and S. Moorthi, "A low jitter wide tuning range phase locked loop with low power consumption in 180nm cmos technology," in *2013 IEEE Asia Pacific Conference on Postgraduate Research in Microelectronics and Electronics (PrimeAsia)*, pp. 228–232, 2013.

[13] D. Boerstler, "A low-jitter pll clock generator for microprocessors with lock range of 340-612 mhz," *IEEE Journal of Solid-State Circuits*, vol. 34, no. 4, pp. 513–519, 1999.

[14] A. Kailuke, P. Agrawal, and R. Kshirsagar, "Design of low power, low jitter pll for wimax application in 0.18µm cmos process," *Procedia Computer Science*, vol. 152, pp. 390–397, 2019. International Conference on Pervasive Computing Advances and Applications- PerCAA 2019.

Single-Layer Wiring Design in UCIe to Realize Low-Cost Interposer Substrate

1st Soshi Shimomura
AI Chip Design
Open Innovation Laboratory
National Institute of Advanced Insdustrial
Science and Technology (AIST)
Tokyo, Japan
soshi.shimomura@aist.go.jp

2nd Yutaka Uematsu
AI Chip Design
Open Innovation Laboratory
National Institute of Advanced Insdustrial
Science and Technology (AIST)
Tokyo, Japan
yutaka.uematsu@aist.go.jp

3rd Katsuya Kikuchi
Semiconductor Frontier
Research Center
National Institute of Advanced Insdustrial
Science and Technology (AIST)
Ibaraki, Japan
k-kikuchi@aist.go.jp

4th Haruo Shimamoto
Semiconductor Frontier
Research Center
National Institute of Advanced Insdustrial
Science and Technology (AIST)
Ibaraki, Japan
haruo.shimamoto@aist.go.jp

5th Yuuki Araga
Semiconductor Frontier
Research Center
National Institute of Advanced Insdustrial
Science and Technology (AIST)
Ibaraki, Japan
yuuki.araga@aist.go.jp

6th Shinichi Ouchi
AI Chip Design
Open Innovation Laboratory
National Institute of Advanced Insdustrial
Science and Technology (AIST)
Tokyo, Japan
shinichi.ouchi@aist.go.jp

Abstract—In recent years, high-performance and chiplet-type CPUs need futher improvements of their performance. It is necessary for the chiplet-to-chiplet communication to increase the throughput. The issue of improving the throughput in UCIe communication is that the large number of required substrate layers makes the interposer expensive. This paper proposes a new design method to implement UCIe wiring, typically using two layers, with a single layer to achieve cost-effective interposers. We adjusted the UCIe form factor rules to the single-layer wiring and identified the need for fine-pitch wiring as 10 μm or less. In addition, to reduce crosstalk in the wiring between dies, we proposed a wiring layout that increases the wiring pitch only between dies. We modeled the interposer, conducted a signal integrity analysis, and confirmed that communication at the transmission rate of 32 Gbps is possible over die-to-die distances of 1–10 mm. We concluded that the interposer cost reduction of about 40 % is feasible using single-layer UCIe wiring.

Index Terms—Chiplet, UCIe, Interposer, Package, Interconnect

I. INTRODUCTION

In recent years, high-performance processors that use chiplets have emerged for servers and laptops. The advantage of chiplet-based processors lies in the improvement of the performance without shrinking the manufacturing process by combining two or more CPU chiplets [1]. However, when chiplets are used, the performance of the chiplet-to-chiplet interconnects becomes critical. Currently, standardization of the interconnects is progressing, and also Universal Chiplet

This work is based on results obtained from a project, JPNP23009, commissioned by the New Energy and Industrial Technology Development Organization (NEDO).

Interconnect Express (UCIe) has become the mainstream communication standard [2].

When we try to improve the throughput of the UCIe interface, increasing the number of lanes is the simplest approach. The UCIe standard also discusses methods to increase the number of lanes. In "stacked" modules, physical layers (PHYs) are stacked vertically along the die edge; in "unstacked" modules, PHYs are horizontally arranged along the die edge (TABLE I). The stacked-type have the advantage of a higher edge bandwidth density as a result of the stacked PHYs; however expensive interposers are necessary because stacked modules require four substrate layers for signal wiring.

This paper proposes a new design to reduce the number of substrate layers with maintaining the edge bandwidth density of the stacked-type. The UCIe standard defines the advanced package based on silicon interposers and the standard package based on organic interposers. This paper targets the standard package that needs cost reduction through layer reduction.

II. PROPOSAL

Two challenges can be identified for implementing single-layer wiring. The first challenge is the feasibility of the form factor for the lead-out wiring section. Fig. 1 shows an example layout in a lead-out wirings area. In typical two-layer wiring, sufficient space is available for single wire to be drawn out between pins; however, single-layer wiring requires more space for three wires to be drawn out between pins. Thus, high-density implementation with fine-pitch wiring is necessary. The second challenge is the increase in crosstalk. Single-layer wiring requires greater wiring density than in two-layer wiring, which results in finer-pitch wiring and larger

TABLE I: Tradeoffs between stacked and unstacked modules.

Type name	Stacked	Unstacked	Proposal
Structure image			
Edge bandwidth density (32Gbps)	**224 Gbps/mm**	112 Gbps/mm	**224 Gbps/mm**
Number of signal wiring layers	4 (Substrate: 8-2-8 or 9-2-9)	2 (Substrate: 4-2-4 or 5-2-5)	2 (Substrate: 4-2-4 or 5-2-5)
Substrate manufacturing cost	1.0	**0.6**	**0.6**

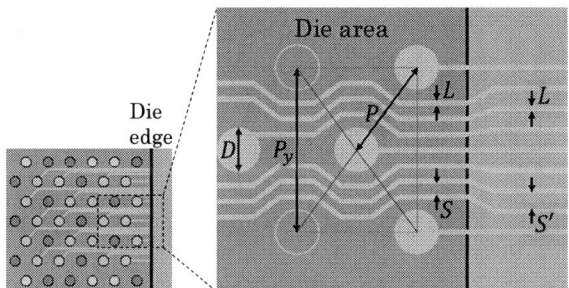

Fig. 1: Example of wiring layout and new form-factor limitation for single-layer wiring design.

crosstalk. The approaches to address these two challenges are described below.

A. High-density implementation of lead-out wiring

We here define new form factor rules and implementable conditions for single-layer wiring. Fig. 1 shows an enlarged view of the lead-out wiring layout in the die edge area. According to the UCIe standard, the form factor rules for the lead-out wiring are defined by the distances between pins in the diagonal and vertical directions (Fig. 1). For the diagonal direction between pins, three wires need to be drawn out; this condition is expressed by Equation (1). For the vertical direction between pins, seven wires need to be drawn out; this condition is expressed by Equation (2).

$$P = D + 3L + 4S \qquad (1)$$
$$P_y = D + 7L + 8S \qquad (2)$$

Consider the application of this new form factor rule with a pin pitch of 130 μm. In this case, the diagonal pin pitch is $P = 130$ μm and the vertical pin pitch is $P_y = 190.5$ μm as defined in the UCIe standard. Assuming that the pin diameter is $D = 50$ μm and that the wiring width L and wiring spacing S are equal, we obtain $L = S = 11.4$ μm from Equation (1) and $L = S = 10.5$ μm from Equation (2). Therefore, high-density wiring with a line-and-space of approximately 10 μm or less is required.

B. Cross-talk reduction in fine-pitch wiring

We here explain a substrate layout to reduce the effect of crosstalk in single-layer wiring. Crosstalk in multi-bit parallel wiring occurs mainly in the area where the wiring runs in parallel between dies [3]. Therefore, we focused on reducing crosstalk in the inter-die wiring. Our design increases the spacing between the wiring in the inter-die area as shown in Fig. 1. The wiring spacing S' in the inter-die area should be $S' > S$, and the condition for line-and-space is expressed by Equation (3). According to the UCIe standard, $P_y = 190.5$ μm is constant,; thus, the wiring spacing S' can be widened to 15 μm with a wiring width $L = 10$ μm.

$$P_y = 7L + 8S' \qquad (3)$$

III. ANALYSIS AND RESULTS

we verified whether UCIe communication is feasible with single-layer wiring by our design of fine-pitch lead-out wiring and crosstalk reduction.

A. Interposer substrate model and simulation conditions

To conduct the analysis, we modeled the UCIe channel of single-layer wiring as shown in Fig. 2. Following the UCIe standard, 16-bit data signal wires, CLP, CKN, TRK, VID, DATASB, and CKSB were laid out as shown in Fig. 2. The form parameters for this model are presented in TABLE II. The bump pitch was $P = 110$ μm, the line-and-space was $L = S = 5$ μm, and the via diameter was $D = 30$ μm. These conditions satisfy Equations (1) and (2). In addition, the wiring spacing between dies is set to $S' = 16$ μm, widening the spacing according to Equation (3). The distance between dies was varied as 1/5/10 mm. Following the UCIe standard, and the transmission rates were adjusted to 4/8/12/16/24/32 Gbps for the analysis.

B. Simulation results

To verify the communication performance of the proposed substrate wiring model, we conducted analysis of s-parameter and eye pattern. In the s-parameter analysis, we examined the

TABLE II: Parameters used in simulation model

Parameter	Value
P	110 µm
P_y	190.5 µm
D	30 µm
L	5 µm
S	5 µm (min.)
S'	16 µm

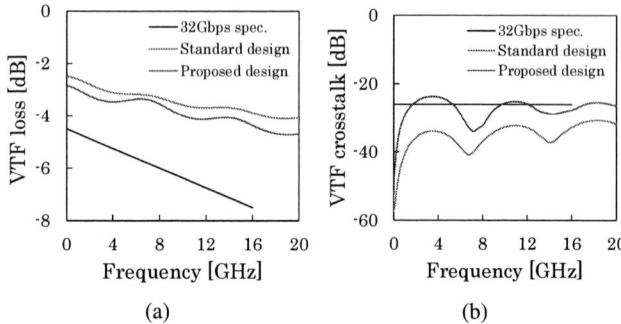

Fig. 3: (a) Insertion loss and (b) crosstalk characteristics of transfer voltage function defined in the UCIe standard.

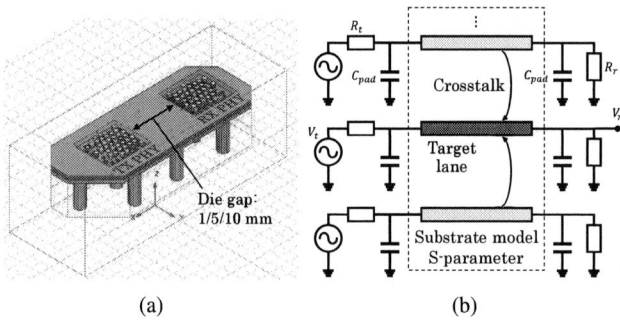

Fig. 2: (a) Substrate model using single-layer wiring design and (b) circuit model of UCIe communication. Substrate characteristics can be simulated in (a), and s-parameter and eye pattern can be analyzed in (b).

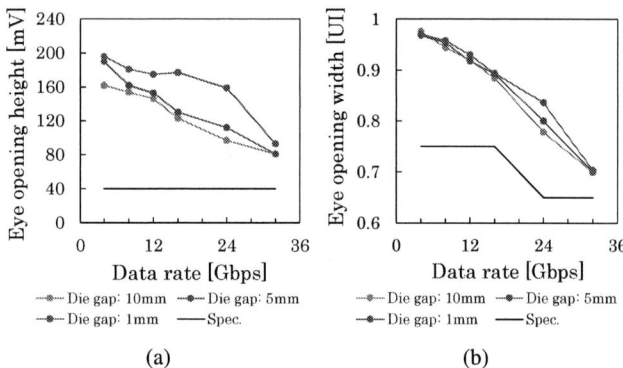

Fig. 4: Data rate characteristics of (a) eye opening height and (b) eye opening width.

influence of insertion loss and far-end crosstalk. Fig. 3 shows frequency characteristics of voltage tranfer function (VTF), which are defined in the UCIe standard, with a die-to-die distance of 10 mm. The red and blue lines represent the results for a standard two-layer wiring design and the single-layer wiring design respectively. Even through our design increases the space between the wiring for corsstalk reduction, far-end crosstalk characteristics don't meet the recommanded spec. of the UCIe standard (black line). The result indicates that only the proposed layout is not enough and even finer-pitch wiring is necessary to widen the space between the wiring. We also verified eye patterns of every lane. We ran the drivers of the victim wirings and the aggressor wiring located in the right two lanes and left two lanes simultaneously to add crosstalk. Fig 4 illustrates the eye opening characteristics of a wiring with the poorest signal quality among all the investigated wirings, as the transmission rate was varied from 4 to 32 Gbps. The green, red, and blue lines represent the results for a die-to-die distance of 10 mm, 5 mm, and 1 mm respectively. The black line is the eye mask requirement of the UCIe standard. The conditions corresponding to the lowest-signal-quality were die-to-die distance of 10 mm and a transmission rate of 32 Gbps. However, even under these conditions, the signal quality was confirmed to be greater than the criteria (black line). Therefore, we conclude that UCIe communication is feasible with single-layer wiring for die-to-die distances of 10 mm or less.

IV. CONCLUSION

This paper discusses a method to implement UCIe wiring, typically using two layers, with a single layer to achieve cost-effective interposers. For single-layer wiring, we adjusted the UCIe form factor rules and identified the need for fine-pitch wiring of 10 µm or less. In addition, to reduce crosstalk in the wiring between dies, we proposed a wiring layout that increases the wiring pitch only between dies. We confirmed that communication at a transmission rate of 32 Gbps is possible over die-to-die distances of 1–10 mm in our simulation. We concluded that the interposer cost reduction of about 40 % is feasible using single-layer UCIe wiring. Further research is required to extend the die-to-die distances by optimizing form factor parameters and verify our design through measurements using prototype boards.

REFERENCES

[1] S. Naffziger et al. Pioneering Chiplet Technology and Design for the AMD EPYC™ and Ryzen™ Processor Families : Industrial Product. In *2021 ACM/IEEE 48th Annual International Symposium on Computer Architecture (ISCA)*, pages 57–70, 2021.

[2] UCIe Consortium. Universal Chiplet Interconnect Express (UCIe) Specification Rev. 1.1. https://www.uciexpress.org/, December 2023.

[3] Samuel H. Russ. *Signal Integrity: Applied Electromagnetics and Professional Practice*, pages 127–137. Springer International Publishing, Cham, 2022.

979-8-3503-5124-8/24 $31.00 © 2024 IEEE

PCIe Gen 6.0 SSD Receiver PAM-4 SI Analysis Based on End-Port Time-domain Measurements for Unknown System Channel

Jinwook Song, Jinan Lee, Jonghee Jeong, Seokwoo Hong, Sungwoo Jin, Hyunwoo Kim, Sungwon Roh, Chorom Jang, Youngjun Ko, Taehyun Shim, Juneyoung Kim, Dongho Choi, Kyungsuk Kim and Sunghoon Chun
Solution Development Team (Memory Business), Samsung Electronics, Hwaseong-si, 18448 South Korea
E-mail: j87.song@samsung.com

Abstract—**This paper presents a system-level PCI Express (PCIe) Gen 6.0 PAM-4 receiver (Rx) signal integrity (SI) analysis using time-domain measurements at a system channel-end test point (TP) without any end-to-end passive channel information such as a S-parameter model. Insertion loss (IL), return loss (RL), and near-end/far-end crosstalk (NEXT/FEXT) for the entire interconnect path at each Nyquist frequency for PCIe Gen 3.0, 4.0, and 5.0 are calculated as a result of measuring voltage waveforms at the TP when the host as the transmitter (Tx) transmits compliance bit pattern. We demonstrated for the first time that reliable Rx PAM-4 SI analysis is possible for entire end-to-end PCIe Gen 6.0 interface channel using a mimic channel model based on the calculated IL, RL, NEXT, and FEXT values.**

Keywords—*Peripheral Component Interconnect Express (PCIe), PAM-4 (Pulse Amplitude Modulation 4-level), Solid State Drive (SSD)*

I. INTRODUCTION

As generative AI continues to evolve, customer requests for high-capacity SSDs are rapidly increasing in both training and inference fields. The learning data capacity increases in proportion to the number of training and parameters, and accordingly, customer requests for SSD products with higher memory capacities are rapidly increasing. In proportion to these increasing nonvolatile memory capacity in the SSD device, the operating speed of the high-speed serial PCIe SerDes interface between the customer CPU and the SSD device is also constantly increasing, and the world's first PCIe Gen 6.0 SSD product is about to be released by Samsung.

For the first time in PCIe interface history, PAM-4 signaling is introduced in the Gen 6.0 generation, and the Nyquist frequency is maintained at 16 GHz, the same as Gen 5.0, but the bandwidth is doubled to 64 GT/s. Based on PCIe base specifications, the loss budget of all channels from the host die to the die of the SSD device is promised to be -32 dB at the Nyquist frequency of 16 GHz, and the height and width of the Rx PAM-4 eye mask are 6 mV and 3.125 ps (= 0.1 unit

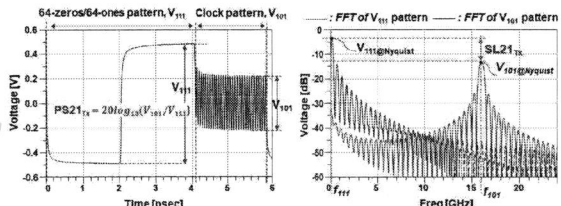

Fig. 2. $PS21_{Tx}$ parameter calculation method using the time-domain measurement data at TP when the host transmits the compliance bit pattern introduced in [1] and the proposed $SL21_{Tx}$ calculation method by using the same data in the frequency-domain.

interval, UI), respectively. As the signal integrity (SI) requirement itself reaches its physical limit on the order of mV and ps, SI robust design for this high-speed serial interface has become very important.

However, as shown in Fig. 1, channels outside of the SSD device account for majority of the entire channel, but most device design company like Samsung is rarely able to obtain models of system interconnection from customers purchasing SSDs. Each system company is competing to build a high-end server environment in the AI market and they are very reluctant to provide their interconnect information. Therefore, it is nearly impossible to predict the overall SI margin or diagnose any risk with only 27 % of the channel information of the SSD. It is an environment that is very unpredictable and difficult for device manufactures to analyze for full-path SI margin, such as a cruise ship sailing without knowing where the iceberg is.

In this paper, we present an alternative to overcome this limitation of the absence of system-side channel model information by time-domain measurements at the test point (TP) the end of the system interconnect where the SSD is connected. We succesfully calculated differential insertion loss (IL), return loss (RL) and crosstalk values at Gen 3.0, 4.0,

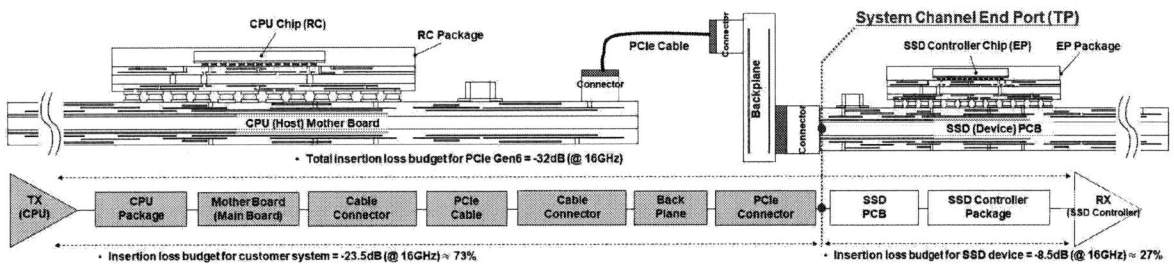

Fig. 1. A conventional system-level PCIe interface interconnect topology for end-to-end (CPU-to-SSD) high-speed serial communication. The system-side channel accounts for the majority of the entire interconnect for PCIe Gen 6.0 communication, and the only measurable TP is at the end port where the SSD is connected. Channel information on the system side that is high-lighted in dark color on the block diagram is usually not available.

979-8-3503-5124-8/24 $31.00 © 2024 IEEE

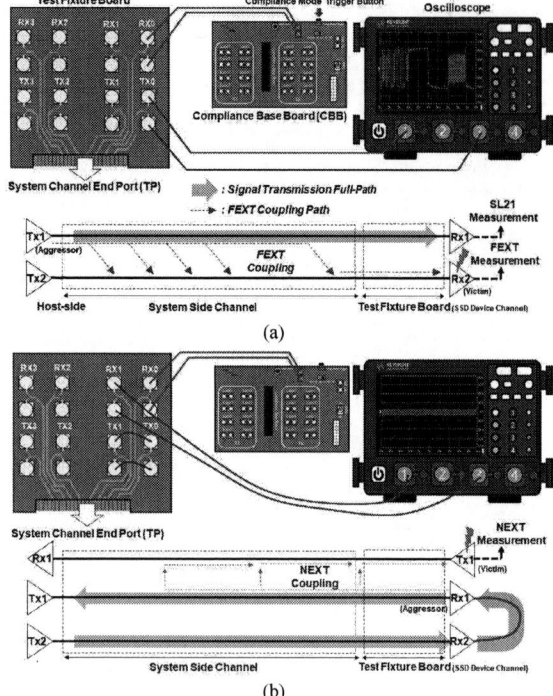

(a)

(b)

Fig. 3. (a) A proposed system insertion loss and FEXT measurement method. The compliance pattern transmitted from the host can be activated with the trigger button on CBB, and multiple lanes can operate simultaneously if the neighboring lane is terminated with a 50 ohm resistor in the CBB. (b) A proposed NEXT measurement method by bypassing host signal to the target aggressor lane, then NEXT coupling to a victim lane can be measured.

and 5.0 frequencies, respectively from measurement data when the promised compliance bit pattern is transmitted from the host. The magnitude of IL is linearized and modeled from DC to the frequency range of 3 times the operating frequency. This assumption has the limitation of reducing the accuracy of SI analysis if a sharp resonance occurs in the region between Nyquist frequencies. To prove the methodology of this process, we extracted S-parameters directly for known DUT channels through VNA measurement and then compared them with the estimated channel models using the proposed methodology. Finally, sucessful Rx PAM-4 SI analyses have been demonstrated by applying the proposed method to the unknown system channel fabricated for PCIe Gen 6.0 SSD product SI margin test.

II. FULL-PATH MIMIC CHANNEL MODEL GENERATION USING SYSTEM END-PORT TIME-DOMAIN MEASUREMENTS

Since the securable TP is only at the end of the system level or at the entrance of SSD device, $PS21_{Tx}$ parameter which is a package loss (including silicon driver bandwidth) is introduced for this limited measurement environment in the specifications [1]. $PS21_{Tx}$ is measured by comparing the 64-zeros/64-ones voltage swing (V_{111}) against a 1010 pattern (V_{101}) as described in Fig. 2. In this paper, unlike $PS21_{Tx}$ calculation comparing voltage amplitudes in the time domain, we propose (1) that calculates the system-level loss, $SL21_{Tx}$ by FFTing the V_{111} and V_{101} pattern periods respectively and comparing the spectral magnitude at the target frequency. However, the measurement method at the TP to calculate both of $PS21_{Tx}$ and the proposed $SL21_{Tx}$ is same as depicted in Fig. 3 (a).

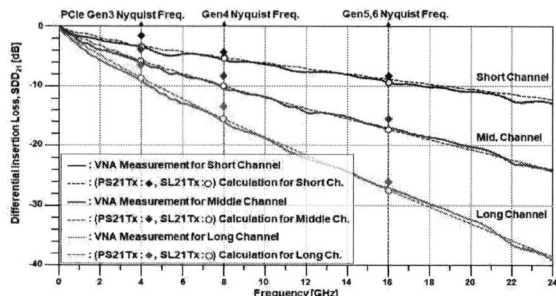

Fig. 4. Verification of the proposed SL21Tx calculation with comparison of the measured S-parameter based SDD21 for 3-types of the DUT channel.

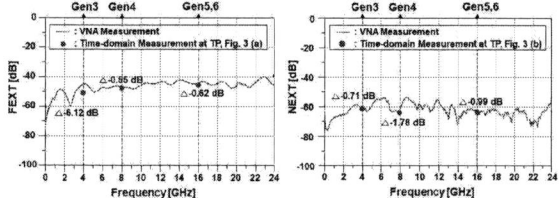

Fig. 5. Verification of the proposed FEXT/NEXT measurement method.

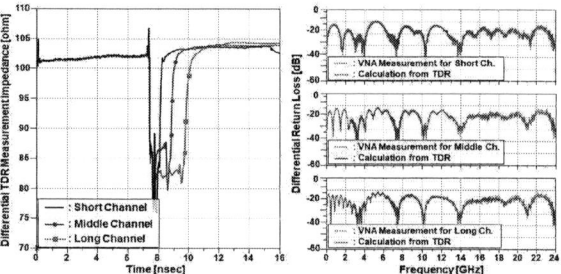

Fig. 6. Verification of the proposed RL calculation method.

$$SL21_{Tx} = FFT(V_{111})|_{dB, f_{111}} - FFT(V_{101})|_{dB, f_{101}} \qquad (1)$$

Since the PCIe interface allows multiple lane operation for bandwidth expansion, differential crosstalk model between lanes is also required to accurately diagnose the Rx SI margin. As shown in Fig. 3 (a), FEXT on the victim lane can be obtained by measuring the magnitude of the noise floor of the oscilloscope when the victim lane is not operating and the coupled voltage when the adjacent aggressor lanes are operating, and calculating the difference between these magnitudes. By terminating 50 ohm to the aggressor lanes of the compliance base board (CBB), simultaneous operation of the aggressor lanes and the victim lane is possible with the toggle signal control. Fig. 3 (b) shows the NEXT measurement environment which adjacent Rx lanes are directly interconnected by cable to bypass compliance bit pattern from host side, then NEXT coupling between Tx and Rx can be measured without any toggle signal injection.

Lastly, the remaining SI design parameter to create a system mimic channel is the RL, and the differential RL, $S_{DD}11$ looking into the host-side from the TP can be calculated based on the differential TDR impedance (TDR_Z_{diff}) measurement data. Since TDR measurement is a well-known verification method and RL is equivalent to analyzing the reflection coefficient in the frequency domain, detailed explanations and illustrations of the process are excluded from this paper [2]. We obtained N reflection coefficient data from TDR_Z_{diff} measurement data, then differentiated the reflection coefficient to obtain the impulse response, then used FFT to calculate $S_{DD}11$ following (2). $S_{DD}11$ and $S_{DD}22$ are actually

979-8-3503-5124-8/24 $31.00 © 2024 IEEE

different, but we assumed that they are the same and generated the system mimic channel by combining differential IL, RL, and crosstalk values calculated by the proposed measurement methods. All measurements are conducted under 100 ohm differential termination ($Z_{diff,term}$). IL is imitated by linearly interpolating 3-point $SL21_{Tx}$ values at 4, 8, and 16 GHz which are PCIe Gen 3.0, 4.0, and 5.0 Nyquist frequencies, respectively.

$$S_{DD}11 = FFT\left\{\frac{d}{dt}\left(\frac{TDR\_Z_{diff}-Z_{diff,term}}{TDR\_Z_{diff}+Z_{diff,term}}\right)\right\}/N \quad (2)$$

To prove the proposed methodology explained so far, differential channels with three different physical lengths (short, middle, long) is manufactured as a DUT, and each S-parameter for the through pattern is measured using a VNA, and IL, RL, and crosstalk values are calculated using the proposed method. As shown in Fig. 4, $SL21_{Tx}$ based imitated IL models show very good correlated results rather $PS21_{Tx}$ values show distinctive difference between measured S-parameter. As shown in Fig 5, compared to the NEXT and FEXT values from the measured S-parameter using a VNA, the crosstalk values by the proposed measurement method described in Fig. 3 yield fairly reliable results except for low frequency region of NEXT results due to open terminated VNA measurement on port 2. In addition, as shown in Fig. 6, the RL extracted from the TDR measurements is analyzed to be at a similar level to the VNA measurement results with almost no error.

III. DEMONSTRATION OF THE PROPOSED METHOD FOR PCIE GEN 6.0 SSD RX PAM-4 SI ANALYSIS

Samsung is about to launch the world's first PCIe Gen 6.0 mass produced SSDs and Fig. 7 shows the test environment to evaluate the SI performance of the fabricated PCIe Gen 6.0 Samsung SSD PoC product. We successfully developed a replica device that can substitute the customer's CPU as a host, and established an evaluation environment that can scalably change the SI parameters (IL, RL, crosstalk) of the passive channel between the SSD controller and the host CPU. Assuming that there is no information on the customer's system interface channel model at all, we conducted SSD Rx PAM-4 simulation analysis by creating IL, RL mimic models through the proposed time domain measurements at the TP for each of the five different channels as described in Fig. 7.

Fig. 8 (a) illustrates the SSD Rx-side BER bathtub curves for each case written with measured $SL21_{Tx}$ values at the Gen 6.0 Nyquist frequency of 16 GHz. Due to confidential concerns, it is expressed in alphabets instead of the actual value of SL21. The y-axis value is based on the code value of the ADC output at the Rx-end, and the voltage margin of the PAM-4 eyes including top, center, and bottom eye could be calculated from the extrapolated target BER of 10^{-6}. When $SL21_{Tx}$ is E dB at 16 GHz, the center bathtub failed to satisfy the target BER and was closed. Fig. 8 (b) shows PAM-4 eye-diagram simulation results using the generated mimic system channels by measured time-domain values such as $SL21_{Tx}$ and TDR values using SEASIM tool provided by PCI-SIG. For end-to-end channel simulation, the rest of the channel inside the SSD used S-parameters extracted through 3D EM simulations. To learn a significant lesson between the simulated Rx PAM-4 eye results based on the mimic channels and measured RX ADC code-based bathtub curves, we defined the PAM-4 eye

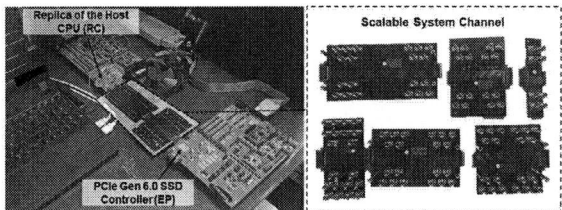

Fig. 7. PCIe Gen 6.0 SSD system-level SI margin test environment.

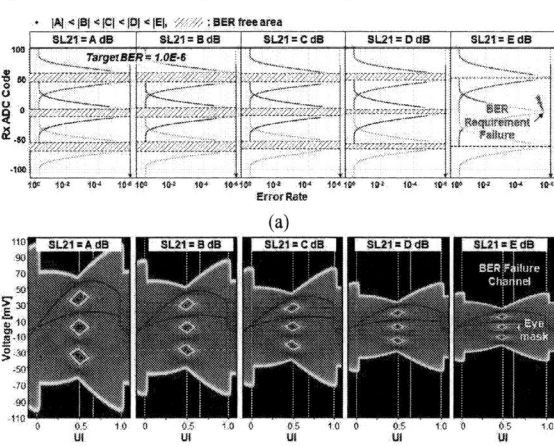

Fig. 8. (a) Measured SSD Rx ADC code based BER bathtub curves for the different channels in the SI margin test environment as shown in Fig. 7. (b) PCIe Gen 6.0 Rx PAM-4 eye-diagram simulation results based on the end-to-end system mimic channels calculated by the proposed methods.

opening in the channel where the bathtub curve is closed as a mask that separates SI success from failure. Based on the simulation PAM-4 eye mask defined in this way, it is possible to analyze the PCIe Gen 6.0 SI margin by simulating the system channel using the proposed measurement methods in any customer environment where Samsung's SSD is used.

IV. CONCLUSION

In this paper, we first proposed the system-level mimic channel model generation method when measurable TP for the PCIe interface is only at the end of the system interconnect and any channel model is limited. We have verified the mimic channel created based on time domain data measured at the TP, when the Tx transmits the compliance bit pattern, showed almost same SI values at PCIe operating Nyquist frequencies compared to the measured S-parameter model based results for the fabricated DUT samples. In addition, by applying the proposed method to the Samsung SSD PCIe SI margin test system, we quantitatively validated system-level channel requirements that Samsung Gen 6.0 SSD product can operate normally without BER, and were also able to secure an eye mask as a result of PAM-4 simulation based on the end-to-end channel model generated by the proposed method. Therefore, we have secured the method to determine whether the channel conditions allow the Samsung SSD to operate normally in PCIe Gen 6.0 mode by the proposed method although it cannot perfectly replace real channel characteristics.

REFERENCES

[1] PCI Express Base Specification Revision 6.2, Jan 25, 2024.

[2] David M. Pozar, Microwave Engineering, 4th ed., John Wiley & Sons, 2011.

979-8-3503-5124-8/24 $31.00 © 2024 IEEE

Using Generative AI to Predict DC Electrical Performance

J. Eric Bracken
ANSYS Inc.
Canonsburg, PA (USA)
https://orcid.org/0009-0007-0051-9637

Abstract—**This paper explores the application of deep learning models to the DC power distribution problem in electronic packaging. A generative model based on an implicit neural representation (INR) is trained with meshed geometry and electromagnetic field data computed by traditional numerical methods. The model is constructed in a way that allows it to make predictions for previously unseen problem geometries.**

Keywords—deep learning, generative design, power distribution

I. INTRODUCTION

Recent advances in artificial intelligence (AI) and machine learning (ML), including deep learning and generative AI, have made the numerical simulation community wonder if and how these methods can be applied to physics problems. It would be very attractive if these methods could speed up or reduce the need for time-intensive numerical computations.

AI/ML models based on artificial neural networks are usually developed with a supervised learning approach. This involves collecting a very large set of training data and curating it so that the feature space of interest is covered in a complete and balanced way. Then a training process follows. This is often a time-consuming task that involves heavy computations with high-end GPU hardware. The trained model must then be evaluated with a reserved set of data to validate its performance. Frequently, some deficiencies are found, and then the model's hyper-parameters must be adjusted and the model re-trained until the desired accuracy is achieved.

Several issues become apparent with this approach when developing a model to predict electrical performance. First, depending on the problems considered, a single electromagnetic simulation can be a lengthy process. When one considers that hundreds of thousands of cases are typically required for training a deep learning model, it can be difficult to produce the training data in a reasonable length of time.

Second, the nature of the feature data may limit the model's applicability and generality. For example, one could take a simple geometry such as a set of coupled transmission lines and generate training data by varying a set of geometric parameters (such as the line spacing, width and length, thickness of the substrate, etc.) and running simulations at each data point. A model trained to make predictions based only on these high-level features of the problem as inputs may do well in predicting the performance of very similar structures, but it will struggle or fail completely when asked to generalize to even slightly different geometries. It is impossible for such a model to generalize because it does not know the underlying geometric details or the physical rules that leads from that geometry to the results. It is simply a high-order curve fit mapping from a small set of abstract parameters to another set of numbers.

To overcome this second problem, it is possible to instead train an AI/ML model using geometry and 2D or 3D field solution data. The advantage of this approach is that the model becomes aware of the geometric structure of the problem and the impact that this has on the electromagnetic fields. A model so trained should have much greater ability to generalize and make predictions on unseen geometries.

A less obvious advantage of this approach is that the number of cases required for training may be significantly lower than what is expected for most deep learning models. Because each set of field simulation data consists of many thousands of values representing electromagnetic behavior in different regions of the problem domain, the training data is much richer in information. The model is better able to "learn" the underlying physical relations from such data.

In the following sections of the paper, we provide some more background information on this class of machine learning methods and describe how they can be used to build a predictive model. We then show some examples to demonstrate their promise for predicting electrical performance.

II. PREDICTING ELECTROMAGNETIC FIELDS

A. Implicit Neural Representations

Implicit neural representations (INRs) are a class of deep learning AI/ML models that represent a continuous function of position with a neural network [1]. These models take a spatial coordinate as their input and map it to the function's value at that position. Samples of the function at discrete points are used for

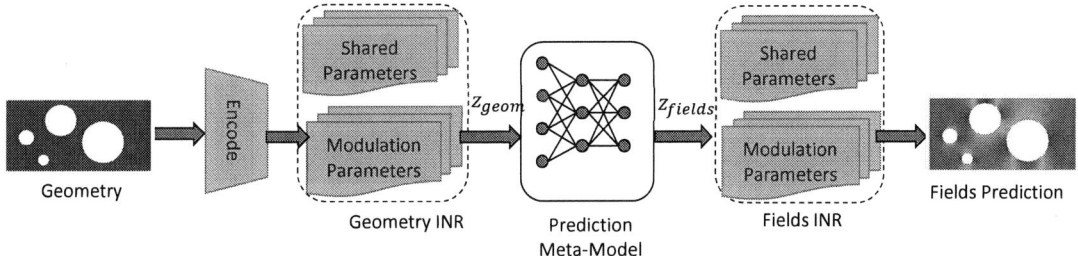

Figure 1. Diagram of the fields prediction process.

training, but the trained model can predict the function's value at any point within the function's domain.

The "function" modeled by the INR can be either a physical field or the problem geometry. The geometry can be represented indirectly with a distance function. The distance function is defined as the distance of any point in the domain from the surface of the geometry. Hence, when the value of the function is zero at a point, that point lies on the surface.

While the INR is useful, it does not immediately lead to a system that is capable of predicting electromagnetic fields, because a basic INR can only evaluate the fields for the single instance used to train it. A generalizable INR can be created through a procedure called modulation (see, e.g. [2].) This involves adjusting some of the internal parameters of the model to allow it to represent multiple data instances. A modulated INR essentially has two sets of internal parameters: one set captures the shared characteristics of the training data, while another smaller set, the modulation parameters, adapts the shared model to represent a single instance.

B. A System for Predicting Electromagnetic Fields

Using modulation, INR models that span the full training data set can be created both for the input geometry and the fields. To develop a predictive model, it is then necessary to establish a mapping between the geometry and the fields. This can be accomplished by training yet another neural network using the modulation parameters z_{geom} of the geometry INR as the inputs and the modulation parameters z_{fields} of the fields INR as the desired outputs. Training a prediction "meta-model" using these compact encodings is a much more manageable task than attempting to train a model with the full descriptions of the fields and geometry.

One more piece is needed: given a new problem geometry, the parameters describing it are not immediately known and must be computed somehow. This can be achieved by iteratively adjusting the z_{geom} parameters until the predicted distance function closely matches the given geometry [3].

With the trained models, the process of predicting the field solutions then becomes:

1. Accept a new problem geometry from the user.

2. Encode a set of modulation parameters z_{geom}.

3. Input the computed geometry parameters to the trained meta-model and perform inference to predict the field modulation parameters z_{fields}.

4. Apply the predicted z_{fields} to the fields INR.

5. Query the fields prediction INR at the desired points in space to extract the estimated fields for the new problem geometry.

The diagram in Figure 1 illustrates the prediction process.

It should be noted that training the models used in this approach can still be a time-consuming process, depending on the number of training cases, the complexity of the meshes, and the computing hardware used. However, the inference process is typically on the order of a few seconds. Thus, it is often considerably faster than performing a full new numerical solution.

III. EXAMPLES

The approach described here has been implemented in a software package called Ansys SimAI [4]. The following examples will demonstrate its effectiveness.

A. Ground Planes with Circular Holes

The first example is a rectangular copper ground plane with 1 amp of current entering at the left and leaving at the right. The model was trained with 50 sample cases. In each training example, one to five circular antipads with random radii are punched out of the plane at random locations. Some examples of the training geometries are shown in Figure 2. A finite element field solver was used to mesh each of these geometries and compute the vector DC current density $\vec{J}_{dc}$. This field data was then used to train the AI/ML model. The resulting resistance values ranged from 350 to 600 μΩ.

Of the 50 cases constructed, 45 were used for training and tuning the model, while 5 were used for unbiased testing of the predictions. The worst error among the test cases for resistance prediction was 4%.

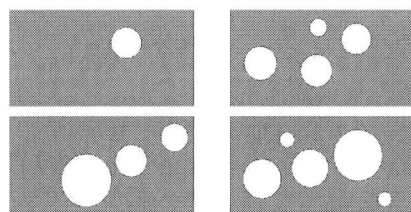

Figure 2. Sample of the geometries used in training the first example.

Figure 3 below shows the ground truth (finite element solution) and predicted current density fields for a manually constructed case not seen in training. The field solutions are

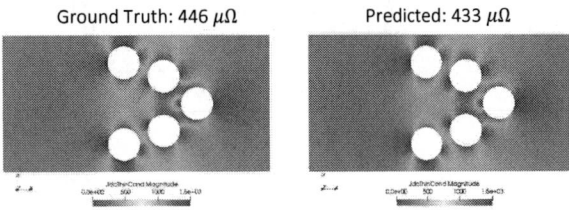

Figure 3. Prediction for a geometry not seen in training.

almost indistinguishable. The actual resistance was 446 $\mu\Omega$ while the predicted resistance was 433 $\mu\Omega$ (-2.9% error).

B. Square Hole

In Figure 4, the same model was used to predict current flow around a square hole. The predicted current density does not capture the singular behavior near the corners of the hole, because only circular holes were seen in training. Nonetheless, the fields are reasonable in the rest of the domain and the predicted resistance (377 $\mu\Omega$) is within 1.3% of the FEM-computed resistance (382 $\mu\Omega$).

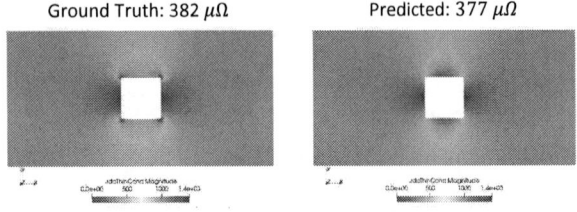

Figure 4. Prediction vs. ground truth for a square hole. (Only circular holes were used in training.)

C. Gridded Power Supply Network

The second example consists of power grids constructed on a rectangular pattern. A number of connections were removed at random from the grids to simulate avoidance of routing obstacles. The widths of the connecting traces were chosen at random (either 0.3 or 0.5 mm) when generating the geometry. In total, 125 different cases were created. The point-to-point current conduction solutions between two terminals at opposite ends of the grid were calculated using finite element analysis and these current distributions were then used to train the prediction model.

A training case that originally used a line width of 0.5 mm was redrawn with 0.3 mm line width, and this new geometry was then input to the model to test its predictive power. Figure 5 shows the resulting current distribution. The resistance of the original network was 5.40 mΩ. After shrinking the line width,

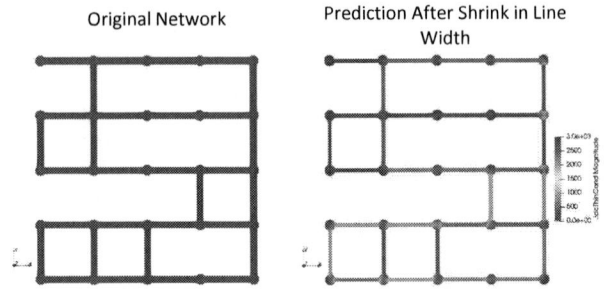

Figure 5. Original power grid and the predicted current distribution after shrinking the line width from 0.5 to 0.3 mm.

the resistance computed by the finite element solver increased by about 60% to 8.65 mΩ. The model predicted a resistance value of 8.37 mΩ, in good agreement with the numerical solution (-3.2% difference.)

IV. CONCLUSIONS AND FUTURE WORK

The model described here shows good predictive power for DC current distributions. Space prevents us from exploring the many possible applications of this technique to other areas, such as high frequency electromagnetics, thermal conduction, mechanical deformation, etc. However, it should be clear from our description of the method that there is nothing specific to the physics in its assumptions, and it has been applied to these other areas successfully. Given the quality of the predictions, this method should serve well for creating surrogate models in applications such as optimization and rapid design exploration.

Future work will focus on improved training techniques, both to increase the prediction accuracy and speed. Other questions to be explored are how much training data must be used and how the chosen data affects the model's ability to generalize successfully.

ACKNOWLEDGMENT

The author would like to thank P. Han, P. Krenz, H. Rashid, and P. Yser, and the entire SimAI development team at Ansys.

REFERENCES

[1] V. Sitzmann, J. Martel, A. Bergman, D. Lindell, and G. Wetzstein, "Implicit neural representations with periodic activation functions." Advances in Neural Information Processing Systems, 33:7462-7473, 2020.

[2] C. Kim, D. Lee, S. Kim, M. Cho, W.-S. Han, "Generalizable implicit neural representations via instance pattern composers," 2023 IEEE/CVF Conf. on Comp. Vision and Pattern Recognition (CVPR), 2023.

[3] L. Serrano, L. Migus, Y. Yin, J. A. Mazari, and P. Gallinari, "INFINITY: Neural field modeling for Reynolds-averaged Navier-Stokes equations," Workshop on Synergy of Scientific and Machine Learning Modeling (ICML 2023), Honolulu,, HI, USA, Jul. 2023.

[4] ANSYS Inc., "Explaining SimAI: How AI is applied to numerical simulation in practice", https://www.ansys.com/blog/explaining-simai, (accessed June 25, 2024.)

PCIe Gen 6.0 SSD PSIJ Estimation Based on Early Design Stage Jitter Sensitivity Measurements

Youngjun Ko, Jinwook Song, Seokwoo Hong, Jinan Lee, Jonghee Jeong, Sungwoo Jin, Hyunwoo Kim, Chorom Jang, Sungwon Roh, Dongho Choi, Kyungsuk Kim and Sunghoon Chun
Solution Development Team (Memory Business), Samsung Electronics, Hwaseong-si, 18448 South Korea
E-mail: yj0531.ko@samsung.com

Abstract— This paper estimates and analyzes the power supply noise induced jitter (PSIJ) caused by system-level power noises in high-capacity PCI express (PCIe) Gen 6.0 SSDs. To guide optimal design of the hierarchical power distribution network (PDN) with consideration of PSIJ before mass production, a measurement-based jitter sensitivity function (JSF) is employed during the PCIe PHY IP chip-level verification stage. For accurate JSF measurement, it must be ensured that the 64 Gbps PAM-4 eye of the PCIe interface is open during operations. In this paper, we propose a test architecture that can measure signal jitter with PAM-4 eyes opened while removing decoupling capacitors (de-caps) at the off-chip level so that single tone power noise applied to the off-chip test point (TP) can be transmitted safely to the on-chip stage without filtering. Using a hierarchical PDN model of the proposed test architecture, JSF was extracted in the 300 kHz to 300 MHz frequency range, considering the voltage transfer ratio (VTR) from TP to the target die bump. We confirmed, for the first time, that PSIJ caused by worst-case system power noise in a high-capacity PCIe Gen 6.0 SSD operating at full performance is estimated to be 1.32 ps, which corresponds to 42.2 % of the eye width specification of the PCIe Gen 6.0 base specification.

Keywords— *Jitter Sensitivity Function (JSF), Power Distribution Network (PDN), Power Supply Noise Induced Jitter (PSIJ), Peripheral Component Interconnect Express (PCIe), Solid State Drive (SSD)*

I. INTRODUCTION

With the transition to PCIe Gen 6.0, the data rate has doubled from 32 Gbps to 64 Gbps with the change in signaling scheme from NRZ to PAM-4. While the Nyquist frequency remains constant at 16 GHz, the transitions between adjacent levels results in a 33 % reduction in eye width. This reduces eye timing margin by about 20.8 ps, making it more vulnerable to system-level jitter. Additionally, as shown in Fig. 1, designing high-capacity, high-performance SSDs in limited spaces increases design complexity and results in more system-level power noise. This system power noise can convert to signal jitter, further reducing the PAM-4 eye timing margin. Moreover, the decline in operating power rail voltage levels due to IP process scaling down deteriorates the jitter sensitivity of the PHY IP. In the case of Samsung's PCIe PHY, as the generation progresses from Gen 5.0 to Gen 6.0, the operating power level decreases and the jitter sensitivity also deteriorates. Consequently, early-stage prediction and management of PSIJ are crucial for optimizing PDN design in high capacity PCIe Gen 6.0 SSDs to maintain overall system performance and reliability in high-speed interfaces.

To estimate and analyze PSIJ in an early design stage, an accurate JSF is required, which can be obtained through measurement. JSF is calculated by applying single-tone power noise of varying frequencies to the power rail and then dividing the increased signal jitter by the applied noise amplitude at the die bump level. Typically, to apply the desired amplitude of sinusoidal waveforms to the PDN using

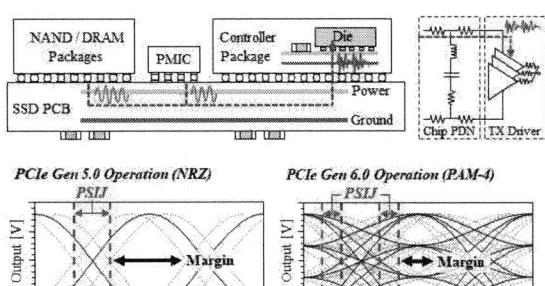

Fig. 1. A conceptual figure of the impacts of system-level power noise on the signal in a high-capacity PCIe Gen 6.0 SSD. The PAM-4 signal has a narrower eye width margin compared to the traditional NRZ signal, making it crucial to analyze SI degradation caused by system-level power noise to meet overall system performance specifications.

signal generator equipment with 50 ohm output impedance, measurement architecture excludes off-chip de-caps. However, since the timing margin of high-speed PAM-4 signal is very small, the removal of de-caps can lead to significant eye deterioration due to self-simultaneous switching noise (SSN). In such environments, measurements do not reflect actual SSD operation. Additionally, because direct measurement at the die bump is physically impossible, it is necessary to measure the applied sinusoidal waveform at off-chip TP and convert it to die bump level. In this case, securing the bandwidth of the off-chip TP to the die bump VTR up to targeted frequency is essential. Additionally, the calculation VTR utilizes z-parameter metrics derived from the hierarchical PDN model [1]. Therefore, ensuring the accuracy of the hierarchical PDN model is important.

In this paper, we estimated and analyzed the system-level PSIJ for high-capacity PCIe Gen 6.0 SSD design at an early design stage. We successfully obtained precise measurement-based JSF within the frequency range of 300 kHz to 300 MHz employing the proposed measurement architecture. After identifying the most severe power noise that could occur in a high-capacity Gen 6.0 SSD, we finally analyze impacts of PSIJ on 64 Gbps PAM-4 signals. The JSF and the power noise at the die bump level extraction process are referenced in [1].

II. EARLY DESIGN STAGE JSF EXTRACTION METHOD

For early stage PDN design guide considering PSIJ, JSF should be extracted during the PHY IP chip-level verification stage. Fig. 2 shows the (a) top view and (b) cross-sectional view of the proposed measurement architecture for JSF. To accurately obtain the JSF of a 64 Gbps PAM-4 PHY, the measurement architecture is modified as follows: all de-caps on the EVB and package are removed, and a 1 ohm resistor and two 22 uF capacitors are mounted in series on the pad where the package de-cap was removed. By mounting the 1

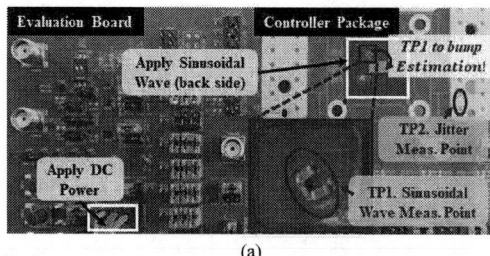

(a)

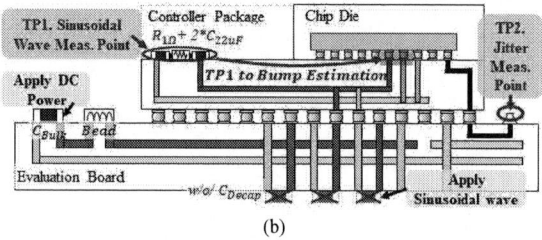

(b)

Fig. 2. (a) Top view and (b) cross-sectional view of the measurement architecture for accurate JSF extraction at the early design stage. A sinusoidal wave of varying frequencies is applied to the power rail, measured at TP1 on the package, and then converted to the bump level using the TP1 to the bump VTR. Signal jitter is measured at TP2 on the EVB.

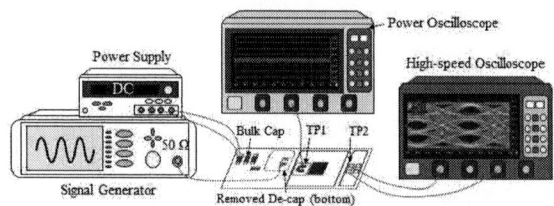

Fig. 3. Proposed Measurement environment to obtain accurate JSF. The input end of each oscilloscope corresponds to the marked TPs in Fig. 2.

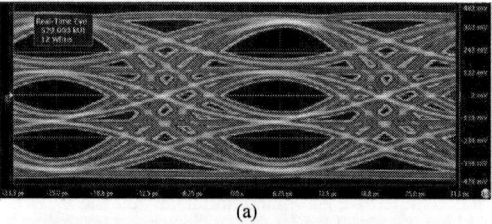

(a)

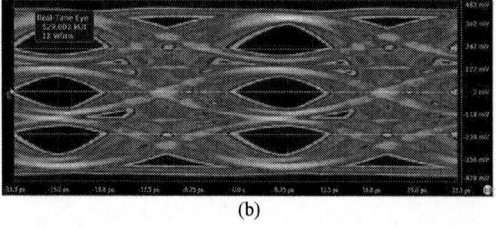

(b)

Fig. 4. PCIe Gen 6.0 PAM-4 eye diagram measured at TP2: (a) without applied power noise, and (b) with 1MHz VDDH power noise applied. The pattern follows the 52-UI jitter measurement pattern defined in the PCIe Gen 6.0 base specification [3]. The increased jitter is determined by the difference of total jitter between (a) and (b).

ohm resistor on the package, large hierarchical PDN self-Z resonance caused by PDN inductance and on-chip capacitance can be reduced. This prevents eye degradation due to SSN for high-speed signals. The bulk capacitor is retained to suppress low-frequency noise from the power supply equipment, and

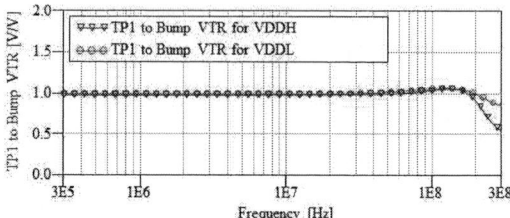

Fig. 5. Package TP1 to bump VTR by hierarchical PDN model for each power rail: VDDH and VDDL. The VTR values are approximately 1 up to 300 MHz, and used for bump-level power noise estimation.

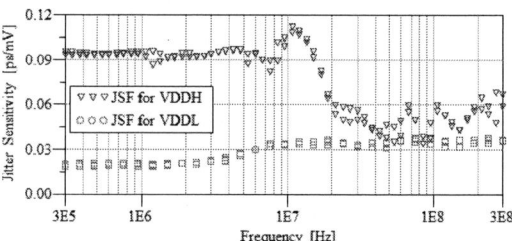

Fig. 6. JSF extracted through the proposed measurement method for the PCIe Gen 6.0 PHY for each power rail: VDDH and VDDL

the bead is retained to prevent PDN impedance reduction due to the bulk capacitor in the measured frequency region.

Fig.3 shows proposed measurement environment to obtain accurate JSF. The measurement process is as follows: DC power is supplied at the bulk capacitor. The PCIe PHY is then speed-changed to Gen 6.0, and a sinusoidal waveform at each frequency is applied to the pad where the de-cap on the bottom side of the EVB has been removed. The amplitude of the applied sinusoidal waveform is set to tens of mV, as defined in the IP vendor's PI ripple specification. The waveform amplitude is measured at TP1, where the 1 ohm resistor and 22 uF capacitor are located on the package, and the signal jitter is measured at TP2 on the EVB. Fig. 4 shows the 64 Gbps PAM-4 eye diagram measurement under two conditions: (a) without power noise and (b) with 1MHz VDDH power noise (assumed to be a sinusoidal waveform) applied. Due to the low hierarchical PDN impedance achieved by mounting a 1 ohm resistor on the package, the eye diagram of Fig. 4 (a) remains open at a normal operational level. To measure PAM-4 jitter, the 52-UI jitter test pattern defined in the PCIe Gen 6.0 base specification is used [3]. Increased jitter is determined by comparing the total jitter between Fig. 4 (a) and Fig. 4 (b). Additionally, because of the high-speed signaling, the measured jitter value can vary depending on the oscilloscope's clock and data recovery (CDR) bandwidth setting [1]. Therefore, the CDR loop transfer function should be applied to the measured jitter value for accurate jitter sensitivity estimation.

Since JSF is an on-chip characteristic, it is necessary to convert the sinusoidal waveform measured at TP1 to the die bump level. To calculate this, the TP1 to bump VTR, derived from hierarchical PDN model-based Z-parameter metrics as described in [1] and [2], can be employed. The targeted frequency range for JSF extraction is up to 300 MHz, determined by considering the noise sources to the die bump VTR. Since VTR is related to the PDN inductance between TP1 and the die bump, which depends on the physical distance, TP1 must be positioned near the bump to broaden the measurement bandwidth. Fig. 5 shows the VTR plot for each

979-8-3503-5124-8/24 $31.00 © 2024 IEEE   68

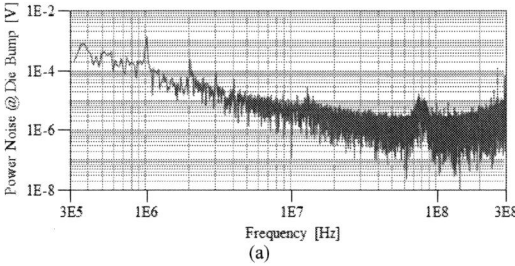

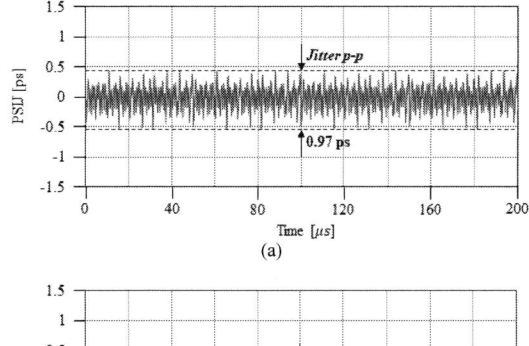

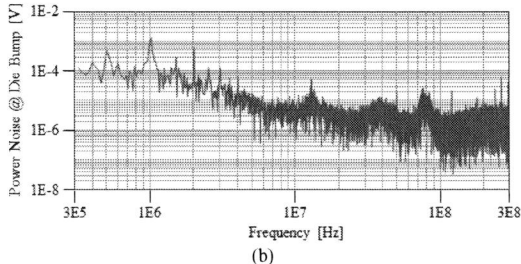

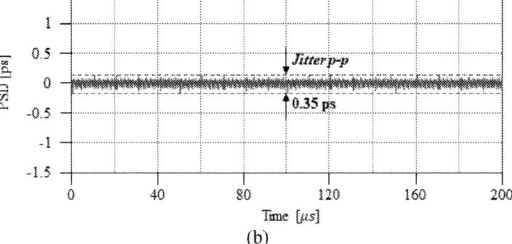

Fig. 7. Worst-case system power noise at die bump in a high-capacity PCIe Gen 6.0 SSD for each power rail: (a) VDDH and (b) VDDL

Fig. 8. Estimated PSIJ caused by worst-case system power noise in a high-capacity Gen 6.0 SSD for each power rail: (a) VDDH and (b) VDDL

power rail from TP1 to the die bump. Because TP1 is positioned adjacent to the die bump, VTR values remain close to 1 across the frequency range up to 300 MHz. For accurate VTR calculation, ensuring the precision of the package and EVB PDN models is crucial [1]. Additionally, modeling of the package balls and the die bumps must be conducted, along with accurate characterization of on-chip capacitance values under applied DC bias (power rail level). Fig. 6 shows the JSF plot for each power rail, calculated with consideration of the VTR values shown in Fig. 5.

III. PSIJ ESTIMATION AND DISCUSSIONS

PSIJ can be estimated by multiplying the JSF and the power noise at the die bump level as follows:

$$PSIJ = \sum_{i=0}^{N} JSF(f_i) \cdot V_{noise}(f_i) \qquad (1)$$

where N is the total harmonic frequencies [3]. Measurement-based JSF is obtained from II. Therefore, for the PSIJ-targeted high-capacity SSD pre-design guide, it is crucial to extract off-chip power noise for various operating conditions that can occur in the system and predict its impact on the signal. In this paper, to analyze the maximum PSIJ for PCIe Gen 6.0 SSD, the worst-case power noise was extracted based on measurements. This extraction assumes the condition of merging power rails that share the same power level and the system condition where all high-performance chips operate at full performance, using the previous Samsung SSD as a reference. Since the power noise cannot be measured directly at the die bump like JSF extraction, the measured power noise at the off-chip level is also converted to the bump level using the VTR calculated from the hierarchical PDN model. Fig. 7 shows the frequency domain power noise plots of the two power rails: (a) VDDH and (b) VDDL. These values are estimated by measuring the worst power noise at the off-chip for each power rail and multiplying it by the VTR from each noise source to the die bump.

Fig. 8 (a) and (b) show the estimated PSIJ for the VDDH and VDDL power rails resulting from system-level power noise in the high-capacity PCIe Gen 6.0 SSD. The values, up

to 0.97 ps for VDDH and 0.35 ps for VDDL. Since the noise in each power rail is independent, the maximum jitter that can occur in the signal is 1.32 ps, obtained by summing these values under the assumption of indefinite data accumulation. Given that the eye margin specification defined in the PCIe Gen 6.0 base specification [4] is 0.1 UI, 3.125 ps, the estimated PSIJ represents 40.7 % of the allowed margin. This accounts for a significant portion of the total system jitter. Therefore, to reduce system power noise, it is necessary to review power rail separation and noise filter designs targeted to the frequency values of the each power noise source.

IV. CONCLUSTION

In this paper, we proposed a novel measurement method to accurately obtain the JSF of high-speed 64 Gbps PAM-4 Serdes PHY. By extracting the worst-case system-level power noise that may occur in high-capacity PCIe Gen 6.0 SSDs based on measurement, we estimated the worst-case PSIJ to be 1.32 ps, which corresponds to 40.7 % of the PCIe Gen 6.0 test specification. This analysis indicates that the risk due to PSIJ cannot be negligible and warrants careful consideration when designing the hierarchical PDN for SSDs. In future work, to reduce PSIJ, we will analyze the power noise sources in the frequency domain and establish optimal guidelines for de-cap quantities and noise filter design, considering cost and area constraints, prior to mass production of SSDs.

REFERENCES

[1] Y. Ko et al., "PCB-level Jitter Sensitivity Measurement and Hierarchical PDN-Z based PSIJ Estimation for PCIe Gen5 SSD," 2023 IEEE 32nd Conference on Electrical Performance of Electronic Packaging and Systems (EPEPS), Milpitas, CA, USA, 2023, pp. 1-3.

[2] J. Song et al., "Novel Target-Impedance Extraction Method-Based Optimal PDN Design for High-Performance SSD Using Deep Reinforcement Learning," in IEEE Transactions on Signal and Power Integrity, vol. 2, pp. 1-12, 2023.

[3] D. Oh, "System Level Jitter Characterization of High Speed I/O Systems," in Proceedings on IEEE International Symposium on EMC, Pittsburgh, PA, USA, Aug. 2012.

[4] PCI SIG Org. (2024), PCI Express® Base Specification Revision 6.2. Available: https://pcisig.com/specifications

979-8-3503-5124-8/24 $31.00 © 2024 IEEE

Application of CAMM2 Connector on PCIe Gen 6.0 SSD Host Interface for Low Near-End Crosstalk

Sungwoo Jin, Jinwook Song, Seokwoo Hong, Youngjun Ko, Hyunwoo Kim, Sungwon Roh, Chorom Jang, Dongho Choi, Kyungsuk Kim
and Sunghoon Chun
Solution Development Team (Memory Business), Samsung Electronics, Hwaseong-si, 18448 South Korea
E-mail: sw.jin@samsung.com

Abstract—The M.2 connector has been widely adopted in the host interface of client Solid State Drives (SSDs) for high-speed data storage in consumer electronics. The pin arrangement of the M.2 connector, where transmitter (TX) and receiver (RX) lines are positioned adjacent, can lead to issues such as near-end crosstalk (NEXT). Particularly, with the upcoming PCI Express (PCIe) Gen 6.0 standard, concerns arise over the potential degradation of NEXT in M.2 connectors, which could negatively affect SSD performance and reliability. This paper analyzes the signal integrity (SI) requirements for connectors to ensure the host interface of SSDs operates reliably at PCIe Gen 6.0 speed. We investigate whether the NEXT in existing M.2 form factors for client SSDs meets these requirements and experimentally examine its impact on PCIe Gen 6.0 64 Gb/s PAM-4 eye diagram at the system level. Furthermore, by applying Compression Attached Memory Module 2 (CAMM2) to the SSD PCIe interfaces, we experimentally demonstrate the capability to reduce NEXT by over 50 % compared to M.2 and limit PAM-4 eye degradation to less than 1 %. This illustrates how CAMM2 connector can overcome the limitations of M.2 connector and enhance performance in PCIe Gen 6.0.

Keywords—*CAMM2, CEM, Gen 6.0, M.2, NEXT, PAM-4, PCI Express, SSD*

I. INTRODUCTION

In recent years, the demand for high-speed data storage solutions has significantly increased with the rapid rise in the number of consumer electronic devices such as personal computers, smartphones, and tablets. Client SSDs have become central to meeting this demand due to their superior performance, reliability, and efficiency compared to traditional hard disk drives (HDDs). Among various form factors, the M.2 has become a popular choice in client SSDs due to its compact size and versatility.

However, the emergence of PCIe Gen 6.0, which supports data speeds of up to 64 GT/s using PAM-4 signaling, has introduced new challenges for SI in M.2 connectors. The PCIe Card Electromechanical (CEM) specification for connectors operating at Gen 6.0 speed is being established, with one of the primary concerns being crosstalk. As shown in Fig. 1, crosstalk generated in the connector can affect the RX of the SSD Controller (CTRL). Particularly, since the SSD channel is generally much shorter than the system channel, severe NEXT in the connector can directly transmit noise to the SSD RX, significantly degrading SI. Additionally, PAM-4 signaling, with its narrower level spacing compared to NRZ signaling, is more susceptible to signal distortion caused by NEXT, which can easily degrade the eye diagram, leading to data integrity issues and overall performance degradation.

The biggest issue with M.2 connectors, compared to other connectors like Enterprise and Datacenter Standard Form Factor (EDSFF) is their poor NEXT characteristics due to the

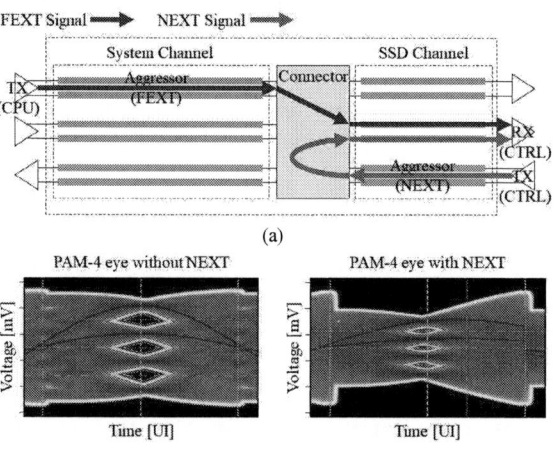

Fig. 1. (a) Possible crosstalk propagation path caused by a connector in the entire host interface channel. (b) Comparison of PAM-4 eye with and without NEXT. When NEXT negatively affects SI, SSDs may experience issues such as bit error rate, data loss, and performance degradation.

side-by-side placement of TX and RX lines. This presents a specification crisis for M.2 connectors and highlights the increasing need for potential redesigns or the introduction of new connectors to ensure stable operation at high data transfer rates.

In this paper, we analyze the host interface connector requirements for ensuring reliable operation of SSDs at PCIe Gen 6.0 speed at the system level. We examine whether the NEXT of M.2 connector meets the necessary requirements and investigate how it can degrade PCIe Gen 6.0 PAM-4 eye diagrams. Furthermore, we propose alternative connector that can meet the stringent SI requirements of PCIe Gen 6.0. Through an example application of CAMM2 connector, commonly used in DRAM modules, to PCIe interfaces, we demonstrate that the connector meet the NEXT SI requirements of the PCIe CEM specification and effectively reduce PAM-4 eye degradation compared to traditional M.2 connector.

The remainder of this paper is structured as follows: Section II outlines the PCIe CEM specifications and the requirements for Gen 6.0 connectors, as well as examining the NEXT-induced eye degradation observed when using M.2 connector. Section III assesses the NEXT levels and corresponding eye diagrams for CAMM2 connector applied to PCIe interface. Finally, Section IV summarizes the findings, discusses their implications, and proposes directions for future research.

II. ANALYSIS OF NEXT ISSUES IN CONVENTIONAL M.2 CONNECTOR FOR PCIE GEN 6.0

Component Contribution Integrated Crosstalk Noise (ccICN) is a method for directly calculating the crosstalk noise

979-8-3503-5124-8/24 $31.00 © 2024 IEEE

that a specific individual component within a channel induces in the RX, taking into account the insertion loss of the pre-channel and post-channel [1]. This approach provides a more intuitive and accurate explanation of the impact of crosstalk compared to limit line-based specifications, which is why the PCIe CEM specification includes requirement values based on this method.

To verify the eye degradation due to NEXT in connectors at the system level, a series of arbitrary connector models with different NEXT values but otherwise identical characteristics were first generated, using a simple s-parameter scaling method, as shown in Fig. 2 (a). The $ccICN_{NEXT}$ values for each connector range from 197.5 uV to 1224.3 uV. Using these connector models, the eye degradation due to NEXT was examined in three different system channels with varying insertion loss. Fig. 2 (b) shows the normalized values compared to the case without NEXT ($ccICN_{NEXT}$ of 0), indicating that as NEXT increases, the eye height and width decrease. Particularly, in the long system channel with higher insertion loss, the degradation due to NEXT is more severe. When the $ccICN_{NEXT}$ exceeds 1200 uV, the PAM-4 eye degrades by more than 50 %, with the current Gen 5.0 M.2 connector's $ccICN_{NEXT}$ specification being 1200 uV [2].

As mentioned in the previous section, M.2 is a widely used connector in client SSDs. In the structure of the M.2 shown in Fig. 3 (a), the SSD is inserted into the connector, with the TX and RX lines positioned next to each other, resulting in high NEXT. To verify the Gen 6.0 PAM-4 eye degradation caused by this high NEXT, channel simulation was conducted at the system level. The comparison of the eye diagrams with and without NEXT for an actual M.2 connector model with a $ccICN_{NEXT}$ of 1100 uV, a system channel with -

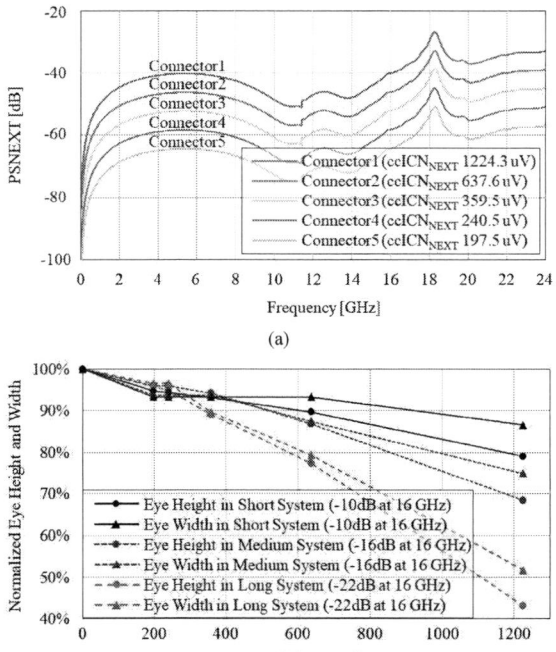

Fig. 2. (a) A series of arbitrary connector models generated with different $ccICN_{NEXT}$ values. (b) Degradation of the eye height and width in SSD RX due to the NEXT of the connector in short, medium, and long systems. The PCIe CEM 6.0 specification proposes a 100 uV $ccICN_{NEXT}$ specification for the Gen 6.0 connector, resulting in an approximate 5 % degradation in the eye diagram [3].

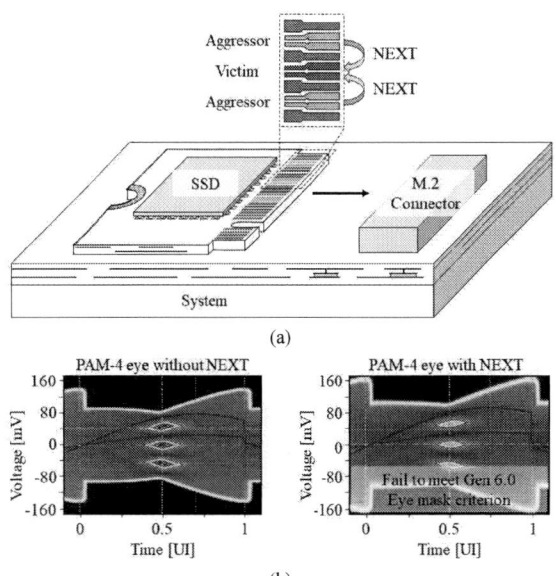

Fig. 3. (a) Structure of the SSD connected to the M.2 connector. (b) Degradation of the PAM-4 eye in the SSD RX due to the presence or absence of NEXT in the M.2 connector within a -22 dB at 16 GHz long system. The $ccICN_{NEXT}$ value of the M.2 is around 1100 uV, resulting in approximately 38 % eye width degradation in Gen 6.0 as shown in Table I.

TABLE I.   PAM-4 EYE HEIGHT AND WIDTH COMPARISON BETWEEN THE WITHOUT AND WITH NEXT FROM M.2 CONNECTOR

Case	M.2	
	w/o NEXT	*w/ NEXT*
EH_{top} [mV]	20.02	10.29
Degradation	**48.65 %**	
EW_{top} [UI]	0.145	0.090 (Fail)
Degradation	**37.93 %**	

PCIe CEM 6.0 Eye mask for Gen 6.0: Eye height 6 mV, Eye width 3.125 ps (Top Eye) [3].

1 Unit Interval (UI) = 31.25 ps.

22 dB at 16 GHz meeting the Gen 6.0 CEM specification, and an SSD channel with -3 dB at 16 GHz is shown in Fig. 3 (b). As shown in Table 1, the eye height and width decreased by 49% and 38%, respectively, with the eye width failing to meet the PCIe Gen 6.0 eye mask criterion of 3.125 ps. Since the PCIe CEM eye mask criteria are based on the top eye, only the top eye was compared [3]. To mitigate the eye degradation caused by NEXT in connectors at PCIe Gen 6.0, it is crucial to implement NEXT reduction designs.

III. PROPOSAL OF APPLYING LOW NEXT CAMM2 CONNECTOR TO GEN 6.0 SSD PCIE INTERFACE

The CAMM2 connector enables fast data transfer between DRAM modules and the system. The connector is located at the bottom of the module and connects to the system. A conceptual diagram of applying the CAMM2 connector to an SSD is shown in Fig. 4 (a), where the connector is similarly positioned at the bottom of the SSD. To verify the NEXT characteristics of the CAMM2 connector, the pin arrangement was modified to fit the PCIe interface and a Design of Experiments (DOE) was conducted on three cases of TX, RX, and ground pin placements to find the optimal NEXT reduction solution, as shown in Fig. 4 (b). The NEXT results for each DOE can be observed in Fig. 4 (c). In DOE1, where

979-8-3503-5124-8/24 $31.00 © 2024 IEEE   71

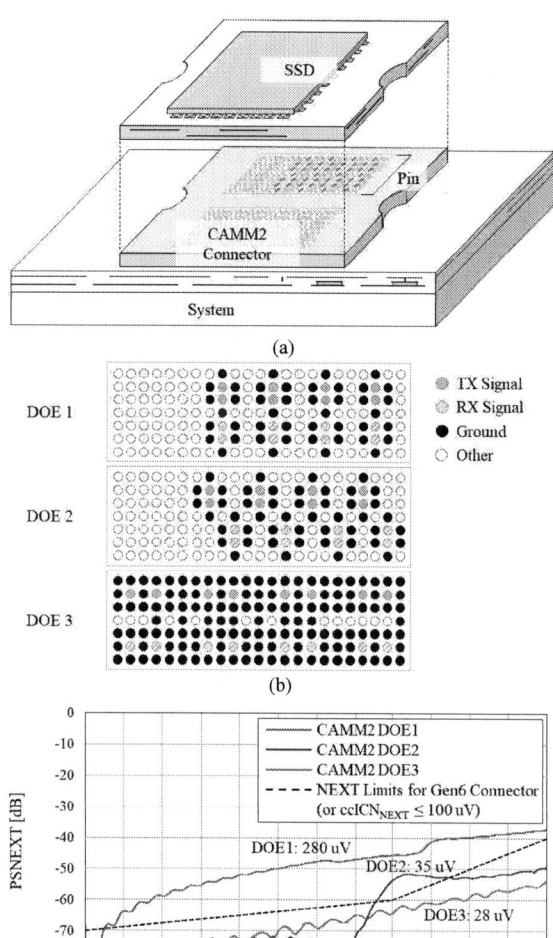

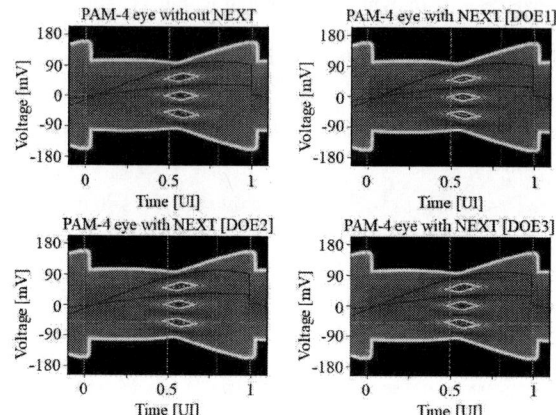

Fig. 5. PAM-4 eye result with or without NEXT for the three DOEs in Fig. 4 (b).

TABLE II. PAM-4 EYE HEIGHT AND WIDTH COMPARISON BETWEEN THE WITHOUT AND WITH NEXT FROM CAMM2 CONNECTORS

Case	DOE1		DOE2		DOE3	
	w/o NEXT	w/ NEXT	w/o NEXT	w/ NEXT	w/o NEXT	w/ NEXT
EH_{top} [mV]	22.64	20.00	22.65	22.46	22.75	22.58
Degradation	11.66 %		0.84 %		0.75 %	
EW_{top} [UI]	0.150	0.140	0.150	0.150	0.150	0.150
Degradation	6.67 %		0.00 %		0.00 %	

*PCIe CEM 6.0 Eye mask for Gen 6.0: Eye height 6 mV, Eye width 3.125 ps (Top Eye) [3].

6.0. This aims to reduce NEXT through the optimized pin arrangement of the CAMM2 connector. Simulations at the system level confirmed that the poor NEXT of the existing M.2 connector could reduce the PAM-4 eye by half. We demonstrated for the first time that applying the CAMM2 connector to the PCIe interface with appropriate pin arrangements reduced eye degradation to less than 1 %. The proposed CAMM2 connector is expected to be widely used in SSD connectors at high speeds beyond PCIe Gen 5.0. Future research will focus on designing SSD PCBs and packages suitable for the CAMM2 connector to analyze SI characteristics at the device level and develop solutions for PCIe Gen 6.0 operation.

ACKNOWLEDGMENT

We would like to express our sincere gratitude to Amphenol for providing technical support and insightful discussions that were crucial for this study. Their support and collaboration significantly contributed to the successful completion of this research.

REFERENCES

[1] S. -J. Moon, Z. Wu and M. Mazumder, "Generalized ccICN (component contribution Integrated Crosstalk Noise) for PAM-N," 2021 IEEE 25th Workshop on Signal and Power Integrity (SPI), 2021.

[2] PCI Express M.2 Specification Revision 5.1.1, Jan 17, 2024.

[3] PCI Express Card Electromechanical Specification Revision 6.0, Draft 0.7, Feb 5, 2024.

Fig. 4. (a) Concept of applying the CAMM2 connector to an SSD. (b) Three DOEs of pin arrangement for the CAMM2 connector aligned with the PCIe interface. (c) Corresponding PSNEXT graphs for each DOE. The ccICN_NEXT is under 100 uV in DOE2 and DOE3.

the TX and RX pins are arranged parallel to each other, a NEXT of 280 uV occurred. In DOE2 and DOE3, where the TX and RX pins are staggered or additional ground pins are added, met the ccICN_NEXT specification.

For the three DOEs, PCIe Gen 6.0 channel simulations are conducted under the same conditions as the previous M.2 analysis. In all DOE, the eye mask criteria are met. As shown in Fig. 5 and Table II, DOE1 with a ccICN_NEXT of 280 uV exhibited approximately 10 % eye degradation, whereas DOE2 and DOE 3 with ccICN_NEXT values below 100 uV showed less than 1 % degradation. This demonstrates that by using CAMM2 connectors instead of M.2 connectors and optimizing the pin arrangement, NEXT can be significantly reduced, thereby effectively preventing SI degradation in the SSD RX due to NEXT at PCIe Gen 6.0 speed.

IV. CONCLUSION

In this paper, we proposed applying the CAMM2 connector into the host interface of client SSDs for PCIe Gen

An Efficient SPICE-compatible Model for Fast Co-simulation of Signal and Power Integrity on Multilayer PCB with Arbitrary Shape

Hyunwoo Kim, Dongryul Park, Seunghun Ryu, Seonghi Lee, Sanguk Lee, Jinwook Lee,
Dongkyun Kim, and Seungyoung Ahn
Korea Advanced Institute of Science and Technology (KAIST), Daejeon, Republic of Korea
E-mail: kimhyunwoo@kaist.ac.kr

Abstract—This paper presents the SPICE-compatible model to efficiently simulate and analyze the multilayer PCBs with arbitrarily shaped plates, signal/power/ground vias, and multi-coupled traces. This model is compared and evaluated using full-wave simulation. The results demonstrate that the proposed model significantly reduces computational time while maintaining high accuracy for complex PCB designs, achieving high efficiency.

Index Terms—Multilayer structure, Signal/Power integrity, SPICE-compatible model

I. INTRODUCTION

As digital systems become complex and operate at higher data rates, multilayer structures with arbitrary shapes are widely used in electronic packages and printed circuit boards (PCBs). However, multilayer packages and PCBs suffer from serious noise caused by crosstalk, signal transitions between different layers, and fast-switching operation of the chip, which leads to degradation of the signal and power integrity (SI/PI) in the packages and PCBs. 3D electromagnetic (EM) solvers like Ansys HFSS provide an accurate solution to analyze these problems. Despite their high accuracy, these software tools are typically time-consuming and demand substantial memory resources.

Several studies have developed physical-based models for rapid simulation of multilayer packages and PCBs to address these challenges. An analytical model of via and traces was introduced to simulate multilayer interconnects at the package and PCB levels, but it only focused on SI analysis [1]. Another study [2] presented a combined method to simulate both SI and PI of multilayer PCBs simultaneously, but did not consider multi-coupled traces. Therefore, a novel model including all critical components of the multilayer PCBs is required for fast and comprehensive SI/PI simulation and analysis.

In this paper, we propose the SPICE-compatible model for the multilayer PCBs with arbitrarily shaped plates, signal/power/ground (S/P/G) vias, and multi-coupled traces. This model was implemented using the test board and validated through comparisons with the 3D EM solver in S-parameter and Z-parameter simulations. The results successfully demonstrated the effectiveness of the proposed model, indicating that it can be readily adapted to the various PCB designs.

II. AN EFFICIENT SPICE-COMPATIBLE MODEL FOR MULTILAYER PCB WITH ARBITRARY SHAPE

In this section, we model the components constituting the complex PCBs and implement the SPICE-compatible model for efficient SI/PI co-simulation of multilayer PCBs. Each component is modeled using the contour integral method (CIM), physics-based via model, and modal decomposition, respectively. Subsequently, the overall SPICE-compatible model is constructed by combining these individual components in Keysight ADS.

A. Arbitrarily Shaped Parallel Plate Model using CIM

The CIM approach, primitively developed for planar circuits analysis, has been effectively utilized to calculate the impedance of power distribution networks having arbitrary shapes [3]. Fig. 1 shows a pair of arbitrarily shaped P/G planes, where the field variation in the z-direction can be ignored due to the smaller distance between the plates compared to the wavelength of the interested frequency. C^p, C^v, and D represent the plate contour, via barrel contour, and computation area, respectively. θ_{ij} denotes the angle formed by the unit normal vectors $\overrightarrow{n}$ of contours C^p, C^v and distance vector $\overrightarrow{r_{ij}}$ between observation point (i) and source point (j).

For the numerical analysis, contours C^p and C^v are divided into N, M segments with width W_i smaller than the wavelength. Assuming constant voltage and current across each

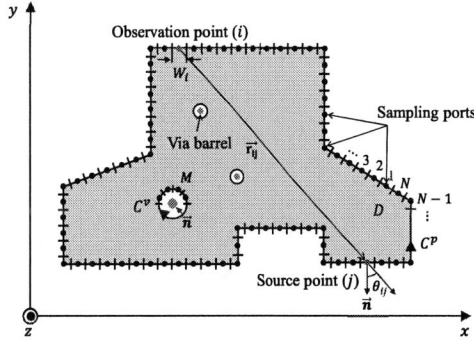

Fig. 1. Definition of the computation area D and parameters for the numerical analysis of the contour integral method.

979-8-3503-5124-8/24 $31.00 © 2024 IEEE

segment, the integral equation results in the following linear matrix equation on each sampling port [3].

$$[\mathbf{Z}] = [\mathbf{U}]^{-1}[\mathbf{H}] \qquad (1)$$

where $[\mathbf{U}]$ and $[\mathbf{H}]$ represent $(N+M) \times (N+M)$ matrices, and $[\mathbf{Z}]$ denotes the impedance matrix. The elements of the $[\mathbf{U}]$ and $[\mathbf{H}]$ matrices are defined by (2) and (3)

$$u_{ij} = \delta_{ij} - \frac{k}{2j} cos\theta_{ij} H_1^{(2)}(kr_{ij})W_j \qquad (2)$$

$$h_{ij} = \begin{cases} \frac{\omega\mu d}{4} H_0^{(2)}(kr_{ij}) & (i \neq j) \\ \frac{\omega\mu d}{4}\left[1 - \frac{2j}{\pi}\left(\ln\frac{kW_i}{4} - 1 + \gamma\right)\right] & (i = j) \end{cases} \qquad (3)$$

where δ_{ij} is the Kronecker delta and $\gamma = 0.5772$ is Euler's constant. $H_0^{(2)}$ and $H_1^{(2)}$ are the zero-order and first-order Hankel functions of the second kind. k denotes the complex wavenumber; ω is the angular frequency; and μ is the permeability of the substrate; and d is distance between the plates.

B. Physics-based Via Model for Signal/Power/Ground Vias

The vias facilitate the signal transition between different layers and connect components to P/G planes in multi-layered boards. Signal vias can be represented by a π-model comprising via capacitance C_{via} and parallel plate admittance Y^{pp} as shown in Fig. 2(a). This model can also be extended for P/G vias, considering the connectivity of via-to-plate illustrated in Fig. 2(b). C_{via} consists of coaxial capacitance C_{coax} and barrel-plate capacitance C_{barrel} meaning the capacitive coupling between via and parallel plate, while Y^{pp} serves as the current return path. The capacitance C_{coax} and C_{barrel} are given by [4]

$$C_{coax} = \frac{2\pi\varepsilon t}{\ln\left(\frac{r_{antipad}}{r_{via}}\right)} \qquad (4)$$

$$C_{barrel} = \frac{8\pi\varepsilon}{d\ln\left(\frac{r_{antipad}}{r_{via}}\right)}\left[\frac{d}{4k}\tan(\frac{kd}{2}) - \sum_{n=1,3,5\cdots}^{2N-1} F_n\right] \qquad (5)$$

where F_n is

$$F_n = -\frac{1}{k_n^2}\frac{H_0^{(2)}\left(k_n r_{antipad}\right)}{H_0^{(2)}\left(k_n r_{via}\right)} \qquad (6)$$

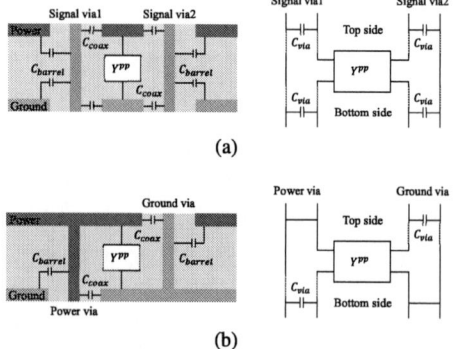

(a)

(b)

Fig. 2. Physics-based via model and circuit implementation: (a) via capacitance between signal vias and P/G planes. (b) via capacitance between P/G vias and P/G planes.

r_{via} and $r_{antipad}$ represent the radius of the via barrel and the antipad; t denotes the thickness of the plate; ε is permittivity of the substrate; k_n is radial wavenumber for TM modes. The admittance Y^{pp} can be derived from the inverse of the impedance matrix of parallel plate Z^{pp}, which is obtained through the CIM approach described in Section II-A.

C. Multi-coupled Trace Model using Modal Decomposition

The stripline is a widely used signal trace scheme in multi-layered boards due to its low radiation and crosstalk. When two signal vias are connected to both sides of the stripline, two modes can be identified within the cavity by applying the modal decomposition, as shown in Fig. 3(a): parallel plate mode and stripline mode. Fig. 3(b) presents the equivalent π-model for the multi-coupled stripline. Y^{pp}_{strip}, Y^{top}_{strip}, and Y^{bot}_{strip} derived as (7) denote the admittance matrices of stripline decomposed by upper and lower plates [5].

$$\begin{aligned} Y^{pp}_{strip} &= \left(\alpha^2 + \alpha\right)Y_{strip} + Y^{pp} \\ Y^{top}_{strip} &= -\alpha Y_{strip} \\ Y^{bot}_{strip} &= (\alpha + 1)Y_{strip} \end{aligned} \qquad (7)$$

where Y_{strip} denotes the admittance matrix for the stripline mode, which was extracted using a 2D solver in this paper. The factor α is defined as $\alpha = -h_l/(h_u + h_l)$, where h_u and h_l denote the height from the stripline to upper and lower plates, respectively.

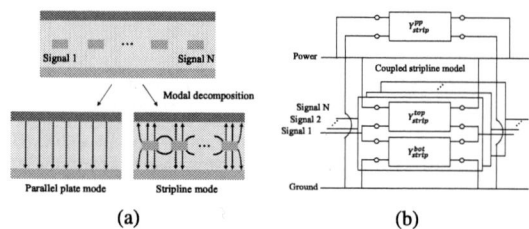

(a)

(b)

Fig. 3. (a) Modal decomposition of the multi-coupled stripline: parallel plate mode and stripline mode. (b) Equivalent π-model for the multi-coupled stripline.

D. Implementation of the SPICE-compatible Model

Fig. 4(a) presents a cross-section view of an exemplary board to describe the combination process for the SPICE-compatible model. A single P/G via pair was employed to consolidate the P/G planes of different layers, forming multiple cavity structures. Additionally, two signal vias were used for the signal transition between the microstrip line on the top and bottom layers and the stripline within i-th cavity. The SPICE-compatible model for the exemplary board was realized by connecting the modeled components, as depicted in Fig. 4(b). Here, $Z_{\mu-strip}$ represents the impedance matrix for the microstrip line, which was also extracted using the 2D solver. Furthermore, this model can easily be extended to cover the various multilayer PCBs by adding additional cavities, vias, and traces as required by specific board designs.

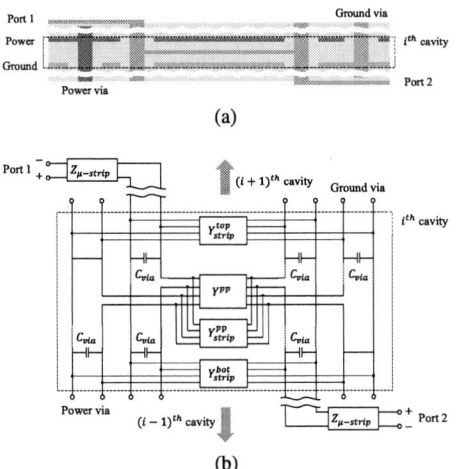

(a)

(b)

Fig. 4. (a) The cross-section view of an exemplary board to describe the combination process. (b) The proposed SPICE-compatible model for the exemplary board.

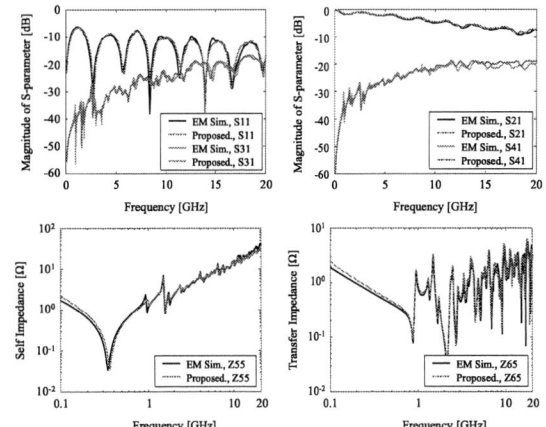

Fig. 6. S-parameter and Z-parameter simulation results for the test board.

III. VERIFICATION OF THE PROPOSED MODEL

The proposed model was evaluated using a 3-cavity arbitrarily shaped board with ten signal vias connected by microstrip lines and striplines, and four P/G via pairs as illustrated in Fig. 5. The board boundaries were approximated as a perfect magnetic conductor (PMC) because the parallel plate model using CIM assumes open boundaries at edges. The six ports were assigned for the simulation. A detailed description of the board's configuration, including its dimensions and material properties, is indicated in the figure.

Fig. 6 shows the S-parameter and Z-parameter simulation results for the test board discussed in Fig. 5. The proposed model shows good agreements with the 3D EM simulation (Ansys HFSS). However, it is observed that this correlation reduces in the high-frequency range, specifically above 10 GHz. This decrease in correlation is attributed to the introduction of analytical assumptions such as neglecting conductor thickness and fringing field, and to the increased parasitic effects such as high-order mode at higher frequencies, resulting in reduced accuracy. Nonetheless, the calculated results remain within acceptable errors compared to the simulated results.

The computation times were compared to verify the efficiency of the proposed model. The 3D EM simulation required approximately 13h 41m 50s (~49,310s) for 1991 frequency points on the computer with a 3.4 GHz CPU and 16 GB

RAM. In contrast, the proposed model took only 73s on the same computer resources, which includes the calculation time of MATLAB for parallel plates and via model, as well as the simulation time of the 2D solver for the trace model. Consequently, the proposed model demonstrated high efficiency for SI/PI co-simulation of complex PCBs, achieving significantly reduced computation time over the 3D EM solver while maintaining high accuracy.

IV. CONCLUSION

In this paper, we introduced the SPICE-compatible model for the multilayer PCB with an arbitrary shape. The results exhibited short computation time in the S-parameter and Z-parameter simulation compared to the 3D EM solver while maintaining high accuracy. This model provides a powerful tool for rapid SI/PI simulation and analysis of complex PCBs, facilitating time and cost savings in practical PCB developments.

ACKNOWLEDGMENT

This work was supported in part by the Institute of Information & communications Technology Planning & Evaluation (IITP) grant funded by the Korea government (MSIT) (No.2022-0-00986, Development of artificial intelligence-based base station electromagnetic wave human exposure prediction algorithm)

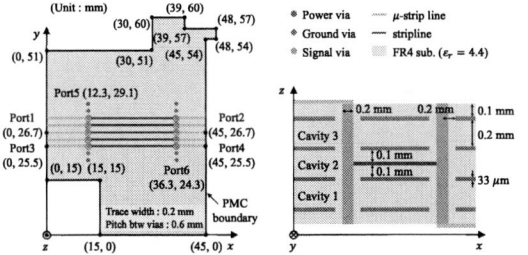

Fig. 5. Top and side view of the arbitrarily shaped test board with three cavities, five coupled traces, ten signal vias, and four power/ground vias.

REFERENCES

[1] R. Rimolo-Donadio, X. Gu, Y. H. Kwark, M. B. Ritter, B. Archambeault, F. de Paulis, Y. Zhang, J. Fan, H.-D. Bruns, and C. Schuster, "Physics-based via and trace models for efficient link simulation on multilayer structures up to 40 ghz," *IEEE Transactions on Microwave Theory and Techniques*, vol. 57, no. 8, pp. 2072–2083, 2009.

[2] X. Duan, R. Rimolo-Donadio, H.-D. Brüns, and C. Schuster, "A combined method for fast analysis of signal propagation, ground noise, and radiated emission of multilayer printed circuit boards," *IEEE Transactions on Electromagnetic Compatibility*, vol. 52, no. 2, pp. 487–495, 2010.

[3] T. Okoshi, *Planar circuits for microwaves and lightwaves.* Springer Science & Business Media, 2012, vol. 18.

[4] S.-P. Gao, F. de Paulis, E.-X. Liu, A. Orlandi, and H. M. Lee, "Fast-convergent expression for the barrel-plate capacitance in the physics-based via circuit model," *IEEE Microwave and Wireless Components Letters*, vol. 28, no. 5, pp. 368–370, 2018.

[5] A. E. Engin, W. John, G. Sommer, W. Mathis, and H. Reichl, "Modeling of striplines between a power and a ground plane," *IEEE Transactions on Advanced Packaging*, vol. 29, no. 3, pp. 415–426, 2006.

979-8-3503-5124-8/24 $31.00 © 2024 IEEE

Latency Insertion Method for Fast Electro-Thermal Simulation of FinFET with Self-Heating Effect

Yi Zhou and José E. Schutt-Ainé
Department of Electrical and Computer Engineering
University of Illinois at Urbana-Champaign
Urbana, IL, USA
{yizhou18, jesa}@illinois.edu

Abstract—Self-heating effect (SHE) is prominent for FinFET devices due to their large currents and compact sizes. With SHE, the power in FinFETs is dissipated into heat, affecting device performance. Thus, fast and accurate electro-thermal analysis of FinFETs incorporating SHE is important. This paper proposes the Latency Insertion Method (LIM) algorithm for fast FinFET electro-thermal simulation with SHE. Based on the BSIM-CMG model, the proposed method leapfrogs between the conventional electrical LIM algorithm and the thermal equations. The LIM results are compared to SPICE-based commercial software to prove the accuracy and the speed.

Index Terms—latency insertion method, FinFET, BSIM-CMG, circuit simulation, electro-thermal simulation, self-heating effect

I. INTRODUCTION

The integrated circuit (IC) industry has been aggressively pursuing higher operating frequencies and higher package densities. As a result, increasing power density causes the dissipated heat to be not negligible. Due to the FinFET's low power assumption and well-suppressed short channel effects, it has been replacing the MOSFET in sub-20 nm integrated circuits [1], [2]. However, the FinFET exhibits prominent self-heating effect (SHE) due to its large drive current and high thermal resistance introduced by the nanoscale 3-dimensional structure [3], [4]. The increasing device temperature caused by SHE can decrease the carrier mobility and saturation velocity [5]. Therefore, SHE can have significant influence on the FinFET's electrical performance. Developing fast and accurate electro-thermal simulation algorithm of FinFET that can capture the SHE is thus important.

Commercial circuit simulators are usually based on the conventional Modified Nodal Analysis (MNA) methods like the SPICE algorithm. SPICE represents the entire circuit by a matrix and relies heavily on matrix calculations and inversions. As the scale of the circuit increases, the computational complexity grows super-linearly, which means simulating large circuits can be computationally expensive. While performing electro-thermal analysis, the SPICE algorithm uses iteration methods to ensure the consistency between the thermal and the electrical results [2]. The iterations can add extra burden to the computation.

The Latency Insertion Method (LIM) was proposed as an alternative algorithm of SPICE [6]. Compared to SPICE, LIM follows time-domain leap-frogging procedure and does not rely on matrix calculations. LIM thus exhibits linear computational complexity, which makes it faster than SPICE for large-scale circuits. In previous works, LIM has been implemented in the electro-thermal analysis of 3-dimentional integrated systems [7].

To simulate non-linear devices like FinFET, a circuit model must be incorporated. For FinFET, the only industry-standard compact model is the BSIM-CMG model [1], [8]. In previous works [9]–[11], it was proven that LIM is a fast and reliable method for FinFET circuit simulation by using the BSIM-CMG model. In this paper, we proposed a LIM algorithm that leapfrogs between electrical and thermal equations. The proposed method can thus conduct electro-thermal simulation of FinFET with SHE.

II. LATENCY INSERTION METHOD

LIM is a time-domain simulation algorithm for circuits [6]. The circuit simulated by LIM must follow the topologies in Fig. 1. A capacitor must be in between each node and the ground. An inductor should be in each branch. If the capacitors and the inductors do not exist in the original circuit, ones with small values must be inserted in.

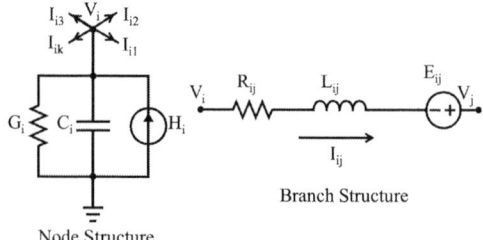

Fig. 1: Node and branch topologies for LIM.

Here, we focus on the Voltage-in-Current (VinC) LIM with improved stability [12]. The VinC LIM formulas for updating the voltages and the currents are

$$V_i^{n+1} = \left(\frac{C_i}{\Delta t} + G_i\right)^{-1} \left(\frac{C_i}{\Delta t} V_i^n - \sum_{k=1}^{M_i} I_{ik}^{n+1} + H_i^{n+1}\right),$$
(1)

979-8-3503-5124-8/24 $31.00 © 2024 IEEE

$$I_{ij}^{n+1} = \cfrac{\cfrac{L_{ij}}{\Delta t}I_{ij}^n + \cfrac{\frac{C_i}{\Delta t}V_i^n - \sum_{k=1,k\neq j}^{M_i} I_{ik}^{n+1} + H_i^{n+1}}{G_i + \frac{C_i}{\Delta t}} - \cfrac{\frac{C_j}{\Delta t}V_j^n - \sum_{k=1,k\neq i}^{M_j} I_{jk}^{n+1} + H_j^{n+1}}{G_j + \frac{C_j}{\Delta t}} + E_{ij}^{n+1}}{\cfrac{L_{ij}}{\Delta t} + R_{ij} + (G_i + \frac{C_i}{\Delta t})^{-1} + (G_j + \frac{C_j}{\Delta t})^{-1}}, \quad (2)$$

where V_i^{n+1} is the voltage of node i at time point t_{n+1}, and I_{ij}^{n+1} is the current between node i and j at t_{n+1}. M_i represents the number of the branches connected to node i. Δt is the time step.

During the transient simulation, at time point t_n, the currents I_{ij}^n are updated first, then V_i^n. The simulation will move on to t_{n+1} and follow the same order.

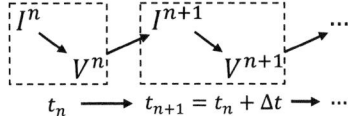

Fig. 2: Leap-frogging procedure of LIM.

III. THE SHE MODEL AND THERMAL LIM EQUATION

Electrical and thermal simulations share similar topologies as shown in Fig. 3, where R_{th} is the thermal resistor, Q is the heat flow, and T is the temperature at the node. Thus, similar LIM equations can be established for thermal problems.

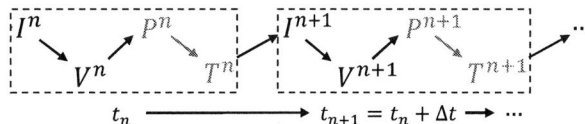

Fig. 3: Electrical-thermal isomorphism.

The equivalent circuit model for FinFET SHE calculation is plotted in Fig. 4. The thermal resistor R_{th} and the thermal capacitor C_{th} are connected in parallel with a current source. The values of R_{th} and C_{th} are calculated through the BSIM-CMG equations [3], [13]. The value of the current source equals the power in the FinFET device which is calculated by $I_{DS} \times V_{DS}$. The voltage value of the lower node equals the environmental temperature of the chip. And the voltage value of the upper node equals the device temperature, which is the target temperature value for the thermal simulation.

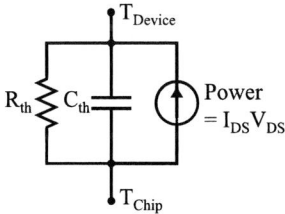

Fig. 4: Model of self-heating effect of FinFET.

While the model is similar to the node structure of LIM, the value of the device temperature can be calculated by

$$T_{Device}^{n+1} = \left(\frac{C_{th}}{\Delta t} + \frac{1}{R_{th}}\right)^{-1} \\ \times \left(\frac{C_{th}}{\Delta t}T_{Device}^n + Power^{n+1} + \frac{1}{R_{th}} \times T_{Chip}\right), \quad (3)$$

where T_{Device}^{n+1} represents the device temperature at the time point t_{n+1}, and $Power$ is the value of the device power which is plugged into the current source. T_{Chip} is the chip temperature at the lower node in Fig. 4. Here, we assume that the chip temperature is constant.

IV. ELECTRO-THERMAL LIM ALGORITHM

The proposed LIM algorithm leapfrogs between electrical and thermal simulation as shown in Fig. 5. After the voltages are achieved, the power in the devices is calculated and inserted into the equation (3) for the temperature calculation. Then both the voltages and the temperatures are inserted into the BSIM-CMG equations as inputs for current calculations.

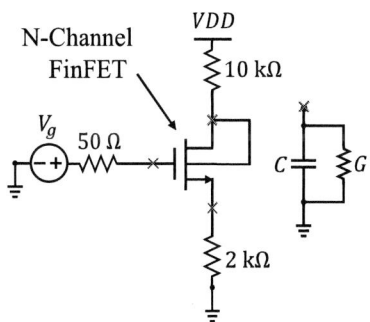

Fig. 5: Leap-frogging procedure of electro-thermal LIM.

For non-linear devices like FinFETs, stable results can be achieved without iteration methods by using small time steps [9], [10]. However, introducing the Newton Raphson iteration method to current calculations can ensure stable results with much bigger time steps, and thus increase the overall simulation speed [14]. It is worth mentioning that LIM only performs iteration to individual devices, while SPICE iterates the entire circuit. Thus, iterations in LIM do not cause as heavy computational burden as in SPICE.

V. SIMULATION RESULTS

The proposed method is tested on an example circuit with single N-channel FinFET as shown in Fig. 6. The simulation results are compared with Cadence Spectre.

Fig. 6: The tested circuit with single N-channel FinFET.

The BSIM-CMG parameters of the transistor is modified from the PTM-MG model cards of 10 nm technologies [15]. The chip temperature and the initial device temperature are set to the room temperature of 300.15 K. The input signal at the gate is set as a periodic pulse with rise and falling time of 1 ps, width of 3 ps, and period of 5 ps.

The simulated voltages at different temperatures are plotted in Fig. 7. Good agreement can be found between LIM and Spectre in electrical simulation at different temperatures.

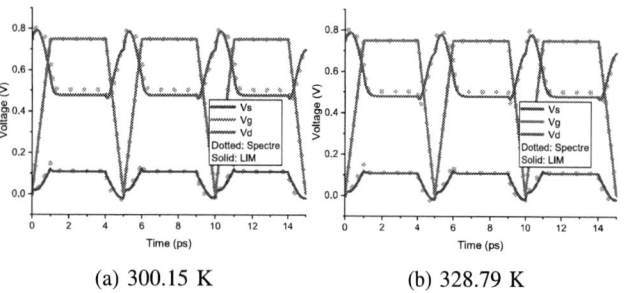

(a) 300.15 K   (b) 328.79 K

Fig. 7: Simulation results of the voltages at different temperatures.

The increase of the temperature with the change of the device power is plotted in Fig. 8. With the input signal turned on continuously, the increasing temperature will eventually saturate as shown in Fig. 9. The simulated results of LIM match with Spectre for both electrical and thermal simulations.

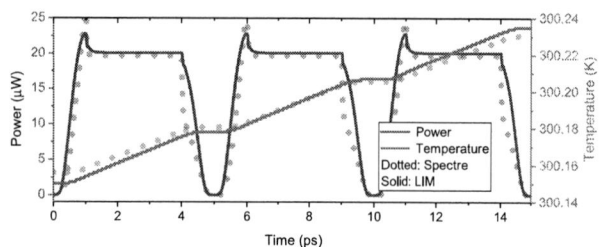

Fig. 8: Simulation results of the device power and temperature.

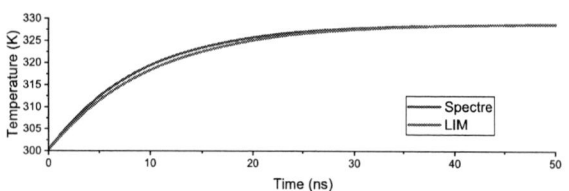

Fig. 9: Simulation device temperature in large time scale.

The simulation time for the proposed LIM algorithm to perform the simulation in Fig. 9 is 8.5 s, while Spectre takes 13.3 s. Thus, the proposed method is 1.56 times faster than the commercial software for the single transistor example. With linear computational complexity, with the scale of the circuit increases, the speed of LIM compared to SPICE should increase correspondingly.

VI. CONCLUSIONS

This paper proposed the LIM algorithm for fast electro-thermal simulation of FinFETs with self-heating effects. The LIM equation for the SHE equivalent circuit model is established. The proposed method leapfrogs between the electrical and thermal equations. Simulation results have good agreement with commercial software for both voltages and temperatures. The proposed method has proven to be faster than the commercial SPICE simulators.

ACKNOWLEDGMENT

This material is based upon work supported by the U.S Army Small Business Innovation Research (SBIR) Program office and the U.S. Army Research Office under Contract No.W911NF-15-P-0005 and by Zhejiang University under the Dynamic Research Enterprise for Multidisciplinary Engineering Sciences (DREMES) program.

REFERENCES

[1] D. Hisamoto *et al.*, "FinFET-a self-aligned double-gate MOSFET scalable to 20 nm," *IEEE Trans. Electron Devices*, vol. 47, no. 12, pp. 2320-2325, Dec. 2000.

[2] Y. G. Chauhan *et al.*, *FinFET modeling for IC simulation and design: using the BSIM-CMG standard*, San Diego, CA, USA: Academic, 2015.

[3] O. Prakash *et al.*, "Transistor self-heating: the rising challenge for semiconductor testing," *2021 IEEE 39th VLSI Test Symposium (VTS)*, San Diego, CA, USA, Apr. 2021.

[4] H. Jiang *et al.*, "The impact of self-heating on HCI reliability in high-performance digital circuits," *IEEE Electron Device Lett.*, vol. 38, no. 4, pp. 430-433, Apr. 2017.

[5] W. Jin *et al.*, "SOI thermal impedance extraction methodology and its significance for circuit simulation," *IEEE Trans. Electron Devices*, vol. 48, no. 4, pp. 730-736, Apr. 2001.

[6] J. E. Schutt-Aine, "Latency insertion method (LIM) for the fast transient simulation of large networks," *IEEE Trans. Circuits Syst. I. Fundam. Theory Appl.*, vol. 48, no. 1, pp. 81-89, Jan. 2001.

[7] D. Klokotov and J. E. Schutt-Ainé, "Latency insertion method (LIM) for electro-thermal analysis of 3-D integrated systems at pre-layout design stages," *IEEE Trans. Compon. Packag. Manuf. Technol.*, vol. 3, no. 7, pp. 1138-1147, July 2013.

[8] J. P. Duarte *et al.*, "BSIM-CMG: standard FinFET compact model for advanced circuit design," *ESSCIRC Conference 2015 - 41st European Solid-State Circuits Conference (ESSCIRC)*, Graz, Austria, Sept. 2015.

[9] Y. Zhou and J. E. Schutt-Ainé, "Latency insertion method for FinFET DC operating point simulation based on BSIM-CMG," *2022 IEEE Electrical Design of Advanced Packaging and Systems (EDAPS)*, Urbana, IL, USA, Dec. 2022.

[10] Y. Zhou and J. E. Schutt-Ainé, "Latency insertion method for FinFET simulation incorporating parasitic source/drain resistances," *2024 IEEE 10th Electronics System-Integration Technology Conference (ESTC)*, Berlin, Germany, Sept. 2024.

[11] Y. Zhou *et al.*, "Signal and power integrity co-simulation of chiplet-to-chiplet channel based on latency insertion method," *2024 IEEE 28th Workshop on Signal and Power Integrity (SPI)*, Lisbon, Portugal, May 2024.

[12] W. C. Chin *et al.*, Tan, K. H., P. Goh, and M. F. Ain, "Voltage-in-current formulation for the latency insertion method for improved stability," *Electron. Lett.*, vol. 52, no. 23, pp. 1904-1906, Nov. 2016.

[13] S. Khandelwal *et al.*, "BSIM-CMG 111.0.0 multi-gate MOSFET compact model technical manual," University of California, Berkely, CA, 2019.

[14] J. Schutt-Aine and P. Goh, "LIM algorithms for MOSFET models," *2019 IEEE 10th Latin American Symposium on Circuits & Systems (LASCAS)*, Armenia, Colombia, Feb. 2019.

[15] S. Sinha *et al.*, "Exploring sub-20nm FinFET design with predictive technology models," *DAC Design Automation Conference 2012*, San Francisco, CA, USA, Jun. 2012.

979-8-3503-5124-8/24 $31.00 © 2024 IEEE

Eye-Diagram Edge Estimation (EEE) Network for Through Silicon Via Design in Next-Generation High Bandwidth Memory

Hyunjun An, Junghyun Lee, Keeyoung Son, Seonguk Choi, Taein Shin, Keunwoo Kim, Jiwon Yoon, Taesoo Kim, Jungmin Ahn, Hyunah Park, Haeseok Suh and Joungho Kim
School of Electrical Engineering, Korea Advanced Institute of Science and Technology (KAIST)
Daejeon, Republic of Korea
E-mail: anhyunjun@kaist.ac.kr

Abstract—In this paper, we propose an eye-diagram edge estimation (EEE) network for fast and accurate eye diagram prediction, which is a novel approach in through silicon via (TSV) design in high bandwidth memory (HBM). As HBM generation is developed, the number of input/outputs (I/Os) and data rate have increased for high bandwidth, causing degradation of signal integrity (SI) in TSV structures. However, SI evaluation takes much time due to time-consuming SI simulations including the eye diagram simulation. Therefore, there is a need for fast eye diagram estimation for TSV structures optimization. The proposed EEE network utilizes only four voltage values at each time step to predict the eye diagram edge. In addition, the EEE network does not require additional time-domain and frequency-domain simulations, enabling a significant reduction in model size and facilitating training with a small dataset and shallow models such as artificial neural network (ANN). As a result, the ANN-based EEE network generated an eye diagram edge with an approximate 1% error in under 1 ms, demonstrating high accuracy and time-efficiency. This novel approach has potential applications in other design scenarios.

Keywords—Artificial neural network, eye diagram, high bandwidth memory, signal integrity, through silicon via

I. INTRODUCTION

With the sharp increase in the parameter size of generative artificial intelligence, the demand for high bandwidth memory (HBM) is also surging. HBM consists of stacked dynamic random access memories (DRAMs), and they are vertically connected by through silicon vias (TSVs). It requires meticulous design because TSV structures are vulnerable to signal integrity (SI) issues such as crosstalk [1]. In the next-generation of HBM, with an increased number of input/outputs (I/Os) and higher data rates, SI becomes more critical. Therefore, SI-aware design becomes even more crucial for TSV channel design, and thereby numerous eye diagram simulations are required.

However, due to high computational cost of eye diagram simulation, it is important to develop the methods to estimate the eye diagram through a fast and accurate scheme. Previous research has explored the methods such as peak distortion analysis (PDA) and machine learning techniques for eye diagram estimation [2], [3], [4]. However, these methods, although successful in improving time efficiency, had the limitation of requiring additional simulation data such as single

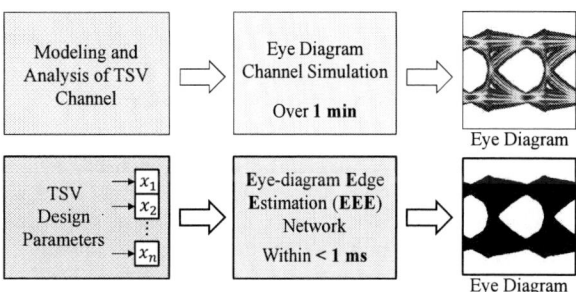

Fig. 1. Eye-diagram edge estimation (EEE) network for fast and accurate eye diagram estimation of through silicon via (TSV) channels.

bit response (SBR), S-parameter and far-end crosstalk (FEXT) simulations.

In this paper, we propose an eye-diagram edge estimation (EEE) network for fast and accurate eye diagram prediction, aiming to improving time efficiency. Unlike previous works, the proposed method omits additional simulations and predicts the eye diagram solely based on physical dimensions of TSV, as illustrated in Fig. 1. We utilized only the four voltage values at each time step to predict the eye diagram edge, which dramatically reduced the model size compared to existing eye diagram prediction models. By focusing on the eye diagram edge alone, we improved model size efficiency because all metrics used to evaluate the eye diagram can be determined from its edge.

II. PROPOSAL OF EYE-DIAGRAM EDGE ESTIMATION (EEE) NETWORK FOR THROUGH SILICON VIA (TSV) DESIGN

In this chapter, the details of the proposed EEE network is presented. The model only utilizes physical dimensions of TSV as input and predicts the voltage values of the eye diagram edge. For the implementation, multivariate polynomial regression and artificial neural network (ANN) are suggested.

A. Eye diagram edge extraction for data compression

The proposed EEE network focuses on the situation that target bit error rate (BER) is fixed during the analyzing eye diagram metrics. In that case, the essential information in an eye diagram lies in its edge, not in the inner part of eye diagram. By analyzing only the edges of the eye diagram, representative evaluation metrics of eye diagram such eye width and height,

979-8-3503-5124-8/24 $31.00 © 2024 IEEE

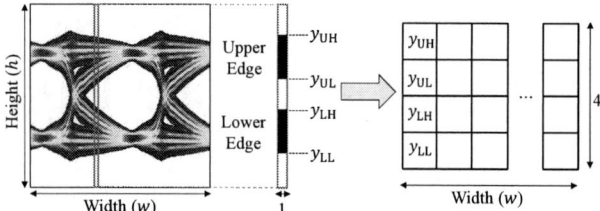

Fig. 2. Eye diagram edge values extraction from eye diagram.

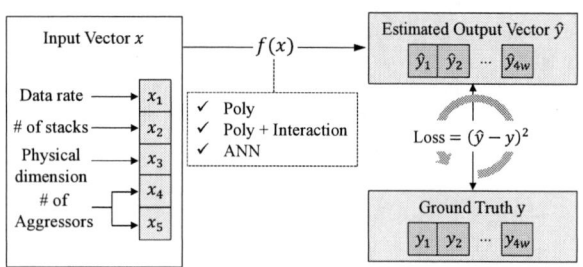

Fig. 3. Overview of EEE network implementation and training process.

overshoot can be determined. Additionally, the eye contour obtained from the eye diagram edge can be used to assess whether it meets the target specifications, allowing for pass/fail judgments. Therefore, we propose the EEE network to predict the positions corresponding to the edge of the eye diagram. This methodology utilizes only four voltage values per time step, determined by the upper and lower edge, as depicted in Fig. 2. This approach offers a revolutionary reduction in computational requirements for eye diagram simulation, which traditionally demands a substantial amount of computing cost due to the need to accumulate multiple symbols.

From the perspective of image processing, the EEE network allows dramatically reduced input data size. By utilizing only the edge of the eye diagram, the input data size is reduced to a $4 \times w$-sized matrix. Additionally, while the eye diagram image possesses a size of 24 bits per pixel, the eye diagram edge requires only 1 bit per pixel by omitting red, green, blue (RGB) color components and their 8-bit intensity information. This approach results in a dramatic image size reduction of approximately 96%. For instance, a 786 KB-sized eye diagram is compressed into a 0.7 KB-sized image through the EEE network, resulting in a significant size reduction of 99.97% for image processing. This compression effectively reduces the model size and leads to a dramatic decrease in inference time.

B. Implementation of the EEE network

In the EEE network, TSV design parameters such as data rate, number of DRAM stacks, radius of TSV, and the number of aggressors are incorporated into the input vector. The output vector is a $4 \times w$ length vector representing the pixel positions of the eye diagram. Therefore, a simple network that maps a vector to a vector is proposed, such as multivariate polynomial regression and artificial neural network (ANN), as illustrated in Fig. 3. Each model can be expressed in mathematical terms as follows:

$$\hat{y}_i = \sum_{\substack{j=1, \\ p \in \{1,2,\cdots,p_{max}\}}}^{n} c_{i,j}^{(p)} x_j^p + b_i \qquad (1)$$

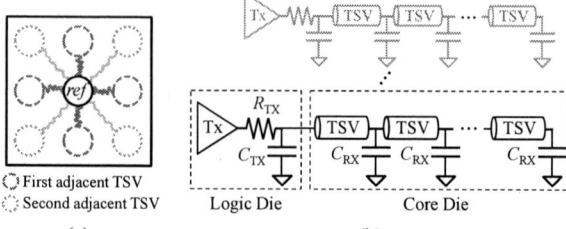

(a)   (b)

Fig. 4. (a) Configuration of TSV array with adjacent aggressors causing crosstalk. (b) Eye diagram simulation setup for HBM TSV interface.

TABLE I
RANGE OF TSV DESIGN PARAMETERS FOR NEXT-GENERATION HBM

Design parameter	Design range	Step size
Data rate (Gb/s)	3-9	3
Number of DRAM stacks	8-16	4
Radius of TSV (μm)	3-5	1
Number of adjacent aggressors	0-2	1
Number of second adjacent aggressors	0-2	1

$$\hat{y}_i = \sum_{\substack{j=1, \\ p \in \{1,2,\cdots,n\}}}^{n} c_{i,j}^{(p)} x_j^p + b_i + \sum_{\substack{j<k, \\ j,k \in \{1,2,\cdots,n\}}} c_{i,j,k} x_j x_k \qquad (2)$$

$$\hat{y}_i = \sum_{j=1}^{h} c_{h_{i,j}} \text{ReLU}(h_j) + b_{h_i}$$

$$\text{where } h_i = \sum_{j=1}^{n} c_{x_{i,j}} x_j + b_{x_i} \qquad (3)$$

As expressed in (1), multivariate polynomial regression focuses on determining coefficients based on the linear relation between the input and output vectors. On the other hand, as shown in (2), incorporating interaction terms between the inputs is a viable approach to achieving non-linearity characteristics. In case of the ANN represented in (3), the model learns the relationship based on the non-linear characteristics provided by rectified linear unit (ReLU). In this paper, we present and analyze the EEE networks composed of these models and conduct a performance comparison, focusing on linearity between the input and output vectors.

When training the EEE network, both input and output vectors were normalized in order to facilitate convergence. Training was conducted using the mean squared error (MSE) of the differences in voltage values as a loss function. After confirming the convergence of the loss, training was terminated. To evaluate model performance, an eye diagram was generated and the errors in eye width and height, overshoot, and eye-opening area were measured against the test dataset.

III. DATA GENERATION

For eye diagram simulation, circuit modeled TSV channel simulation was conducted through ADS SPICE simulation tool [5]. Fig. 4(a) shows 3×3 configuration of TSV array with first and second adjacent aggressors, and Fig. 4(b) illustrates simulation setup for eye diagram generation [6]. The TSVs were figured with a height of 50 μm, a silicon dioxide thickness of 0.5 μm and a pitch of 50 μm. In the eye diagram setup, the resistance of TX driver (R_{TX}) and the capacitance of both TX driver (C_{TX})

979-8-3503-5124-8/24 $31.00 © 2024 IEEE

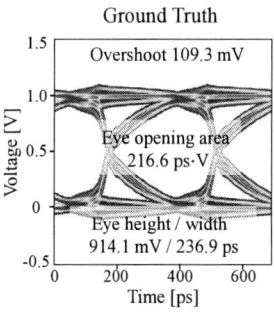 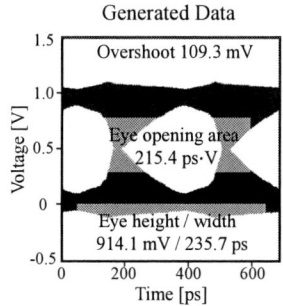

Fig. 5. Ground-truth and estimated eye diagram through ANN.

TABLE II
ESTIMATION ACCURACY OF EACH MODELS FOR EEE NETWORK

Model	Eye width (error)	Eye height (error)	Eye-opening area (error)	Over-shoot (error)
Poly	5.9%	1.1%	6.1%	19.2%
Poly + Interaction	4.8%	1.2%	1.5%	14.5%
ANN	0.3%	0.3%	0.6%	1.9%

TABLE III
MODEL SIZE AND TIME CONSUMPTION OF EACH MODELS FOR EEE NETWORK

Model	Parameter size	Training time (s)	Generation time (ms)
Poly	22.5 K	11.9	0.31
Poly + Interaction	43.0 K	19.9	0.43
ANN	527.9 K	55.1	0.69

and RX driver (C_{RX}) were set to 50 Ω, 0.1 pF and 0.1 pF, respectively.

To generate dataset, five design parameters were swept, namely: data rate, number of DRAM stacks, radius of TSV, number of adjacent aggressors, and number of second adjacent aggressors. Considering the higher data rate, increased number of DRAM stacks, reduced size of TSVs in next-generation HBM, the design ranges were determined as shown in Table I. To incorporate the effects of crosstalk, the number of aggressors was also considered as a parameter. The other fixed parameters assumed the standards of the next-generation HBM, such as low voltage swing, and determined the pitch between TSVs by assuming technological advancements such as hybrid bonding.

In total, 405 eye diagrams were collected. Essentially, only the eye edge pixel information is extracted, resulting in a dataset consisting of a 4×512 size matrix, which is equivalent to a vector of size 2048. Thus, each eye diagram sample is transformed into an input vector of size 5 and an output vector of size 2048, forming the dataset.

IV. VERIFICATION OF THE PROPOSED METHOD

After training, the accuracy of the EEE network was measured using the test dataset. Relative error was obtained using eye width, height, overshoot and eye-opening area as metrics, as indicated in Table II. We employed three models: second-order multivariate polynomial regression and the same with added interaction terms, and an ANN with a hidden layer size of 256. In the case of multivariate polynomials, the best-performing second order regression was employed, and the size of the hidden layer in ANN was similarly determined.

When analyzing the performance of each models, the following observations can be made. Firstly, in the case of models utilizing ANN, it exhibited the highest accuracy, as illustrated in Fig. 5. It is presumed to be attributed to its effective incorporation of nonlinearity, characteristic of neural networks. For polynomial models, the addition of interaction terms to introduce nonlinearity resulted in better performance compared to models lacking such features. For all models, the significant error in over-shoot is attributed to its pronounced nonlinearity and the small scale, leading to a sensitive response to even a 1-pixel error.

When utilizing the Intel i7-9700 CPU, it was observed that for all models, training was terminated within 1 minute, and the inference time was also confirmed to be at a similar level of around 1 ms in all models, as shown in Table III. Considering that eye diagram simulation typically takes about 1 min, the dramatic reduction in time is noteworthy. Furthermore, it's remarkable that despite training on a small dataset, the interpolation of eye diagrams achieves high accuracy for variations in design parameters within a certain range.

V. CONCLUSION

In this paper, we proposed EEE network for eye diagram estimation which significantly enhanced time efficiency while maintaining high accuracy in TSV channels of HBM. Also, we showed EEE network can be trained through small dataset and shallow model such as ANN. Proposed network has a potential to be extended to other transmission line structures with various design parameters. Furthermore, it can be utilized for reward estimation system for iterative optimization algorithm, including reinforcement learning, genetic algorithm and generative flow network.

ACKNOWLEDGMENT

We would like to acknowledge the technical support from ANSYS Korea. This work was supported by Samsung Electronics Co., Ltd (Contract ID: MEM230315_0004).

REFERENCES

[1] K. Kim *et al.*, "Deep Reinforcement Learning-based Through Silicon Via (TSV) Array Design Optimization Method considering Crosstalk," *2020 IEEE Electrical Design of Advanced Packaging and Systems (EDAPS)*, Shenzhen, China, 2020, pp. 1-3.

[2] B. K. Casper, M. Haycock, and R. Mooney, "An accurate and efficient analysis method for multi-Gbps chip-to-chip signaling," in *Proc. IEEE Symp. VLSI Circuits*, pp. 54-57, Jun. 2002.

[3] N. Ambasana et al., "Eye Height/Width Prediction From *S*-Parameters Using Learning-Based Models," in *IEEE Transactions on Components, Packaging and Manufacturing Technology*, vol. 6, no. 6, pp. 873-885, June 2016.

[4] J. Lee *et al.*, "Adaptive Gramian-Angular-Field Segmentation Integration Based Generative Adversarial Network (AGSI-GAN) for Eye Diagram Estimation of High Bandwidth Memory (HBM) Interposer," *2023 IEEE 32nd Conference on Electrical Performance of Electronic Packaging and Systems (EPEPS)*, Milpitas, CA, USA, 2023, pp. 1-3.

[5] J. Cho *et al.*, "Modeling and Analysis of Through-Silicon Via (TSV) Noise Coupling and Suppression Using a Guard Ring," in *IEEE Transactions on Components, Packaging and Manufacturing Technology*, vol. 1, no. 2, pp. 220-233.

[6] H. Kim *et al.*, "Signal Integrity Analysis of Through-Silicon Via (TSV) With a Silicon Dioxide Well to Reduce Leakage Current for High-Bandwidth Memory Interface," in *IEEE Transactions on Components, Packaging and Manufacturing Technology*, vol. 13, no. 5, pp. 700-714, May 2023.

979-8-3503-5124-8/24 $31.00 © 2024 IEEE

Automated Accurate Quadratic Formulation of Nonlinear Circuits

Germin Ghaly
Dept. of Electronics
Carleton University
Ottawa, Canada
germinghaly@cmail.carleton.ca

Emad Gad
Sch. of Electrical Eng. and Computer Science
University of Ottawa
Ottawa, Canada
egad@uottawa.ca

Michel Nakhla
Dept. of Electronics
Carleton University
Ottawa, Canada
michelnakhla@cunet.carleton.ca

Abstract—**A new approach to automate the modelling of general nonlinear circuits in a quadratic type nonlinearity is presented. The proposed quadratic model is characterized by high accuracy when used in numerical simulation such as time marching.**

Index Terms—**Circuit Simulation, Computer-Aided Design, Alternative Circuit Formulation**

I. Introduction

Model-order reduction (MOR) techniques have been used as a methodology to reduce the computational time associated with simulating complex circuits [1]. Those techniques have been well-established in the domain of linear circuits. There has nevertheless been a sustained interest to push MOR into handling the simulation of complex nonlinear circuits. Projection-based reduction approaches have been one of the most widely used in the literature for linear circuit MOR. One of the major impediments to using the projection framework of MOR in nonlinear circuits has been the presence of nonlinear vector (a vector whose entries are nonlinear functions of the circuit state variables $\mathbf{f}(\mathbf{x})$) that captures the contributions of the nonlinear components to the currents and charges of the circuit. In contrast to the linear components, which are always modelled by matrices where the projection yields matrices of reduced sizes, that are used throughout the simulation operations, the projection of the nonlinear vector forces a mapping process from the reduced domain to the original (large) domain and then back again to the reduced domain, in every such operation, typically eroding the efficiency of the MOR.

There are two approaches that have been proposed to handle the above issue in the context of MOR of nonlinear circuits. The first is known as the discrete empirical interpolation method (DEIM) [2]. The key operating principle in DEIM is based on the assumption that the full entries in $\mathbf{f}(\mathbf{x})$ can be represented from a linear combination of few of those entries (interpolation entries). Based on that assumption, DEIM creates an auxiliary projection matrix that projects the full set of entries onto the reduced space spanned by the interpolation entries.

An earlier approach, not based on the DEIM assumption, relies on approximating the entries in $\mathbf{f}(\mathbf{x})$ as multi-dimensional polynomials [3]. However, the limited ability of the polynomial approximation approach to capture a general strong nonlinear behavior has instigated seeking alternatives that represent the nonlinear behavior more robustly. One such alternative was based on an idea that reformulates specific types of nonlinearity into a quadratic form [4]. That approach operates through augmenting the system arising from the modified nodal analysis (MNA) with additional variables selected so that the final form of nonlinearity becomes quadratic in the form $\mathbf{G}(\mathbf{x} \otimes \mathbf{x})$. Although the latter approach is an exact representation for the original nonlinearity that is suitable for projection MOR [5], it remained limited in two issues. The first issue is an inaccuracy observed in the numerical simulation with time-domain solvers. The second issue is that this approach lacks a general framework to automate the process of creating the quadratic formulation. That drawback made it limited in the range of its application to few circuits [6].

The objective of the approach proposed in this paper addresses these issues. First, the problem of the inaccuracy emerging during the numerical simulation is handled by a staged type of quadratic formulation. Second, it proposes the usage of the idea of rooted trees to automate the building process of the quadratic formulation. The rest of the paper is organized as follows: Section II discusses the difficulties of projection based MOR with the general MNA formulation. The existing alternative formulations are shown in Section III, while the proposed automated formulation is demonstrated and verified in Sections IV and V, respectively. Finally, the conclusion is presented in Section VI.

II. Projection-based MOR: Difficulties with the General Formulation

The classical and common method of formulating the general nonlinear circuits in the mathematical domain is based on the MNA which takes the form of a system of a mixed set of differential-algebraic equations,

$$\mathbf{C}\frac{\mathrm{d}\mathbf{x}(t)}{\mathrm{d}t} + \mathbf{G}\mathbf{x}(t) + \mathbf{f}(\mathbf{x}(t)) = \mathbf{b}(t) \quad (1)$$

where $\mathbf{C}$ and $\mathbf{G}$ are matrices in $\mathbb{R}^{N \times N}$ that represent the memory and memoryless elements, respectively, and

979-8-3503-5124-8/24 $31.00 © 2024 IEEE

$\mathbf{f}(\mathbf{x}(t)): \mathbb{R}^N \to \mathbb{R}^N$ is a vector with entries given by non-linear functions that capture the contributions of the nonlinear devices to the circuit currents, charges and fluxes. In addition, $\mathbf{b}(t)$ is a vector that collects the independent stimuli.

The regular course in the projection-based MOR typically involves creating an orthogonal basis $\mathbf{Q} \in \mathbb{R}^{N \times q}$ with $q \ll N$. Next, and through performing a change of variables replacing $\mathbf{x}(t)$ with $\mathbf{Q}\hat{\mathbf{x}}(t)$, and pre-multiplying both sides of (1) with $\mathbf{Q}^\top$, a reduced system of the form

$$\hat{\mathbf{C}}\frac{\mathrm{d}\hat{\mathbf{x}}(t)}{\mathrm{d}t} + \hat{\mathbf{G}}\hat{\mathbf{x}}(t) + \hat{\mathbf{f}}(\hat{\mathbf{x}}(t)) = \hat{\mathbf{b}}(t) \tag{2}$$

is obtained, where $\hat{\mathbf{C}} = \mathbf{Q}^\top \mathbf{C}\mathbf{Q}$, $\hat{\mathbf{G}} = \mathbf{Q}^\top \mathbf{G}\mathbf{Q}$, $\hat{\mathbf{b}}(t) = \mathbf{Q}^\top \mathbf{b}(t)$ and $\hat{\mathbf{f}}(\hat{\mathbf{x}}(t)) = \mathbf{Q}^\top \mathbf{f}(\mathbf{Q}\hat{\mathbf{x}}(t))$. The particular approach of computing the basis $\mathbf{Q}$ and its effect on the accuracy of the reduced system is not the main focus in this work. However, it is the computation of the reduced system that provides the main motivation in this paper. More particularly, and in contrast to the two reduced linear terms in (2), the computation of the reduced nonlinear term, $\hat{\mathbf{f}}(\hat{\mathbf{x}}(t))$, forces back and forth arithmetic operations in the original $\mathbb{R}^N$ and the reduced domain $\mathbb{R}^q$ at every time point t at which numerical simulation is performed. This fact is obvious from the need to use the full vector $\mathbf{f}(\cdot)$ before forming the reduced vector $\hat{\mathbf{f}}(\cdot)$, and represents a major drawback that subtracts from the computational efficiency of MOR.

III. EXISTING ALTERNATIVE FORMULATIONS

Different approaches have been proposed to overcome the above hurdle in MOR of nonlinear circuits. The earlier approaches relied on the approximation of the nonlinear operator $\mathbf{f}(\cdot)$ using multidimensional polynomials [3], [7]. More recent approaches sought to reformulate this nonlinear operator without involving approximation [3], [8]. The main concept of the reformulation approach can be illustrated through a simple example circuit shown in Fig. 1, of which the MNA formulation takes the form,

$$\frac{\mathrm{d}x(t)}{\mathrm{d}t} + x(t) + \exp(40x(t)) - 1 = u(t) \tag{3}$$

The alternative formulation is obtained by defining a new variable $y(t) = \exp(40x(t))$, while noting that

$$\frac{\mathrm{d}y(t)}{\mathrm{d}t} = 40 \exp(40x(t))\frac{\mathrm{d}x(t)}{\mathrm{d}t} = 40\, y(t)\frac{\mathrm{d}x(t)}{\mathrm{d}t}. \tag{4}$$

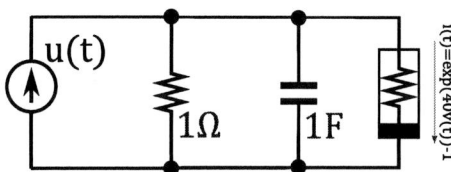

Fig. 1. An illustrative circuit example.

One technique [4] continues by substituting for $\frac{\mathrm{d}x(t)}{\mathrm{d}t}$ using (3) to form a system of quadratic equations

$$\frac{\mathrm{d}x(t)}{\mathrm{d}t} + x(t) + y(t) - 1 = u(t) \tag{5}$$

$$\frac{\mathrm{d}y(t)}{\mathrm{d}t} + 40\left(y(t)x(t) + (y(t))^2 - y(t)\right) = 40\, y(t)u(t) \tag{6}$$

Another technique [8] lumps (3) directly with (4) along with an added variable $z(t) = \frac{\mathrm{d}x(t)}{\mathrm{d}t}$ to form the system

$$
\begin{bmatrix} 1 & 1 & 0 \\ 0 & 0 & 0 \\ 0 & 0 & -1 \end{bmatrix}
\begin{bmatrix} x \\ y \\ z \end{bmatrix}
+
\begin{bmatrix} 1 & 0 & 0 \\ 0 & 1 & 0 \\ 1 & 0 & 0 \end{bmatrix}
\begin{bmatrix} \frac{\mathrm{d}x}{\mathrm{d}t} \\ \frac{\mathrm{d}y}{\mathrm{d}t} \\ \frac{\mathrm{d}z}{\mathrm{d}t} \end{bmatrix}
+
$$
$$
\begin{bmatrix} 0 & 0 & 0 \\ 0 & 0 & -40 \\ 0 & 0 & 0 \end{bmatrix} \mathbf{0}_{3\times 6}
\left(\begin{bmatrix} x \\ y \\ z \end{bmatrix} \otimes \begin{bmatrix} x \\ y \\ z \end{bmatrix} \right)
=
\begin{bmatrix} u+1 \\ 0 \\ 0 \end{bmatrix}
\tag{7}
$$

Notice that both techniques end up replacing the exponential nonlinearity with a quadratic type nonlinearity.

IV. PROPOSED AUTOMATED ALTERNATIVE CIRCUIT FORMULATION

The approach proposed in this paper has been motivated by two main issues. The first issue arises from the observations that numerical simulations of the developed formulation described in the previous section exhibit significant inaccuracies when used in a common marching on in time solvers, such as the backward Euler or trapezoidal rule. Those observations of inaccuracies are reported in Section V. The other issue addressed by the proposed approach is the lack of framework that facilitates the formulation of general (user-defined) circuits in an automated way. This section outlines how these two issues are handled by the proposed approach, while making reference to the simple circuit example of Fig. 1 for illustration purposes.

A. Quadratic Formulation for Accurate Numerical Simulation

The basic idea proposed in this section is to introduce a set of n variables, $y_1(t), \cdots, y_n(t)$. Using the simple example in Fig. 1 for illustration, those variables will be defined as follows,

$$y_1(t) = \exp\left(\frac{40}{2^{n-1}}x(t)\right) \tag{8}$$

$$y_i(t) = y_{i-1}(t)\, y_{i-1}(t), \qquad i = 2, \cdots, n \tag{9}$$

It follows that,

$$\frac{\mathrm{d}y_1(t)}{\mathrm{d}t} = \frac{40}{2^{n-1}}y_1(t)\frac{\mathrm{d}x(t)}{\mathrm{d}t} \tag{10}$$

The new formulation can then be written by combining (3), (8)-(10) and adding a new variable representing $\frac{\mathrm{d}x(t)}{\mathrm{d}t}$

$$\tilde{\mathbf{C}}\frac{\mathrm{d}\tilde{\mathbf{x}}(t)}{\mathrm{d}t} + \tilde{\mathbf{G}}_1\tilde{\mathbf{x}}(t) + \tilde{\mathbf{G}}_2\left(\tilde{\mathbf{x}}(t) \otimes \tilde{\mathbf{x}}(t)\right) = \tilde{\mathbf{u}}(t) \tag{11}$$

where $\tilde{\mathbf{x}}(t) \in \mathbb{R}^M$ denote a new set of variables that includes the original set of variables, i.e., $\mathbf{x}(t)$ augmented with the extra variables introduced above, where M in the case of the illustrative example is given by $M = N + n + 1$. Matrices $\tilde{\mathbf{G}}_1$

979-8-3503-5124-8/24 $31.00 © 2024 IEEE

and $\tilde{\mathbf{C}}$ are of dimensions $M \times M$ while the matrix $\tilde{\mathbf{G}}_2$ is of dimension $M \times M^2$. $\tilde{\mathbf{u}}(t) \in \mathbb{R}^M$ is built from the vector of independent stimuli by adding sparse vector of constants.

It is important to note here that if n is set to $n = 1$, the new approach reverts to the conventional approach in (7).

B. Automated Quadratic Formulation

Automating the process of building the quadratic formulation from the general nonlinear expressions is executed by representing the latter in the form of Rooted Trees (RT). Fig. 2 shows the RT corresponding to a hypothetical expression given by $\exp\left(0.1\, x(t)^2\right)$. Each node in the RT structure receives a call from its parent node prompting it to stamp its contribution on the quadratic formulation. It reacts to the call by inserting the required entries into the matrices $\tilde{\mathbf{C}}$, $\tilde{\mathbf{G}}_1$ and $\tilde{\mathbf{G}}_2$. The process starts at the root node of the nonlinear expression and terminates after the leaf nodes return the call from its parents.

V. NUMERICAL SIMULATIONS

The accuracy of the proposed approach was validated through using it to simulate the time-domain response of the nonlinear transmission line (NLTL) circuit shown in Fig. 3. The NLTL consists of 5 segments of R-C-diode cascaded as shown in the figure. The time-domain simulation of this circuit was performed using the trapezoidal rule (TR). Two sets of experiments were performed corresponding to two values of the amplitude of the sinusoidal input source $J(t) = A\cos(\omega t)$, namely, $A = 0.5\,\text{V}$ and $A = 2.0\,\text{V}$. In each set, three different approaches are used and compared. The first approach is based on applying the TR to the classical MNA formulation; the second and third approaches are based on applying the TR to solve the quadratic formulation for two cases: the first case has $n = 1$, which is equivalent to the previous approach [8], and the second case was for $n = 3$ as proposed in this paper. The results from the two sets of experiments for the voltage of the input node of the circuit are shown in Figs. 4 and 5.

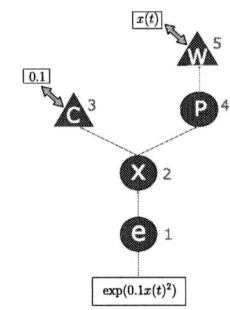

Fig. 2. Nodes of a RT representing the expression $\exp(0.1\, x(t)^2)$.

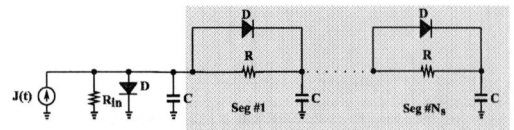

Fig. 3. Nonlinear Transmission Line.

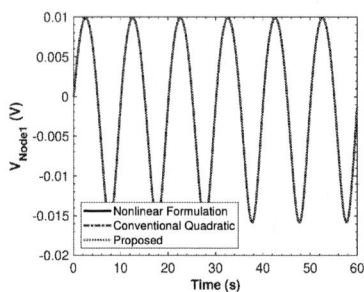

Fig. 4. Time Domain Response at $A = 0.5\,\text{V}$.

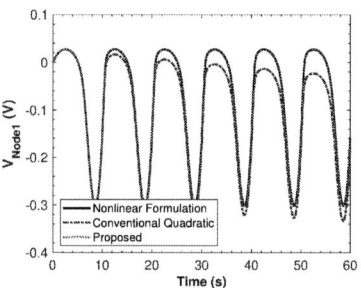

Fig. 5. Time Domain Response at $A = 2.0\,\text{V}$.

Those results show that the proposed quadratic formulation maintained good accuracy, compared with the classical MNA formulation, while those of previous approach lost accuracy for the case $A = 2.0\,\text{V}$.

VI. CONCLUSION

This paper presented a framework and a methodology aimed at casting general nonlinear circuit into a quadratic type suitable for the projection-based MOR.

REFERENCES

[1] P. Benner, W. Schilders, S. Grivet-Talocia, A. Quarteroni, G. Rozza, and L. Miguel Silveira, *Model order reduction: volume 3 applications*. De Gruyter, 2020.

[2] S. Chaturantabut and D. C. Sorensen, "Nonlinear model reduction via discrete empirical interpolation," *SIAM J. Sci. Comput.*, vol. 32, no. 5, pp. 2737–2764, 2010, doi: 10.1137/090766498.

[3] J. R. Phillips, "Projection-based approaches for model reduction of weakly nonlinear, time-varying systems," *IEEE Trans. Comput.-Aided Des. Integr. Circuits Syst.*, vol. 22, no. 2, pp. 171–187, Feb 2003, doi: 10.1109/TCAD.2002.806605.

[4] C. Gu, "QLMOR: A projection-based nonlinear model order reduction approach using quadratic-linear representation of nonlinear systems," *IEEE Trans. Comput.-Aided Des. Integr. Circuits Syst.*, vol. 30, no. 9, pp. 1307–1320, sep 2011, doi: 10.1109/TCAD.2011.2142184.

[5] B. Kramer and K. E. Willcox, "Nonlinear model order reduction via lifting transformations and proper orthogonal decomposition," *AIAA Journal*, vol. 57, no. 6, pp. 2297–2307, 2019.

[6] P. Benner, M. Hinze, and E. J. W. Ter Maten, *Model reduction for circuit simulation*. Springer, 2011.

[7] H. Liu, L. Daniel, and N. Wong, "Model reduction and simulation of nonlinear circuits via tensor decomposition," *IEEE Trans. Comput.-Aided Des. Integr. Circuits Syst.*, vol. 34, no. 7, pp. 1059–1069, Jul. 2015, doi: 10.1109/TCAD.2015.2409272.

[8] A. Elhamshary, Y. Ismail, Y. A. Aziz, and H. Ragae, "A new second order nonlinear formulation for fast-spice circuit simulation," *IEEE Access*, 2024.

Crosstalk Analysis in Add-In Card structure for High-Speed SerDes Channels with PCIe Gen6

Sungjin Yoon, Manho Lee, Kwangho Kim, Hyeongi Lee, Chulhee Cho, Youngjae Lee, Wooshin Choi,
Young-Chul Cho, Jung-Hwan Choi and Young-Soo Sohn
Samsung Electronics Co. Ltd, Hwaseong, South Korea
E-mail: sjin.yoon@samsung.com

Abstract—**With the increasing demand for high-performance computing (HPC), the significance of high-speed SerDes channels is growing. An importance of crosstalk is larger than ever, in particular, peripheral component interconnect express (PCIe) Gen6 uses pulse amplitude modulation of four-level (PAM4), not non-return-to-zero (NRZ). In this paper, minimizing crosstalk Add-in Card (AIC) based on the network parameters and specific example of enhancement with reinforcement grounding is presented. And its effectiveness of that is evaluated in NRZ and PAM4 modulations. As a result, the crosstalk effect caused by the AIC's ground return path in NRZ is negligible, but case of PAM4, it has a significant impact that can directly degrade the signal quality.**

Keywords—Connector, NRZ, PAM4, Crosstalk, AIC, PCIe, Ground return path, Signal integrity (SI)

I. INTRODUCTION

Ensuring stable signal integrity becomes increasingly challenging as data speeds in high-speed SerDes applications increase [1-2]. In particular, for PCIe Gen6, the operating speed has doubled from 32 Gb/s to 64 Gb/s, and PAM4 has been adopted to meet these requirements. PAM4 can transmit twice data at the same frequency as NRZ, with maintaining channel bandwidth. However, a voltage margin is reduced by 1/3, making it more susceptible to noise interference [3].

To achieve high data speeds, the density of system is became high, exacerbating signal quality degradation problems due to crosstalk [4-5]. To mitigate crosstalk, the root cause has to be solved through accurate analysis for whole components such as packages (PKG), printed circuit boards (PCB), connectors, and cables. Previous researches have conducted precise analysis of crosstalk occurring in systems and proposed various design guides to prevent. For instance, it was suggested that changing ground rule of arrangement of via for isolate channel and reinforcement immunity[6,7], and a way to reduce crosstalk using the additional components on connector[8]. It could be seen that the crosstalk of the components of interface studied and well known how to minimize it. However, there are few studies for exact analysis of the crosstalk impact in AIC. In previous study, ground discontinuity in the interconnector region can lead to impedance mismatch, increasing noise, crosstalk, and reflections on AIC[9].

This paper studies the impact of crosstalk generated in AIC on system SI using Enterprise and Data center SSD Form Factor (EDSFF) connectors. The target structure is three pairs of differential channels (1-victim/2-aggressor) by combining an AIC and a receptacle connector as shown in Fig. 1 (a). And the channel signaling is using a differential mode with GSSG

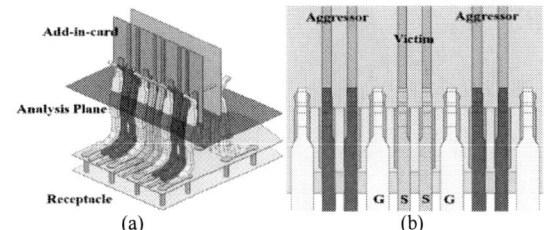

Fig. 1. (a) Design of receptacle and AIC mated in EDSFF connector and (b) description of GSSG array and Aggressors and Victim in the connector.

(Ground-Signal-Signal-Ground) as shown in Fig. 1 (b) for immunity to external noise[10-11].

The crosstalk of this structure is identified by network parameters. And, a transient simulation is performed on the 3-connector topologies to determine the impact of crosstalk on the SI for the Gen6 system. To mitigate crosstalk, design modifications to AIC are conducted and the results are compared to the original design. And also crosstalk between aggressor and victim would be shown with electric field on the analysis plane. And finally, the effectiveness of the proposal is verified by manufacturing a test coupon and performing measurements.

II. CROSSTALK ANALYSIS OF GROUND RETURN PATH IN AIC

A. Identifying the Impact of AIC's Crosstalk on PCIe Systems

To identify the impact of crosstalk within the connector on the systems, the eye opening of NRZ and PAM4 signaling is performed for a typical 3-connector PCIe topology as depicted in Fig. 2. The system operates in a differential mode, with a total end-to-end insertion loss (IL) greater than -32 dB at 16 GHz, and all connectors use EDSFF connectors. Within the connector, the AIC and receptacle are combined to transmit signals in GSSG mode. During this process, the adjacent channels share the same ground conductor, which can form a crosstalk path.

Fig. 3 (a) illustrates an AIC design with the minimum number of ground via, and it was analyzed using a 3D electromagnetic (3D-EM) simulation tool by using the AIC as part of the receptacle, as depicted in Fig. 1. The bandwidth of crosstalk was focused on up to 16 GHz, which is the fundamental frequency of PCIe Gen5 and Gen6. The crosstalk

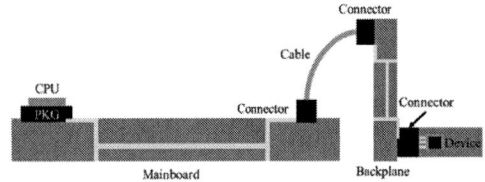

Fig. 2. Typical 3-connector PCIe topology and physical structure illustration.

979-8-3503-5124-8/24 $31.00 © 2024 IEEE

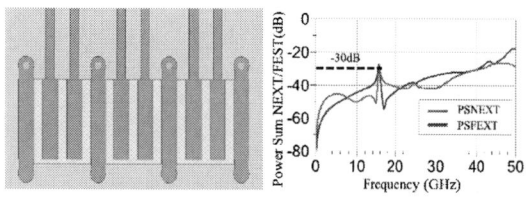

(a)   (b)

Fig. 3. Top view of (a) original AIC design, and (b) crosstalk results of original AIC.

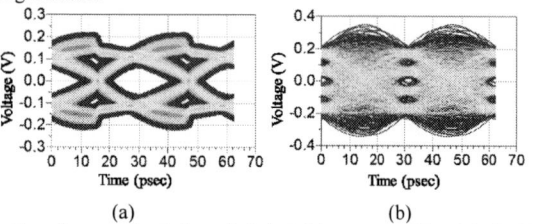

(a)   (b)

Fig. 4. Eye-diagram simulation of whole PCIe systems with original AIC : (a) NRZ signaling and (b) PAM4 signaling.

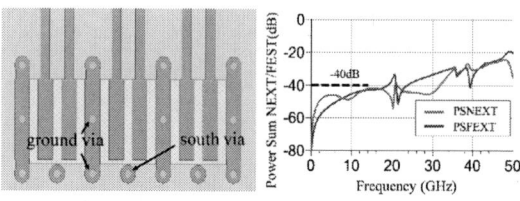

(a)   (b)

Fig. 5. Top view of (a) modified AIC design with adding ground via on pad and additional via, and (b) crosstalk results of revised AIC.

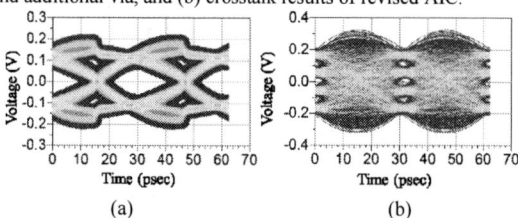

(a)   (b)

Fig. 6. Eye-diagram simulation of whole PCIe systems with modified AIC : (a) NRZ signaling and (b) PAM4 signaling.

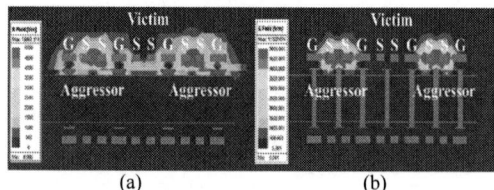

(a)   (b)

Fig. 7. Comparison of electric field on analysis plane : (a) original AIC pad design and (b) revised design.

factors can be expressed to power sum near-end crosstalk (PSNEXT) and power sum far-end crosstalk (PSFEXT), and the results are shown in Fig. 3 (b). At the 16 GHz, two factors are -30 dB or higher.

Using AIC-inclusive connector, eye diagram simulation was conducted based on PCIe Gen5/Gen6 requirement[12]. The simulation includes both Tx and Rx equalization, and three channels, consist of one victim and two aggressors, are used in simulation. Fig. 4 (a) shows the eye opening of PCIe Gen5, which meets the criteria, however, the eye closed in PCIe Gen 6 as in Fig. 4 (b). It can be seen that the crosstalk effect is negligible when using NRZ signaling, but it can severely degrade the eye opening when using PAM4 signaling with current AIC structure.

B. Adding Ground Vias in AIC pad for Crosstalk Mitigation

To mitigate crosstalk inside the connectors, ground reinforcement design is performed. Fig. 5 (a) illustrates the AIC design with three ground via and an additional south via added to each ground pad. This structure was analyzed 3D EM simulation, and its results are shown in Fig. 5 (b). At 16 GHz, the two factors are below -40 dB. It can be seen that the added ground via on AIC effectively mitigates crosstalk at 16 GHz.

Similarly, transient simulation of whole system was conducted. Fig. 6 (a) shows eye diagram of PCIe Gen5 with NRZ signaling, and the eye opening is similar to original AIC pad. In PAM 4 signaling case, the eye opening meets agreement the criteria as in Fig. 6 (b). It is a significant improvement in the PCIe Gen6 compared to original AIC pad. An enhancement also can be confirmed in the electric field on space. To identify the crosstalk paths more intuitively, electric field analysis at 16 GHz was performed for both types of AIC design. Field analysis was conducted on analysis plane which is expressed in Fig. 1 (a). It is indicated that an electric field is concentrated between the ground pad and the ground plane. Conversely, the electric field intensity is few on the victim channel as in Fig. 7 (b). Consequently, the crosstalk occurring at 16 GHz is transmitted through the ground return path and effectively mitigated through the ground via.

III. EVALUATION EFFECTIVENESS OF MITIGATING CROSSTALK ON AIC

To evaluate the previous result, in this chapter, the two types of test coupons were fabricated and eye-diagram of those two types were measured using NRG and PAM4 signaling.

A. Measurement S-parameter for test coupons

In Fig. 8 (a), (b), and (c) are corresponding to original plug, revised plug, and a receptacle board, respectively. Each test coupon consists of 3 channels which are 2 aggressors and 1 victim. By using VNA, the S-parameter for the test coupons was measured using a 4-port setup, and the measurement frequency was set from 10 MHz to 50 GHz in 10 MHz steps.

Even if the ground via are added in the AIC ground pad and lower side of differential signal pads, it is confirmed that the insertion loss (S_{21}) and return loss (S_{11}) are make few difference between two types up to 50 GHz. However, according to Fig. 9, both the PSNEXT and PSFEXT factors are above -35 dB in the original plug and below -50 dB in the revised plug at 16 GHz. It has similar tendency of the previous simulation.

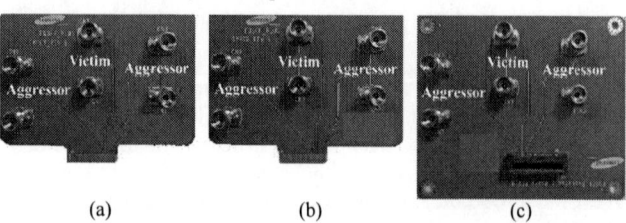

(a)   (b)   (c)

Fig. 8. Manufactured test coupon : (a) plug board of original, (b) plug board of revised and (c) receptacle board.

979-8-3503-5124-8/24 $31.00 © 2024 IEEE

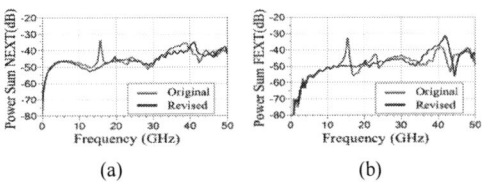

(a)   (b)

Fig. 9. Comparison with a original plug to a revised plug for (a) PSNEXT and (b) PSFEXT.

B. Eye-diagram of two type plugs using NRZ and PAM4

To verify effectiveness of mitigating crosstalk using both NRZ and PAM4, measurement for two cases of test coupons is demonstrated. A Bit Error Rate Tester (BERT) and an oscilloscope are applied for the eye diagram measurement. The BERT generates two signals to propagate two aggressors and other one signal for victim channel. On the receptacle board, only the victim channel is connected to the oscilloscope, and broadband 50 Ω was terminated opposite side of the aggressor channels. In the measurement, an equalizer was not applied because of evaluating channel characteristics only.

Fig. 10 shows the eye-diagram measurement results with enabling aggressors, and Fig. 10 (a) and (b) are corresponding to original plug and revised plug with NRZ modulation, respectively. And also, Fig. 10 (c) and (d) corresponding to original plug and revised plug with PAM4 modulation using, respectively. Notably, while results of eye-opening of two cases are similar in NRZ modulation, however, the eye diagram improves in the revised plug in the case of PAM4 modulation.

Table I presents eye-diagram measurement results for two types of plugs, it can be seen that the eye height degraded 6 % in NRZ modulation, but in PAM4 signal processing, the eye height was 24 % worse than the case of ignore crosstalk.

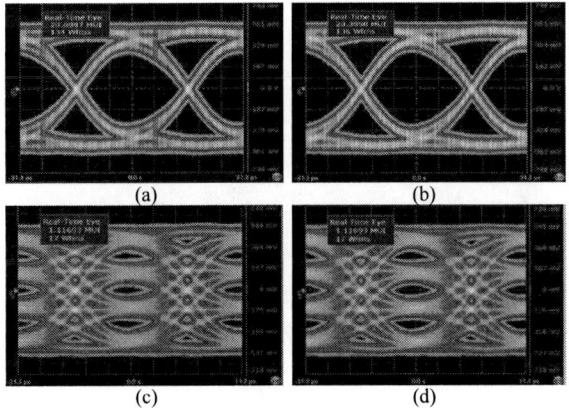

(a)   (b)

(c)   (d)

Fig. 10. Results of transient measurement : the eye diagram plot using NRZ signaling for (a) original plug and (b) revised plug test coupons, and using PAM4 signaling for (c) original plug and (d) revised plug test coupons.

TABLE I.   EYE-DIAGRAM MEASUREMENT RESULTS

		NRZ			PAM4		
		w/o aggr	w/ aggr	Delta	w/o aggr	w/ aggr	Delta
height (mV)	ORG	263	**247**	6 %	54	**37**	24 %
	REV	263	**262**	0 %	55	**51**	7 %
width (ps)	ORG	26.3	**25.9**	2 %	6.8	**5.5**	19 %
	REV	26.1	**26.0**	0 %	6.9	**6.5**	6 %

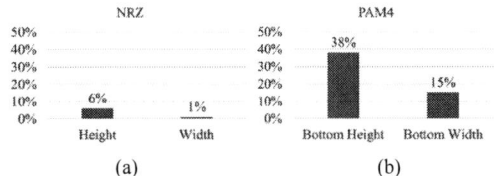

(a)   (b)

Fig. 11. Comparison of performance improvement of revised plug compared to original plug in (a) NRZ and (b) PAM4

For the case of revised AIC pad, each eye height and width were 7 % and 6 % degraded thus an immunity from crosstalk is enhanced. Fig. 11 shows the relative improvement of eye height and width in case of NRZ and PAM4 modulation, and it also can be confirmed that in NRZ modulation. In contrast, the bottom height improves by 38 % and the bottom width by 15 % in PAM4 modulation.

IV. CONCLUSION

This paper demonstrates the importance of crosstalk in high-speed serial data transmission, particularly when using PAM4 signaling. To mitigate this issue, additional vias were used to reduce ground coupling, resulting in a significant reduction in eye diagram degradation. The effectiveness was verified in experiments using actual connectors, emphasizing the importance of a ground-robust design in AIC design for high-speed serial data transmission using PAM4 based on the comparison of NRZ and PAM4 signals.

REFERENCES

[1] Vasa, Mallikarjun, et al. "Pcie gen-5 design challenges of high-speed servers." 2020 IEEE 29th Conference on Electrical Performance of Electronic Packaging and Systems (EPEPS). IEEE, 2020.

[2] Chew, Li Wern, et al. "PCB channel optimization techniques for high-speed differential interconnects." 2022 International Conference on Electronics Packaging (ICEP). IEEE, 2022.

[3] G Chen, Sherman Shan, et al. "Crosstalk Performance Analysis: ENRZ, NRZ, PAM3, and PAM4." IEEE Transactions on Signal and Power Integrity, vol. 2, pp. 53-63, 2023.

[4] Gong, Ouyang, et al. "Switching voltage regulator noise coupling to connector signal pins through near field radiation." 2014 IEEE International Symposium on Electromagnetic Compatibility (EMC). IEEE, 2014.

[5] Halligan, Matthew Scott. Maximum crosstalk estimation and modeling of electromagnetic radiation from PCB/high-density connector interfaces. Missouri University of Science and Technology, 2014.

[6] Zhang, Mu-Shui. "A high-density and low-crosstalk differential pin map for 112 Gb/s PAM4 applications." IEEE Microwave and Wireless Components Letters vol. 32, no. 6, pp. 635-638, 2022,.

[7] Kotzev, Miroslav, et al. "Crosstalk analysis in high density connector via pin fields for digital backplane applications using a 12-port vector network analyzer." 3rd Electronics System Integration Technology Conference ESTC. IEEE, 2010.

[8] Hsu, Rung-Bin, et al. "High frequency and high speed connector design for 5G applications." 2021 IEEE International Symposium on Radio-Frequency Integration Technology (RFIT). IEEE, 2021.

[9] Dsilva, Hansel Desmond, et al. "Impact of Discontinuity in the Ground on the Signal Integrity of 100G PAM4 Ethernet." 2023 Joint Asia-Pacific International Symposium on Electromagnetic Compatibility and International Conference on ElectroMagnetic Interference & Compatibility (APEMC/INCEMIC). IEEE, 2023.

[10] E. Bogatin, Signal and Power Integrity—Simplified, 2nd ed. Upper Saddle River, NJ, USA: Prentice-Hall, 2009.

[11] Montrose, Mark I. EMC and the printed circuit board: design, theory, and layout made simple. John Wiley & Sons, 2004.

[12] PCI Express Base Specification Revision 6.0, Version 1.0, January 2022.

Agile Analysis for Worst-Case Eye-Diagrams in Multi-Line Links of CoWoS Packaging

Zhu-Chen Chang, Chien-Min Lin, *Senior Member, IEEE*, and Ruey-Beei Wu, *Fellow, IEEE*

Department of Electrical Engineering and Graduate Institute of Communication Engineering, National Taiwan University, Taipei, Taiwan

r12942122@ntu.edu.tw, cmlin@ieee.org, rbwu@ntu.edu.tw

Abstract — **High-dense wires on CoWoS's redistribution layers suffer from the crosstalk to degrade the eye-opening margin. A single-line equivalent-circuit methodology with Peak Distortion Analysis is proposed to assess HBM-link performance with accuracy and efficiency.**

Keywords — Eye diagram, multi-line HBM- link, worst bit.

I. INTRODUCTION

In recent years of high-speed and high-bandwidth designs, the advanced Chip-on-Wafer-on-Substrate (CoWoS) packaging has become the most promising technology for multi-line links in High Bandwidth Memory (HBM) applications [1]. The complex wiring structure is built on the redistribution layer (RDL) within the limited space of the interposer to transmit the signals. The high-loss characteristic of the narrow-size lines and the crosstalk effects on each other degrades the signal integrity (SI).

Regarding Fig. 1 for the generic HBM links [2], the SI-performance metric is assessed on their worst-case eye diagrams through the interconnection with Pseudo Random Binary Sequence (PRBS) stimuli. The worst-case eye height indicator in the multi-line HBM links ensures all the signal transmissions across the interconnection; nonetheless, the tedious PRBS analysis is very time-consuming. An agile analysis for the eye opening margin with computational efficiency and reasonable prediction is thereof proposed.

II. ANALYTIC MODEL OF A SINGLE TRANSMISSION-LINE

An exemplary analysis for the step response of a single short-length transmission line was developed by using the equivalent lumped circuit model [3].

A. Equivalent Lumped Circuit for RDL Transmission-Line

As shown in Fig. 2, R_S and C_L are the source resistance and capacitive load, respectively, of the circuit, R_{RDL} is the series resistance, L_{RDL} is the series inductance, G_{RDL} is the parallel admittance, and C_{RDL} is the parallel capacitance. These parasitic parameters are extracted by Ansys Q2D for the transmission lines [4], while the small value G_{RDL} is ignorable when constructing the second-order RLC circuit.

Considering the step response of this RLC circuit to V_{in}, its analytic solution for V_L at the load end is attributed to the normalized resistance ζ as given by

$$\zeta = \frac{R_S + R_{RDL}}{2}\sqrt{\frac{C_L + C_{RDL}}{L}} \tag{1}$$

along with the specific time delay $T_d = \sqrt{L(C_L + C_{RDL})}$, where $L = L_{RDL}$.

This study is supported in part by MediaTek Inc., Hsinchu, Taiwan, ROC, under the Grant: 112HZA38003.

In acknowledgement of the technical support and advice from Dr. Ting-Li Yang, Cheng-Lin Huang, and Kai-Chieh Yang associated with MediaTek Inc., Hsinchu, Taiwan, ROC.

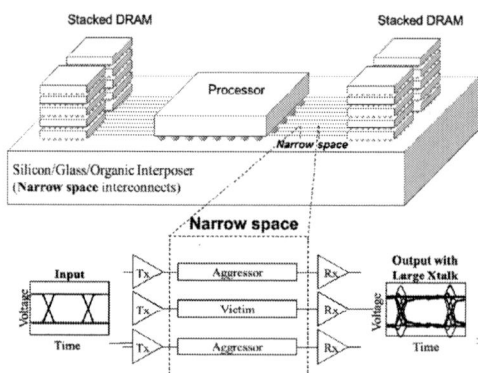

Fig. 1. Eye diagram analysis for the multi-line links [1].

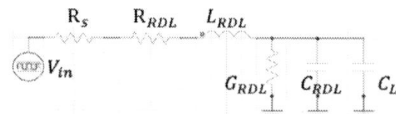

Fig. 2. Equivalent lumped circuit model of a transmission-line.

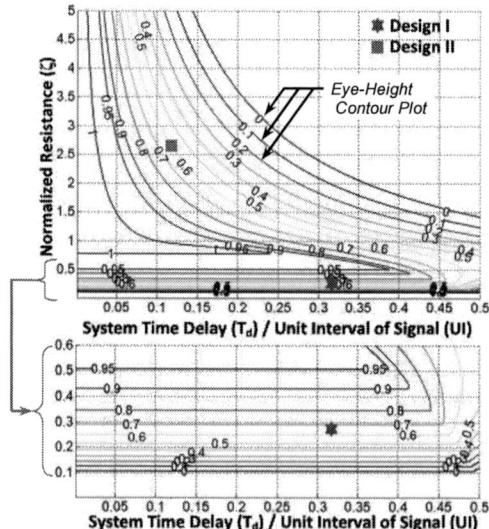

Fig. 3. Eye height solution-space contours over (T_d / UI) versus ζ [3].

For the unit-interval signals with the zero rising time at the varying data rates over the circuit-element parameters for ζ's, the single-line eye height contours are derived as shown in Fig. 3. It can be seen from (1) that adjusting R_S and C_L can significantly change ζ and T_d. The eye height trend can be intuitively predicted from Fig. 3. Therefore, while the interposer structure remains unchanged, the circuit designer can adjust the internal resistance and load according to fit the eye height requirements.

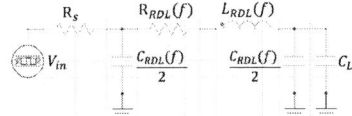

Fig. 4. Equivalent *RLC* π- model of a transmission-line.

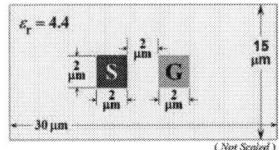

Fig. 5. Cross-sectional view of two 5-mm signal (S) and ground (G) lines.

B. Single-Line Eye Height of Improvably Simplified Model

The worst bit is referred to the input digital signal that contributes to the worst eye height [5] while the worst bit pattern is acquired by the Peak Distortion Analysis (PDA). As shown in Fig. 3 for $\zeta > 0.9$, the signal response behaves a monotonically rising trend with the corresponding worst bits of "0000000010" by the PDA analysis, where "1" represents the main cursor.

Concerning the characteristic of transmission lines with a comprehensive impedance for the more accurate eye height analysis, the aforementioned lumped model is improved by the *RLC* π-model as presented in Fig. 4. It accounts for the frequency dependency of the distributed parasitics and skin depths over a frequency range of interests so that one section of the π-model can simulate the length of the wire up to about 1/6 of the wavelength and two sections of the π-model can simulate up to 1/3 of the wavelength. For this circuit with $R_S = 50 \ \Omega$ and $C_L = 1$ pF, two 5-mm transmission-lines are designed to be analyzed as depicted in Fig. 5.

In this example with $\zeta > 0.9$, the worst eye heights are predicted by invoking the worst bits of "0000000010" at the data rates of 4 and 8 Gbps, respectively. Notice that "SnP Model" refers to the full-wave analysis for S-parameters of transmission lines as extracted by Ansys Q2D. Both the analysis results of using the *RLC* π-models without and with the frequency dependency are thus compared with that of "SnP Model" for accuracy. Their corresponding eye heights are thus summarized in Table I as well.

III. ANALYSES FOR MULTI-LINE LINKS

By observing whether the worst bit is in-phase or out-of-phase with reference to the bit response of the main line in the multi-line links, this coupling system can be equivalent to a single-line circuit to facilitate the performance analysis for the worst-case eye height.

A. Peak Distortion Analysis for Multi-Line Links

Considering that exemplary 20-line links as depicted in Fig. 6, Line #1 is taken as the main signal line to be analyzed for all the crosstalk effects from the other 19 aggressor lines. The specific bit-pattern of "0100000000" is fed in Line #1 while another bit-pattern of "0000000000" is given to all the other 19 lines such that the step-pulse response appears on Line #1 as demonstrated in Fig. 7. The worst bit-responses of all the lines are listed in Table II with the corresponding phase-identity, respectively. Those first ten bits of the worst bit responses correspond to the worst upper waveforms for the worst eye heights and the next successive ten bits correspond to the worst lower waveforms for the worst eye heights.

TABLE I. EYE HEIGHT MEASUREMENTS OF ANALYSIS MODELS

Data Rate	Model	Eye Height	Error
4 Gbps	SnP Model	0.833 volt	
	Frequency-Independent *RLC* π-model	0.801 volt	-3.8%
	Frequency-Dependent *RLC(f)* π-model	0.821 volt	-1.4%
8 Gbps	SnP Model	0.383 volt	
	Frequency-Independent *RLC* π-model	0.307 volt	-19.8%
	Frequency-Dependent *RLC(f)* π-model	0.376 volt	-1.8%

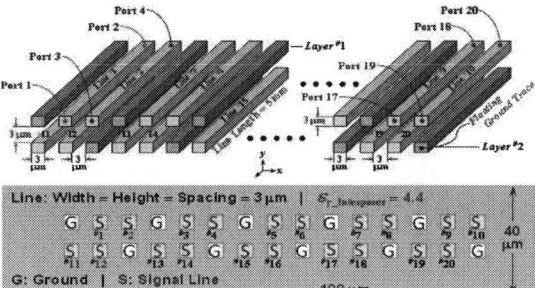

Fig. 6. Layout parameters and structure for the exemplary 20-line links.

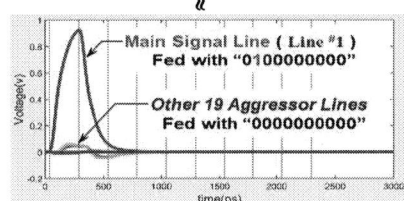

Fig. 7. Step-pulse response of the 20-line links with the input bits.

TABLE II. WORST BIT-RESPONSE AND PHASE OF 20-LINE LINKS

Line #	Worst Bit-Response	Phase Identity
Line #1	"0 0 0 0 0 0 0 0 1 0 1 1 1 1 1 1 1 1 0 1"	
Lines #2, #11, #12	"0 0 0 0 0 0 0 1 0 0 1 1 1 1 1 1 1 0 1 1"	Out-of-Phase
Other Lines	"0 0 0 0 0 0 0 0 1 0 1 1 1 1 1 1 1 1 0 1"	In-Phase

Note that the ninth bit-response is the main cursor of Line #1. If the eighth and ninth bit-responses on an aggressor line are opposite to that of the main cursor, it is realized with the out-of-phase identity as Lines #2, #11, and #12. For the aggressor lines with the eighth and ninth bit-responses with the same phase identity as that of the main cursor, they are realized as in-phase.

B. Agile Analysis with Single-Line Equivalent Cirucuit

If the worst bit on an aggressor line is in phase with that of the main line to have a positive response voltage, the mutual-inductance $Lm_{in\text{-}phase}$ is added to its self-inductance L_{self} whereas the mutual-capacitance $Cm_{in\text{-}phase}$ is subtracted from its self-capacitance C_{self}. If out-of-phase with the main line has a negative response voltage, $Lm_{out\text{-}of\text{-}phase}$ is then subtracted from L_{self} but $Cm_{out\text{-}of\text{-}phase}$ is added to C_{self}. To sump up all the coupling effects among the aggressors in the multi-line links, the equivalent inductance L_{eq} and equivalent capacitance C_{eq} are given, respectively, by

$$L_{eq} = L_{self} + \sum_i Lm_{in\text{-}phase} - \sum_k Lm_{out\text{-}of\text{-}phase} \quad (2)$$

and

$$C_{eq} = C_{self} - \sum_i |Cm_{in\text{-}phase}| + \sum_k |Cm_{out\text{-}of\text{-}phase}|, \quad (3)$$

where i = 3, 4, 5, 6, 7, 8, 9, 10, 13, 14, 15, 16, 17, 18, 19, and 20 for the in-phase lines and k = 2, 11, and 12 for the out-of-phase lines, respectively.

An efficient method using a single-line equivalent-circuit scheme to assess the eye diagram performance by replacing L_{RDL} and C_{RDL} in Fig. 4 with the derived L_{eq} and C_{eq} in (2) and (3), respectively. Note that L_{eq} and C_{eq} for Line [#]1 are inferred based on the worst bit-responses in Table II.

Nonetheless, it is time-consuming to analyze the single-bit response to all the lines for each worst bit pattern. As shown in Fig. 6 for the 7-mm lines, only Lines [#]2, [#]11, and [#]12 exhibit the significant crosstalk with Line [#]1 due to being out-of-phase while all the other 16 lines have a negligible impact. Those 16 in-phase lines are isolated by the ground lines to keep the crosstalk effects on Line [#]1 ignorable. As for (1) that the normalized resistance ζ is proportional to $[(C_L + C_{eq})/L_{eq}]^{0.5}$, in the out-of-phase case, ζ is increased by adding $\Sigma|Cm_{out-of-phase}|$ into C_{self} and subtracting $\Sigma Lm_{ou-of-phase}$ from L_{self}. Whereas in the in-phase scenario, ζ is decreased by subtracting $\Sigma|Cm_{in-phase}|$ from C_{self} and adding $\Sigma Lm_{in-phase}$ into L_{self}.

At a specific T_d/UI in reference to Fig. 3 for this 20-line example with $\zeta > 0.9$, the predicted eye height becomes smaller if ζ increases due to the more noticeable C_{eq}. On the other hand, for a given $\zeta > 0.9$ but with a faster data rate, the eye height also becomes smaller because the value T_d/UI increases; in other words, the out-of-phase of the worst-bit pattern will make the eye height smaller due to the increase of ζ. When $\zeta > 0.9$, the single-pulse response is a single wave so that the worst bit can only be in-phase or out-of-phase and the crosstalk will reduce the eye height. Hence, the main cause of a decreasing eye height must be out-of-phase. Since those adjacent lines cause the most obvious crosstalk, the worst bit-responses only to the adjacent lines are assumed to facilitate the analysis without considering the other in-phase lines. Thereof, L_{eq} and C_{eq} can be simplified, respectively, as

$$L_{eq} = L_{self} - \sum Lm_{out-of-phase} \qquad (3)$$

and

$$C_{eq} = C_{self} + \sum |Cm_{out-of-phase}|. \qquad (4)$$

To this stage for the 20-line links as depicted in Fig. 6, (4) and (5) are invoked to construct the frequency-dependent RLC π-model with $R_s = 50\ \Omega$ and $C_L = 1$ pF. Table III shows the eye height comparison of those three analysis schemes with "SnP Model," "$RLC(f)$ π-Model with all the Phase cases," and "$RLC(f)$ π-Model with Only Out-of-Phase," respectively, by the worst-bit assessment. Nonetheless, "2^{20}-bit PRBS with SnP Model" is implemented by Keysight Advanced Design System [6] as the run-time reference. By the proposed single-line equivalent-circuit scheme with only the out-of-phase scenarios, the eye height performance can be reliably predicted with an accuracy comparable to that of the traditional analysis method. The simulation efficiency is achieved without analyzing the worst bit response to all the lines but only treating those adjacent lines as out-of-phase.

IV. CONCLUSIONS

For the multi-line links in the CoWoS packaging designs, a single-line equivalent-circuit scheme is proposed to assess the worst eye height performance. Frequency-dependent

RLC π-models are constructed for the RDL layout structure to characterize their crosstalk effects in the links. The step-pulse responses to the lines are verified for the normalized resistance ζ in the solution-space contour plot and the phase identities to derive the equivalent inductance and equivalent capacitance of the short-length high-loss multi-line links.

To improve the computational efficiency with the worst-bit analysis, only those adjacent lines with the out-of-phase response are accounted for the worst eye height prediction. An agile simulation with a single-line equivalent circuit thus achieves the efficient and reliable worst eye height measure comparable to that of the time-consuming PRBS method. To aim at promising multi-line link applications using the novel HBM technology at a higher data-rate, the proposed analysis methodology can be readily applied for the worst-case performance assessment on signal integrity.

REFERENCES

[1] K. Cho, Y. Kim, H. Lee, H. Kim, S. Choi, J. Song, S. Kim, J. Park, S. Lee, and J. Kim, "Signal integrity design and analysis of silicon interposer for GPU-memory channels in high-bandwidth memory interface," *IEEE Trans. Compon., Packag., Manuf. Technol.*, vol. 8, no. 9, pp. 1658–1671, Sept. 2018.

[2] S. Choi, H. Kim, K. Kim, J. Park, D. H. Jung, and J. Kim, "Signal integrity analysis of silicon/glass/organic interposers for 2.5D/3D Interconnects," in *IEEE 67th Electron. Compon. Technol. Conf.*, May 30 – June 2, 2017, Orlando, Florida, USA.

[3] K.-B. Wu, T.-Y. Kuo, C.-C. Hung, B. Lin, I.-H. Peng, M.-T. Yang, and R.-B. Wu, "Novel RDL design of wafer-level packaging for signal/power integrity in LPDDR4 application," *IEEE Trans. Compon., Packag., Manuf. Technol.*, vol. 8, no. 8, pp. 1431–1439, Aug. 2018.

[4] Q2D, ANSYS Inc. [Online]. Available: www.ansys.com

[5] Kai Li, "Analytic method of fast eye-diagram index for multiple coupled lines in DDR," Master's Thesis, Graduate Inst. Comm. Engr., National Taiwan University, Taipei, Taiwan, ROC, Aug. 2023.

[6] Advanced Design System, PathWave Design, Keysight Technologies. [Online]. Available: www.keysight.com

TABLE III. EYE HEIGHT COMPARISON OF ANALYSIS SCHEMES

Data Rate	Analysis Scheme	Eye Height	Error % to SnP Model	Run Time
3.2 Gbps in HBM2E	SnP Model	0.624 volt		
	$RLC(f)$ π-Model with all Phase Cases	0.629 volt	0.8%	0.5 sec.
	$RLC(f)$ π-Model with Only Out-of-Phase	0.610 volt	-2.2%	0.5 sec.
	2^{20}-Bit PRBS with SnP Model	0.635 volt	1.8%	15 hours
6.4 Gbps in HBM3	SnP Model	0.182 volt		
	$RLC(f)$ π-Model with all Phase Sets	0.185 volt	1.6%	0.5 sec.
	$RLC(f)$ π-Model Only Out-of-Phase	0.176 volt	-3.3%	0.5 sec.
	2^{20}-Bit PRBS with SnP Model	0.189 volt	3.8%	8 hours

Worst-bit assessment on:
  SnP Model: Circuit with the full S-parameters extracted by Ansys Q2D.
  $RLC(f)$ π-Model with all Phase Cases: Frequency-dependent RLC π-model using Eqs. (2) & (3) for all the in-phase and out-of-phase cases.
  $RLC(f)$ π-Model Only Out-of-Phase: Frequency-dependent RLC π-model using Eqs. (6) and (7) for only the out-of-phase cases.
Traditional simulation with the PRBS of 2^{20}-bit stimuli:
  PRBS with 2^{20} Bits: Analysis by ADS for 2^{20}-bit PRBS and SnP model.

Operator Inference for Rigid-Flex Printed Circuit Boards Subject to Large Deformations

Pascal den Boef
Mathematics and Computer Science
Eindhoven University of Technology
Eindhoven, The Netherlands
p.d.boef@tue.nl

Diana Manvelyan
Siemens AG
Garching, Germany
diana.manvelyan@siemens.com

Wil Schilders
Mathematics and Computer Science
Eindhoven University of Technology
Eindhoven, The Netherlands
w.h.a.schilders@tue.nl

Joseph Maubach
Mathematics and Computer Science
Eindhoven University of Technology
Eindhoven, The Netherlands
j.m.l.maubach@tue.nl

Nathan van de Wouw
Mechanical Engineering
Eindhoven University of Technology
Eindhoven, The Netherlands
n.v.d.wouw@tue.nl

Abstract—**We use operator inference for the accelerated simulation of the deformation of rigid-flex printed circuit boards (PCBs). Nonlinear behavior associated with large deformation of rigid-flex PCBs is captured by polynomial terms in the model structure.**

Index Terms—**Model reduction, rigid-flex PCBs**

I. Introduction

Printed Circuit Boards (PCBs) which consist of rigid and flexible parts have numerous applications. To name a few: flexible flow sensors [19], microrobotics [9] and high-density interconnect applications [20]. The field of origami robotics uses rigid-flex PCB systems to build foldable robotics with applications in, e.g., planetary exploration [10] and solar tracking [15].

In addition to the materials commonly used in rigid PCBs (such as FR4 and copper), rigid-flex PCBs use materials with distinctive mechanical properties (such as Nomex [6]). In a simulation and modeling context, specialized techniques are needed to properly handle these materials. In [7], a finite element model is developed focused on large-scale deformations of rigid-flex PCBs with Nomex hinges. In [14], the process characterization of rigid-flex PCBs is considered and reliability models are established based on empirical studies. Simulation models for thermo-mechanical reliability analysis of rigid-flex PCBs are studied in [3].

Since the resulting simulation models are often large-scale, they are computationally expensive to run. This poses a challenge especially in the multi-query context, i.e., running the simulation for many different combinations of parameters (such as boundary conditions, material parameters, etc.). In this context, model reduction [2] is a crucial tool to accelerate simulation studies. Linear model reduction methods are applied in the context of (micro)electronics in [22], [21], [11],

[12]. Some works also treat more general models, such as time-varying [13] or nonlinear systems [1]. In the case of rigid-flex PCBs, large deformations are present - which is a nonlinear phenomenon. Hence, this paper proposes a suitable model reduction technique to accelerate the simulation of rigid-flex PCB systems exhibiting nonlinear behavior.

Specifically, the contributions of this paper are:

1) We propose a model reduction technique for the accelerated simulation of the deformation of rigid-flex PCB systems accounting for the nonlinear behavior caused by large deformations.
2) The method is non-intrusive (i.e., requiring no analytical model) and hence can be integrated with ease in workflows involving (commercial) simulation software.
3) We demonstrate the method on a numerical example involving rigid-flex PCB systems.

Our notation is as follows. $A \succeq B$ means that the matrix $A - B$ is positive semidefinite. $\nabla f(x)$ represents the gradient of f at x. $\|\cdot\|_F$ is the Frobenius norm for matrices, and $\|\cdot\|_2$ is the Euclidean norm for vectors.

II. Stable Operator Inference for Static Problems

We consider the following mechanical model of a PCB describing how the displacement $y \in \mathbb{R}^n$ depends on a parameter $\mu \in \mathbb{R}^{n_\mu}$:

$$K^o \phi^o(y) = B^o \mu, \qquad (1)$$

where $K^o : \mathbb{R}^{n \times n_{\phi^o}}$ and $B^o : \mathbb{R}^{n \times n_\mu}$. $\phi^o : \mathbb{R}^n \to \mathbb{R}^{n_{\phi^o}}$ is a possibly nonlinear function. In our setting, the model (1) originates from a discretization (using, e.g., finite element method) of equations describing hyperelasticity. μ represent a parametrization of load terms that are considered in the design and operation of the PCB. For example, elements of μ can be associated with a prescribed force or a temperature (in case of temperature-induced deformation). The model (1) can be computationally expensive if n is large (i.e., many

979-8-3503-5124-8/24 $31.00 © 2024 IEEE

displacement variables). At the same time, often we need to compute y for many values of μ (the so-called multi-query context), for example to evaluate the mechanical robustness of the PCB against a wide range of parameter values. The high computational cost needed for these evaluations motivates the need for a reduced-order model. A complicating factor is that, often, the model (1) is implemented in commercial software. Such software does not allow the user to directly access the operators K^o, ϕ^o and B^o describing the model. Hence, we need an alternative method of obtaining a reduced-order model without requiring these operators. This alternative method is operator inference [18]. The basic ingredient of operator inference is a set of simulation snapshots:

$$\mathcal{Y} := \begin{bmatrix} y_1 & \cdots & y_N \end{bmatrix}, \quad \mathcal{U} := \begin{bmatrix} \mu_1 & \cdots & \mu_N \end{bmatrix}, \quad (2)$$

where N denotes the number of simulations and y_i, μ_i (with $i = 1, \ldots, N$) are the displacement and parameter extracted from the simulation software, respectively. In the context of model reduction, the process of creating the simulation snapshots (2) is termed the *offline phase*. Subsequently, we build a reduced model which can subsequently be used to speed up any further simulations: this is the *online phase*.

y is high-dimensional. As a first step in the reduction, the Proper Orthogonal Decomposition (POD) [8] can be used to approximate y on a low-dimensional manifold:

$$y \approx Vx, \quad (3)$$

with $x \in \mathbb{R}^r$ the coordinates on the r-dimensional manifold (with $r \ll n$) and $V \in \mathbb{R}^{n \times r}$ a basis for the manifold. In POD, V is set to the first r left singular vectors of $\mathcal{Y}$ in (2). The truncation index r should be chosen based on the decay of the singular values of $\mathcal{Y}$ to ensure a good balance between accuracy and speed of the reduced model.

We define the simulation snapshot of the reduced state as $\mathcal{X} := V^{\mathrm{T}}\mathcal{Y}$. To find a model in the reduced state space approximating (1), we propose the model structure

$$K_{\mathrm{lin}}x + K_{\mathrm{nl}}\phi(x) = B\mu, \quad (4)$$

where K_{lin}, K_{nl} are appropriately sized matrices and $\phi : \mathbb{R}^r \to \mathbb{R}^{n_\phi}$ is a vector of nonlinear basis functions. The model structure (4) can be used as a surrogate model for (1). A reasonable choice for the nonlinearity ϕ in (4) is typically generated by exploiting prior knowledge of ϕ^o (for example, from knowledge of the material models). We match the reduced-order model (4) as closely as possible to the dataset by solving

$$\min_{K_{\mathrm{lin}} \succeq I, K_{\mathrm{nl}}, B} \|K_{\mathrm{lin}}\mathcal{X} + K_{\mathrm{nl}}\phi(\mathcal{X}) - B\mathcal{U}\|_{\mathrm{F}}. \quad (5)$$

The optimization problem (5) can be seen as a variant of operator inference [18] for static problems with the constraint $K_{\mathrm{lin}} \succeq I$ ensuring that, in the neighborhood of the undeformed state $x = 0$, the model is physically sound by having a positive definite strain energy function.

III. NUMERICAL EXAMPLE

We consider a discretized 2D large deformation model of two rigid PCB segments joined by a flexible hinge using $n = 5,364$ degrees-of-freedom (Figure 2a), implemented in DOLFINx 0.7.0 [4] on a computer with 32 GB of memory and an Intel i7-9750H CPU. The material of the rigid segment is FR4 (Young's modulus $E = 24\mathrm{GPa}$, Poisson's ratio $\nu = 0.13$), while the material of the flexible hinge is natural rubber ($E = 6\mathrm{MPa}$, $\nu = 0.49$). The system is fixed on the left edge with a homogeneous Dirichlet boundary condition. On the right edge we impose a distributed force μ for which the magnitude α_μ and angle β_μ are parametrized (Figure 2b), i.e.:

$$\mu = \alpha_\mu \begin{bmatrix} \cos(\beta_\mu) \\ \sin(\beta_\mu) \end{bmatrix} \quad (6)$$

with $\alpha_\mu \in [0, 500]\mathrm{N\,mm}^{-1}$ and $\beta_\mu \in [0, \pi/3]\mathrm{rad}$.

For the rigid segments, we use the St. Venant-Kirchoff material law to account for the large deformation [5]. For the flexible hinge, we use a Neo-Hookean material law [17], for which the strain energy is

$$\psi(u) = \frac{\mu}{2}(\mathrm{tr}(F^{\mathrm{T}}F) - 2) - \mu\ln(\det(F)) + \frac{\lambda}{2}(\ln(\det(F)))^2, \quad (7)$$

with $F = \nabla y + I$ and μ, λ the Lamé parameters.

We define a 12-point uniform gridding of the loading angle β_μ from 0 to $\pi/3$. For each point, we perform a static simulation by increasing α_μ uniformly over 2,000 steps, leading to $N = 24,000$ snapshots as in (2). For each of the 6 pairs of grid points of β_μ, we use the snapshots of the first one to solve the operator inference (5) and the snapshots of the second one to validate the model accuracy.

We reduce the dimension using POD on the inference dataset to obtain the basis V as in (3). The first 100 singular values of $\mathcal{Y}$ are shown in Figure 1. We compare 3 reduced-order models (ROMs) of dimensions $r \in \{2, 3, 4\}$. For each ROM, the accuracy is given by

$$\frac{\sum_{i=1}^{M} \|y_i - Vx_i\|_2}{\sum_{i=1}^{M} \|y_i\|_2}, \quad (8)$$

where $M = 12,000$. We compute x_i from (4) for each β_μ using the fsolve command in MATLAB, initialized from the previous step in α_μ, with the operators of (4) obtained by solving (5) using YALMIP [16]. For $\phi(x)$ in (4), we use all third-degree monomials of x ($x_1^3, x_1^2x_2, \ldots$). The results are displayed in Table I. In the table, the offline phase consists of simulating the inference dataset (404 seconds), performing the POD and solving (5). The online phase consists of using the ROM for simulation of the validation dataset.

For $r = 2$, the ROM fails to capture the relevant dynamics in both datasets. For $r = 3$, good results are obtained, while further increasing to $r = 4$ increases error on the validation dataset. In all cases, a speed-up is achieved considering that the total duration of offline and online phase is lower (less than 505 seconds) than the time required to simulate the entire dataset using the full-order model (808 seconds).

979-8-3503-5124-8/24 $31.00 © 2024 IEEE

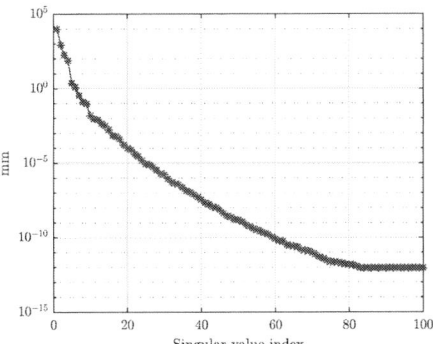

Fig. 1: First 100 singular values of the snapshot matrix $\mathcal{X}$.

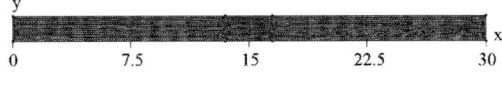

(a) Undeformed (dimensions in mm).

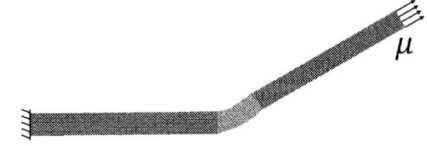

(b) Deformed (with boundary conditions).

Fig. 2: Rigid-Flex PCB model, side-profile.

IV. CONCLUDING REMARKS

We proposed a model reduction technique based on operator inference for the mechanical simulation of rigid-flex PCB systems. The method can handle static mechanical models with nonlinear stifness terms (such as large deformations or hyperelastic material models). We demonstrated the method on a simplified 2D model of a rigid-flex PCB system.

As a next step, the method should be evaluated on a more complex 3D model of a rigid-flex PCB system. Furthermore, the method can be extended to include dynamics models and multiphysics domains.

This article is dedicated to our esteemed colleague and friend, Michel Nakhla, whose work has been of utmost importance to our field of research.

REFERENCES

[1] Ramachandra Achar, Pavan K Gunupudi, Michel Nakhla, and Eli Chiprout. Passive interconnect reduction algorithm for distributed/measured networks. *IEEE Transactions on Circuits and Systems II: Analog and Digital Signal Processing*, 47(4):287–301, 2000.

[2] Athanasios C Antoulas. *Approximation of large-scale dynamical systems*. SIAM, 2005.

TABLE I: Error of and time required for the ROMs.

r	Error (inf.)	Error (val.)	Offline phase	Online phase
2	$1.1 \cdot 10^{+0}$	$1.1 \cdot 10^{+0}$	441 s	17.4 s
3	$6.9 \cdot 10^{-3}$	$5.5 \cdot 10^{-3}$	447 s	24.2 s
4	$3.3 \cdot 10^{-3}$	$8.3 \cdot 10^{-2}$	445 s	59.5 s

[3] Luciano Arruda, Quayle Chen, and Jairo Quintero. Failure evaluation of flexible-rigid PCBs by thermo-mechanical simulation. In *2009 International Conference on Electronic Packaging Technology & High Density Packaging*, pages 1201–1205. IEEE, 2009.

[4] Igor A. Baratta, Joseph P. Dean, Jørgen S. Dokken, Michal Habera, Jack S. Hale, Chris N. Richardson, Marie E. Rognes, Matthew W. Scroggs, Nathan Sime, and Garth N. Wells. DOLFINx: The next generation FEniCS problem solving environment, December 2023.

[5] Jernej Barbič and Doug L James. Real-time subspace integration for St. Venant-Kirchhoff deformable models. *ACM transactions on graphics (TOG)*, 24(3):982–990, 2005.

[6] J Bell, Laura Redmond, K Carpenter, and JP de la Croix. Experimental dynamic characterization of rigid-flex PCB systems. *Experimental Techniques*, 47(2):419–433, 2023.

[7] John Bell, Laura Redmond, Kalind Carpenter, and Jean-Pierre de la Croix. Finite Element Modeling of Rigid-Flex Printed-Circuit-Board Origami Robotics Subjected to Large-Scale Deformations. *AIAA Journal*, 62(4):1264–1280, 2024.

[8] Gal Berkooz, Philip Holmes, and John L Lumley. The proper orthogonal decomposition in the analysis of turbulent flows. *Annual review of fluid mechanics*, 25(1):539–575, 1993.

[9] Erik Edqvist, Niklas Snis, Raimon Casanova Mohr, Oliver Scholz, Paolo Corradi, Jianbo Gao, Angel Diéguez, Nicolas Wyrsch, and Stefan Johansson. Evaluation of building technology for mass producible millimetre-sized robots using flexible printed circuit boards. *Journal of Micromechanics and Microengineering*, 19(7):075011, 2009.

[10] Samuel Felton, Michael Tolley, Erik Demaine, Daniela Rus, and Robert Wood. A method for building self-folding machines. *Science*, 345(6197):644–646, 2014.

[11] Lihong Feng, Peter Benner, and Jan G Korvink. System-Level Modeling of MEMS by Means of Model Order Reduction (Mathematical Approximations)–Mathematical Background. *System-Level Modeling of MEMS*, pages 53–93, 2013.

[12] Francesco Ferranti, Michel S Nakhla, Giulio Antonini, Tom Dhaene, Luc Knockaert, and Albert E Ruehli. Multipoint full-wave model order reduction for delayed peec models with large delays. *IEEE Transactions on Electromagnetic Compatibility*, 53(4):959–967, 2011.

[13] Emad Gad and Michel Nakhla. Efficient model reduction of linear periodically time-varying systems via compressed transient system function. *IEEE Transactions on Circuits and Systems I: Regular Papers*, 52(6):1188–1204, 2005.

[14] Shiqing Huang. *Process characterization and reliability modeling of Rigid-Flex Printed Circuits (RFPCs)*. PhD thesis, Hong Kong Polytechnic University, 2011.

[15] Aaron Lamoureux, Kyusang Lee, Matthew Shlian, Stephen R Forrest, and Max Shtein. Dynamic kirigami structures for integrated solar tracking. *Nature communications*, 6(1):8092, 2015.

[16] Johan Lofberg. YALMIP: A toolbox for modeling and optimization in MATLAB. In *2004 IEEE international conference on robotics and automation*, pages 284–289. IEEE, 2004.

[17] Raymond W Ogden. *Non-linear elastic deformations*. Courier Corporation, 1997.

[18] Benjamin Peherstorfer and Karen Willcox. Data-driven operator inference for nonintrusive projection-based model reduction. *Computer Methods in Applied Mechanics and Engineering*, 306:196–215, 2016.

[19] Anastasios Petropoulos, Dimitrios Goustouridis, Thanassis Speliotes, and Grigoris Kaltsas. Demonstration of a new technology which allows direct sensor integration on flexible substrates. *The European Physical Journal Applied Physics*, 46(1):12507, 2009.

[20] Abdellah Salahouelhadj, Marion Martiny, Sébastien Mercier, L Bodin, D Manteigas, and B Stephan. Reliability of thermally stressed rigid–flex printed circuit boards for high density interconnect applications. *Microelectronics Reliability*, 54(1):204–213, 2014.

[21] M Weninger, J Zündel, T Krivec, M Frewein, S Waschnig, P Fuchs, and C Obst. Evaluation of thermomechanical behavior of electronic devices through the use of a reduced order modelling approach. In *2023 24th International Conference on Thermal, Mechanical and Multi-Physics Simulation and Experiments in Microelectronics and Microsystems (EuroSimE)*, pages 1–5. IEEE, 2023.

[22] Ibrahim Zawra, Jeroen Zaal, Michiel van Soestbergen, Torsten Hauck, Evgeny Rudnyi, and Tamara Bechtold. Compact modelling of wafer level chip-scale package via parametric model order reduction. In *International Conference on Scientific Computing in Electrical Engineering*, pages 217–228. Springer, 2022.

979-8-3503-5124-8/24 $31.00 © 2024 IEEE

A Signal Integrity Comparison of VIPPO Technology for PCIe 5.0 DC Blocking Capacitors

Andrew Page, Matteo Cocchini

IBM Infrastructure, Poughkeepsie, NY 12601, USA
andrew.page@ibm.com, mcocchi@us.ibm.com

Abstract—This work studies the signal integrity effects of VIPPO-style vias and DC blocking surface-mount capacitors required by many high-speed serdes interfaces, focusing on PCIe 5.0 transmitters. A test vehicle was built containing four via-and-capacitor configurations fixtured with coaxial probe contacts, as well as a calibration structure for broadband de-embedding. Passive measurements are discussed, and de-embedding is attempted. A set of 3D full-wave models are built that validate the reliability of de-embedded results after post-processing. The 0201-capacitor seated on VIPPO vias performs the best against the demonstrated tests from a high-speed signal integrity standpoint.

Index Terms—DC block, PCIe 5.0, VIPPO, de-embedding, full-wave modeling

I. INTRODUCTION

DC-blocking (or AC-coupling) capacitors are used on the transmitter side of PCIe channels mainly to obtain DC isolation between the driver and receiver side of the lane [1]. They are a necessary part of a PCIe link, but they take up valuable board space and present discontinuities to high-speed digital signals. Such surface-mounted capacitors would require vias to connect with buried high-speed wiring. Classic "dogbone" traces are typically used to connect the capacitor to the vias. They are easy to manufacture, but the structures they require for proper routing add to footprint size and generate unwanted losses, making for poor signal integrity (SI). Vias made with the via in- pad plated-over (VIPPO) technique allow surface-mount components to be placed on top of pads, eliminating bulky microstrip traces. Much work is being done to characterize the SI of VIPPO structures [2].

This work presents experimental results, based on a test vehicle, comparing use of VIPPO (V) and non-VIPPO (NV) vias to support 0201 and 0402 surface-mount capacitors intended for PCIe 5.0 DC blocking. Section II introduces the structures built on the test board. Section III shows passive measurements of each built structure and provides an initial attempt at broadband de-embedding using an industry tool. Section IV discusses full-wave modeling of the 0201V structure with the goal of validating post-processing techniques that improve de-embedding results, which are applied to the measured data in Section V. Concluding remarks are offered in Section VI.

II. OVERVIEW OF STRUCTURE VARIATIONS

The test vehicle houses four devices-under-test (DUT's) and one calibration structure called the 2xThru. Each DUT is comprised of a surface-mount 220nF monolithic ceramic capacitor

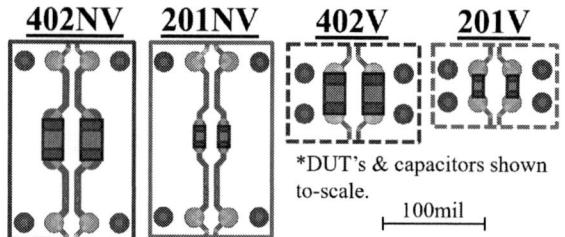

Fig. 1. Top-down diagram of the four DUT variations shown to-scale. 130mil differential stripline leadup included in each DUT but not shown.

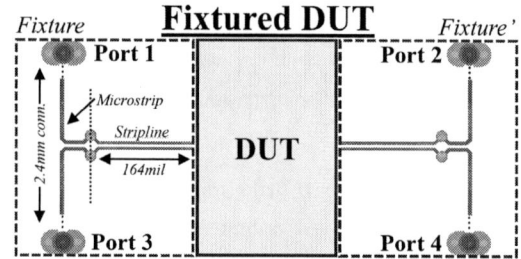

Fig. 2. Top-down diagram of fixture-DUT-transpose fixture (*fixture'*) as implemented on the test vehicle. Not to-scale.

of either the 0201 or 0402 footprint, accompanied by either dogbone- or VIPPO-style vias breaking out into an embedded stripline layer. These DUT's are labelled by capacitor size and V/NV as shown in Fig. 1. A test vehicle printed circuit board (PCB) houses each DUT, which are connected to identical fixtures that interface each DUT with four FMCN1480 2.4mm surface-mount connectors for VNA and TDR probing, as in Fig. 2. Measurements obtained from such probing will include the effects of these fixtures, which may skew conclusions on performance. The 2xThru structure, built on the same test vehicle, is comprised of two fixtures attached to each other. It will serve as a calibration structure used in broadband de-embedding, explored in sections III-C and V. The DUT's and 2xThru were built in class B compliance with the IEEE 370-2020 fixture electrical requirements through 32GHz [3].

III. COMPARISON BY PASSIVE MEASUREMENTS

A. Time-domain reflectometry

Each DUT represents a different disruption in the path of a high-speed digital signal propagating through the PCIe link.

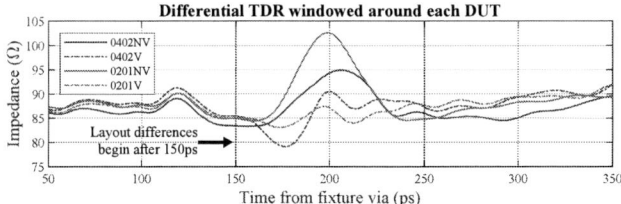

Fig. 3. Measured impedance profile centered on capacitor region. Each fixtured DUT is identical (± manufacturing tol.) until 150ps.

TABLE I
DUT IMPEDANCE DISCONTINUITIES SEEN IN MEASURED TDR

DUT	Z_{ref}	Z_{min}	Z_{max}	δZ_{down}	δZ_{up}	δZ
0402NV	83.41	83.23	94.9	0.18	11.5	11.68
0402V	85.37	79.11	90.53	6.26	5.16	11.42
0201NV	84.89	84.38	102.61	0.51	17.71	18.22
0201V	85.05	83.04	87.49	2	2.44	4.45

*Each value carries units of Ω.

Time-domain reflectometry (TDR) can be used to quantify the disturbance presented by each structure as an impedance ripple. Each fixtured DUT was prepared with open terminations on ports 2 & 4. A 500mVpp differential step with a 12ps rise time was injected into ports 1 & 3 using a Tektronix DSA8300 with an 80E10B sampling module pair. The differential impedance profile was extracted for each DUT and is shown in Fig. 3 after windowing out the fixtures. A difference between the DUT's is apparent, as 0201NV and 0402NV share large inductive spikes while the VIPPO pair ripple evenly up and down. 0201V demonstrates the least disturbance to the TDR wavefront.

The impedance swing for each DUT was calculated by defining a reference impedance and calculating the upward and downward swings, as well as peak-to-peak, as summarized in Table I. The VIPPO DUT's show balanced up- and downswings, while the NV are almost purely inductive. The authors believe the wide-pitched microstrip present on the NV DUT's causes this inductance, especially seen in 0201NV. 0201V shows the lowest peak-to-peak ripple by roughly 7Ω.

B. Scattering parameters

The S-parameters of each fixtured DUT were measured to 50GHz using a P5028A VNA. These measurements relay information of losses and reflections from the fixtured DUT. The results are shown in Fig. 4, calibrated up to but still including each fixture. Any difference in DUT IL here is small, due in part to DUT behavior being obscured by fixture effects. Close inspection shows that 0201V stands out in IL through 20GHz, being either at or tied for the lowest loss. Both V designs have improved RL over their NV counterparts. The V designs trend together, as do the NV designs, until they all gather within roughly ±3dB after 18GHz.

C. Initial de-embedding

The measured S-parameters covered in Section III-B represent the fixtured DUT as in Fig. 2, which makes comparing

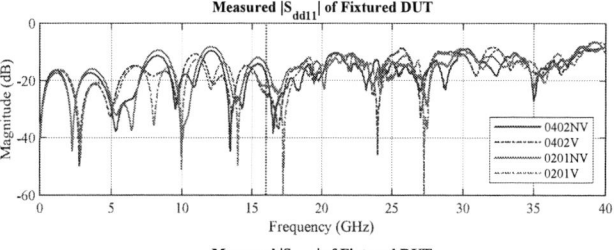

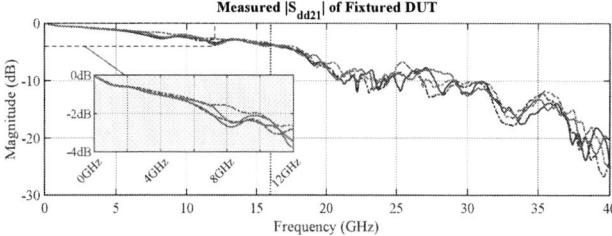

Fig. 4. Differential-mode insertion and return loss (IL/RL) measured with calibration up to connectors. PCIe 5.0 fundamental frequency is highlighted.

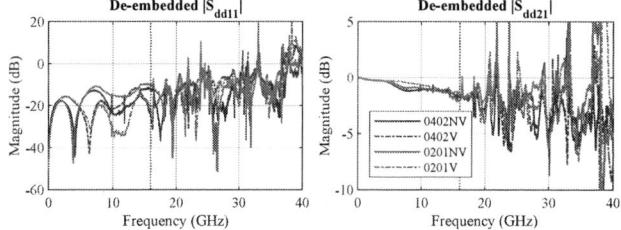

Fig. 5. DUT IL/RL from de-embedding Fig. 4 data with measured 2xThru. Spurious spikes begin past 10GHz and worsen past 16GHz.

DUT's difficult. These issues may be avoided by employing de-embedding, the act of shifting the reference plane from the connectors to the boundaries of the DUT itself. This is done by providing an industry-accessible tool with measurements of the 2xThru, which is two fixtures connected without a DUT. This tool characterizes the fixtures and removes their effects from the full S-parameter measurements, isolating the DUT. The results of this procedure are shown in Fig. 5, where spikes and passivity violations begin below 16GHz and grow with frequency, rendering the data unusable in further simulation. Several de-embedding tools were used, all delivering similar results. It is unclear whether these issues indicate bias in the results, and so no comparisons of dB IL or RL levels should be made among this data without further investigation.

IV. FULL-WAVE MODEL EXPERIMENTS

A. Model-hardware correlation of 0201V

We now attempt to recreate the non-physical spikes seen in the previous de-embedded data using 3D models and find a potential solution. A 3D model of the fixtured 0201V DUT was built to correlate with the passive measurements of section III. A starting model was extracted from the test vehicle board file and updated based on dimensions measured under an optical profilometer. Multipole Debye models were used for

979-8-3503-5124-8/24 $31.00 © 2024 IEEE

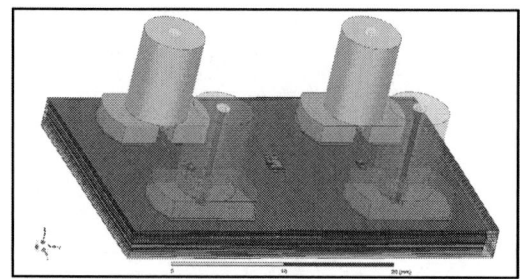

Fig. 6. 3D model of 0201V with "copper brick" capacitors and 3D connectors.

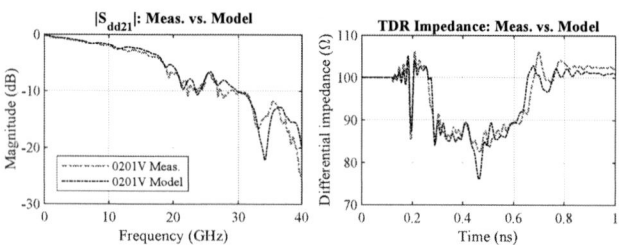

Fig. 7. IL and simulated TDR of 0201V VNA measurement and HFSS model.

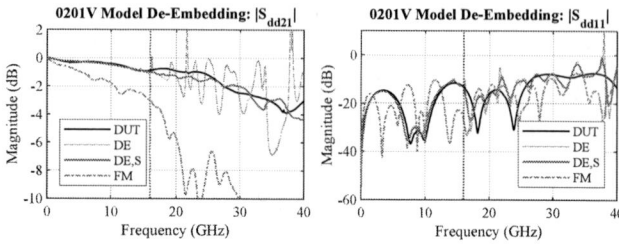

Fig. 8. 0201V models: DUT/full-model (FM)/de-emb. (DE)/smoothed (S).

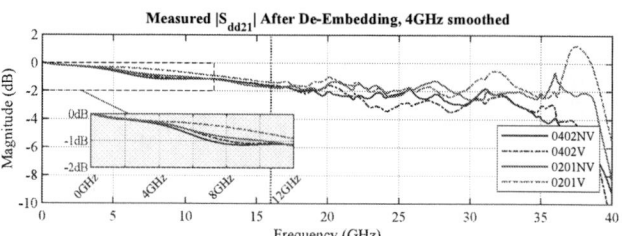

Fig. 9. Measured IL after de-embedding, with 4GHz smoothing.

the low-loss signal layer and the FR-4 layers, and a Djordjevic-Sarkar model was used for the soldermask. A high-impedance high-attenuation boundary-scan corner model was used for the low-loss layer [4]. Renditions of the 2.4mm surface-mount connectors were built into the HFSS model to simulate the transition between the coaxial pin and the microstrip launch. The 3D model is shown in Fig. 6.

The 0201V model is compared to the measured structure both by differential IL and TDR simulated from S-parameters; these are shown in Fig. 7. Features like dB losses, trace impedances, and resonance synchronization are achieved within reasonable margins. This model will serve as a taxonomic match to the test vehicle structure and will be used to test de-embedding in the following section. Improved capacitor modeling will be explored in future work.

B. Comparison of de-embedded and split 0201V 3D models

We now perform de-embedding with the 0201V 3D model. A 2xThru model was prepared in the same manner. An 0201V DUT-only model was created by subtracting the fixtures from the 0201V model with ports at the fixture-DUT interface. The same tool will be used to de-embed the full model using the 2xThru model, which should resemble the DUT-only model.

Fig. 8 shows the de-embedded model fluctuating around the DUT-only model. Good agreement is seen in both IL and RL until 16GHz. A 4GHz and 1GHz rolling-average smoothing was applied to the de-embedded IL and RL respectively, which eliminates small spikes and helps the match extend past 16GHz. The IL discrepancy between smoothed DE and DUT-only is less than 0.08dB until 16GHz, and peaks at 0.61dB until past 31GHz. These results suggest that smoothing the measured IL in Fig. 5 may provide a reasonable estimate of the DUT dB loss until the spikes exceed a few dB and GHz.

V. REVISITING DE-EMBEDDED PASSIVE MEASUREMENTS

The previous section instilled confidence that the de-embedded measurements depict the DUT characteristics in a qualitative sense fit for comparison. Fig. 9 shows the measured IL's after de-embedding and 4GHz smoothing. 0201V is the clear best choice in IL, at its best having a 0.38dB and 0.63dB lead over the other DUT's around 8GHz. RL smoothing did not help in Fig. 8, so Fig. 5 can be re-examined. 0201V and 0402V match well until 8GHz; 0201V remains lowest or tied for lowest RL until past 16GHz. A comparison of the windowed TDR's of Fig. 3 roughly agree with the RL plots, though detailed correlation is nontrivial.

VI. CONCLUSION

The 0201 capacitors placed on VIPPO pads performed the best in all tests performed. Its small footprint produces comparatively little parasitic behavior, low discontinuity, and low losses near the PCIe 5.0 fundamental. Use of 3D models validated advanced de-embedding with post-processing for DUT comparison. This technique allowed estimate of DUT dB losses isolated from fixture effects. TDR remains a practical means of quantifying discontinuity of an embedded structure.

REFERENCES

[1] PCI-SIG, "Pci-sig," 2024, accessed: Jun. 14, 2024. [Online]. Available: https://pcisig.com.

[2] Y. Zhang, J. Tang, X. Duan, S. Datta, P. R. Paladhi, M. Bohra, S. Chun, and D. M. Dreps, "Analyses of via-in-pad plated over (vippo) and dogbone in fanout routing high-density interconnects," in *2023 IEEE 32nd Conference on Electrical Performance of Electronic Packaging and Systems (EPEPS)*, 2023, pp. 1–3.

[3] "Ieee standard for electrical characterization of printed circuit board and related interconnects at frequencies up to 50 ghz," *IEEE Std 370-2020*, pp. 1–147, 2021.

[4] Z. Chen, "Transmission line attenuation-impedance realistic corner modeling by scaled-down tolerance boundary scan," in *2007 IEEE International Symposium on Electromagnetic Compatibility*, 2007, pp. 1–6.

979-8-3503-5124-8/24 $31.00 © 2024 IEEE

Analysis of Interconnects in Multilayer SIW Bandpass Filters Design

Lu Qiu, Xiao-Wei Zhu, *Member, IEEE* and Xian-Long Yang, *Graduate Student Member, IEEE*
School of Information Science and Engineering
Southeast University
Nanjing210096, China
Email: {qiulu, xwzhu, xlyang_mw}@seu.edu.cn

Abstract—Two cross coupling substrate integrated waveguide (SIW) filters designed on multilayer structure, vertically interconnected with screws and prepreg respectively, are presented, analyzed and measured. Spurious signals caused by prepreg is analyzed to propose elimination suggestions.

Keywords—Bandpass filter (BPF), interconnect, substrate integrated waveguide (SIW)

I. INTRODUCTION

With the rapid development of wireless communication systems, substrate integrated waveguide (SIW) technology has emerged as a critical technique for designing compact and high-performance filters [1][2]. Multilayer structures have been commonly utilized in a variety of SIW bandpass filters (BPFs), including wide stopband filters [3][4], high-selectivity filters [5][6], and multiband filters [7][8].

For multilayer filters, achieving vertical layer interconnections while maintaining filter performance and system integration is a crucial issue. The majority of these filters employ screws as vertical interconnection for the layers. While simple and straightforward, this approach is clearly detrimental to the integration of the filter within the system and may even negatively impact the in-band performance of the filter.

Compared to screws, using prepreg for layer interconnection is a more effective yet challenging approach. Prepreg, a composite material pre-impregnated with resin, not only simplifies the interconnect process but also improves overall structural integrity in communication systems, making it suitable for high-frequency filters that have a smaller size. However, the adoption of prepreg introduces its own challenge. Directly bonding layers with prepreg between copper layers can introduce spurious signals for filter, requiring further analysis and improvement.

In this paper, two kinds of SIW multilayer filter with identical topologies are designed, one using screws and the other using prepreg as the vertical interconnect methods. We compare and analyze their performances, particularly regarding spurious issues, and provide recommendations for future filter designs utilizing prepreg. The filter with prepreg is fabricated and measured, showing great agreement with the simulation results.

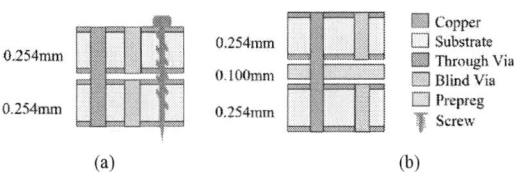

Fig.1. Two vertical interconnect methods for multilayer filters. (a) With screws penetrating the entire structure. (b) With the prepreg between two copper layers.

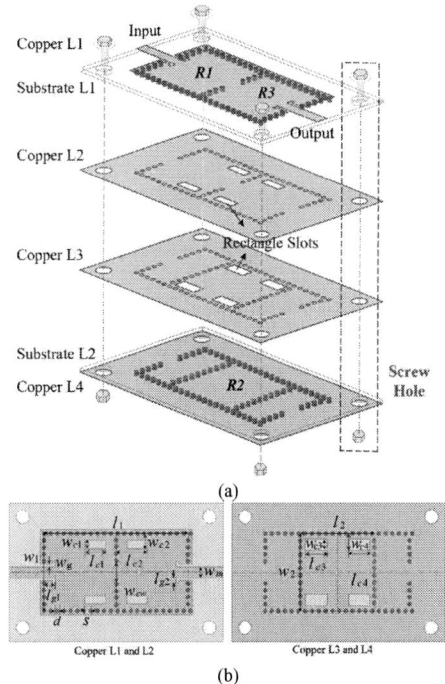

Fig.2. Construction of the cross-coupling BPF with screws. (a) 3-D layered perspective. (b) Top view and details (w_1=5.6, l_1=10.8, w_2=5.6, l_2=5.57, w_g=0.2, l_{g1}=0.85, l_{g2}=0.39, w_m=0.78, w_{cw}=1.4, w_{c1}=0.6, l_{c1}=1.55, w_{c2}=1.1, l_{c2}=2.35, w_{c3}=0.8, l_{c3}=1.6, w_{c4}=1.2, l_{c4}=1.98, d=0.3, s=0.5. Unit:mm.).

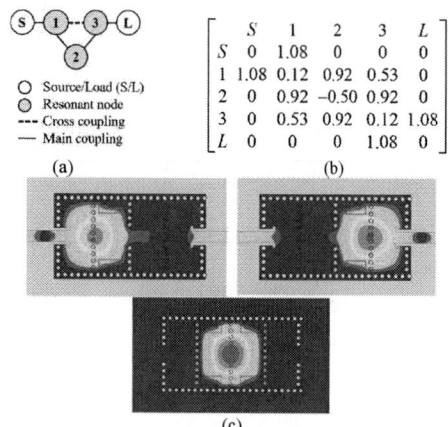

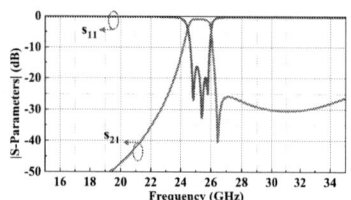

(a)

$$\begin{bmatrix} & S & 1 & 2 & 3 & L \\ S & 0 & 1.08 & 0 & 0 & 0 \\ 1 & 1.08 & 0.12 & 0.92 & 0.53 & 0 \\ 2 & 0 & 0.92 & -0.50 & 0.92 & 0 \\ 3 & 0 & 0.53 & 0.92 & 0.12 & 1.08 \\ L & 0 & 0 & 0 & 1.08 & 0 \end{bmatrix}$$

(b)

○ Source/Load (S/L)
◎ Resonant node
--- Cross coupling
— Main coupling

(c)

Fig.3. Simulated results of the cross coupling BPF with Screws. (a) Coupling topology. (b) Coupling matrix. (c) E-field distributions.

Fig.4. Simulated transmission responses of the cross coupling BPF with Screws.

II. COMPARE AND ANALYSIS

Fig. 1 illustrates two vertical interconnect methods utilized for multilayer filters. The initial approach involves inserting screws through the screw holes and subsequently tightening them, as depicted in Fig. 1(a). Meanwhile, the second method employs prepreg to directly bond the two copper layers, as shown in Fig. 1(b).

A. Multilayer cross coupling BPF with screws

Fig. 2(a) shows the 3-D layered perspective of the cross coupling BPF with screws. The structure comprises two full-mode SIW cavities in the upper layers and one full-mode SIW cavity in the lower layers. The initial size of the full-mode cavity can be calculated based on the following equation [4]:

$$f_{TE_{m0n}} = \frac{c}{2\sqrt{\mu_r \varepsilon_r}} \sqrt{\left(\frac{m}{l}\right)^2 + \left(\frac{n}{w}\right)^2} \qquad (1)$$

where m and n represent the mode indices along the x-axis and y-axis, respectively, and l and w denote the equivalent length and width of the full-mode cavity. ε_r and μ_r are the relative permittivity and permeability of the substrate, respectively. The top view and details are shown in Fig. 2(b).

A coupling window is created by blind vias on the upper layers to establish the main magnetic coupling from resonator

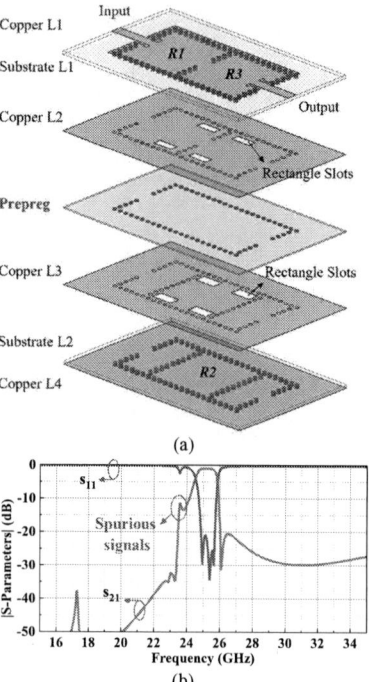

(a)

(b)

Fig.5. Construction of the cross coupling BPF with prepreg. (a) 3-D layered perspective. (w_1=5.6, l_1=10.8, w_2=5.6, l_2=5.52, w_g=0.2, l_{g1}=0.85, l_{g2}=0.39, w_m=0.78, w_{cw}=1.4, w_{c1}=0.6, l_{c1}=1.55, w_{c2}=1.1, l_{c2}=2.35, w_{c3}=0.8, l_{c3}=1.6, w_{c4}=1.2, l_{c4}=1.98, d=0.3, s=0.5. Unit:mm.). (b) Simulated transmission responses.

$R1$ to $R3$. Eight rectangular coupling slots are etched on Copper L2 and L3, facilitating the cross coupling signals from $R1$ to $R2$ and subsequently to $R3$. Based on the configuration, the coupling topology and the coupling matrix can be built as depicted in Fig. 3(a) and 3(b). The main coupling between $R1$ and $R3$, and the cross-coupling between $R1$ and $R2$, as well as $R2$ and $R3$, are all magnetic couplings. Three SIW cavities are all operated on TE_{101} modes with a working frequency of 25 GHz, and the simulated E-field distributions of three resonators are shown in Fig. 3(c).

Fig. 4 illustrates the simulated transmission responses of the cross-coupling bandpass filter (BPF) employing screws. The transmission zero is situated at 26.45 GHz, and the stopband performance is satisfactory. However, this vertical layer interconnect method requires space for screws, and improper tightening during mechanical fastening may create gaps between the two copper layers, potentially affecting the insertion loss of the passband.

B. Multilayer cross coupling BPF with prepreg

Fig. 5(a) shows the 3-D layered perspective of the cross coupling BPF with prepreg. The selected cavity and coupling topology of the filter are identical to the previous one. Notably, this filter eliminates the need for screw holes, with prepreg directly inserted between Copper L2 and L3.

979-8-3503-5124-8/24 $31.00 © 2024 IEEE   98

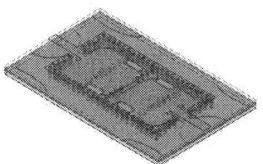

Fig.6. Simulated E-field distributions in the prepreg layer of the cross-coupling BPF with prepreg.

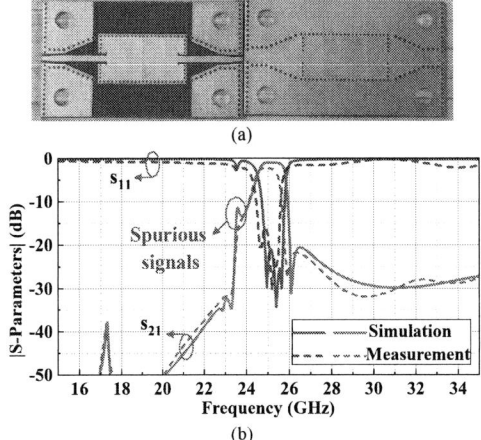

Fig.7. (a) Photograph. (b) Simulated and measured transmission responses of the cross-coupling BPF with prepreg.

Fig. 5(b) illustrates the simulated transmission responses of the filter. The passband performance is satisfactory, with the transmission zero accurately positioned in the upper stopband. However, spurious signals emerge at 23.50 GHz, peaking at -12.3 dB of $|S_{21}|$. A comparison with the transmission response of the previous filter reveals that this issue stems from the prepreg. As shown in Fig. 5(a), the through via forms a resonant cavity, denoted as R_{pp}, within the prepreg. Furthermore, Fig. 6 presents the simulated E-field distributions of the resonator R_{pp} at 23.5 GHz, demonstrating the excitation of the TE_{102} mode.

The filter employing prepreg can achieve vertical layer interconnect directly during the lamination process, eliminating the requirement for subsequent mechanical fastening. This results in improved integration and ensures a more stable in-band performance.

III. RESULTS AND DISCUSSION

The proposed cross coupling SIW BPF with prepreg utilizes a layer stack comprising the substrate *TLY-5* from Taconic with a relative permittivity ϵ_r = 2.2 and a loss tangent of $\tan \delta$ = 0.0009, and the prepreg *FR-25-0021-45* from Taconic with a relative permittivity ϵ_r = 2.43 similar to *TLY-5*. The thickness of Substrate L1, Prepreg, and Substrate L2 is 0.254mm, 0.1mm, and 0.254mm, respectively.

The filter is simulated with ANSYS HFSS software and manufactured using a standard printed circuit board (PCB) process. Subsequently, it is measured utilizing a vector network analyzer and southwest microwave connectors. The photograph of the filter is depicted in Fig. 7(a). The simulation and measurement results are presented in Fig. 7(b). The measurement reveals a center frequency (CF) of 24.95 GHz, 3-dB fractional bandwidth (FBW) of 3.0%, minimum inband insertion loss (IL) of -2.34 dB, in-band return loss (RL) exceeding 15.96 dB, and a TZ at 25.9 GHz. The nearest spurious signal is located at 23.6 GHz. Although the spurious spike is smaller in the measurement compared to the simulation, they still degrade the out-of-band performance of the filter.

In most cases [3]-[8], filters use screws as the vertical layer interconnect method. However, as working frequencies increase and filter sizes decrease, screws become less suitable for integrating components in communication systems. While prepreg can eliminate the need for mechanical tightening, directly bonding two copper layers with prepreg may introduce spurious signals. Therefore, prepreg should be avoided between two copper layers in future designs. Instead, utilizing the prepreg to bond a copper layer to a substrate layer should be considered.

IV. CONCLUSION

In this paper, two vertical layer interconnect methods for multilayer filters—prepreg and screws—are presented and analyzed. Initially, a cross coupling BPF mounted with screws is designed and analyzed. Subsequently, a filter with the same coupling topology but interconnected with prepreg is designed, fabricated, and compared. In conclusion, although the method of using screws is simple and straightforward, it is detrimental to integration. While the prepreg improves integration, the filters should avoid directly bonding two copper layers with prepreg. Instead, prepreg should bond a copper layer to a substrate directly in future designs.

REFERENCES

[1] A. Iqbal, J. J. Tiang, S. K. Wong, M. Alibakhshikenari, F. Falcone and E. Limiti, "Miniaturization Trends in Substrate Integrated Waveguide (SIW) Filters: A Review," *IEEE Access*, vol. 8, pp. 223287-223305, 2020.

[2] K. Wu, M. Bozzi and N. J. G. Fonseca, "Substrate Integrated Transmission Lines: Review and Applications," *IEEE Journal of Microwaves*, vol. 1, no. 1, pp. 345-363, Jan. 2021.

[3] P. Chu et al., "Using Mixed Coupling to Realize Wide Stopband Multilayer Substrate Integrated Waveguide Filter," *IEEE Transactions on Circuits and Systems II: Express Briefs*, vol. 70, no. 8, pp. 2744-2748, Aug. 2023.

[4] P. Chu, J. Feng, L. Guo, L. Zhang, L. Liu and K. Wu, "Multilayer Substrate Integrated Waveguide Filter With Multimode Suppression and Wide Stopband," *IEEE Transactions on Circuits and Systems II: Express Briefs*, vol. 69, no. 11, pp. 4553-4557, Nov. 2022.

[5] L. Gu and Y. Dong, "Compact Half-Mode SIW Filter With High Selectivity and Improved Stopband Performance," *IEEE Microwave and Wireless Components Letters*, vol. 32, no. 9, pp. 1039-1042, Sept. 2022.

[6] W. Shen and H. -R. Zhu, "Vertically Stacked Trisection SIW Filter With Controllable Transmission Zeros," *IEEE Microwave and Wireless Components Letters*, vol. 30, no. 3, pp. 237-240, March 2020.

[7] K. Zhou, C. -X. Zhou and W. Wu, "Substrate-Integrated Waveguide Dual-Mode Dual-Band Bandpass Filters With Widely Controllable Bandwidth Ratios," *IEEE Transactions on Microwave Theory and Techniques*, vol. 65, no. 10, pp. 3801-3812, Oct. 2017.

[8] G. Zhang, X. Zhou, J. Hu, K. Tam, D. Li, Z. Zhang, T. Ma, and J. Yang, "Multilayer Packaged Single- and Dual-Band Dual-Channel Bandpass Filters Based on Substrate Integrated Waveguide Cavities," *IEEE Transactions on Components, Packaging and Manufacturing Technology*, vol. 14, no. 1, pp. 114-121, Jan. 2024.

Yield-Aware Interposer Design for UCIe Interconnects

Ram Krishna*, Ashita Victor†, Srujan Penta†, Xu Chen*, Muhannad S. Bakir†, Nam Sung Kim*, and Elyse Rosenbaum*

*School of Electrical and Computer Engineering, University of Illinois Urbana-Champaign, Urbana, USA
†School of Electrical and Computer Engineering, Georgia Institute of Technology, Atlanta, USA
Email: ramk3@illinois.edu, avictor8@gatech.edu

Abstract—This work presents an interposer design methodology for UCIe die-to-die interfaces that maximizes yield given a set of signal integrity specifications. Four different routing configurations on a silicon interposer are considered, and the cost, performance, and yield tradeoffs are elucidated. The fabrication steps for the D2D interconnects are outlined and sources of yield-limiting variability are identified. The yield analysis is expedited by the use of a Gaussian Process Regression surrogate model.

Index Terms—UCIe, heterogeneous integration, interposer, yield, signal integrity, insertion loss, crosstalk

I. INTRODUCTION

Heterogeneous integration (HI) of separately manufactured chiplets may provide significant benefits over the monolithic system-on-chip (SoC) in terms of flexibility, cost, and yield. The scaling down of chiplet dimensions and the ability to integrate chiplets of different technology nodes aids in cost reduction and yield improvement. HI systems require die-to-die (D2D) interfaces for communicating data among the chiplets.

Universal Chiplet Interconnect Express (UCIe) is an open-source industry-standard interconnect technology that is receiving significant attention because of its several favorable properties, such as low cost, low latency, high bandwidth, and high interoperability between chiplets [1]. Recent works consider architectural and high-level implementations of UCIe [2], [3].

However, no prior work has delved into the interposer architecture and routing strategy required to meet the UCIe signal integrity specifications. Those specifications include voltage transfer function (VTF) loss, VTF crosstalk, and eye height [1]. Additionally, there is a need to consider manufacturing ("process") variations while designing the interposer. In this work, the design point that maximizes yield is termed the optimal design.

This work is supported in part by PRISM and CHIMES, two of seven centers in JUMP 2.0, a Semiconductor Research Corporation (SRC) program sponsored by DARPA. It is also supported in part by Global Research Collaboration (GRC), an SRC program. This work was performed in part at the Georgia Tech Institute for Electronics and Nanotechnology, a member of the National Nanotechnology Coordinated Infrastructure (NNCI), which is supported by the National Science Foundation under Grant ECCS-2025462.

UCIe specifications dictate that 19 neighbors be considered when calculating crosstalk [1]. An electromagnetic (EM) simulation to extract S-parameters of a layout with 19 neighbors takes 20 minutes on an Intel Xeon CPU with 8 cores and 32 GB memory. Thus, a complete design space exploration is infeasible. This work proposes using a Gaussian Process Regression (GPR) [4] surrogate model to expedite the design space exploration. The model is demonstrated in the context of a silicon interposer technology.

II. DESIGN METHODOLOGY

A. Routing Configurations

The four different routing configurations depicted in Fig. 1 were investigated. There are four design variables: conductor height (C_H), dielectric thickness (D_T), line width (L), and spacing (S). Each UCIe module has a fixed beachfront length of 388.8 μm [1]. This imposes a constraint on the pitch of the D2D interconnects. This work assumes an x32 UCIe module, and the corresponding pitch is 4.8 μm for configurations 1 and 4. That is the maximum distance between two wires and was calculated based on the number of signal lines that need to pass through the fixed beachfront length in an x32 UCIe module. The pitch for configurations 2 and 3 is 9.6 μm. There are cost and performance tradeoffs among the configurations. Configurations 1 and 2 are the least expensive because they require only three layers, while configuration 3 is the most costly because it has the largest number of layers and fabrication steps. Routing density is defined as the number of wires that are accommodated in one layer. Configuration 1 has the highest routing density because all the signal lines are routed in one layer. Configuration 4 has same routing density as configuration 1 but an additional layer is required to route all the signal lines. Configuration 4 is a checkerboard pattern with lateral shielding, which is supposed to provide better crosstalk immunity than configuration 1.

B. Interposer Properties

The interposer technology considered in this work is based on a wafer-scale packaging technology called Integrated Chiplet Encapsulation (ICE) [5]. The key feature of this technology is that the chiplets are encapsulated in low-temperature

979-8-3503-5124-8/24 $31.00 © 2024 IEEE

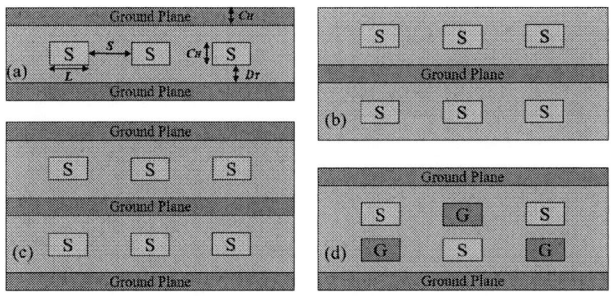

Fig. 1. Illustration of interposer architectures considered for the study, viz., (a) Configuration 1 with one routing and two ground layers, (b) Configuration 2 with two routing and one ground layers, (c) Configuration 3 with two routing and three ground layers, and (d) Configuration 4 with two layers of checkerboard routing and two ground layers.

silicon dioxide to form a reconstituted-SiO_2 chiplet tier. These chiplet tiers can then be stacked in 3D to enable high-density heterogeneous integration. The D2D interface for a single tier of this packaging technology resembles a silicon interposer. The results presented in this paper reflect the minimum achievable feature size and the process control for this specific interposer technology.

A summary of the fabrication process flow for a structure that resembles configuration 1 from Fig. 1 is as follows. First, the bottom ground plane is fabricated by evaporating 30-nm/1-μm of Ti/Cu, followed by the deposition of approximately 1-μm-thick low-temperature (100 °C) SiO_2. The SiO_2 interlayer dielectric is deposited using ICP-PECVD (Inductive Coupled Plasma Enhanced Chemical Vapor Deposition) at a rate of approximately 1 μm/hr. Next, signal traces are fabricated using a lift-off process. For a recent testbed, signal traces with an L/S of 2.8/2 μm and a conductor thickness of approximately 1 μm (Ti/Cu of 30-nm/1-μm) were fabricated, showing the current capabilities of this technology. 2-μm-thick ICP-PECVD SiO_2 is deposited to ensure complete oxide infill between the signal traces, as shown in Fig. 2(a). The excess oxide is then planarized using a benchtop Bruker Tribolab CMP. A silica-based slurry is used to remove the excess oxide at a removal rate of 100 nm/min. Next, an additional 1-μm-thick SiO_2 layer is deposited prior to depositing the top ground plane, as shown in Fig. 2(b). The top ground plane is fabricated by evaporating 30-nm/1-μm/30-nm of Ti/Cu/Au.

C. Design Space Exploration Setup

Fig. 3 shows the design flow used to train a GPR model of VTF loss and VTF crosstalk. The GPR model uses a radial basis function (RBF) kernel; the Broyden-Fletcher-Goldfarb-Shanno (BFGS) algorithm was used to optimize the kernel hyperparameters. The feasible design space was estimated based on the UCIe pitch and the smallest feature size allowed by the fabrication process. EM simulation was performed for 200 feasible design points for each of the four configurations of Fig. 1. Each of the simulated structures contains 20 signal lines, and the worst-case VTF loss and crosstalk among the the 20 signal lines were extracted. 100 arbitrarily-selected design points were used to train a GPR model for each of the four

(a)

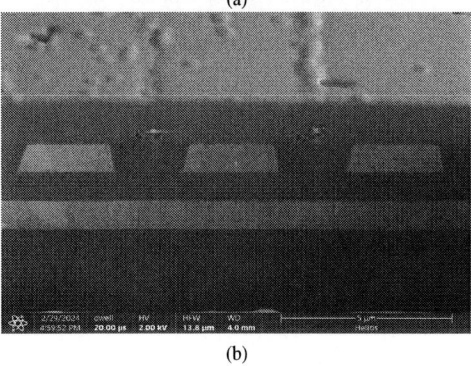

(b)

Fig. 2. FIB-SEM images of the fabricated structures. (a) Oxide deposition over the signal traces (L/S = 2.8/2 μm) and (b) Oxide planarization using CMP, followed by additional oxide deposition.

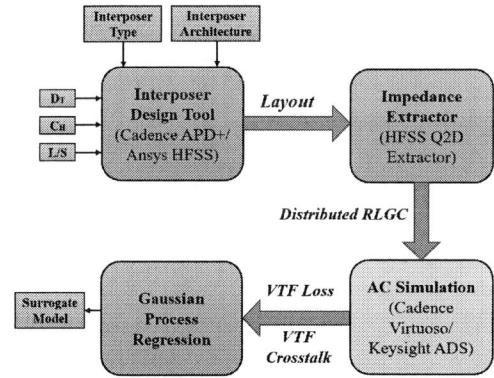

Fig. 3. Training flow for a GPR model of a D2D interface.

configurations, and the rest are used to test the model accuracy. Average accuracy (AA) is defined as:

$$AA(\%) = 100 - \frac{1}{N}\sum_{i=1}^{N}\left(\left|\frac{(AV)_i - (PV)_i}{(AV)_i}\right| \times 100\right) \quad (1)$$

AA is calculated using the actual value (AV) obtained from the EM simulation and the predicted value (PV) obtained from the trained GPR model. N is the number of samples in the test set. The trained GPR model gave an average accuracy of 96% for VTF loss and 91% for VTF crosstalk. The prediction accuracy and the confidence interval for VTF loss and VTF crosstalk for configuration 1 are shown in Fig. 4.

On the basis of test structure measurements, the distri-

979-8-3503-5124-8/24 $31.00 © 2024 IEEE   101

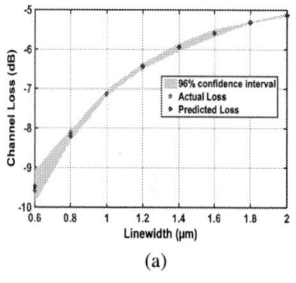 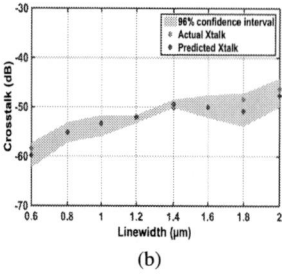

(a)   (b)

Fig. 4. (a) VTF loss and (b) VTF crosstalk of 2-mm long UCIe interconnects with 4.8-μm pitch, 1-μm conductor height and 1-μm dielectric thickness.

bution for each design variable was fitted to a Gaussian distribution. For each configuration, 10,000 design points were sampled from the four Gaussian distributions using Monte Carlo sampling [6]. The performance metrics at those points were obtained from the trained GPR model and subsequently used to calculate the yield and its confidence interval. Here, yield is defined as the percentage of samples that meet the UCIe channel specifications.

III. Results and Analysis

Design space exploration for one configuration with 10,000 design points takes 120 minutes. Thus, the GPR model provides significant improvement in runtime over EM simulations.

Table I summarizes the results of the design space exploration for each of the four routing configurations. It shows the VTF loss and VTF crosstalk values of the design point with the highest yield. Configuration 2 has the lowest yield because it has only one ground plane, leading to higher values of crosstalk. Yield is highest for configurations 3 and 4. The lateral shielding and checkerboard pattern of configuration 4 reduces the crosstalk considerably, thereby increasing the number of samples that meet the specifications. Configuration 3 uses an additional layer of routing, and because each signal is shielded by a ground plane at the top and bottom, the yield is similarly improved due to the reduction in crosstalk.

As demonstrated by the results of Table I, the yield is less than 100% due to the within-wafer variation of the design parameters C_H, D_T, L, and S. Typically, variations in L and S occur during the photoresist development stage, and variability of C_H and D_T is affected by the chamber pressure and the deposition rate, among other factors. By improving the process control, one can reduce the variability of the design parameters across a wafer, which has a beneficial effect on the achievable yield. Fig. 5 depicts the effect of reducing the variance of all design variables simultaneously. It demonstrates that the yield is a function of both the process variations and the design point, thereby highlighting the importance of the design point selection. As shown in the figure, if the selected design point is that which results in the least crosstalk while satisfying the other specifications, reducing the variability provides only a marginal improvement in the yield. This is because the minimum-crosstalk design point lies at the edge of the VTF loss specification. However, if the selected design

TABLE I. Maximum yield across the feasible design space for different configurations. VTF Loss and VTF Crosstalk values at 16 GHz are listed.

Configurations	VTF Loss (dB)	VTF Crosstalk (dB)	Yield (%)
1	-3.78	-30.86	62.02
2	-3.64	-34.37	49.88
3	-2.79	-50.23	99.33
4	-1.30	-37.36	99.87

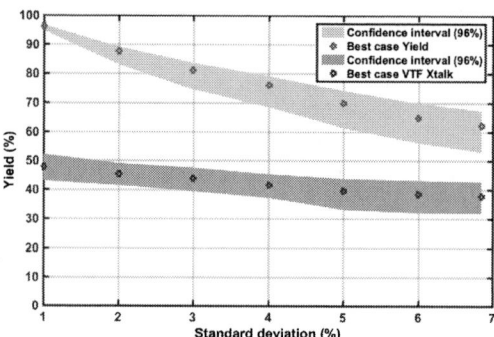

Fig. 5. Effect of process control at two design points (one which provides maximum yield and the other which provides the least crosstalk). Configuration 1.

point is that which maximizes the yield given the nominal process variations, the yield will be improved significantly by a tightening of the variance, as neither the loss nor crosstalk are at the edge of the specification range.

IV. Conclusion

A novel approach for yield-aware design space exploration for a D2D interface is presented. The Gaussian process regression-based surrogate model used to predict the channel performance metrics provided a 1000× runtime improvement with a reasonable accuracy of 90-96%. With the GPR surrogate model, it is possible to identify the design point with the highest yield, and quantify how much further the yield could be improved by reducing the interposer fabrication process variations. Future work includes the testing and verification of fabricated test structures.

References

[1] Universal Chiplet Interconnect Express (UCIe) Specification, 7 2023, rev. 1.1.

[2] D. Das Sharma, G. Pasdast, Z. Qian, and K. Aygun, "Universal Chiplet Interconnect Express (UCIe): An open industry standard for innovations with chiplets at package level," IEEE Transactions on Components, Packaging and Manufacturing Technology, vol. 12, no. 9, pp. 1423–1431, 2022.

[3] D. D. Sharma, "System on a package innovations with universal chiplet interconnect express interconnect," IEEE Micro, vol. 43, no. 2, pp. 76–85, 2023.

[4] C. Williams and C. Rasmussen, "Gaussian processes for regression," in Advances in Neural Information Processing Systems, D. Touretzky, M. Mozer, and M. Hasselmo, Eds., vol. 8. MIT Press, 1995.

[5] M.-J. Li and M. S. Bakir, "3-D Integrated Chiplet Encapsulation (3-D ICE): High-density heterogeneous integration using SiO2-Reconstituted Tiers," IEEE Transactions on Components, Packaging and Manufacturing Technology, vol. 11, no. 12, pp. 2242–2245, 2021.

[6] W. Hu, Z. Wang, S. Yin, Z. Ye, and Y. Wang, "Sensitivity importance sampling yield analysis and optimization for high sigma failure rate estimation," in 2021 58th ACM/IEEE Design Automation Conference (DAC), 2021, pp. 895–900.

Efficient Thermal Analysis for Heat Dissipation in Three-Dimensional Chip-Stacking Packaging

Cheng-Yuan Lu, Chien-Min Lin, *Senior Member, IEEE*, and Ruey-Beei Wu, *Fellow, IEEE*
Graduate School of Advanced Technology, National Taiwan University, Taipei, Taiwan, ROC
r12k43015@ntu.edu.tw, cmlinster@gmail.com, rbwu@ntu.edu.tw

Abstract — **Advances in high-performance computing are driving thermal analysis for 3D chip-stacking packaging to assess its heat dissipation. Rather than the finite-element method, an efficient and accurate semi-analytic estimation method of temperature distribution is proposed.**

Keywords— *Chip-stacking packaging, Fourier transform, thermal analysis.*

I. INTRODUCTION

With an advancement of the prevailing high-speed, high-bandwidth, and high-power applications, the development of high-performance and heterogeneous packaging has become increasingly important. To integrate all the edge devices, the planar space is limited to accommodate the multi-chips with the diverse functionalities and the chip-stacking architecture is a promising trend [1]. Since the high-power consumptions within those high-performance multi-chips are significant, the thermal analysis for the heat dissipation over the modules emerges the very attention [2].

To analyze the temperature distribution in complex structures, the finite-element method (FEM) is often used but time-consuming. Early in 1974, Kokkas developed a Fourier series model to characterize the frequency response of the surface temperature in a three-layer structure [3]. Albers thereof extended the recursion relation technique in 1995 to solve for the steady-state surface temperature of general multilayer structures [4].

For a multi-layered packaging as exemplified in Fig. 1, the Fast Fourier Transform (FFT) on its equivalent stacked structure is derived for a recursive relation of the thermal conductance on account of the uniform heat sources within the multi-layered structure. The various thermal conductivity and thickness of layered materials are the primary concerns in the thermal distribution through this chip-stacking system. Compared to the FEM-based simulations using Ansys Icepak [5], the proposed methodology achieves computational efficiency and consistent accuracy. Moreover, the FFT scheme accomplishes faster computation than the traditional Fourier series method.

II. ANALYTICAL MODEL FOR HEAT DISTRIBUTION

To efficiently analyze the thermal distribution, the heat dissipation over the stacking-structure as depicted in Fig. 2 is assessed. The infinitely large and uniform layers over the x-y plane are assumed with the material variations of thermal conductivity $\kappa(z)$ along the z direction. A two-dimensional uniform heat-flux magnitude q_0 extents on both the x and y axes to be confined within the zero-thickness plane size of W_x

This study is supported in part by MediaTek Inc., Hsinchu, Taiwan, ROC, under the Grant: 112HZA38003.

In acknowledgement of the technical support and advice from Ting-Li Yang, Cheng-Lin Huang, and Kai-Chieh Yang associated with MediaTek Inc., Hsinchu, Taiwan, ROC.

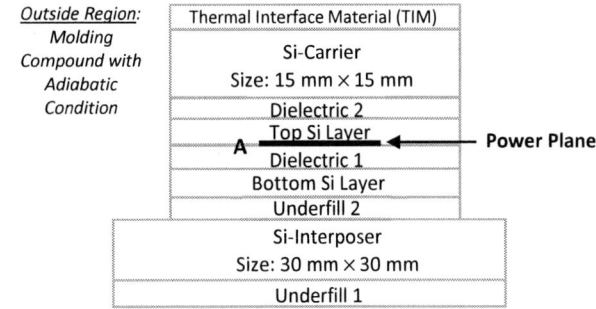

Fig. 1. Layered Structure for the analysis model

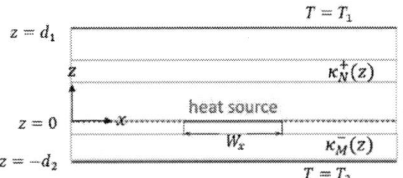

Fig. 2. A multilayer structure with the uniform heat-source

$\times W_y$ at $z = 0$. The heat transfer mechanism thus governs the temperature distribution along the z axis.

A. Governing Equation of Thermal conduction

For a multi-layered structure with the uniform heat-source setting as depicted in Fig. 2, the governing equation of the steady-state thermal conduction in the region outside of the heat source is given by Laplace equation

$$\nabla \cdot (\kappa \nabla T) = 0 \qquad (1)$$

where T represents the temperature distribution and $\kappa(z)$ is the thermal conductivity of the specific layer materials. The boundary conditions are assumed as

$$T(x, y, z = d_1) = T_1 \quad \text{and} \quad T(x, y, z = -d_2) = T_2 \qquad (2)$$

where d_1 and d_2 are the distances between the source and top/bottom-boundaries, respectively.

At the heat source plane $z = 0$, the heat flux $\vec{q} = -\kappa \nabla T$ is referred to the Fourier's law of heat conduction. Considering the heat flow in both the upward and downward directions, the continuity condition at $z = 0$ is given by

$$\vec{q}(x, y) = -\kappa^+(z) \left. \frac{\partial T}{\partial z} \right|_{z=0^+} + \left(\kappa^-(z) \left. \frac{\partial T}{\partial z} \right|_{z=0^-} \right) = q_0 \quad (3)$$

over $|x| \leq \frac{W_x}{2}$ & $|y| \leq \frac{W_y}{2}$; whereas, zero elsewhere. Note that q_0 is a constant corresponding to the uniform heat-flux and $W_x \times W_y$ is the plane size.

B. Computational Decomposition for Stacked Layers

By the superposition theorem, the thermal conduction in a multilayer structure can be decomposed into the two parts:

979-8-3503-5124-8/24 $31.00 © 2024 IEEE

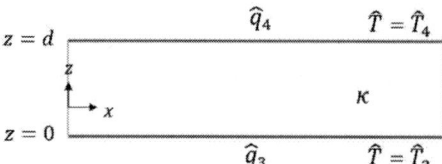

Fig. 3. Spectral domain analysis for the sides of a single uniform layer.

1) No Heat Source: Without the heat source, i.e., $q_0 = 0$, Eq. (1) is thus solved for the temperature distribution along its vertical axis as a one-dimensional problem. With Eq. (2), the temperature at the source posistion $z = 0$ is accordingly found as

$$T_0(z=0) = \frac{\int_{-d_2}^{0} dz/\kappa(z)}{\int_{-d_2}^{d_1} dz/\kappa(z)} T_1 + \frac{\int_{0}^{d_1} dz/\kappa(z)}{\int_{-d_2}^{d_1} dz/\kappa(z)} T_2 \quad (4)$$

or in a more physical expression for a multilayer structure,

$$T_0(z=0) = \frac{R^-}{R^+ + R^-} T_1 + \frac{R^+}{R^+ + R^-} T_2 \quad (5)$$

where

$$R^- = \int_{-d_2}^{0} \frac{dz}{\kappa(z)} = \sum_{i=1}^{M} \frac{l_i^-}{\kappa_i^-}; \quad R^+ = \int_{0}^{d_1} \frac{dz}{\kappa(z)} = \sum_{i=1}^{N} \frac{l_i^+}{\kappa_i^+}$$

indicates the sum of thermal resistance of lower layers ($z < 0$) and upper layers ($z > 0$), respectively.

2) Zero-Temperature Boundary with Heat Source: Thermal effect of the heat source with a zero-temperature on both the top and bottom boundaries is further analyzed. The heat transfer equation is thus solved with the heat source present but with the zero-temperature imposed on the boundaries.

To solve Eq. (1) for the temperature distribution $T(x, y, z)$, Fourier transform is employed to convert spatial domain into spectral domain by

$$\mathcal{F}(T) = \hat{T}(k_x, k_y, z)$$
$$= \frac{1}{2\pi} \iint_{-\infty}^{\infty} T(x, y, z) e^{-jk_x \cdot x - jk_y \cdot y} \, dx \, dy \quad (6)$$

Considering κ as a constant within each homogeneous layers, Eq. (1) becomes an ordinary differential equation as

$$\kappa^{\pm} \left(\frac{d^2 \hat{T}}{dz^2} - k_\rho^2 \hat{T} \right) = 0; \text{ for } z \gtrless 0 \quad (7)$$

where $k_\rho^2 = k_x^2 + k_y^2$. Note that the spectral solution of Eq. (7) is readily available as

$$\hat{T}(z) = c_1 \cosh(k_\rho z) + c_2 \sinh(k_\rho z) \quad (8)$$

along with the heat flux derived as

$$\hat{q}(z) = -\kappa^{\pm} \frac{d\hat{T}}{dz} = -c_1 \kappa^{\pm} k_\rho \sinh(k_\rho z) - c_2 \kappa^{\pm} k_\rho \cosh(k_\rho z) \quad (9)$$

For a multilayer structure, the spectral temperature $\hat{T}(z)$ and its derivative $\hat{q}(z)$ should be continuous across the layer boundary, except at the heat source plane $z = 0$ to satisfy Eq. (3). The two constants c_1 and c_2 on each layer can be solved iteratively.

To be more specific, a layer of uniform material shown in Fig. 3, which temperature distribution above is $\hat{T}_4$ and below

Fig. 4. Spectral thermal conductance recursion for the multilayer structure.

is $\hat{T}_3$, while the heat flux $\hat{q}_4$ above and $\hat{q}_3$ below. Then, from Eq. (8), the temperature inside the layer can be written as

$$\hat{T}(z) = \hat{T}_3 \cosh\left(k_\rho(d - z)\right) + c_2 \sinh\left(k_\rho(d - z)\right) \quad (10)$$

and the heat flux from Eq. (9) as

$$\hat{q}(z) = \kappa k_\rho \hat{T}_3 \sinh\left(k_\rho(d - z)\right) + \kappa k_\rho c_2 \cosh\left(k_\rho(d - z)\right) \quad (11)$$

This gives $\hat{q}_4 = \kappa k_\rho c_2$ since $\hat{q}(z) = \hat{q}_4$ at $z = d$.

Substituting $z = 0$ to Eq. (10) and (11) yields $\hat{T}_4$ and $\hat{q}_4$ in the left-hand-side, respectively. To infer the spectral-domain thermal conductance $G(k_\rho)$ as

$$\frac{\hat{q}_3}{\hat{T}_3} = \frac{\kappa k_\rho \hat{T}_4 \sinh(k_\rho d) + \hat{q}_4 \cosh(k_\rho d)}{\hat{T}_4 \cosh(k_\rho d) + \frac{\hat{q}_4}{\kappa k_\rho} \sinh(k_\rho d)}$$
$$= \frac{\frac{\hat{q}_4}{\hat{T}_4} + \kappa k_\rho \tanh(k_\rho d)}{1 + \frac{\tanh(k_\rho d)}{\kappa k_\rho} \frac{\hat{q}_4}{\hat{T}_4}} \quad (12)$$

In Eq. (12), $G(k_\rho)$ between each layer can be written as

$$G_{n-1}(k_\rho) = \frac{G_n + \kappa_n k_\rho \tanh(k_\rho d_n)}{1 + \frac{\tanh(k_\rho d_n)}{\kappa_n k_\rho} G_n} \quad (13)$$

It is noteworthy that in the uppermost layer, $\hat{T}_4 = 0$ in Eq. (12) and the initial values $G(k_\rho) = \kappa k_\rho \coth(k_\rho d)$.

A similar relation is applied for the lower layers. Starting from both the sides to further derive it toward the specific heat source, $G(k_\rho)$ can be calculated as depicted in Fig. 4.

Given the total heat flux $\hat{q}$ at $z = 0$, which is the Fourier transform of the spatial heat-source distribution, the spectral-domain thermal distribution is thus given by $\hat{T} = \hat{q}/G(k_\rho)$ where $G(k_\rho) = G_1^+(k_\rho) + G_1^-(k_\rho)$ from Fig. 4. Performing an inverse Fourier transform (IFFT) on the $\hat{T}$ can acquire the desired temperature distribution in the real space, accordingly.

979-8-3503-5124-8/24 $31.00 © 2024 IEEE

C. Temperature Distribution by Fast Fourier Transform

To facilitate the computational efficiency, the generic Fast Fourier Transform (FFT) is invoked in aid of the tools with MathWorks Matlab [6] for the thermal analysis.

To invoke the 2^N-stage algorithm in FFT, a periodicity of $G(k_\rho)$ should be ensured to match the characteristic of $\hat{q}$ for an accurate computation. The temperature distribution due to the heat source in the layered model is thereof acquired as

$$T_s(z = 0) = IFFT\{\hat{T}(z = 0)\} \tag{14}$$

The actual temperature distribution on the heat-source layer is summed up by $T = T_0 + T_s$ with the superposition of T_0 for "*No Heat Source*" and T_s for "*Zero-Temperature Boundary with Heat Source.*"

III. CALCULATE ACCURACY COMPARISON

To verify the accuracy of the proposed model and analysis scheme, the temperature distribution data are compared to the simulation results by using Ansys Icepak. A multi-layered structure in Fig. 1 along with the material settings and design parameters in Table I is assumed. The upper portion above $z = 0$ consists of the stacked chips and the lower portion is the silicon interposer. 100 watts of the heat source is uniformly distributed on a plane of $W_x \times W_y = 5 \times 5$ mm^2 at the center of the chip. As being a sensitive parameter with the varying thickness of Layer Dielectric 2, the maximum temperature observed at heat source center point is analyzed.

In Fig. 5, the maximum temperature increases from 66°C to 72° C while the thickness of Layer Dielectric 2 is increased from 1.5 μm to 5.0 μm. The analysis result by the proposed method is compared with that by Ansys Icepak to have an error less than 1%. Moreover, the computational time by the proposed method achieves a 100-fold efficiency against that by Ansys Icepak.

As a function of the position along the x axis across the power plane, the temperature distribution is shown in Fig. 6. The temperature profiles within the x-axis size of 15 mm for a chip-stacking model are consistent with the analysis results by both Ansys Icepak and the proposed method. Outside the modeling region, the discrepancy arises because the molding compound is not accounted in the proposed method.

IV. CONCLUSIONS

The promising technologies toward 3D chip-stacking heterogeneous integration post significant challenges in the thermal and heat-dissipation resolution. Due to the analysis

TABLE I.   DESIGN PARAMETERS IN THE LAYERED STRUCTURE

Layer name	Design Parameter	
	κ *(watt/m·K)*	*thickness (um)*
TIM	10	100
Si-Carrier	180	750
Dielectric 2	1.5	1.5~5.0
Top Si Layer	180	5~10
Dielectric 1	1.5	5
Bottom Si Layer	180	10~20
Underfill 2	2.8	40
Si-Interposer	180	50
Underfill 1	2.8	80

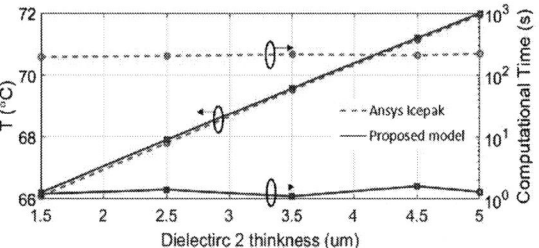

Fig. 5.   Maximum temperature and computatuon time (thickness of Top/Bottom Si Layer are 5/10 um, respectively)

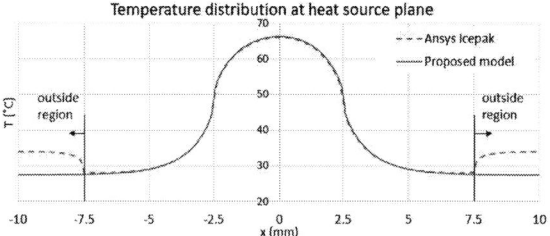

Fig. 6.   Temperature distribution result using Icepak (finite element method) and proposed analytical equation along x-axis

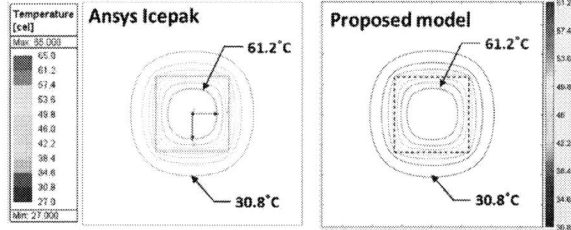

Fig. 7.   Temperture distribution contour on the x-y power plane at z = 0.

efficiency, a feasible model for the layered structure along with the fast computational algorithm is proposed to enable a reliable assessment of temperature distribution with the sensitive design parameters.

The model's simplification of thermal conductivity as a constant within each layer may lead to inaccuracies when there are significant variations or a large number of unevenly distributed components. Additionally, the Fourier transform's assumption of infinite and homogeneous x and y directions can introduce errors when the heat source size approaches the wafer size. Despite these limitations, this developed scheme is expected to analyze the thermal integrity of prevailing chiplet-packaging applications as a cost-effective tool.

REFERENCES

[1] J. H. Lau, "Recent advances and trends in advanced packaging," *IEEE Trans. Compon., Packag., Manuf. Technol.*, vol. 12, no. 2, pp. 228–252, Feb. 2022.

[2] L. Su and S. Naffzige, "Innovation for the next decade of compute efficiency," in *2023 IEEE Int'l Solid-State Circuits Conf.*, Feb. 19-23, 2023, San Francisco, California, USA.

[3] A. G. Kokkas, "Thermal analysis of multiple-layer structures," in IEEE Transactions on Electron Devices, vol. 21, no. 11, pp. 674-681, 1974.

[4] J. Albers. "An exact recursion relation solution for the steady-state surface temperature of a general multilayer structure." IEEE Transactions on Components, Packaging, and Manufacturing Technology: Part A., vol 18, no 1, 1995, pp 31-38.

[5] Icepak, ANSYS Inc. [Online]. Available: www.ansys.com

[6] MATLAB, MathWorks. [Online]. Available: www.mathworks.com

Design of an Ultra-High-Speed Digital Interface Based on a Coplanar Stripline

Mun-Ju Kim
School of Electronic and Electrical Engineering
Kyungpook National University
Daegu, Republic of Korea
dranswn@knu.ac.kr

Byung-Cheol Min
School of Electronic and Electrical Engineering
Kyungpook National University
Daegu, Republic of Korea
minbc4658@knu.ac.kr

Hyun-Chul Choi
School of Electronic and Electrical Engineering
Kyungpook National University
Daegu, Republic of Korea
hcchoi@ee.knu.ac.kr

Kang-Wook Kim
School of Electronic and Electrical Engineering
Kyungpook National University
Daegu, Republic of Korea
kang_kim@ee.knu.ac.kr

Abstract—**A design for an ultra-high-speed digital interface, providing autonomous signal integrity improvement, is proposed. The proposed interface is on a coplanar stripline and verified to perform from DC to 30 GHz for 5G/6G communications.**

Keywords—*ultra-high-speed interface, vertical transition, ultra-high-speed digital transmission*

I. INTRODUCTION

Nowadays, technologies capable of transmitting large amounts of digital data at high speeds of hundreds of Gbps are required for advancements in next-generation autonomous driving, artificial intelligence (AI), and 6G communications. A high-speed digital signal, composed of a main frequency component and its harmonics, occupies a significant frequency bandwidth, often reaching up to mm-wave frequencies, necessitating the application of microwave technology in designing digital signal transmission structures [1]. A digital interface, consisting of multiple digital signal transmission structures, connects between a circuit board and an external module, and plays a key role in the overall performance of digital signal transmission systems, requiring the development of practical digital interfaces for ultra-high-speed digital signal transmission.

High-speed digital interfaces are utilized in various fields such as cloud computing systems for large-volume data transmission, high-speed signal transmission between CPUs and GPUs in AI chips, and connection between external memory devices. A universal serial bus (USB) is one of the high-speed digital interfaces, facilitating the link from a circuit board to an external device through several vertical pins and vias, and exists in multiple versions depending on the transmission speed. For instance, the USB 3.2 offers the transmission speed of 5 Gbps for Gen 1 and 10 Gbps for Gen 2, while the USB 4 can reach up to 20 Gbps per lane [2]. For high-speed digital signal transmission, the transmission speed is often increased through parallel circuits and digital signal processing techniques. However, for ultra-high-speed digital transmission, requiring over hundreds of Gbps, the conventional structures show obvious limits on the maximum frequency bandwidth (~10 GHz), and securing an alternative transmission structure is crucial for the success of next-generation technologies.

Up to now, for high-speed digital signal transmission, a differential line (DL) consisting of two parallel microstrip lines with opposite polarities, supporting differential signaling, has been widely used. The DL offers advantages of operating at low voltages and being less susceptible to external noise compared to a single-ended line. However, as digital signal speed increases over 10 Gbps, the DL, within a complex circuit board with curves and vertical transitions through vias, is exposed to various unfavorable effects such as signal skew, uneven EM interference, and discontinuous line impedance [3]. Particularly for sections containing vertical vias, it is very difficult to cope with impedance discontinuity due to the abrupt change in electric field line shapes [4]. Therefore, the digital interfaces utilizing DL-based via structures are inherently limited in frequency bandwidth of around 10 GHz, making them unsuitable for ultra-high-speed digital transmission.

Research efforts have been made to enhance the performance of DL-based interfaces, and guidelines for signal line placement and via structures have been provided [5], [6]. Additionally, a study has focused on expanding the operating bandwidth by analyzing parasitic capacitance in DL-based via structures used in the interfaces [7]. Line coding schemes have also been devised for faster-speed interfaces [8]. Despite these guidelines and various efforts, DL-based interfaces have not been able to surpass the inherent performance limits, leading to the need for developing of a new ultra-high-speed digital transmission structure.

Planar balanced lines (BLs), such as coplanar stripline (CPS) and parallel stripline (PSL), accompanied by an ultra-wideband DL-to-BL transition, have been proposed for digital signal transmission with a frequency bandwidth exceeding 40 GHz [3]. A CPS comprises two signal lines placed in parallel on the same side of the substrate without an additional ground plane, and a PSL also consists of two parallel signal lines, with one on the top and the other on the bottom of the substrate. BLs can transmit differential signals, with each line acting as the ground for the other, resulting in strong EM coupling between the lines. This enables autonomous phase recovery in the presence of curved sections, and minimizes EM interference, making it suitable for ultra-high-speed digital transmission lines [3].

For the implementation of an ultra-high-speed digital interface, a CPS-based vertical structure can be utilized,

979-8-3503-5124-8/24 $31.00 © 2024 IEEE

having a much simpler via structure and providing ultra-wide bandwidth compared to conventional DL-based via structures [4]. In this paper, a practical and efficient implementation of an ultra-high-speed digital interface based on CPS is proposed. The proposed CPS-based interface is designed to maintain consistent impedance throughout the transitional structures, using analytical design formulas obtained through conformal mappings.

II. DESIGN OF AN ULTRA-HIGH-SPEED DIGITAL INTERFACE

A configuration of the proposed ultra-high-speed digital interface is illustrated in Fig. 1, consisting primarily of a receptacle and a plug securely coupled through a housing. The interface connects three pairs of differential signal lines from the receptacle to the plug, and the pairs are separated from the other lines by a gap of 50 mil.

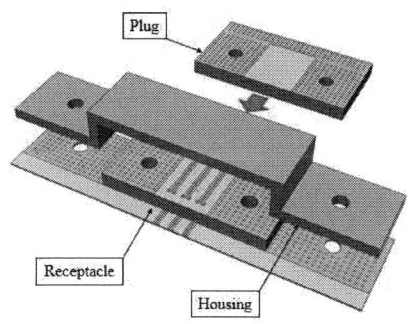

Fig. 1. Perspective view of the proposed ultra-high-speed digital interface.

In Fig. 2, layouts and design parameters for a single pair are presented. A top view of the receptacle and its design parameters are illustrated in Fig. 2a. For the analysis, the receptacle is divided into five sections. The AA'-BB' section is a DL section for connection to a typical digital PCB, where each signal pair consists of two parallel signal lines with a width of w_d spaced apart by a gap of g_d. The width and gap of the DL can be adjusted to fit with the connected digital circuits. The BB'-CC' section depicts a DL-to-CPS transition, where the width of the signal lines widens while maintaining a constant line gap, and the ground aperture s beneath the signal lines opens. The CC'-DD' section is a typical CPS line, consisting of two signal lines with a width of w_{c1}, spaced by a gap of g_{c1}. The DD'-EE' section is a CPS-based vertical transition, with two cylindrical vias of a same radius r separated by a center-to-center distance of g_v. The vertical vias are connected to the upper CPS line in the EE'-FF' section with a width of w_{c2} and a gap of g_{c2}.

A top view of the plug and its design parameters are shown in Fig. 2b, where it is divided into three sections for the analysis. The FF'-GG' section represents the CPS that interfaces with the EE'-FF' section in Fig. 2a, with two signal lines having the same width w_{c2} and gap g_{c2}. The GG'-HH' section is the CPS-to-DL transition. The width of the signal lines narrows while maintaining the constant gap, and the ground aperture s beneath the signal lines closes, transitioning to the DL in the HH'-II' section. The two signal lines have the width of w_d and the gap of g_d.

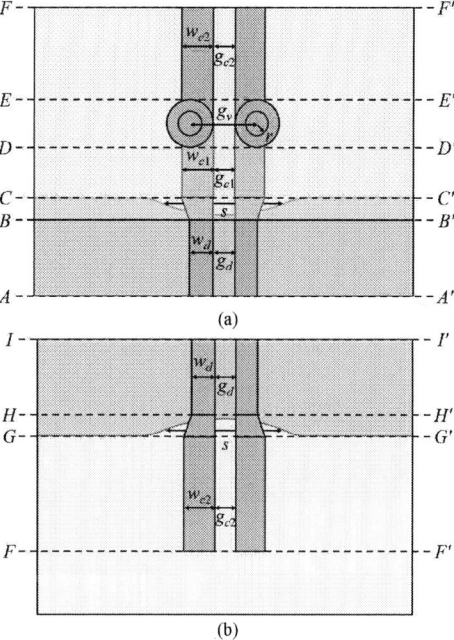

Fig. 2. Configurations and design parameters of the proposed ultra-high-speed digital interface (w_d=11.4 mil, w_{c1}=13 mil, w_{c2}=12 mil, g_d=g_{c1}=g_{c2}=5 mil, r=6 mil, g_v=29.1 mil): (a) receptacle, (b) plug.

The proposed CPS-based interface is designed using the Rogers 4003C substrate with a relative permittivity ε_r of 3.38. For the receptacle, one 8 mil substrate for the DL and DL-to-CPS transition, and two 32 mil substrates along with one 8 mil substrate for the CPS-based vertical transition are used to have a total height of 72 mil. For the plug, two 32 mil substrates and one 8 mil substrate are used to ensure a stable contact with the receptacle.

III. SIMULATION AND MEASUREMENTS

In order to efficiently design the CPS-based interface structure, the analytical design formulas are derived using conformal mapping in [4]. To verify the accuracy of the design formulas, the characteristic line impedances calculated with the analytical formulas are compared with the EM simulated results as shown in Fig. 3. For these EM simulations, a commercial 3D EM simulator (CST Microwave Studio) is used. The characteristic line impedances for the DL-to-CPS transitions of the receptacle and plug, as a function of the ground aperture width s, are shown in Figs. 3a and 3b. The deviations between the calculated characteristic line impedances and the EM simulated values are 2.2% and 4.8%, respectively. Fig. 3c represents the characteristic line impedances as a function of the center-to-center distance g_v of the CPS-based vertical transition, with the via radius of 6 mil. The deviations between the calculated characteristic line impedances and the EM simulated values are within 3.0%. Using the analytical design formulas for the transitions of the proposed ultra-high-speed digital interface, all design parameters are chosen to match the characteristic line impedance of 100 Ω (Fig. 2).

979-8-3503-5124-8/24 $31.00 © 2024 IEEE

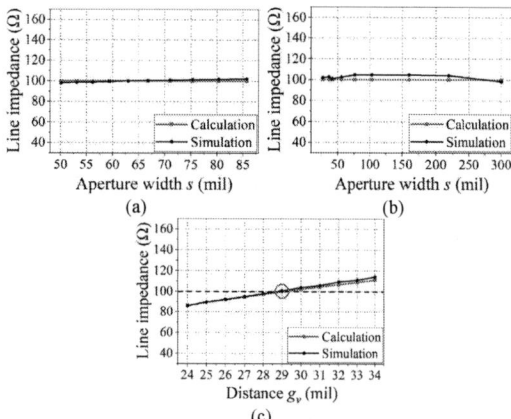

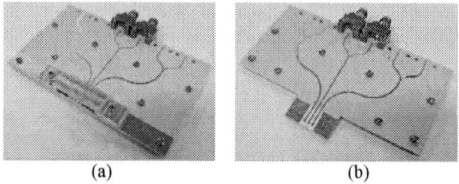

Fig. 3. Calculated and EM simulated characteristic line impedances of the DL-to-CPS transitions and CPS-based vertical transition: (a) DL-to-CPS in the receptacle, (b) DL-to-CPS in the plug, (c) CPS-based vertical transition.

The images of the fabricated ultra-high-speed digital interface are presented in Fig. 4. The differential signal transmission characteristics of the proposed CPS-based interface are measured using a 4-port vector network analyzer (Rohde & Schwarz ZNB40). To measure the 4-port S-parameters, an auxiliary structure, which gradually transforms from the two signal lines of the DL into two microstrip lines, is additionally connected at the input and output ends. The measured 4-port S-parameters are then converted into mixed-mode S-parameters using the AITT (Advanced Interconnect Test Tool) software [9], obtaining calibrated results.

Fig. 4. Pictures of the fabricated ultra-high-speed digital interface: (a) receptacle, (b) plug.

The measured S-parameters of the fabricated ultra-high-speed digital interface are compared with the EM simulated S-parameters in the frequency range of DC to 30 GHz. The transmission and isolation characteristics of the central pair among the three pairs in the fabricated interface are depicted in Fig. 5. The EM simulated results show a maximum insertion loss of 1.1 dB, a minimum return loss of 14.2 dB, and an average isolation of 41.3 dB. The measured results provide a maximum insertion loss of 3.5 dB, a minimum return loss of 14.1 dB, and an average isolation of 41.4 dB. The discrepancy between the measured and the EM simulated insertion loss can be attributed to slight misalignment between the signal lines of the receptacle and plug, and the fabrication tolerances of the multi-layered interface (manually-assembled). Therefore, the proposed interface demonstrates ultra-wideband characteristics, potentially enabling digital signal transmission of over 120 Gbps per pair using 4-level pulse amplitude modulation (PAM4).

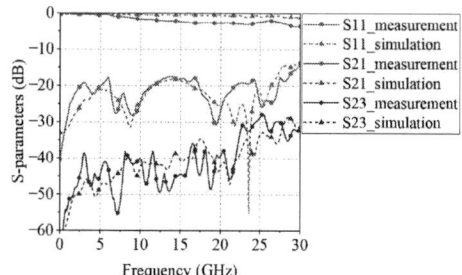

Fig. 5. Measured and EM simulated S-parameters of the fabricated ultra-high-speed digital interface.

IV. CONCLUSION

In this paper, an ultra-high-speed digital interface based on CPS is proposed. The proposed CPS-based interface structure offers a much simpler configuration and provides three times the operating bandwidth compared to the conventional DL-based interface structure. The interface design process is very efficient using accurate analytical design formulas derived with conformal mapping. Both measured and EM simulated results of the proposed interface provide a return loss of over 10 dB in the frequency range from DC to 30 GHz. The proposed interface can transmit a vast amount of digital data at ultra-high-speed, required for next-generation communications.

ACKNOWLEDGMENT

This research was supported in part by the National Research Foundation of Korea (NRF) grant funded by the Korea government (Ministry of Science and ICT) (No. RS-2023-00281275), and in part by the BK21 FOUR project funded by the Ministry of Education (No. 4199990113966).

REFERENCES

[1] E. Bogatin, Signal and Power Integrity-Simplified, 2nd ed.. Boston, MA, USA: Prentice Hall, 2010.

[2] "USB: Port Types and Speeds Compared," Tripp Lite by Eaton, [Online]. Available: https://tripplite.eaton.com/products/usb-conn ectivity-types-standards?q=USB%3A+Port+types+and+speeds+comp ared. [Accessed: May 31, 2024].

[3] B. C. Min et al., "Ultra-wideband differential line-to-balanced line transitions for super-high-speed digital transmission," Sensors, vol. 22, pp. 6873, September 2022.

[4] M. -J. Kim et al., "Ultra-wideband vertical transition in coplanar stripline for ultra-high-speed digital interfaces," Sensors, vol. 24, pp. 3233, May 2024.

[5] High-Speed Interface Layout Guidelines, Texas Instruments, 2018.

[6] AN 672: Transceiver Link Design Guidelines for High-Gbps Data Rate Transmission, Intel, 2020.

[7] J. Han, S. Hao, Z. Wang, and P. He, "Analysis of system level signal integrity for high speed interface design based on GTY transceivers," in Proceedings of the 2020 International Conference on Microwave and Millimeter Wave Technology (ICMMT), Shanghai, China, 2020.

[8] J. P. Bak, T. J. An, Y. W. Kim, and J. K. Kan, "An overhead-reduced key coding technique for high-speed serial interface," IEEE Access, vol. 10, pp. 21187-21192, February 2022.

[9] Y. Chen, "De-embedding method comparisons and physics based circuit model for high frequency D-probe," M. Eng. thesis, Department of Electrical and Computer Engineering, Missouri University of Science and Technology, Rolla, MO, USA, 2018.

Design and Analysis of High-Density Silicon Interposer Channel and Power Distribution Network

Haeyeon Kim, Joonsang Park, Hyunah Park, Keeyoung Son, Hyunsik Kim*, Taeil Bae*, Haekang Jung*
and Joungho Kim

School of Electrical Engineering (EE), Korea Advanced Institute of Science and Technology (KAIST)
SK hynix Inc.
haeyeonkim@kaist.ac.kr

Abstract—**This paper presents a design guide for silicon interposer channels and power distribution networks (PDN) in post-HBM3 high-density I/O implementations. We introduce a novel reference-paired shielding (RPS) scheme that mitigates crosstalk and reduces metal layer count. Our optimal RPS pattern, which takes into account crosstalk, simultaneous switching noise (SSN), and redistribution layer effects, achieves a maximum datarate of 7.1 Gbps for high-density I/O implementation. This represents a 31.5% improvement over conventional interposer channels while using one fewer metal layer. We also analyze the trade-off between I/O channel and PDN allocation within a limited interposer area, examining crosstalk and SSN effects relative to channel dimensions and PDN area.**

Index Terms—**High bandwidth memory, silicon interposer, signal integrity, power integrity, power distribution network, crosstalk mitigation, simultaneous switching noise**

I. INTRODUCTION

With the rising significance of high data bandwidth for AI model training and applications, high bandwidth memory (HBM) has emerged as the one and only memory solution. The 3D Stacking of DRAMs with TSVs and 2.5D integration of HBM and SoCs with silicon interposer are the two key technologies that made the AI era come true.

Silicon interposer enables fine-pitch interconnections that achieve terabyte-per-second scale bandwidth as illustrated in the signal path of Fig. 1. A single HBM3 module has 1024 I/O count and operates at a speed of 6.4Gbps, delievering a bandwidth of 819GB/s, which is 12.8 times larger than the bandwidth of GDDR6, which has 32 I/O count and operates at 16Gbps, resulting in 64GB/s.

Nevertheless, the size of large language models (LLMs) has already exceeded 1 trillion parameters and is exponentially growing. As a result, their training and inference are increasingly memory-bound. To keep pace with the expanding size of AI models, memory bandwidth must continue to increase ,and increasing the I/O count is considered a promising solution.

Higher I/O density on silicon interposer is anticipated to increase the bandwidth. However, merely adding more metal layers will increase the cost of production and degrade the signal integrity and speed due to amplified crosstalk. Thus, a novel high-density I/O channel design scheme is necessary to mitigate crosstalk and reduce the metal layer count.

Furthermore, metal layers in silicon interposer are used not only for channel routing but also for the power distribution network (PDN). The signal integrity of the interposer channels

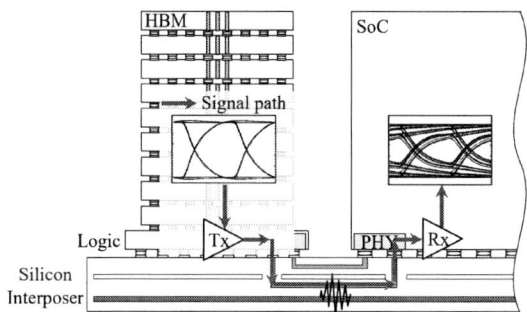

Fig. 1: Side-view of HBM-SoC Module and its I/O signal path.

benefits from increased channel width and spacing, while the power integrity of the PDN improves with a larger allocated area. Therefore, due to the limited area available for metal layers, a trade-off exists between optimizing the interposer channels and the PDN in terms of area allocation.

In this paper, we propose a reference-paired shielding (RPS) structure for implementing high-density I/O channels on a silicon interposer to reduce crosstalk and minimize the number of metal layers. Additionally, given the limited area available for metal layers, we examine various scenarios to co-optimize the high-density I/O channel dimensions and PDN area, aiming to achieve maximum bandwidth.

II. SILICON INTERPOSER CHANNEL DESIGN PARAMETERS

Silicon interposer channel design parameters include signal-ground channel patterns and physical dimensions such as channel width and space. To optimize the channel design, we propose a novel reference-paired shielding (RPS) structure and ablate its various patterns with different signal-to-ground ratio and channel dimensions.

A. Reference-Paired Shielding (RPS) Structure

Reference-paired shielding(RPS) structure refers to the interposer stack-up with interleaving signal and ground channel pairs and diagonally stacked metal layers as shown in Fig. 2. Motivated by the ground-referenced signaling for improved simultaneous switching output (SSO) and reference voltage margin in PAM-3 transceiver [1], RPS was devised to reduce crosstalk by shielding the signals by the surrounding ground channels and maintain their solitary return path.

979-8-3503-5124-8/24 $31.00 © 2024 IEEE

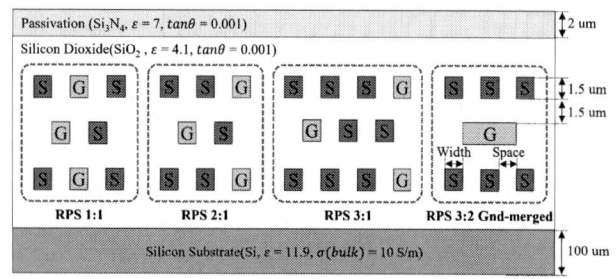

Fig. 2: Proposed reference-paired shielding (RPS) scheme for interposer channel design. RPS pattern can vary depending on signal-to-ground ratio.

TABLE I: RPS Pattern Ablation Cases

Case	S:G Ratio	Gnd-Merged	No. Layers	Width [um]	Space [um]	Eye Aperture @9Gbps [UI]
1	1:1	X	2.99 (3)	1.5	1.5	0.575
	2:1	X	2.24 (3)			0.520
	3:1	X	1.99 (2)			0.405
	3:2	X	2.49 (3)			0.635
	3:2	O	2.49 (3)			0.571
2	2:1	X	3	1.5	2.510	0.585
	3:1	X			3.010	0.550
	3:2	X			2.110	0.595
3	2:1	X	3	2.005	2.005	0.635
	3:1	X		2.255	2.255	0.590
	3:2	X		1.805	1.805	0.615
	3:2	O		1.805	1.805	**0.655**
3 Aligned	1:1	X	3	1.500	1.500	0.581
	2:1	X		2.005	2.005	0.609
	3:1	X		2.255	2.255	0.525
	3:2	X		1.805	1.805	0.575

B. Signal Integrity Analysis of Various RPS Patterns

Depending on the signal-to-ground ratio, there exist various RPS patterns and allowed channel dimensions with limited interposer area and metal layer count. To figure out the optimal RPS pattern, we ablated 16 different RPS channel patterns with different signal-to-ground ratios and channel dimensions. Furthermore, we also compared the RPS patterns with diagonally stacked metal layers to conventional ground-signal-ground co-planar patterns with aligned metal layers.

Given the limited interposer area and fixed channel length of $5mm$, Case 1 in Table I has five RPS patterns with fixed channel width and space. Case 2 has a fixed layer count and channel width while case 3 has a fixed layer count and equally distributed channel width and space. Lastly, an additional case for aligned metal layers with case 3 channel dimensions was studied as a reference that reflects the conventional ground-signal-ground coplanar channel structure.

For the 16 cases, we evaluated the worst channel eye aperture with HBM mask spec. of $120mV$ and 0.6 UI at 9 Gbps. According to the Table I, only 5 cases satisfied the 0.6UI spec. and RPS 3:2 ground merged of case 3 achieved the best eye aperture. Furthermore, the more the layer count, the better the signal integrity, and with the same layer count, equally distributed channel width and space were more effective in mitigating crosstalk.

III. SILICON INTERPOSER CHANNEL AND PDN CO-OPTIMIZATION

Silicon interposer not only includes the high-density I/O channels but also includes power distribution network (PDN) within the limited interposer area and metal layers. To examine the trade-off relationship between signal integrity and power integrity driven by the interposer area allocation of channel and PDN, we formulated three scenarios as shown in Table II. The optimal RPS pattern accordinig to Table I, signal-to-ground ratio of 3:2 with merged-ground pattern was implemented. Within the fixed interposer area according to the HBM JEDEC ballmap spec. and fixed metal layer count to 3 layers, the larger the channel dimension, the smaller the interposer PDN area.

TABLE II: SI/PI Co-Simulation Scenarios

Scenario	Width/Space [um]	Interposer PDN Area
1	1.5	1x
2	1.2	2x
3	1.0	2.6x

A. SI/PI Co-Simulation Setup

To analyze the effect of crosstalk, we modeled RPS channels and HBM I/O drivers with IBIS model. To analyze the effect of simultaneous switching noise (SSN) along the hierarchical PDN, we modeled hierarchical PDNs composed of PCB, package, interposer, and on-chip layers as described in [2] and simultaneous switching current (SSC) at 32 channels in VD-DQL domain. Furthermore, we also modeled the aluminium redistribution layer (RDL) on top of silicon interposer metal layers to evaluate the signal degradation from escape routing.

B. Design Guide for High-Density Silicon Interposer

The effect of crosstalk, SSN and RDL for each scenario was evaluated by the maximum datarate satisfying the HBM eye mask, $120mV$ and 0.6UI. According to the Table III, scenario 3, which has the smallest channel width and space, was the most susceptible to crosstalk. Furthermore, as shown in Fig. 5, the addition of SSN similarly degraded the max. datarate in all scenario. This is explained by the impedance profile of PDNs in Fig. 4. The Off-chip hierarchical PDNs excluding on-chip PDN have different impedance profile depending on the interposer PDN area. However, the impedances of full hierarchical PDNs with the on-chip PDN were suppressed and converged to the same level, indicating that the size effect of interposer PDNs was negligible. Also, the PDN impedance and consequent SSN can be controlled by proper decoupling capacitor placement [3].

TABLE III: SI/PI Co-Simulation Conditions and Results

Simulation Condition			Max. Datarate [Gbps]			
Crosstalk	SSN	RDL	Conventional	Scen.1	Scen.2	Scen.3
O	X	X	6.5	8.3	6.2	4.6
O	O	X	5.9	7.5	5.4	4.0
O	X	O	-	8.0	6.0	4.5
O	O	O	5.4	7.1	5.0	3.9

979-8-3503-5124-8/24 $31.00 © 2024 IEEE

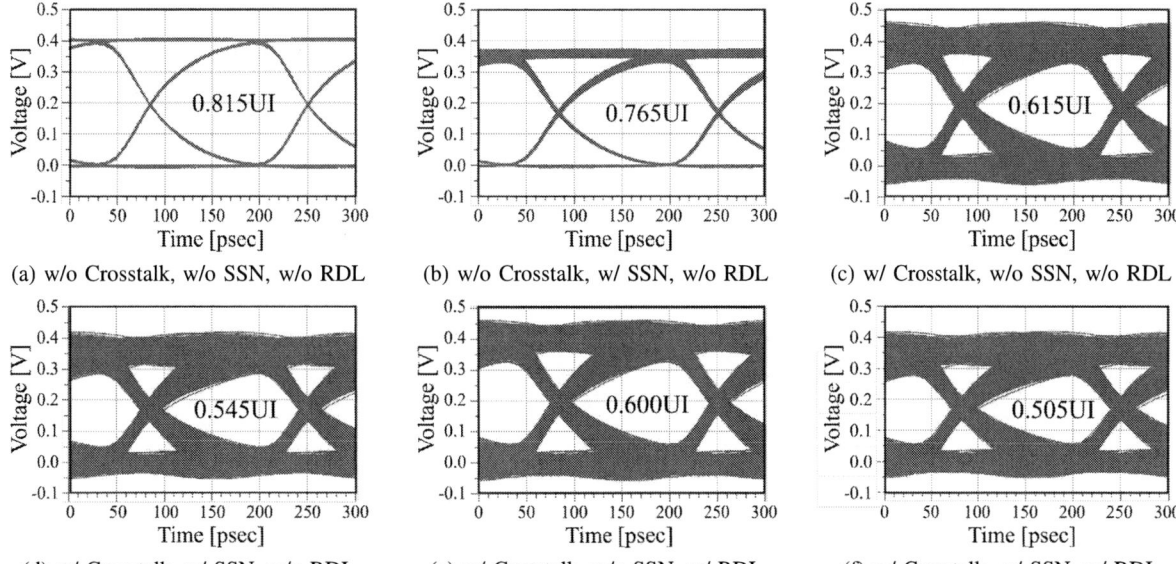

Fig. 3: Eye diagram of scenario 2 at the datarate of 6Gbps depending on the presence of crosstalk and SSN and inclusion of RDL layer in evaluation. Eye aperture at 6Gbps is measured and labeled on each diagram.

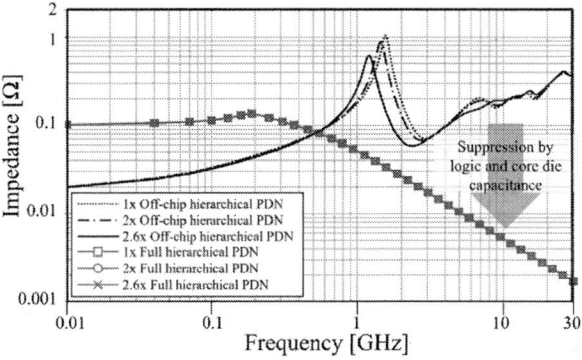

Fig. 4: Impedance profile of off-chip hierarchical PDN and full hierarchical PDN depending on interposer PDN area.

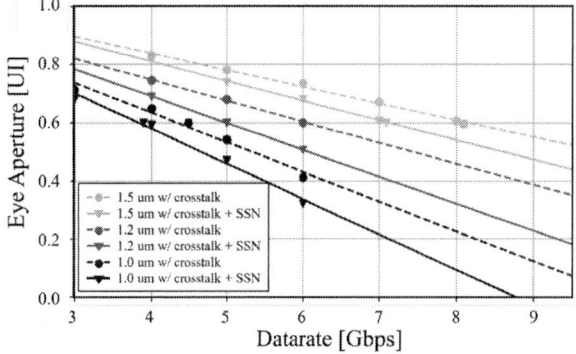

Fig. 5: Eye aperture of HBM eye mask $120mV$ depending on datarate for each scenarios in Table II.

Moreover, the addition of RDL also degraded the max. datarate mainly due to the increased loss from increased channel length. To illustrate the effect of each component, eye diagrams for each simulation setup for scenario 2 are plotted in Fig. 3. It was verified that the PDN impedance caused not only SSN but also voltage droop as the maximum voltage shifted down for the cases with SSN.

IV. CONCLUSION

In conclusion, this paper proposed a novel RPS scheme for high-density silicon interposer channel design that mitigates corsstalk and reduces metal layer count. We ablated various RPS patterns and figured out the optimal signal-to-ground ratio and channel dimensions. Due to the limited area in silicon interposer, there exists a trade-off relationship between the allocation of channel and PDN. We set three scenarios and evaluated the effect of crosstalk and SSN depending on the distribution of interposer area to channel and PDN. Crosstalk was the most dominant factor and the effect of interposer PDN area was negligible. Finally, the optimal 3 layer RPS pattern achieved max. datarate of 7.1Gbps, which is 31.5% greater than the 4 layer conventional channel with 5.4Gbps.

ACKNOWLEDGMENT

This research was funded by SK hynix Inc. We would also like to acknowledge the technical support of Ansys Korea.

REFERENCES

[1] Y. Kwon, H. Park, Y. Choi, J. Sim, J. Choi, S. Park, K.-M. Kim, C. Choi, H.-K. Jung, and C. Kim, "A 33-gb/s/pin 1.09-pj/bit single-ended pam-3 transceiver with ground-referenced signaling and time-domain decision technique for multi-chip module memory interfaces," *IEEE Journal of Solid-State Circuits*, vol. 58, no. 8, pp. 2314–2325, 2023.

[2] J. Park, S. Kim, K. Son, H. Kim, H. Kim, H. Kim, S. Choi, J. Kim, and J. Kim, "Design and analysis of an irregular-shaped power distribution network (pdn) for high bandwidth memory (hbm) interposer," in *2023 IEEE 32nd Conference on Electrical Performance of Electronic Packaging and Systems (EPEPS)*, 2023, pp. 1–3.

[3] H. Kim, H. Park, M. Kim, S. Choi, J. Kim, J. Park, S. Kim, S. Kim, and J. Kim, "Deep reinforcement learning framework for optimal decoupling capacitor placement on general pdn with an arbitrary probing port," in *2021 IEEE 30th Conference on Electrical Performance of Electronic Packaging and Systems (EPEPS)*, 2021, pp. 1–3.

Signal Integrity Analysis of PCIe Channel with Floating Board-to-Board Connectors in Automotive Infotainment System

Junghyun Lee[1], Keeyoung Son[1], Junho Park[1], Joonsang Park[1], Keunwoo Kim[1], Hyunjun An[1], Seonguk Choi[1], Jihun Kim[1], Hyunah Park[1], Sumi Choi[2], Sanghyuk Son[2] and Joungho Kim[1]

[1]*School of Electrical Engineering, Korea Advanced Institute of Science and Technology (KAIST),* Daejeon, Republic of Korea
[2]*Korea Electric Terminal (KET),* Incheon, Republic of Korea
E-mail: junghyunlee@kaist.ac.kr

Abstract— In this paper, we analyzed signal integrity (SI) of peripheral component interconnect express (PCIe) channels in an infotainment system utilizing floating board-to-board (BtoB) connectors for automotive environments. The floating BtoB connector is designed to maintain reliable electrical performance at high frequencies, making it compatible with various I/O protocols. Thus, the floating BtoB connectors can be effectively used for PCIe interconnect between the CPU on the main board and the ECU on the daughter board. To accommodate numerous components, including connectors, packages, and chips within a single system, the placement of the ECU package varies across different scenarios. Accordingly, we explored PCIe channel configurations by altering the ECU package's position, both on top and bottom-side of the daughter board. By designing PCIe interconnects within the system, we conducted electromagnetic (EM) simulations to analyze differential insertion loss (DIL), considering the impact of routing paths containing floating BtoB connectors. To verify compatibility with PCIe, we achieved eye diagrams at 16 Gb/s via equalization. Finally, we suggest design guidance for the proposed system considering cost and SI.

Keywords— Floating board-to-board (BtoB) connector, infotainment system, signal integrity (SI), simulation

I. INTRODUCTION

In high-speed systems, multiple printed circuit boards (PCB) connected through connectors and cables are widely used [1]. To enable high-speed link paths between multi-board PCBs, a board-to-board (BtoB) connectors have been designed in various applications, like automotive systems and mobile devices [2], [3]. Specifically, a BtoB connector for automotive infotainment systems must be mechanically stable while supporting high-speed interfaces. A floating BtoB connector, featuring elastic terminal pins and a floating housing, mitigates horizontal misalignment and reduces stress from vibration, as illustrated in Fig. 1(a). Additionally, it offers a shorter interconnection length for high-speed differential signaling compared to cable-type BtoB connectors. As shown in Fig. 1(b), repetitive ground-signal-signal-ground (GSSG) differential pairs are configured while separating power pins from terminal pins to minimize return current path discontinuity.

Previous research [2] focused on optimizing the mechanical design to enhance the electrical performance at high frequencies, thereby ensuring the floating BtoB connector's

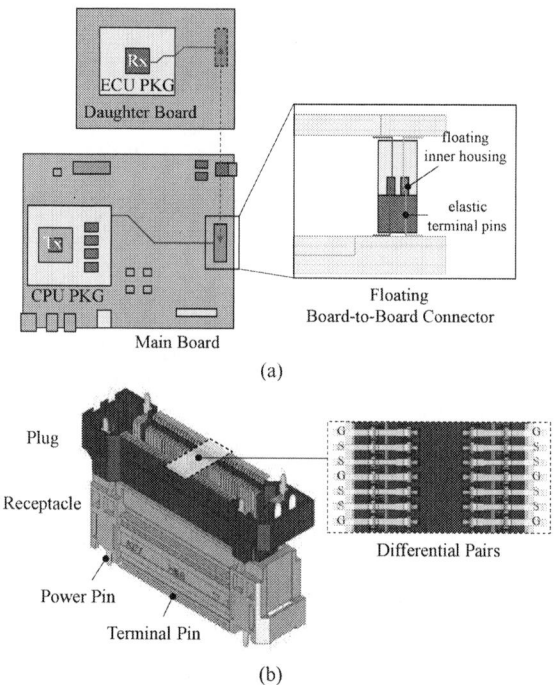

Fig. 1. (a) Infotainment system with a floating board-to-board (BtoB) connector. (b) Designed BtoB connector with adjacent ground-signal-signal-ground (GSSG) differential pairs for high-speed signaling.

compatibility with various high-speed I/O protocols. However, evaluation of the connector within a comprehensive system configuration was not conducted. Peripheral component interconnect express (PCIe) is widely used high-speed serial link due to its efficient bandwidth expansion capabilities. In previous work [4], PCIe channels are designed and evaluated including BtoB interconnects. However, the design did not consider the use of multiple floating BtoB connectors in a compact automotive environment, nor did it account for the impacts of PCB vias on signal integrity (SI).

In this paper, we design and analyze PCIe channels with the floating BtoB connector specifically designed for high-speed differential signaling in an infotainment system. The system comprises a CPU package, main board, floating BtoB connectors, daughter board and ECU package on the daughter

---

979-8-3503-5124-8/24 $31.00 © 2024 IEEE

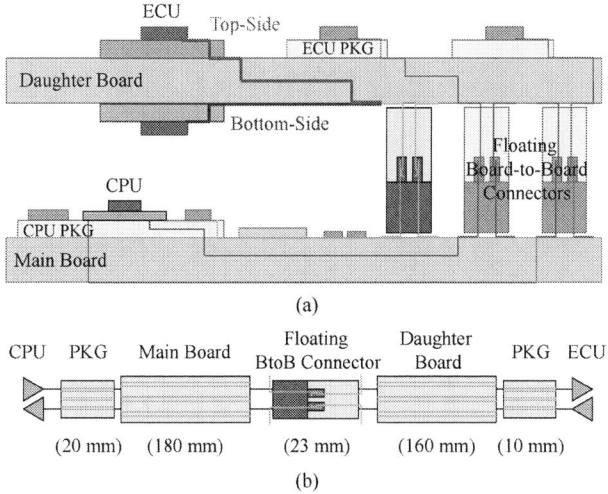

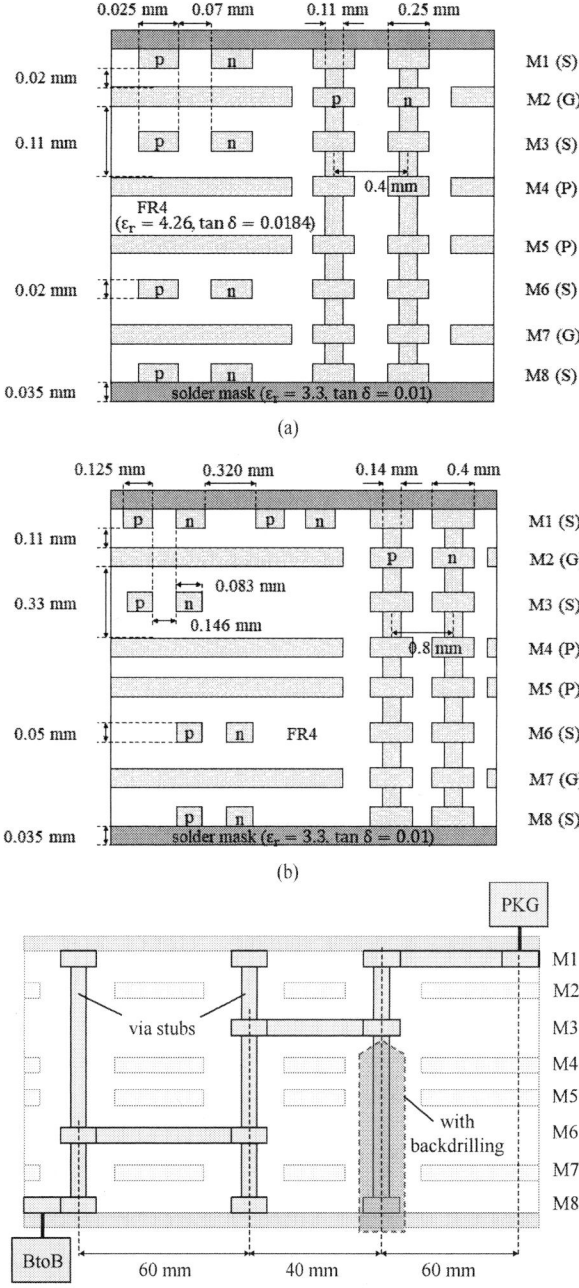

Fig. 2. (a) Infotainment system with ECU package placed on top and bottom side of the daughter board. (b) PCIe channel configuration between CPU and ECU.

board. We configured two different PCIe channel designs by varying the position of the ECU package, placing it on either the top or bottom side of the daughter board. Assuming a data rate of 16 Gb/s for PCIe Gen 4.0, we simulated and analyzed differential insertion loss (DIL) and eye diagrams with receiver (Rx) equalization. The results indicate that the bottom-side design is preferred over the top-side design. The top-side design necessitates an additional backdrilling process to improve SI, whereas the bottom-side design, which is routed solely with microstrip lines, exhibits a wider eye opening.

II. PROPOSAL OF INFOTAINMENT SYSTEM WITH FLOATING BOARD-TO-BOARD CONNECTORS

A. Configuration of Infotainment System

The infotainment system is designed to handle various functions, making it essential to incorporate numerous chips, within a confined space, as depicted in Fig. 2(a). The dense arrangement of these chips necessitates the use of multiple floating BtoB connectors to ensure all components are properly connected and function seamlessly. Because many packages are densely placed on the daughter board for a compact design, the ECU connected to the connector closest to the CPU is designed to have the longest channels on the daughter board. Therefore, we designed and analyzed the full-system to assure compatibility with PCIe. The designed floating BtoB connectors have a mating height of 20 mm, which ensures sufficient vacant space between the main and daughter boards. Consequently, ECU packages can be placed either on the top or bottom layer of the daughter board.

B. High-speed PCIe Channel Design

As shown in Fig. 2(b), PCIe channels are designed with a total length of approximately 390 mm. Fig. 3(a) and (b) shows the designed stack-ups and design parameters of packages and PCBs. While the microstrip line on the PCB has limited space per PCIe channels, resulting in a close pitch between differential pairs, the stripline is designed with sufficient pitch to minimize crosstalk. Fig. 3(c) illustrates PCIe channels of top-

Fig. 3. Stack-ups of designed (a) CPU and ECU packages, and (b) main board and daughter board printed circuit boards (PCBs). (c) The PCIe channels of top-side design with multiple via transitions.

side design on daughter board, between the pins of floating BtoB connector to ECU package. The routing is done through striplines with multiple via transitions to avoid interrupting the routing of other chips and packages on daughter board. All vias are formed as plated-through-holes (PTHs) to minimize manufacturing costs. Because via stubs can significantly impact SI, we also employed the backdrilling process. However, the additional backdrilling process incurs significant costs. The bottom-side design, on the other hand, is routed only with microstrip lines on the bottom layer, without any via transitions.

979-8-3503-5124-8/24 $31.00 © 2024 IEEE   113

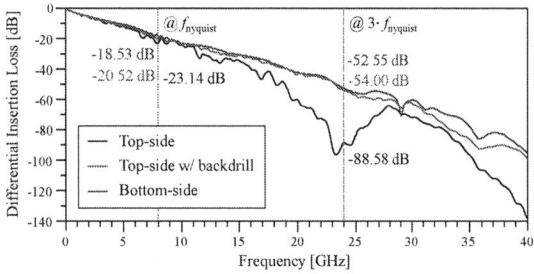

Fig. 4. Differential insertion loss (DIL) of the designed PCIe channels.

III. SIGNAL INTEGRITY ANALYSIS OF PCIe CHANNEL IN INFOTAINMENT SYSTEM

In this chapter, we analyze full path PCIe channels through differential insertion loss (DIL) and eye diagram simulations. The data rate is set at 16 Gb/s, corresponding to PCIe Gen 4.0 [5]. The electrical characteristics of the floating BtoB connector are obtained through 3-D EM simulation, including the PCB pads, where the attached pins of the connector exhibit high capacitance. We compared the DIL up to 40 GHz for all cases, as depicted in Fig. 4. The bottom side design shows consistent attenuation across all frequencies, demonstrating -18.53 dB of loss at $f_{nyquist}$.

The impact of via stubs is evident, as the DIL of top-side design have broad dips around 24 GHz and 40 GHz, which correspond to the third and fifth harmonics of $f_{nyquist}$. Altering the stack-up to adjust the length of via stub can shift resonance frequencies, but this could adversely affect other channels operating at different data rates. Therefore, to place ECU on the top layer, it is essential to eliminate stubs through the backdrilling process. With backdrilling, broad dips at high frequencies are eliminated, showing less than 2 dB difference with bottom-side design at the first and third harmonics of $f_{nyquist}$.

For eye diagram simulation, we set CPU as the transmitter (Tx), and ECU as the Rx. Differential peak-to-peak voltage (V_{pp}) is set at 800 mVPP. We implemented first-order continuous time linear equalization (CTLE) with -6 dB A_{dc} and 2-tap decision feedback equalization (DFE) at the Rx, without Tx equalization to reduce power consumption. Fig. 5 shows the simulated eye diagrams. The results demonstrate that both the top-side design with backdrilling and the bottom-side design have sufficient eye openings, while the design with via stubs failed to achieve an open eye contour. However, the bottom-side design provides the widest eye contour than the top-side design with backdrilling. Therefore, considering both cost and performance, the bottom-side design is the most recommended configuration for the proposed system.

IV. CONCLUSION

In this paper, we designed and analyzed the PCIe channels of an infotainment system utilizing floating BtoB connectors. We examined the worst-case scenario by placing the ECU package far from the BtoB connector, resulting in the longest PCIe channels. For channel configuration, we positioned the ECU on both the top side of the daughter board, using striplines with multiple via transitions, and on the bottom side, routing only with microstrip lines. Consequently, the bottom-side design not only enhances overall system performance but also

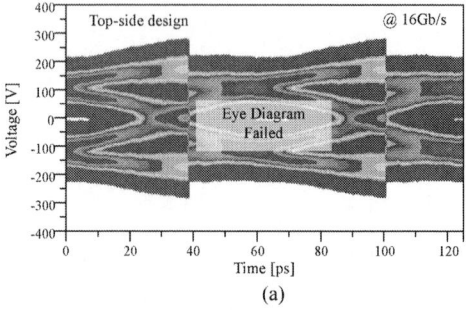

(a)

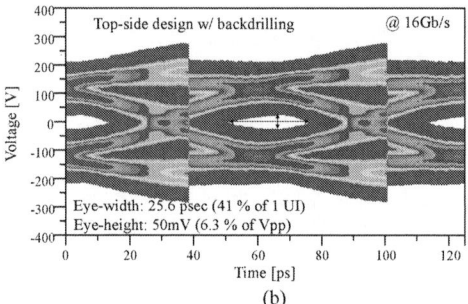

(b)

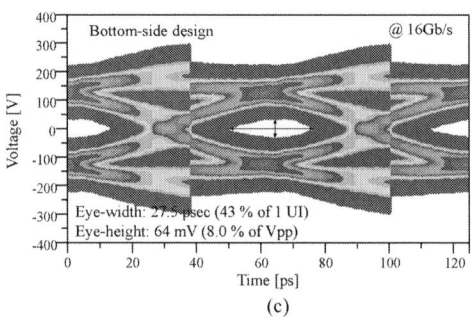

(c)

Fig. 5. Eye diagrams of (a) top-side design, (b) top-side design with backdrilling, and (c) bottom-side design, each equalized with -6 dB CTLE and 2-tap DFE.

provides a cost-effective solution, avoiding via transitions and additional backdrilling process.

ACKNOWLEDGMENT

We would like to acknowledge the technical support from ANSYS Korea. This work was supported by Samsung Electronics Co., Ltd (MEM230315_0004).

REFERENCES

[1] M. Mondal et al., "Electrical Analysis of Multi-board PCB Systems with Differential Signaling Considering Non-ideal Common Ground Connection," *2007 IEEE Electrical Performance of Electronic Packaging*, Atlanta, GA, USA, 2007, pp. 37-40.

[2] S. Park et al., "Signal integrity analysis of high-speed board-to-board floating connectors for automotive systems," *2017 IEEE 26th Conference on Electrical Performance of Electronic Packaging and Systems (EPEPS)*, San Jose, CA, USA, 2017, pp. 1-3.

[3] K. Kim et al., "A Low EMI Board-to-board Connector Design for 5G mmWave and High-speed Signaling," *2022 IEEE 31st Conference on Electrical Performance of Electronic Packaging and Systems (EPEPS)*, San Jose, CA, USA, 2022, pp. 1-3.

[4] S. Kim et al., "Signal Integrity Analysis of High-speed PCIe Channel with Board-to-Board Interconnect for High-Performance Server," *2023 IEEE Electrical Design of Advanced Packaging and Systems (EDAPS)*, Rose-Hill, Mauritius, 2023, pp. 1-3.

[5] NCB-PCI Express Base 4.0, 2014.

Design and Analysis of Extended Scale Cache (ESC) Stacked-GPU-HBM Module Architecture Considering Power Integrity (PI)

Hyunah Park, Seonguk Choi, Haeyeon Kim, Taein Shin, Keeyoung Son, Jiwon Yoon, Junghyun Lee,
Haeseok Suh, Taesoo Kim, Jungmin Ahn, Hyunjun An and Joungho Kim
School of Electrical Engineering (EE), Korea Advanced Institute of Science and Technology (KAIST)
hyunah.park@kaist.ac.kr

Abstract—This paper proposes the extended scale cache (ESC) stacked-graphics processing unit (GPU)-high bandwidth memory (HBM) module architecture. The main concept of the proposed architecture involves stacking an ESC, which is an L2 cache, atop a GPU to decrease off-chip data movement, thereby reducing power consumption and latency. However, the simultaneous switching noise (SSN) of the proposed architecture is increased due to additional switching noise transferred from the ESC to the GPU cores, leading to degraded power integrity (PI) performance. To ensure stable PI performance of the proposed architecture, we design and analyze the hierarchical power distribution network (PDN) of the proposed architecture and the simultaneous switching current (SSC) for both the GPU and ESC. Additionally, we conduct modeling and analysis of the SSN of the proposed architecture. By employing decoupling capacitors (decaps) with a total capacity of 1,152 nF, the SSN of the proposed architecture is reduced to 7.65 mV, which is 0.695% of the supply voltage (VDD), demonstrating that the proposed architecture guarantees reliable PI performance.

Index Terms—Extended scale cache (ESC), graphics processing unit (GPU), high bandwidth memory (HBM), power integrity (PI), simultaneous switching noise (SSN)

I. INTRODUCTION

Recently, with the emergence of artificial intelligence (AI) models with extremely large numbers of parameters, graphics processing unit (GPU)-high bandwidth memory (HBM) module has been widely used for training and inference of AI models. It offers high memory capacity and high bandwidth up to TB/s range between memory and processor. However, in conventional GPU-HBM module, high power consumption and increased latency occur when data move through off-chip interconnections [1], presenting a bottleneck in the overall system.

Therefore, to reduce data movement through off-chip interconnections, it is essential to expand the on-chip cache capacity by implementing on-chip network. To increase on-chip cache capacity, several new architectures have been introduced, including accelerators using static random access memory (SRAM), commonly employed as caches, as the main memory and central processing unit (CPU) with stacked L3 cache [2]. recently released NVIDIA GPUs have increased the capacity of their L2 cache. However, despite these advancements, the on-chip cache capacity still remains insufficient.

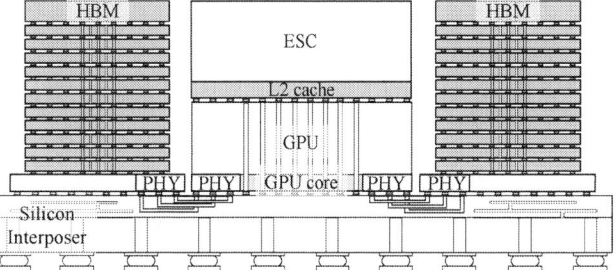

Fig. 1: Conceptual view of the proposed extended scale cache (ESC) stacked-GPU-HBM module architecture.

In this paper, we propose the extended scale cache (ESC) stacked-GPU-HBM module architecture which involves stacking SRAM-based L2 caches on top of a GPU, as illustrated in Fig. 1. By utilizing the stacked ESC, data movement through off-chip interconnections is effectively reduced by increasing data movement through shorter on-chip interconnections. Thus, the proposed architecture is a promising solution for achieving high performance and energy efficiency in AI computing.

However, in the proposed architecture, the simultaneous switching noise (SSN) is increased due to additional switching noise transferred through the hierarchical power distribution network (PDN) from the ESC to the GPU cores. This SSN, which degrades the power integrity (PI) performance, causes voltage fluctuations, potentially leading to logic failures in the operations of GPU cores. Thus, to ensure the stable PI performance of the proposed architecture, we analyze the SSN of the proposed architecture in the time domain at the VDD power domain.

II. DESIGN OF THE ESC STACKED-GPU-HBM MODULE ARCHITECTURE

As shown in Fig. 1, the main concept of the proposed ESC stacked-GPU-HBM module architecture involves stacking an ESC, which consists of L2 cache, on top of a GPU. Through this implementation, the reduction in latency results in higher throughput, while the decreased power consumption leads to improved energy efficiency. To facilitate this architecture, the GPU undergoes back grinding, enabling the height of

979-8-3503-5124-8/24 $31.00 © 2024 IEEE

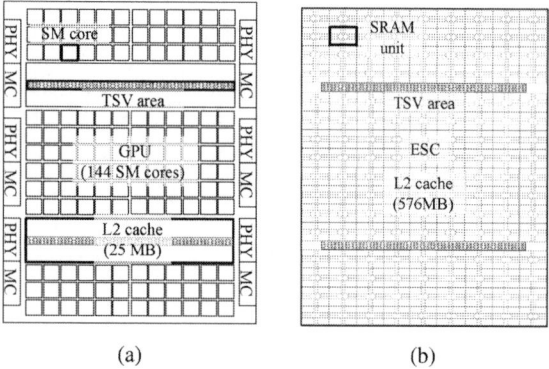

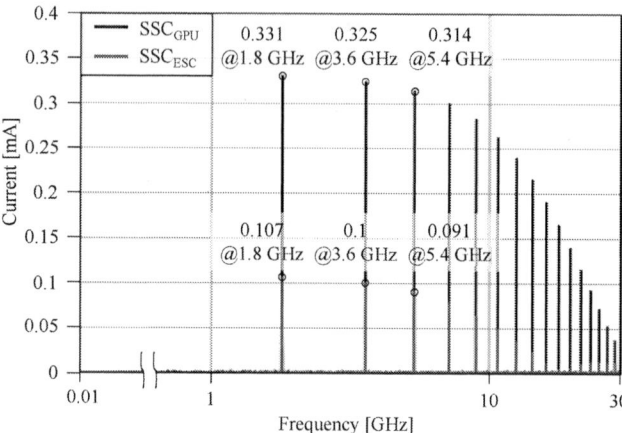

Fig. 2: Designed floorplans and configurations of the (a) GPU and (b) ESC.

Fig. 3: SSC spectrum each for GPU and ESC in the frequency domain.

the ESC stacked-GPU to match that of the HBM, which adheres to the $720\mu m$ set by the JEDEC specification [3]. This implemented ESC stacked-GPU integrates six next-generation HBMs, featuring with high-density I/Os, designed to deliver high bandwidth in the range of tens of TB/s.

Fig. 2 illustrates the designed floorplans of GPU and ESC. Both ESC and GPU have identical total areas. The GPU features 144 streaming multiprocessor (SM) cores and includes a total of 50 MB L2 cache, which also incorporates through silicon via (TSV) areas for connectivity with the ESC. In the case of the ESC, it is configured with a 576 MB L2 cache, arranged in 9 × 16 SRAM units, where each SRAM unit is 4 MB in size. Similarly to the GPU, it includes TSV areas to ensure connectivity with the GPU L2 cache, thereby maintaining the memory hierarchy.

III. PI ANALYSIS OF THE PROPOSED ARCHITECTURE

A. Modeling and Analysis of the SSC of the Proposed Architecture

We model the two SSC spectra one each for the GPU and ESC. To model the SSC of the GPU, the chip power model (CPM) is employed. CPM is a recognized method for modeling the power behavior of chips in PI analysis [4]. This is achieved by implementing a piecewise linear (PWL) current source to reflect the dynamic operations of the chip. The PWL current source generates a current profile of the GPU with peak current (I_{peak}) of 7.1 mA, calculated based on its 700 W peak power and the number of bumps. The current profile features a period (T) of 0.55 ns, and rising and falling times of 0.05 T. For the ESC, a SRAM circuit is designed using SPICE simulations, which generates a current spectrum during repeated read operations, reaching a I_{peak} of 0.4 mA. Both the GPU and ESC operate at a supply voltage (VDD) of 1.1 V and a clock frequency (f_{clock}) of 1.8 GHz.

As shown in Fig. 3, the SSC spectra of the GPU and ESC in the frequency domain are derived through fast fourier transform (FFT). Both SSC spectra exhibits peak currents predominantly in the high frequency domain, particularly at the harmonic frequencies of the f_{clock}. Comparing the peak currents of the GPU and ESC, the GPU exhibits larger peak currents than the ESC, which is attributable to its higher peak

power. This increased peak power in the GPU is primarily due to the high power consumption of its SM cores during operation.

B. Modeling and Analysis of the Hierarchical PDN of the Proposed Architecture

We model the hierarchical PDN, which consists of multiple components including package PDN, the interposer PDN, the GPU PDN and the ESC PDN. The dimensions of the package PDN are 50.6 mm × 35.6 mm. The interposer PDN, which is a meshed type, has dimensions of 6.8 mm × 6.2 mm. The GPU PDN and ESC PDN, both employing a grid type, have dimensions of 3.4 mm × 3.4 mm and 2.8 mm × 1.9 mm, respectively. Each unit cell of both meshed and grid-type PDNs are modeled as an equivalent circuit using the balanced transmission line method (TLM). For the interposer PDN, each unit cell features a 75% metal density and a length of $20\mu m$. For the on-chip PDNs in the GPU and ESC, each unit cell has an 80% metal density and a length of $5\mu m$. The PDN stack-up and metal density parameters are adopted from [5]. Finally, all modeled components are cascaded using the segmentation method.

Fig. 4 presents the impedance profile of the hierarchical PDN across a frequency range of 0.01 to 30 GHz. The self-impedance (Z_{self}) measured at the center of the GPU indicates that the overall capacitance increases with the additional ESC on-chip PDN (C_{ESC}). This increase is attributed to the high metal density in the unit cells of the on-chip PDN, which results in increased areas of overlap between the power and ground planes. Regarding the transfer impedance ($Z_{transfer}$) between GPU and ESC port, it aligns with the Z_{self} at low frequencies. However, it varies at high frequencies due to variance of perceived distances between the ports [6]. The sum of $Z_{transfer}$ across the ports is considerably high, which necessitates a detailed analysis of SSN.

C. Modeling and Analysis of the SSN of the Proposed Architecture

In this section, we model and analyze the SSN of the proposed architecture by multiplying the SSC with the PDN

979-8-3503-5124-8/24 $31.00 © 2024 IEEE

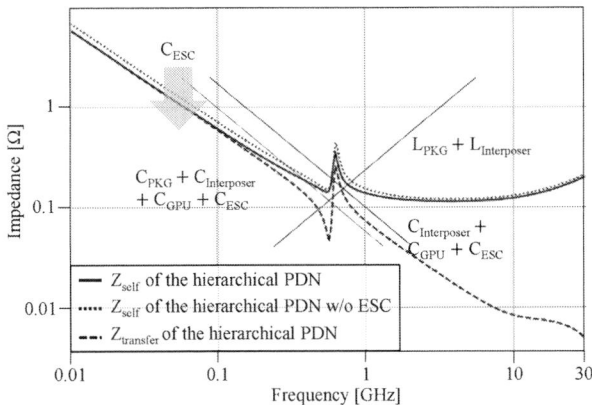

Fig. 4: Self and transfer PDN impedance of the proposed architecture probed at the GPU center port.

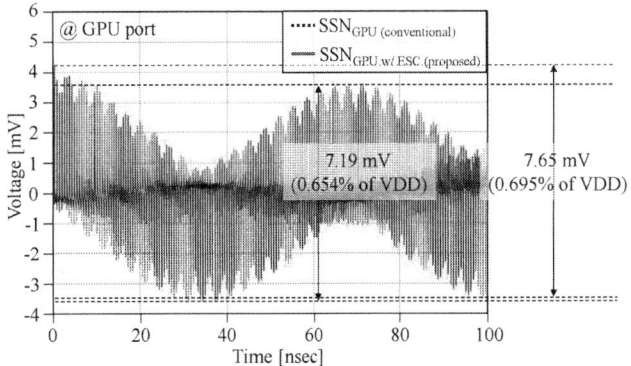

Fig. 5: Comparison of SSN between conventional GPU-HBM module and the proposed architecture after decap placement.

impedance, followed by an inverse fast fourier transform (IFFT) for time domain analysis. The SSN for the conventional GPU-HBM module is derived, incorporating both self switching noise and sum of transfer switching noise. In the proposed architecture, we also consider the additional switching noise transferred from the ESC, as detailed in Eq. (1).

$$
\begin{aligned}
\text{SSN} = {}& Z_{\text{GPU,self}} \cdot \text{SSC}_{\text{GPU}} + \sum \left(Z_{\text{GPU,transfer}} \cdot \text{SSC}_{\text{GPU}} \right) \\
& + \sum \left(Z_{\text{ESC,transfer}} \cdot \text{SSC}_{\text{ESC}} \right)
\end{aligned}
\tag{1}
$$

To analyze the SSN of the proposed architecture, probing is conducted at the center of the GPU. Modeled SSN value significantly exceeds the nominal operating threshold voltage of less than 5% of VDD, leading to essential need for SSN suppression through the placement of decoupling capacitors (decaps). Among the various types of decoupling capacitors, such as metal-insulator-metal (MIM) capacitors and N-type metal-oxide-semiconductor (NMOS) capacitors, MIM capacitors are frequently used in high-frequency applications due to their low resistivity and minimal parasitic capacitance, despite having a lower capacitance density than NMOS capacitors. Additionally, MIM capacitors are placed in the overlapping areas of power and ground planes, thereby not consuming extra areas on the chip. Thus, we employ MIM capacitors for decaps.

Fig. 5 presents the SSN of both the conventional GPU-HBM module (SSN$_{\text{GPU}}$) and proposed architecture (SSN$_{\text{GPU w/ ESC}}$) after placing decaps with a total capacity of 1,152 nF within the GPU and ESC PDN. The SSN$_{\text{GPU}}$ is reduced to 7.19 mV, which is 0.654% of VDD and SSN$_{\text{GPU w/ ESC}}$ is reduced to 7.65 mV, which is 0.695% of VDD. The SSN$_{\text{GPU w/ ESC}}$ is similar to SSN$_{\text{GPU}}$, and its value is significantly below 5% of VDD, indicating that the proposed architecture guarantees reliable PI performance.

IV. CONCLUSION

We propose the ESC stacked-GPU-HBM module architecture, which incorporates an L2 cache from the ESC stacked atop the GPU. To implement this architecture, we design the floorplans for both the GPU and the ESC. To analyze the PI performance of the proposed architecture, we model and analyze the hierarchical PDN of the proposed architecture and SSC for each the GPU and ESC. Furthermore, we conduct modeling and analysis of the SSN of the proposed architecture. After placing decaps of 1,152nF, the SSN of the proposed architecture is similar to that of conventional GPU-HBM modules. Furthermore, its value is 7.65 mV, which constitutes 0.695% of VDD, indicating that the proposed architecture ensures reliable PI performance.

ACKNOWLEDGMENT

We would like to acknowledge the technical support from ANSYS Korea. This research was supported by National RD Program through the National Research Foundation of Korea(NRF) funded by Ministry of Science and ICT (NO. 2022M3I7A4072293). Also, this research was funded by SK hynix Inc.

REFERENCES

[1] Y.-H. Chen, T. Krishna, J. S. Emer, and V. Sze, "Eyeriss: An energy-efficient reconfigurable accelerator for deep convolutional neural networks," *IEEE journal of solid-state circuits*, vol. 52, no. 1, pp. 127–138, 2016.

[2] J. Wuu, R. Agarwal, M. Ciraula, C. Dietz, B. Johnson, D. Johnson, R. Schreiber, R. Swaminathan, W. Walker, and S. Naffziger, "3D V-cache: the implementation of a hybrid-bonded 64MB stacked cache for a 7nm x86-64 CPU," in *2022 IEEE International Solid-State Circuits Conference (ISSCC)*, vol. 65. IEEE, 2022, pp. 428–429.

[3] Standard, JEDEC, "High bandwidth memory DRAM (HBM3)," JESD238, 2022.

[4] J. Xu, S. Bai, B. Zhao, K. Nalla, M. Sapozhnikov, J. L. Drewniak, C. Hwang, and J. Fan, "A novel system-level power integrity transient analysis methodology using simplified CPM model, physics-based equivalent circuit pdn model and small signal vrm model," in *2019 IEEE International Symposium on Electromagnetic Compatibility, Signal & Power Integrity (EMC+ SIPI)*. IEEE, 2019, pp. 205–210.

[5] H. Park, M. Kim, S. Kim, K. Kim, H. Kim, T. Shin, K. Son, B. Sim, S. Kim, S. Jeong *et al.*, "Transformer network-based reinforcement learning method for power distribution network (PDN) optimization of high bandwidth memory (HBM)," *IEEE Transactions on Microwave Theory and Techniques*, vol. 70, no. 11, pp. 4772–4786, 2022.

[6] K. Cho, Y. Kim, H. Lee, H. Kim, S. Choi, S. Kim, and J. Kim, "Design and analysis of power distribution network (PDN) for high bandwidth memory (HBM) interposer in 2.5 d terabyte/s bandwidth graphics module," in *2016 IEEE 66th Electronic Components and Technology Conference (ECTC)*. IEEE, 2016, pp. 407–412.

979-8-3503-5124-8/24 $31.00 © 2024 IEEE

Explainable Reinforcement Learning(XRL)-based Decap Placement Optimization for High Bandwidth Memory (HBM)

Keunwoo Kim[1], Hyunwook Park[2], Keeyoung Son[1], Seonguk Choi[1], Taein Shin[1], Junghyun Lee[1], Jiwon Yoon[1], Hyunjun An[1], Haeyeon Kim[1], Wooshin Choi[3], Jung-Hwan Choi[3] and Joungho Kim[1]

[1]School of Eletrical Engineering, Korea Advanced Institute of Science and Technology (KAIST),
Daejeon, Republic of Korea
[2]MST, Missouri, United States
[3]Samsung Electronics Co. Ltd, Hwasung, South Korea
E-mail: keunwookim@kaist.ac.kr

Abstract— In this paper, for the first time, we propose an explainable reinforcement learning (XRL)-based decap placement optimization method for high bandwidth memory (HBM) considering power integrity (PI). The proposed XRL-based method enhances explainability by transforming the sum of various types of rewards into a vector sum operation for the trained model. A CNN-based network was used for training, with each reward considered from a multi-objective RL perspective. To verify the proposed method, we applied it to solve the problem of placing decaps at VDDQ domain of HBM3 module. In this paper, rewards were set as the suppression of self-impedance and transfer-impedance at each probing port. The proposed method achieved improvements of 2.8% compared to usage of general scalar sum reward. Ultimately, the vector differences in the Q-value for different actions provided grounds for action taken and allowed for the evaluation of whether the model was well-trained.

Keywords— Power integrity (PI), high-bandwidth memory (HBM), explainable reinforcement learning (XRL)

I. INTRODUCTION

The growing computational demands of large language models (LLMs) have accelerated the development of advanced memory technologies like high bandwidth memory (HBM) [1]. Designing HBM requires integrating AI and machine learning (ML) techniques to optimize the complex architecture and performance of modern semiconductor devices. In signal integrity (SI) and power integrity (PI) domains, AI and reinforcement learning (RL) are used for optimization [2]-[4]. While RL excels in complex tasks, it often faces the black-box problem, where the decision-making process is opaque as shown in Fig. 1, limiting trust and transparency. Explainable reinforcement learning (XRL) addresses this by making the decision process more transparent, enhancing its applicability in critical SI/PI tasks.

In this paper, we present a novel approach to decap placement optimization for HBM modules using an XRL agent. We define the decap placement problem as a Markov decision process (MDP), a mathematical framework essential to RL that models decision-making in environments where outcomes are partly random and partly under the control of a decision-maker, tailored to the XRL framework. Q-values, which represent the expected future rewards associated with specific actions in a given state, are also fundamental to reinforcement learning. Specifically, we vectorize the reward function, allowing the Q-values to be expressed as vectors, which facilitates the interpretation and analysis of the underlying causes for the

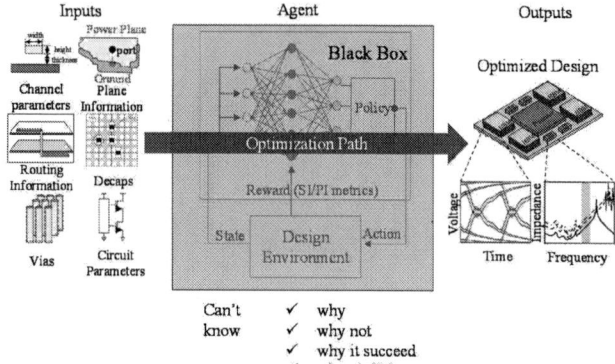

Fig. 1. The concept of reinforcement learning (RL)-based optimization in signal and power integrity (SI/PI) domain and its black-box problem.

agent's decisions. This methodology not only optimizes the decap placement within the HBM module but also provides a clear explanation of the optimal results. Through this approach, we demonstrate significant improvements in decap placement efficiency and provide a detailed explanation of the optimal placement strategy derived by the XRL agent.

II. PROPOSAL OF EXPLAINABLE REINFORCEMENT LEARNING-BASED OPTIMAL DECAP PLACEMENT METHOD

Fig. 2 shows the concept of the proposed XRL-based decap placement optimization method. The MDP parameters are defined as follows: State s is a 3-D array representing the hierarchical power distribution network (PDN) and includes information about ports and the presence or absence of decaps; action a is mapped to the index of the location where a decap will be assigned; reward $\vec{r}$ is the degree of suppression of self-impedance and transfer-impedance compared to the initial state. The rewards are measured at a total of four ports, and the specific equation is as follows:

$$\vec{r}(a|s) = \alpha_r \sum_p \overrightarrow{\Delta Z_{pp}}/4 + (1 - \alpha_r) \sum_p \overrightarrow{\sum_q \Delta Z_{pq}}/6 \quad (1)$$

, where α_r is coefficient of self-impedance suppression. In this work, we used 0.7 for α_r. This method is similar to the decap placement method using deep Q-network (DQN) presented in [5]. However, this study uses a multi-objective DQN agent composed of several DQNs, allowing each reward to be learned separately.

979-8-3503-5124-8/24 $31.00 © 2024 IEEE

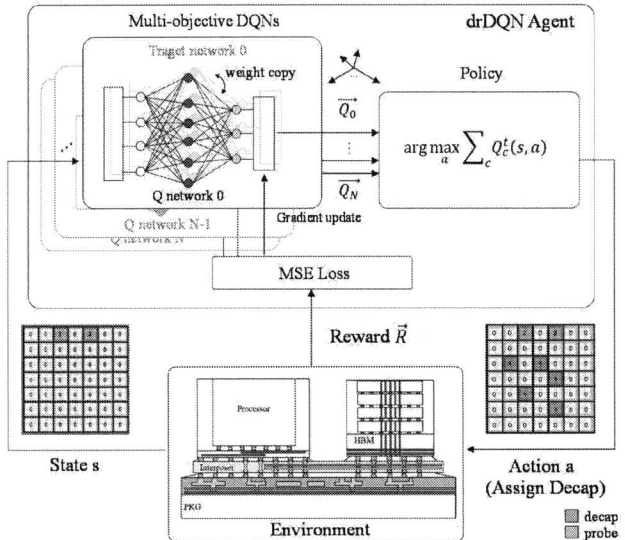

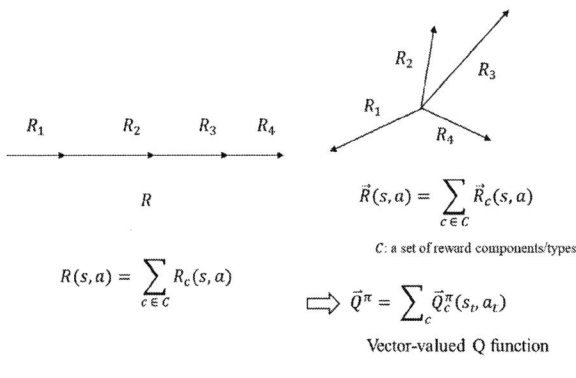

Fig. 2. Explainable reinforcement learning (XRL)-based decap placement for high bandwidth memory (HBM).

Fig. 3. The concept of reward and Q-value decomposition using vector.

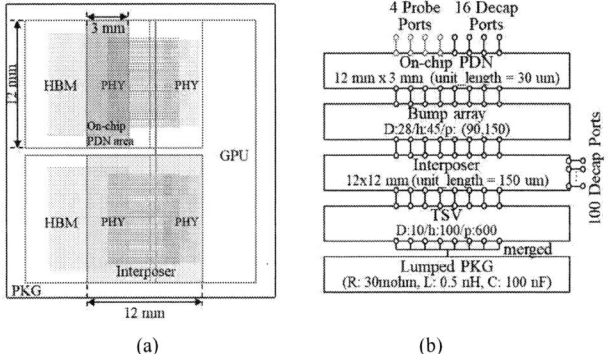

Fig. 4. (a) Physical dimension of HBM3 VDDQ domain and (b) RLGC equivalent circuit modeling-based hierarchical power distribution network (PDN) of HBM3 VDDQ domain.

A. Reward and Q-value Vectorization for Explanation

In scenarios where rewards are determined by various factors, traditional MDP formulations typically combine these diverse rewards into a scalar sum, which aggregates the influences of each factor. However, this approach results in scalar-defined Q-values, making it impossible to discern the contribution of each reward component to the overall Q-value after training. To address this issue, we combined the rewards from each component using a vector sum, as depicted in Fig. 3. This methodology allows us to obtain vectorized Q-values, enabling us to identify which reward components influence the Q-value. Ultimately, this provides a clearer understanding of the factors driving the agent's decisions, enhancing the explainability of the RL process [6].

The vector-valued Q-value learned by the agent allows us to provide various explanations. The reason why the agent prefers action 2 over action 1 in the same state can be explained using $\Delta(s, a_1, a_2) = \vec{Q}(s, a_1) - \vec{Q}(s, a_2)$, which we refer to as reward difference explanation (RDX). Each vector component of $\Delta(s, a_1, a_2)$ can be either positive or negative, indicating whether a_1 has an advantage or disadvantage over a_2 for that specific component. This framework allows us to explain the critical disadvantages or advantages of particular actions, providing a clearer understanding of the agent's decision-making process.

B. Training XRL Agent through Multi-objective DQN

However, a conventional DQN cannot handle vector-valued rewards or Q-values. To address this, we employed the concept of a multi-objective DQN to train each component of the reward and Q-value separately. Specifically, we created separate networks for each reward component to derive the component-specific Q-values. The policy was then defined as:

$$\pi(s) = \operatorname{argmax}_a \sum_{c \in C} Q_c(s, a) \qquad (2)$$

, where C is a set of reward components. The loss for each component was computed as the mean squared error (MSE) between the target Q-value and the current Q-value for that

component. This approach allows each reward component to be learned independently, facilitating a more explainable RL process where the influence of individual reward components on the overall Q-value can be clearly understood.

III. VERIFICATION OF THE PROPOSED METHOD

We applied the proposed method to the decap placement in the HBM3 VDDQ domain to demonstrate its optimality and to explain the optimal design obtained by the trained agent. The physical dimensions of the elements constituting the hierarchical PDN of the HBM are shown in Fig. 4(a). Additionally, to employ RL effectively, we required a fast impedance estimator. We modeled the PKG, TSV array, interposer, bump array, and on-chip PDN using an RLGC equivalent circuit, with specific values detailed in Fig. 4(b). The modeling method used is identical to that described in [7].

We solved the problem of placing a total of 8 decaps in the interposer and on-chip regions. The on-chip area has 16 decap candidates excluding the 4 ports, while the interposer contains 100 decap candidates. The material information and decap specifications used for verification are also consistent with those in [7].

To train the network, we leveraged the fact that the input state is a 3-D matrix similar to an image. Thus, we used a network consisting of two convolutional layers (CNN) followed by two linear layers. ReLU was used as the activation function for each layer. Detailed information about the network and hyperparameters is provided in Table I. The learning rate is set as 0.001, gamma as 0.5. batch size as 64, and total training

979-8-3503-5124-8/24 $31.00 © 2024 IEEE

TABLE I
NEURAL NETWORK ARCHITECTURE OF THE PROPOSED METHOD

Convolution Layer 1 & 2	Filter	3×3 filter 256 / 512
	Stride	1
	Padding	Zero padding
	Activation	ReLU
Fully-connected Layer 1 & 2	# of node	512
	Activation	ReLU

episode as 1000. For comparison with a conventional DQN using scalar sum rewards, we used the same network. However, to train our proposed method, we used a bundle of 8 networks corresponding to the 8-dimensional rewards; self and transfer-impedance at 4 ports.

A. Optimality Verification

We observed that the total reward of the optimal design by the trained agent increased as the number of training episodes grew. Fig. 5 depicts the optimal decap design at episodes 100, 500, and 1000. As the number of training episodes increased, the interposer decaps were increasingly placed directly beneath the ports, while the on-chip decaps were positioned near the ports. Additionally, Fig. 6 illustrates the self-impedance at port 3 and the transfer-impedance at port 2 as a function of the training episodes, along with the results from a DQN agent trained for 1000 episodes for comparison. As the number of episodes increased, the degree of impedance suppression, compared to the state without decaps, also increased. Compared to the reward sum of 1.818 achieved by the DQN agent, our proposed method yielded a reward sum of 1.869, indicating a performance improvement of 2.8%.

B. Explanation using Vectorized Q-value

To clarify the decision-making process of the agent trained for 1000 episodes when placing the 5th decap, denoted as s_{t_5}, we use counterfactual reasoning to interpret the agent's decisions. The Q-value vector is structured as follows: Self-impedance for ports 1, 2, 3, and 4, followed by transfer-impedance for the same ports. For the state s_{t_5} with action [0, 6, 1], $\vec{Q}(s_{t_5}, [0, 6, 1])$ is [2.87, 2.20, 2.21, 2.45, 1.92, 2.41, 2.38, 2.24]. For the state s_{t_5} with action [0,4,1], $\vec{Q}(s_{t_5}, [0, 4, 1])$ is [2.79, 2.14, 2.31, 2.39, 1.85, 2.43, 2.47, 2.26]. The preference for placing the decap at on-chip [0, 6] over [0, 4] is given by the difference: [0.08, 0.06, -0.11, 0.06, 0.07, -0.03, -0.09, -0.03]. This difference indicates that placing the decap at on-chip [0, 6] is more advantageous in terms of self-impedance at ports 1, 2, and 4, and transfer-impedance at port 1. Despite the negative impact on some components, particularly the self-impedance at port 3, the overall impedance suppression is enhanced when the decap is placed at [0, 6].

IV. CONCLUSION

In this paper, we proposed XRL-based method for optimizing decap placement in HBM, enhancing model explainability through vectorized reward. Validated on the VDDQ domain of an HBM3 module, the proposed method achieved a 2.8% improvement over conventional DQN methods. The analysis of

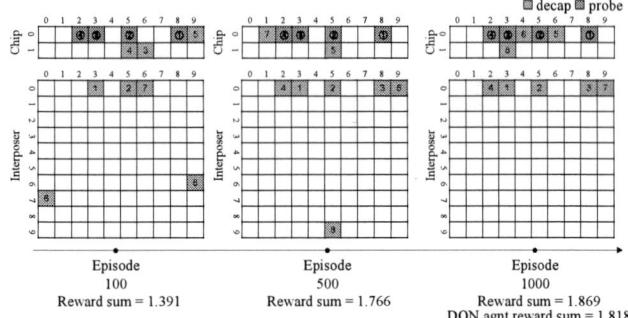

Fig. 5. Optimal decap placement by the proposed method according to the training episode.

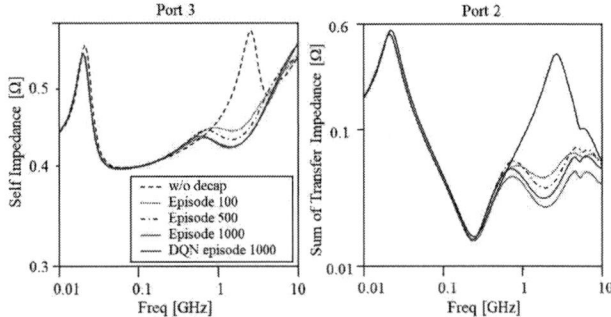

Fig. 6. (a) Self-impedance at port 3 and (b) sum of transfer-impedance at port 2 of the proposed method according to the training episode and DQN agent.

vector differences in Q-values provided clear explanations for the agent's actions, confirming the effectiveness of the XRL approach.

ACKNOWLEDGMENT

We would like to acknowledge the technical support from ANSYS Korea. This work was supported by Samsung Electronics Co., Ltd(Contract ID: MEM230315_0004).

REFERENCES

[1] J. Lee et al., "13.4 A 48GB 16-High 1280GB/s HBM3E DRAM with All-Around Power TSV and a 6-Phase RDQS Scheme for TSV Area Optimization," 2024 IEEE International Solid-State Circuits Conference (ISSCC), San Francisco, CA, USA, 2024, pp. 238-240

[2] H. Park et al., "Policy Gradient Reinforcement Learning-based Optimal Decoupling Capacitor Design Method for 2.5-D/3-D ICs using Transformer Network," 2020 IEEE Electrical Design of Advanced Packaging and Systems (EDAPS), Shenzhen, China, 2020, pp. 1-3

[3] S. Choi et al., "Sequential Policy Network-based Optimal Passive Equalizer Design for an Arbitrary Channel of High Bandwidth Memory using Advantage Actor Critic," 2021 IEEE 30th Conference on Electrical Performance of Electronic Packaging and Systems (EPEPS), Austin, TX, USA, 2021, pp. 1-3

[4] K. Kim et al., "Policy-Based Reinforcement Learning for Through Silicon Via Array Design in High-Bandwidth Memory Considering Signal Integrity," in IEEE Transactions on Electromagnetic Compatibility, vol. 66, no. 1, pp. 256-269, Feb. 2024

[5] H. Park et al., "Deep Reinforcement Learning-Based Optimal Decoupling Capacitor Design Method for Silicon Interposer-Based 2.5-D/3-D ICs," in IEEE Transactions on Components, Packaging and Manufacturing Technology, vol. 10, no. 3, pp. 467-478, March 2020

[6] Juozapaitis, Zoe, e t al. "Explainable reinforcement learning via reward decomposition." IJCAI/ECAI Workshop on explainable artificial intelligence. 2019.

[7] H. Park et al., "Transformer Network-Based Reinforcement Learning Method for Power Distribution Network (PDN) Optimization of High Bandwidth Memory (HBM)," in IEEE Transactions on Microwave Theory and Techniques, vol. 70, no. 11, pp. 4772-4786, Nov. 2022

Nonlinear macromodeling of voltage-regulated power delivery networks

Antonio Carlucci, Stefano Grivet-Talocia

Dept. Electronics and Telecommunications, Politecnico di Torino, Italy

antonio.carlucci@polito.it

Abstract—We introduce a frequency-domain macromodeling approach that generalizes Vector Fitting to mildly nonlinear systems, such as power delivery networks including integrated voltage regulators described by averaged models. The proposed approach overcomes the limitation of linearized descriptions, and leads to a black-box nonlinear macromodel of the entire power distribution network with drastically enhanced accuracy.

I. INTRODUCTION

Modern power delivery architectures for high-end microprocessors rely on integrated buck converters to improve efficiency [1]. A typical power delivery network (PDN) topology, shown in Fig. 1, includes power converter stages between the *input* and *output network* blocks, which are both linear and passive. The power switches are driven by the duty cycle signals $d(t)$ produced by feedback compensators based on the deviation between the sensed load voltages $y(t)$ and a desired reference y_{ref}.

In system-level power integrity (PI) analyses, the linear blocks are EM-accurate representations of all components and parasitics in the system (interconnects at board and package level, decaps, etc.), resulting in highly complex large-scale networks. This makes the transient analyses that are required in PI verification computationally demanding or even non-convergent. However, since only the load voltage responses $y(t)$ to current excitations $u(t)$ are of interest, a typical approach is to build a compact behavioral model of the u-y relation to reproduce the same *external* behavior as the original system, at a fraction of the computational cost. In absence of the voltage regulators, such u-y relation is linear and represented by the impedance matrix $Z(s)$ seen by the load. In such situation, compact models are typically obtained in a data-driven manner via off-the-shelf rational approximation algorithms such as Vector Fitting (VF) [2], [3].

The presence of DC-DC converters for local voltage regulation poses a challenge, because it introduces a weak nonlinearity, thus ruling out standard frequency-domain macromodeling approaches based on transfer function samples. This calls for a generalization to account for such nonlinear effects.

II. PROBLEM STATEMENT AND CONTRIBUTION

Following a standard practice in PI analysis, the power switches are here represented via averaged models. Under this

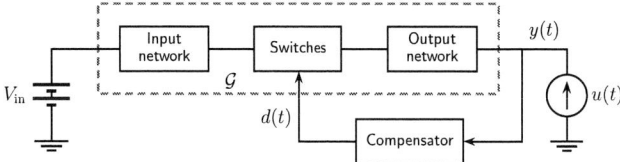

Fig. 1. Power Delivery Network topology. The dashed box encloses the target $\mathcal{G}$ of the proposed nonlinear macromodeling approach.

assumption, we aim at building a behavioral model of the PDN including the converter stage and the linear blocks ($\mathcal{G}$, the dashed box in Fig. 1) and excluding the compensator. The input excitations to $\mathcal{G}$ are both $d(t)$ and $u(t)$. The feedback loop with the compensator will be reintroduced at runtime to evaluate $d(t)$ in real time.

We choose Wiener-Hammerstein (W-H) model structure [4], [5] to represent the nonlinearity due to the power switches. Our main contribution is a novel modeling technique that combines VF with Harmonic Balance (HB) analysis [6] to identify a W-H model that accounts for nonlinear effects induced by $d(t)$. This can be seen as a nonlinear extension of VF for the adopted W-H topology. As for standard VF, model identification is performed in the frequency domain.

III. FORMULATION

A. Model structure

The proposed model $\mathcal{G}$ is built by augmenting a linear model $\mathcal{G}_{\mathrm{lin}}$ with a nonlinear correction $\mathcal{G}_{\mathrm{nl}}$ so that the overall model response is the sum of two components $y(t) = y_{\mathrm{lin}}(t) + y_{\mathrm{nl}}(t)$. The output $y_{\mathrm{lin}}(t)$ of $\mathcal{G}_{\mathrm{lin}}$ is the superposition of the effects due to d and u

$$y_{\mathrm{lin}}(t) = g_{\mathrm{lin},u}[u](t) + g_{\mathrm{lin},d}[d](t),$$

where the notation $g[u](t)$ indicates the response of the linear system g to the input u at time t, and capitalized G denotes its transfer function (TF). The linear block $\mathcal{G}_{\mathrm{lin}}$ has the following two-component TF

$$\mathbf{G}_{\mathrm{lin}}(\mathrm{j}\omega) = \big(G_{\mathrm{lin},u}(\mathrm{j}\omega) \quad G_{\mathrm{lin},d}(\mathrm{j}\omega)\big).$$

As for $\mathcal{G}_{\mathrm{nl}}$, we assume the W-H model structure shown in Fig. 2. Both u and d are first filtered by n linear systems to produce the signals $w_i[u]$, $w_i[d]$, $i = 1, \ldots, n$. Using vector notation, we define

$$\boldsymbol{w}[\star](t) = \big(\star(t) \quad w_1[\star](t) \quad \cdots \quad w_n[\star](t)\big)^T \in \mathbb{R}^{n+1}$$

979-8-3503-5124-8/24 $31.00 © 2024 IEEE

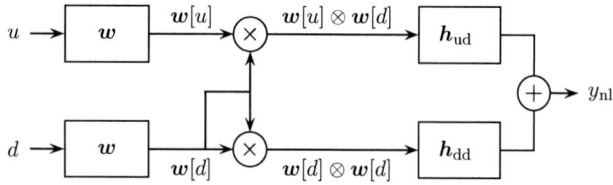

Fig. 2. Block schematic of the W-H model $\mathcal{G}_{\mathrm{nl}}$.

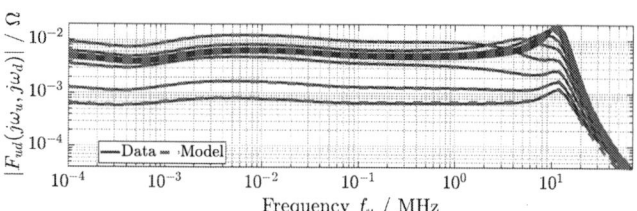

Fig. 3. Model-data comparison of the bi-variate function $F_{ud}(\mathrm{j}\omega_d, \mathrm{j}\omega_u)$. Different curves correspond to several fixed values of ω_d. Similar results hold for F_{dd}.

where $\star$ stands for either u or d. Next, the static nonlinearity in the middle forms the Kronecker products $\boldsymbol{w}[u] \otimes \boldsymbol{w}[d]$ and $\boldsymbol{w}[d] \otimes \boldsymbol{w}[d]$, giving rise, respectively, to a) an intermodulation product between d and u and b) second-order effects in d. Finally, the two linear blocks $\boldsymbol{h}_{ud}$ and $\boldsymbol{h}_{dd}$ map $(n+1)^2$ inputs into a single output, yielding

$$y_{\mathrm{nl}}(t) = \boldsymbol{h}_{ud}[\boldsymbol{w}[u] \otimes \boldsymbol{w}[d]] + \boldsymbol{h}_{dd}[\boldsymbol{w}[d] \otimes \boldsymbol{w}[d]].$$

In continuity with VF-based macromodeling, we adopt a frequency-domain approach in the characterization of $\mathcal{G}_{\mathrm{nl}}$ so as to obtain a model representation that is compatible with rational approximation tools. In particular, we consider the steady-state response of $\mathcal{G}_{\mathrm{nl}}$ under harmonic excitation as in [7], to define rational functions suitable to describe nonlinear effects. By setting $u(t) = e^{\mathrm{j}\omega_u t}$ and $d(t) = e^{\mathrm{j}\omega_d t}$ and propagating them through the blocks in Fig. 2, we see that the resulting $y_{\mathrm{nl}}(t)$ contains, among others, the intermodulation component $F_{ud}(\mathrm{j}\omega_u, \mathrm{j}\omega_d)e^{\mathrm{j}(\omega_u+\omega_d)t}$, with

$$F_{ud}(\mathrm{j}\omega_u, \mathrm{j}\omega_d) = \mathbf{H}_{ud}(\mathrm{j}\omega_u + \mathrm{j}\omega_d)\mathbf{W}_{ud}(\mathrm{j}\omega_u, \mathrm{j}\omega_d) \quad (2)$$

where $\mathbf{W}_{ud}(\mathrm{j}\omega_u, \mathrm{j}\omega_d) = \mathbf{W}(\mathrm{j}\omega_u) \otimes \mathbf{W}(\mathrm{j}\omega_d)$. Similarly, to isolate the second-order effects in d, consider $u(t) = 0$ and $d(t) = e^{\mathrm{j}\omega_1 t} + e^{\mathrm{j}\omega_2 t}$. In this case the output component at frequency $\omega_1 + \omega_2$ is $F_{dd}(\mathrm{j}\omega_1, \mathrm{j}\omega_2)e^{\mathrm{j}(\omega_1+\omega_2)t}$, with

$$F_{dd}(\mathrm{j}\omega_1, \mathrm{j}\omega_2) = \mathbf{H}_{dd}(\mathrm{j}\omega_1 + j\omega_2)\cdot$$
$$[\mathbf{W}(\mathrm{j}\omega_1) \otimes \mathbf{W}(\mathrm{j}\omega_2) + \mathbf{W}(\mathrm{j}\omega_2) \otimes \mathbf{W}(\mathrm{j}\omega_1)] \quad (3)$$

This simple analysis motivates using the bi-variate rational functions F_{ud} and F_{dd} to represent the W-H model $\mathcal{G}_{\mathrm{nl}}$.

B. Data collection

Regarding the linear part, we extract samples $\breve{\mathbf{G}}_{\mathrm{lin}}^{(k)} = \breve{\mathbf{G}}_{\mathrm{lin}}(\mathrm{j}\omega^{(k)})$, for $k = 1, \ldots, K$, using small-signal AC analysis around the nominal operating point. Note that, in the following, the accented symbol $\breve{x}$ indicates the observed or measured value of the quantity x. As for $\mathcal{G}_{\mathrm{nl}}$, two-tone HB analysis allows to extract the samples

$$\breve{F}_{ud}^{(k)} = \breve{F}_{ud}(\mathrm{j}\omega_u^{(k)}, \mathrm{j}\omega_d^{(k)}), \quad k = 1, \ldots, K_{ud}$$
$$\breve{F}_{dd}^{(k)} = \breve{F}_{dd}(\mathrm{j}\omega_1^{(k)}, \mathrm{j}\omega_2^{(k)}), \quad k = 1, \ldots, K_{dd}$$

defined over two-dimensional frequency domains. In particular, $\breve{F}_{ud}^{(k)}$ is found by setting $u(t) = \cos(\omega_u^{(k)}t)$ and $d(t) = d_0 + \alpha \cos(\omega_d^{(k)}t)$, with α chosen to keep the duty cycle in $[0, 1]$. The resulting harmonic component at $\omega_u^{(k)} + \omega_d^{(k)}$ gives the result. Similarly, $\breve{F}_{dd}^{(k)}$ is extracted by HB with $d(t) = d_0 + \alpha \cos(\omega_1^{(k)}t) + \alpha \cos(\omega_2^{(k)}t)$.

C. Rational approximation

Using the above-defined data, the behavioral model can be built using rational approximation. In fact, a direct application of VF to $\breve{\mathbf{G}}_{\mathrm{lin}}^{(k)}$ gives a TF model in pole-residue form

$$\mathbf{G}_{\mathrm{lin}}(s) = \sum_{\ell=1}^{\nu} \mathbf{R}_{\mathrm{lin},\ell}(s - p_\ell)^{-1}$$

where $p_\ell \in \mathbb{C}$ are ν poles and $\mathbf{R}_{\mathrm{lin},\ell} \in \mathbb{C}^{1\times 2}$ the corresponding residues. Once the (linearized) system poles p_ℓ have been identified through VF, we assume these are sufficient to describe the whole system dynamics. This simplifies the problem of building models of $\mathbf{W}(s)$, $\mathbf{H}_{ud}(s)$, $\mathbf{H}_{dd}(s)$ because we take them to be rational functions with the same poles. The next observation is that there is no further loss of generality in assuming that $\mathbf{W}(s)$ coincides with the partial fraction basis

$$\mathbf{W}(s) = \begin{pmatrix} 1 & (s - p_1)^{-1} & \cdots & (s - p_\nu)^{-1} \end{pmatrix}^T.$$

Hence, the only quantities to be estimated are the residues of $\mathbf{H}_{ud}(s)$, $\mathbf{H}_{dd}(s)$, which have pole-residue expansions

$$\mathbf{H}_\star(s) = \sum_{\ell=1}^{\nu} \mathbf{T}_{\star,\ell}(s - p_\ell)^{-1}, \quad \star = \{dd, ud\} \quad (4)$$

where $\mathbf{T}_{\star,\ell} \in \mathbb{C}^{1\times(n+1)^2}$. Then, (2) and (4) imply that matching F_{ud} to the sampled data $\breve{F}_{ud}$ translates in the approximation

$$\sum_{\ell=1}^{\nu}[\mathrm{j}(\omega_u^{(k)} + \omega_d^{(k)}) - p_\ell]^{-1}\mathbf{T}_{ud,\ell}\mathbf{W}_{ud}(\mathrm{j}\omega_u^{(k)}, \mathrm{j}\omega_d^{(k)}) \approx \breve{F}_{ud}^{(k)}$$

that is a linear least-squares problem in the unknowns $\mathbf{T}_{ud,\ell}$. The residues $\mathbf{T}_{dd,\ell}$ are found analogously by optimizing the approximation $F_{dd} \approx \breve{F}_{dd}$ and using (3)–(4).

IV. NUMERICAL RESULTS

The proposed model is tested on a single-core version of the multicore PDN described in [8], which refers to an Intel-based mobile architecture. We sampled all transfer functions by sweeping the frequency of the load current excitation u and that of the duty cycle d up to $0.5\,\mathrm{GHz}$. In total, $K = 120$, $K_{dd} = 10^4$, $K_{ud} = 1.2 \cdot 10^4$ samples were collected. We used $\nu = 12$ poles to fit the bi-variate models, obtaining the results reported in Fig. 3.

In time domain, the model was tested with two load current signals. The first is a $3\,\mu\mathrm{s}$ long ramp-up analysis using a

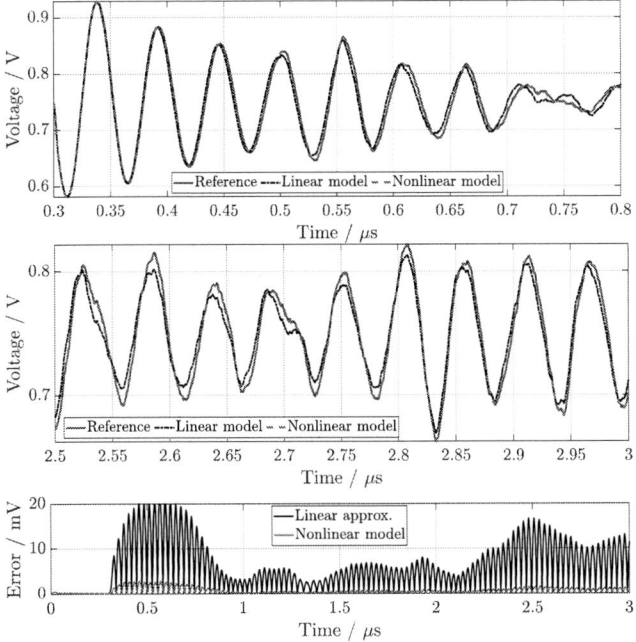

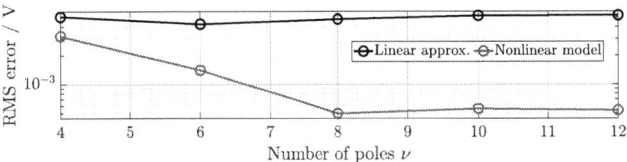

Fig. 5. RMS error comparison between a small-signal (linear) model and the proposed nonlinear model, reported by increasing the number of model poles.

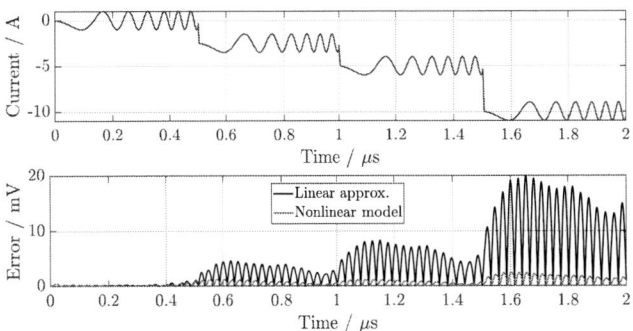

Fig. 6. Transient simulation using a chirp signal around several bias levels. Top panel: load current excitation. Bottom panel: instantaneous error.

Fig. 4. Transient response comparing the reference with linear and nonlinear models. Top: subinterval $t \in [0.3, 0.8]\,\mu s$, i.e. response to noisy current step. Middle: response in the interval $[2.5, 3]\,\mu s$. Bottom: instantaneous error along the entire simulation.

10 A current step occurring at $t = 0.3\,\mu s$ with 3 ns rise time. This is superimposed to a filtered white noise with bandwidth $[4, 16]$ MHz (close to the peak in Fig. 3) and peak amplitude equal to 2 A. The model response is compared to the reference in Fig. 4, showing the detail of the step response in the top panel, and the response to noise in the middle panel. A quantitative error analysis comparing both linear and nonlinear model responses with the reference along the entire simulation interval is also reported in Fig. 4 (bottom panel). Solving the proposed model in MATLAB for 5×10^4 fixed-length timesteps takes 2.7 s, whereas the same simulation of the original system requires 191 s. The proposed algorithm runs in about 63 s, whereas data collection via HB takes 1.2 hours in our experiments. It is important to remark that the benefit of this modeling approach is also that it captures nonlinear effects that would be impossible to model through linear macromodeling. This is evident in Fig. 5, which shows that the RMS error of the linear model remains almost constant as the number of poles ν is increased, while the nonlinear model provides higher accuracy because the additional model poles improve the approximation of higher-order effects that are structurally neglected with a linear approximation.

The second test signal is reported in Fig. 6 (top panel). It consists of a sequence of steps that sequentially shift the operating point of the load current from 0 A to 10 A every $0.5\,\mu s$, in discrete steps. This is superimposed to a chirp signal of amplitude 1 A that repeatedly sweeps from 5 to 25 MHz. As the average load current departs from zero (that is the operating point used to construct the small-signal model $\mathcal{G}_{\mathrm{lin}}$),

the linear approximation becomes increasingly inaccurate. Adding the second-order nonlinear corrections decreases this error by at least one order of magnitude.

V. CONCLUSION

This paper has demonstrated that, in presence of integrated voltage regulators, linearized models of the PDN impedance may be inadequate to reproduce with sufficient accuracy the transient load voltages. The proposed approach, which extends Vector Fitting to account for the nonlinear terms that are induced by the regulator switches, is able to correct the linearized impedance model to represent, as a black-box, the entire voltage-regulated PDN structure.

REFERENCES

[1] E. A. Burton et al., "FIVR — Fully integrated voltage regulators on 4th generation Intel® Core™ SoCs," in *2014 IEEE Applied Power Electronics Conference and Exposition*, Mar. 2014, pp. 432–439.

[2] B. Gustavsen and A. Semlyen, "Rational approximation of frequency domain responses by vector fitting," *IEEE Transactions on Power Delivery*, vol. 14, no. 3, pp. 1052–1061, 1999.

[3] S. Grivet-Talocia and B. Gustavsen, *Passive macromodeling: Theory and applications*. John Wiley & Sons, 2015.

[4] E.-W. Bai and F. Giri, *Introduction to Block-oriented Nonlinear Systems*. London: Springer London, 2010, pp. 3–11.

[5] J. A. Oliver, R. Prieto, J. A. Cobos, O. Garcia, and P. Alou, "Hybrid wiener-hammerstein structure for grey-box modeling of DC-DC converters," in *2009 Twenty-Fourth Annual IEEE Applied Power Electronics Conference and Exposition*, 2009, pp. 280–285.

[6] R. J. Gilmore and M. B. Steer, "Nonlinear circuit analysis using the method of harmonic balance—a review of the art. part i. introductory concepts," *International Journal of Microwave and Millimeter-Wave Computer-Aided Engineering*, vol. 1, no. 1, pp. 22–37, 1991.

[7] S. Baumgartner and W. Rugh, "Complete identification of a class of nonlinear systems from steady-state frequency response," *IEEE Transactions on Circuits and Systems*, vol. 22, no. 9, pp. 753–759, 1975.

[8] A. Carlucci, S. Grivet-Talocia, S. Kulasekaran, and K. Radhakrishnan, "Structured model order reduction of system-level power delivery networks," *IEEE Access*, vol. 12, pp. 18 198–18 214, 2024.

Efficient parametric assessment of worst-case voltage droop in power delivery networks

Tommaso Bradde, Antonio Carlucci, Riccardo Trinchero, Paolo Mandredi, Stefano Grivet-Talocia

Dept. Electronics and Telecommunications, Politecnico di Torino, Italy

tommaso.bradde@polito.it

Abstract—Power Delivery Network (PDN) optimization is crucial for guaranteeing adequate power integrity performance in modern microprocessor systems. In this work, we introduce a novel surrogate modeling workflow for efficiently predicting the worst-case voltage droop occurring at the loading points of a PDN including a set of free design parameters. We apply the proposed approach for modeling the impact of a set of decoupling capacitors on the performance of a template PDN structure.

Index Terms—Power Delivery Networks, Decoupling Capacitors, Machine Learning

I. INTRODUCTION

Power Integrity (PI) optimization is crucial for achieving target performance in terms of efficiency and reliability of modern microprocessor systems. With current technologies, careful assessments of the Power Delivery Network (PDN) electrical performance must be necessarily performed at the system level. In this view, one challenging task in PI is the minimization of the PDN power noise, and in particular, of the voltage droop resulting from the dynamic activity of its loads. Most commonly, this performance specification is met by designing the PDN with the objective of bounding its impedance magnitude below a (possibly frequency-dependent) threshold, known as target impedance [1] [2] [3]. Optimizing a PDN to achieve minimal voltage droops can be challenging and time consuming, as any design configuration must be ultimately verified in time domain, possibly interlacing extremely costly transient analyses with the impedance shaping routine, in order to actually verify the entity of the droop under the operating conditions of interest.

This contribution introduces a surrogate modeling approach thought to alleviate the above-mentioned computational issues. In particular, we propose an efficient approach for directly predicting the worst-case voltage droop (WCVD) of a PDN as a function of the relevant design parameters, assuming that the load currents are bounded in amplitude and slew-rate. The approach expands on the recent numerical tools introduced in [4], that allow to compute directly the voltage droop without relying on any costly transient simulations. Exploiting these tools, we compute the WCVD of the PDN for a discrete number of design parameters configurations, and we exploit this information as input for generating a surrogate model which predicts the desired performance index efficiently, throughout the whole design space. The surrogate is obtained by applying Gaussian Process Regression (GPR),

which is a machine learning approach that allows reproducing a complex functional relationship using a relatively low amount of training data [5]. Numerical experiments based on a template PDN description provide a proof of concept for the efficacy and efficiency of the proposed approach.

II. PROBLEM STATEMENT

We consider a generic Linear Time Invariant (LTI) PDN structure, subject to independent current stimuli acting in correspondence of P well-defined electrical ports, representing its loading points. We allow the PDN description to include a number ρ of free design parameters, collected in the vector $\boldsymbol{x} = [x_1, \ldots, x_\rho]^T \subset \mathcal{X}$, where $\mathcal{X}$ is a hyperectangle defining the allowed domain of variation. We denote as $\mathbf{Z}(s, \boldsymbol{x}) \in \mathbb{C}^{P \times P}$ the parameterized output impedance matrix of the PDN defined at the loading points, and we assume that $\mathbf{Z}(s, \boldsymbol{x})$ can be sampled at discrete frequency-parameter configurations, via real or virtual measurements.

Let us denote the impulse response as $\mathbf{z}(t, \boldsymbol{x}) = \mathcal{L}^{-1}\{\mathbf{Z}(s, \boldsymbol{x})\}$, the vector of load current sources entering the PDN as $\boldsymbol{i}(t)$, and the corresponding port voltages as $\boldsymbol{v}(t)$. Then, the convolution integral

$$\boldsymbol{v}(t, \boldsymbol{x}) = \int_0^t \mathbf{z}(\tau, \boldsymbol{x})\boldsymbol{i}(t - \tau)d\tau = (\mathbf{z}(\boldsymbol{x}) \star \boldsymbol{i})(t), \quad (1)$$

provides the instantaneous voltage droop of the PDN for any admissible design and load profile. Our objective is to generate a surrogate model for efficiently predicting the WCVD of the PDN as a function of the parameters, assuming load currents that are bounded in amplitude and slew rate. Formally, we desire a surrogate representation for the following function

$$y(\boldsymbol{x}) = \sup_{t \geq 0, \boldsymbol{i}(t) \in \mathcal{I}} ||\boldsymbol{v}(t, \boldsymbol{x})||_\infty, \quad (2)$$

where $\mathcal{I}$ denotes the set of admissible load current stimuli

$$\mathcal{I} = \{\boldsymbol{i}(t) : 0 \leq |i_j(t)| \leq I_{j,\max}, \left|\frac{di_j(t)}{dt}\right| \leq \Delta_{\max},$$
$$\forall t \geq 0, j = 1, \ldots, P\}. \quad (3)$$

Notice that for step transitions $0 \leftrightarrow I_{\max}$ the slew-rate constraint implies a minimum rise time $\tau_r = I_{\max}/\Delta_{\max}$.

979-8-3503-5124-8/24 $31.00 © 2024 IEEE

III. SURROGATE MODEL GENERATION

The proposed modeling workflow is based on two main steps. The first consists in the generation of a dataset of pairs $\mathcal{D} = \{(\boldsymbol{x}_k, y_k)\}_{k=1}^K$, with $y_k = y(\boldsymbol{x}_k)$, obtained by evaluating (2) at discrete parameter values $\boldsymbol{x}_k \in \mathcal{X}$. The second exploits this dataset to generate the GPR model.

A. Dataset Extraction

1) Voltage Droop Computation: Given a design parameter configuration $\boldsymbol{x}_k$, the exact computation of the corresponding bound $y(\boldsymbol{x}_k)$ would require constructing the worst-case load current signal, as outlined in [6], and simulating the PDN response in the time domain applying such load profile. To avoid this expensive procedure, in this contribution we apply the simplified approach described in [4], which is equivalent for practical purposes. Specifically, we evaluate (2) as

$$y(\boldsymbol{x}_k) = \max_{i=1,\dots,P} v_{i,\max}(\boldsymbol{x}_k), \tag{4}$$

$$v_{i,\max}(\boldsymbol{x}_k) = \sum_{j=1}^P I_{j,\max} \int_0^\infty |(z_{ij}(\boldsymbol{x}_k) \star g_{\tau_r})(t)|_+ dt \tag{5}$$

where $g_{\tau_r}(t)$ is a unit-area square pulse having width τ_r and $|a|_+$ is equal to a if $a \geq 0$ and is 0 otherwise. See [4, Sec.II-B] for technical details. Assuming that a closed form expression for $\boldsymbol{z}(t, \boldsymbol{x}_k)$ is available, the target value $y(\boldsymbol{x}_k)$ is obtained via numerical integration of (5). Most commonly, the numerical or experimental characterization of the PDN behavior is available only terms of measurements of the corresponding impedance matrix $\mathbf{Z}(s, \boldsymbol{x})$. In this scenario, a closed form approximation for $\boldsymbol{z}(t, \boldsymbol{x}_k)$ can be retrieved via rational fitting, using a set of measurements of the kind $\mathcal{V}_k = \{(j\omega_m, \mathbf{Z}(j\omega_m, \boldsymbol{x}_k))\}_{m=1}^M$, with $\omega_m = 2\pi f_m$, $f_m \in [f_{\min}, f_{\max}]$. The measurements are used as input for the Vector Fitting iteration [7] to generate a stable (yet not necessarily passive) approximation $\tilde{\mathbf{Z}}_k(s)$ for the PDN impedance, with structure

$$\tilde{\mathbf{Z}}_k(s) = \sum_{\ell=1}^{\bar{\ell}} \frac{\mathbf{R}_\ell}{s - p_\ell} \approx \mathbf{Z}(s, \boldsymbol{x}_k). \tag{6}$$

An approximation for the required impulse response is obtained via analytical inverse Laplace transform

$$\tilde{\boldsymbol{z}}_k(t) = \mathcal{L}^{-1}\{\tilde{\mathbf{Z}}_k(s)\} = \sum_{\ell=1}^{\bar{\ell}} \mathbf{R}_\ell\, e^{p_\ell t} \approx \boldsymbol{z}(t, \boldsymbol{x}_k), \tag{7}$$

and can be used in place of $\boldsymbol{z}(t, \boldsymbol{x}_k)$ in (5) to compute the desired sample $y(\boldsymbol{x}_k)$.

B. Modeling via Gaussian Process regression

Once the dataset $\mathcal{D}$ is available, we use it to train a GPR model. We choose GPR because it offers a good trade-off between model accuracy and complexity, providing a flexible model with limited training cost. We consider a standard implementation with a constant prior trend β_0 and an anisotropic Matérn 5/2 covariance function, i.e.,

$$k(\boldsymbol{x}, \boldsymbol{x}') = \sigma^2 \left(1 + \sqrt{5}u + \frac{5}{3}u^2\right) \exp\left(-\sqrt{5}u\right), \tag{8}$$

where

$$u = \sqrt{\sum_{j=1}^\rho \frac{(x_j - x_j')^2}{\theta_j^2}} \tag{9}$$

and $\theta_1, \dots, \theta_\rho$ are unknown scale parameters. The covariance function describes the correlation between two points in the input space, $\boldsymbol{x}$ and $\boldsymbol{x}'$, which translates into the model smoothness. Owing to the complexity of the target function, an anisotropic kernel (in which the scale parameter differs for each input parameter) turns out to be substantially more accurate, at the expense of a slight increase of the training cost. The WCVD is predicted as

$$\hat{y}(\boldsymbol{x}) = \beta_0 + \sum_{k,m=1}^K \left[\tilde{\mathbf{K}}^{-1}\right]_{km} (y_m - \beta_0) k(\boldsymbol{x}, \boldsymbol{x}_k). \tag{10}$$

In (10), $\tilde{\mathbf{K}}$ is a matrix with entries $\tilde{K}_{km} = k(\boldsymbol{x}_k, \boldsymbol{x}_m) + \sigma_n^2 \delta_{km}$, for $k, m = 1, \dots, K$, where δ_{km} is the Kronecker's delta and σ_n^2 is a noise parameter that acts as a regularizer. It should be noted that the size of $\tilde{\mathbf{K}}$ is determined by the available data samples and usually does not need to be substantially increased, even for larger input spaces [5]. The coefficient β_0 is obtained via a generalized least-square estimate as

$$\beta_0 = \frac{\boldsymbol{e}^\mathsf{T} \tilde{\mathbf{K}}^{-1} \boldsymbol{y}}{\boldsymbol{e}^\mathsf{T} \tilde{\mathbf{K}}^{-1} \boldsymbol{e}}, \tag{11}$$

where $\boldsymbol{e} \in \mathbb{R}^K = (1, \dots, 1)^\mathsf{T}$ is a column vector of ones and $\boldsymbol{y} = (y_1, \dots, y_K)^\mathsf{T}$ is the vector of data samples. The scale parameters $\{\theta_j\}_{j=1}^\rho$, the noise variance σ_n^2, and the kernel variance σ^2 are estimated via likelihood maximization.

IV. NUMERICAL RESULTS

We provide a proof of concept for the proposed approach considering a 2-D distributed structure, representative of a template PDN on a printed circuit board. The board consists of two parallel square planes with side length $l = 7.5$ cm, separated by a dielectric material of width $d = 0.5$ mm. The dielectric has relative permittivity $\epsilon_r = 5.5$ and loss tangent $\tan\delta = 0.01$. Five ideal (lumped) ports are defined between top and bottom planes at coordinates $\{(0, 3.8), (3, 3), (2.7, 3), (3, 2.7), (3.3, 3)\}$, defined in cm taking the bottom left corner of the board as the origin. We close port #1 on a Voltage Regulator Module (VRM), modeled as a RL series circuit with $\mathrm{R_{VRM}} = 2$ mΩ and $\mathrm{L_{VRM}} = 1$ nH. Port #2 is considered as the loading point of the PDN, and closed on a RC series circuit representing a simplified silicon die model (with $\mathrm{R_{die}} = 50$ mΩ and $\mathrm{C_{die}} = 5$ nF) in parallel with a decoupling capacitor. With the remaining ports left open, the impedance of the structure is shown in Fig. 1 (blue line). We assume ports #2 to #5 be shunted with four decoupling capacitors. The latter are modeled as series RLC circuits, whose lumped element values are the design parameters of interest. Table I reports the admissible parameter ranges, using subscripts to identify the components reference ports. The PDN impedance obtained with one random design choice is shown in Fig. 1 (red line).

979-8-3503-5124-8/24 $31.00 © 2024 IEEE

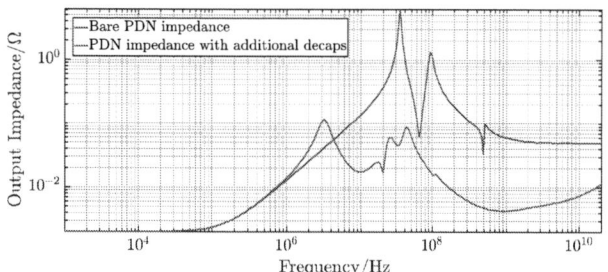

Fig. 1. The output impedance of the considered template PDN. The blue line shows the bare impedance without decoupling capacitors. The red line shows the impedance loaded by with a random admissible choice of such capacitors.

TABLE I
VARIABILITY RANGES OF DESIGN PARAMETERS

C_2 /nF	C_3 /μF	C_4 /μF	C_5 /μF	R_{2-5} /mΩ	L_{2-5} /pH
[50, 150]	[1, 100]	[0.1, 10]	[0.01, 1]	$25 \pm 30\%$	$1 \pm 90\%$

Our objective is to predict the WCVD of the PDN as a function of these parameters when the loading point is subject to a load current with maximum amplitude $I_{\max} = 1$ A and minimum rise time $\tau_r = 3$ ns. To this aim, we generate the dataset $\mathcal{D}$ with $K = 2500$, defining the sampling points $\{\boldsymbol{x}_k\}_{k=1}^{K}$ using a Sobol sequence. For each sampling point, we retrieve the rational model (6) approximating the output impedance of the PDN and we compute the corresponding worst-case droop according to (4), (5). The data are then divided into disjoint training and test sets, denoted as $\mathcal{D}_{\text{train}} = \{(\boldsymbol{x}_j, y_j)\}_{j=1}^{K_{\text{train}}}$ and $\mathcal{D}_{\text{test}} = \{(\boldsymbol{x}_i, y_i)\}_{i=1}^{K_{\text{test}}}$, where we set $K_{\text{train}} = 1000$ and $K_{\text{test}} = 1500$. Using $\mathcal{D}_{\text{train}}$, the GPR model is trained in 40 s on a common laptop.

The scatter plot of the GPR predictions against the test samples is shown in Fig. 2 and proves the remarkable accuracy of the surrogate, exhibiting a coefficient of determination of

$$R^2 = 1 - \frac{\sum_{i=1}^{K_{\text{test}}} (y_i - \hat{y}(\boldsymbol{x}_i))^2}{\sum_{i=1}^{K_{\text{test}}} (y_i - \bar{y})^2} = 0.98, \tag{12}$$

where $\bar{y}$ is the dataset mean. Using the model, a batch of 1000 test samples is computed in 66 ms, against the 54 s required by directly computing (5). This confirms that also in case of the considered academic example the proposed approach provides major improvements in terms of efficiency. To conclude, following the approach presented in [4], for each test sample, we build the worst-case current giving rise to the WCVD, and compute the associated voltage responses (1), over a timespan of 5 μs. Figure 3 shows the ensemble of these responses over a restricted time window. The solid red line marks the WCVD for the best design among the test configurations, and shows that the bound is actually attained by the corresponding voltage droop waveform (blue solid line); the GPR prediction deviates by 0.5% from the reference.

V. CONCLUSIONS

This work introduced a novel approach for fast parametric assessment of the WCVD occurring in a PDN under con-

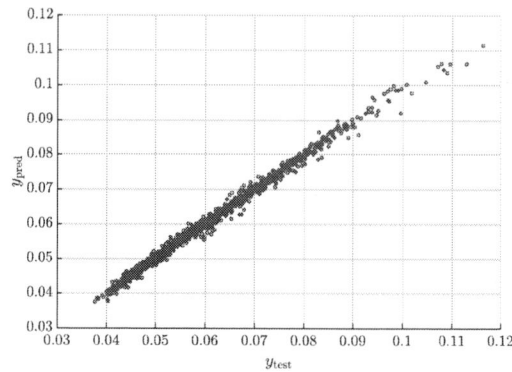

Fig. 2. Correlation plot of the reference data against the predictions obtained via the GPR model built with K_{train}=1000 training samples.

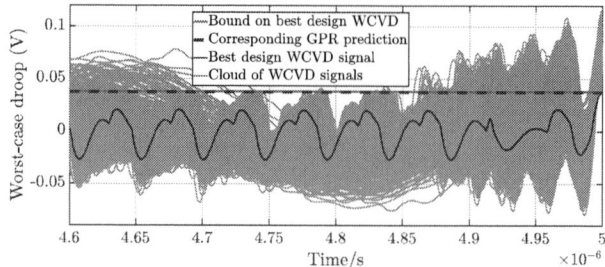

Fig. 3. Grey lines: WCVD waveforms associated to the test samples. The blue line is the voltage droop signal corresponding to the best design. Red and black dashed lines are the corresponding exact and surrogate WCVD bounds.

strained load current profiles. The proposed method combines GPR surrogate modeling with established numerical tools, enabling efficient prediction of the considered performance index for the sake of design verification and optimization.

REFERENCES

[1] J. Chen and M. Hashimoto, "A frequency-dependent target impedance method fulfilling voltage drop constraints in multiple frequency ranges," *IEEE Transactions on Components, Packaging and Manufacturing Technology*, vol. 10, no. 11, pp. 1769–1781, 2020.

[2] Y. Kim, K. Kim, J. Cho, J. Kim, K. Kang, T. Yang, Y. Ra, and W. Paik, "Power distribution network design and optimization based on frequency dependent target impedance," in *2015 IEEE Electrical Design of Advanced Packaging and Systems Symposium (EDAPS)*, 2015, pp. 89–92.

[3] S. Liang, B. Zhao, S. Bai, S. Connor, M. Cocchini, S. Scearce, D. Becker, M. Cracraft, M. S. Doyle, A. Ruehli, and J. Drewniak, "Decoupling capacitor optimization to achieve target impedance in pcb pdn design," in *2021 IEEE International Joint EMC/SI/PI and EMC Europe Symposium*, 2021, pp. 967–972.

[4] A. Carlucci, T. Bradde, and S. Grivet-Talocia, "Fast prediction of worst-case voltage droops in power distribution networks," in *2024 IEEE 28th Workshop on Signal and Power Integrity (SPI)*, 2024, pp. 1–4.

[5] P. Manfredi and R. Trinchero, "A probabilistic machine learning approach for the uncertainty quantification of electronic circuits based on Gaussian process regression," *IEEE Transactions on Computer-Aided Design of Integrated Circuits and Systems*, vol. 41, no. 8, pp. 2638–2651, 2021.

[6] W. Reinelt, "Maximum output amplitude of linear systems for certain input constraints," in *Proceedings of the 39th IEEE Conference on Decision and Control*, vol. 2, 2000, pp. 1075–1080.

[7] S. Grivet-Talocia and B. Gustavsen, *Passive Macromodeling: Theory and Applications.* New York: John Wiley and Sons, 2016.

979-8-3503-5124-8/24 $31.00 © 2024 IEEE

Limit of the Impact of the Via Stub Length on the Via Impedance in Printed Circuit Boards

Katharina Scharff
IBM Deutschland
Research & Development GmbH
Germany
Email: katharina.scharff1@ibm.com

Xiaomin Duan
IBM Deutschland
Research & Development GmbH
Germany
Email: xduan@de.ibm.com

Dierk Kaller
IBM Deutschland
Research & Development GmbH
Germany
Email: dkaller@de.ibm.com

Abstract—This paper investigates the impedance discontinuities due to residual via stubs for a differential via pair. Different stub lengths are investigated. The influence of the stub is comparable to the influence of the via pad.

Index Terms—PCB, Via Stub, Via Pad

I. INTRODUCTION

Higher bitrates put high demands on all parts of the interconnect. The printed circuit board (PCB) technology is an important element that enables high bitrates. The via stub is long known as a major contributor to a poor performance [1]–[3]. It is necessary to backdrill to remove the long stubs and mitigate their detrimental effect. However, a small residual stub remains which can still negatively influence the signal propagation [4].

In current PCB technologies, the via stubs are residuals of the backdrilling process. This removes the majority of the stub but a small stub, called the residual stub, is left. Technologies are improving and it is possible to reduce the length of the residual stub. However, this increases the manufacturing cost. Microvia technology is another alternative, but this technology is also expensive. While the general assumption is that a shorter stub will significantly improve the performance, it is not obvious if the additional cost and effort are necessary. Fig. 1 shows a drawing of a differential via pair in a multilayer PCB. The pair is surrounded by ground (GND) vias. The signal vias are backdrilled but a small part of the stub cannot be removed and remains.

In this work we study the impact of the via stub based on the model in Fig. 1 and a cross-section with state-of-the-art technology. We investigate the impact of reducing the via stub by using more advanced technology. The via pad as a second major impedance discontinuity is also taken into consideration. We can show that a reduction of the via stub improves the via impedance but the via pad discontinuity is equally influential.

II. VIA LAYOUT

To evaluate the effect of the via stub, a simple differential via structure was modeled and simulated. This layout can be found in Fig. 1. A single differential via pair is surrounded by 10 GND vias. The pitch between the vias is 1.060 mm. On

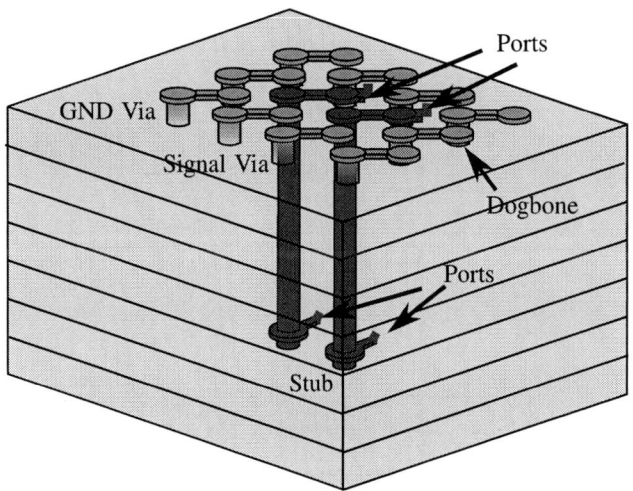

Fig. 1. Printed circuit board with a differential via pair surrounded by GND vias. The GND vias extend to the bottom but only the upper parts of the GND vias are shown here. The signal vias are backdrilled with a small residual stub. The pitch between the vias is 1.060 mm.

the top layer the vias are connected to dogbones. A schematic of the cross-section is shown in Fig. 2. The cross-section is a multilayer stackup with multiple signal layers. The dielectric material is a standard FR4 in the upper part and a ultra-low loss (ULL) dielectric with a small loss tangent in the lower part, where the high-speed signal layers are routed. The differential via pair is surrounded by ground vias which extend to the bottom of the stackup (only the upper part is shown in Fig. 1). Two via models with different via lengths were designed. The short via is 1.65 mm long, and the long via is 3.26 mm long. The signal pad and a short stripline is connected to the vias in the signal layer. The via diameter is 305 μm (12 mil) and the standard pad diameter is 20 mil. The stub length is varied between 14, 12, 8, and 4 mil, which represent different technologies with increasing backdrilling abilities. The pad diameter was also varied between 18 and 22 mil. Additionally, three more examples, which are not representative of current technology, were generated. There either the stub was removed completely or the pad in the signal layer was removed. All model variations can be found in Table I.

979-8-3503-5124-8/24 $31.00 © 2024 IEEE   127

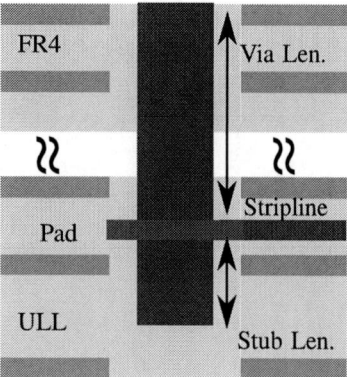

Fig. 2. Cross-section and side view of the via. The upper part of the stack-up is a standard FR4 material. The lower layers are made of an ultra-low loss dielectric (ULL). A short stripline is attached to the via.

TABLE I
MODEL VARIATIONS OF THE SHORT AND LONG VIA MODEL.

Via length	Stub Length	Pad diam.
1653 μm	101.6 μm(4 mil)	508 μm(20 mil)
	203.2 μm(8 mil)	508 μm(20 mil)
	304.8 μm(12 mil)	508 μm(20 mil)
	355.6 μm(14 mil)	508 μm(20 mil)
	203.2 μm(8 mil)	0 μm
	203.2 μm(8 mil)	457.2 μm(18 mil)
	203.2 μm(8 mil)	558.8 μm(22 mil)
	0 μm	508 μm(20 mil)
	0 μm	0 μm
3264 μm	101.6 μm(4 mil)	508 μm(20 mil)
	203.2 μm(8 mil)	508 μm(20 mil)
	304.8 μm(12 mil)	508 μm(20 mil)
	355.6 μm(14 mil)	508 μm(20 mil)
	203.2 μm(8 mil)	0 μm
	0 μm	508 μm(20 mil)
	0 μm	0 μm

The models were simulated with the full-wave finite elements tool Ansys HFSS [5]. The lumped ports are placed at the top of the vias at the dogbones and at the end of the striplines (see Fig. 1). The model is surrounded by an airbox with a radiation boundary.

III. SIMULATION RESULTS

The model variations in Table I were simulated with Ansys HFSS. The resulting return losses for the short and the long via and a variation of the stub length are shown in Fig. 3. Fig. 3(a) shows the differential reflection for the short via and Fig. 3(b) the reflection for the long via. The resonance at 35 GHz is due to the dogbone structures and their associated ports. For both via lengths the reflection increases with the stub length.
Fig. 5 shows the time-domain reflectometry (TDR). This is the reflected voltage normalized to $85\,\Omega$ based on the input step with a rise time of 15 ps (see Fig 4). The TDR shows the impedance variation along the channel for the short (see Fig. 5(a)) and the long via (see Fig. 5(b)) for different stub lengths. The results show that the impedance of the via is less than $85\,\Omega$. The via stub is another impedance discontinuity and with a longer stub the impedance decreases further and

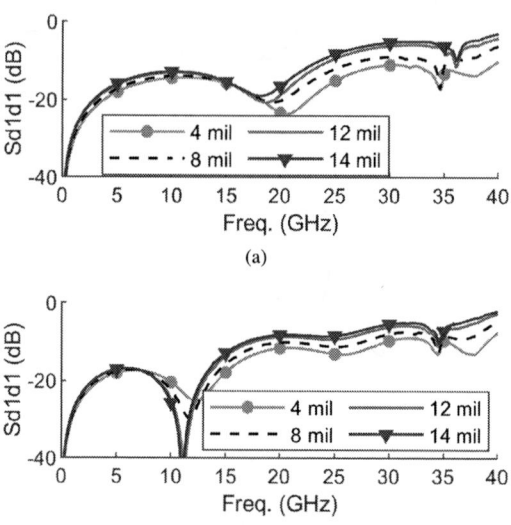

Fig. 3. Return loss for the short and the long via with different stub lengths. (a) Short via. (b) Long via.

the drop becomes steeper. In the longer via the signal takes longer to propagate, thus the stub impedance discontinuity is shifted in time compared to the short via. The impedance discontinuity is greater with a longer stub. Reducing the stub size to 8 mil improves the impedance but the impedance drop is still present.

To investigate how much improvement would be achievable in an ideal case, additional models were simulated. Here, either the stub length is set to zero, or the pad is removed, or both. The pad has a large influence on the impedance and to estimate whether a smaller stub is necessary, it should be considered that the discontinuity of the pad will still impact the impedance. Fig. 6 shows the S-parameters for the short (see Fig. 6(a)) and long via (see Fig. 6(b)). Based on the reflection it appears that changing the pad diameter to 0 mil has the same impact as changing the stub length to 0 mil. In the TDR plot for long and short via (see Fig. 7(a) and (b)) it can be seen that a stub length of 4 mil is almost comparable to a stub length of 0 mil. The impact of removing the via pad is slightly larger than completely removing the stub. The lowest impedance discontinuity could theoretically be achieved by removing both stub and via pad. Though this is not a realistic technology, this result was included as a theoretical upper limit.
Fig. 7(c) shows additional simulation results for the short via with different pad diameters and a stub length of 8 mil. The pad diameter of the result without a stub is 20 mil. These results are comparable to the via with 18 mil pad diameter. Overall, both pad and stub contribute to the impedance discontinuity and the influence of the pad will still be significant even if the stub is very small.

IV. CONCLUSION

In this paper the influence of a small stub in a multi-layer PCB is analyzed. We compare the influence on the via impedance of different stub lengths both in frequency

979-8-3503-5124-8/24 $31.00 © 2024 IEEE

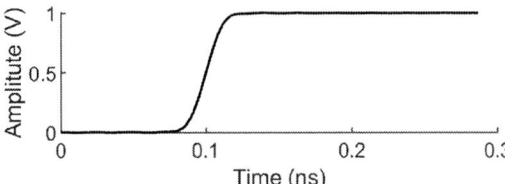

Fig. 4. Exciting step of the TDR simulation with a rise time of 15 ps.

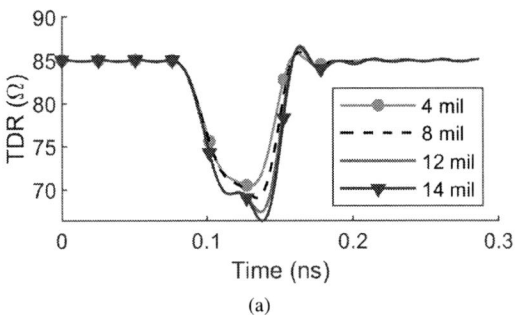

(a)

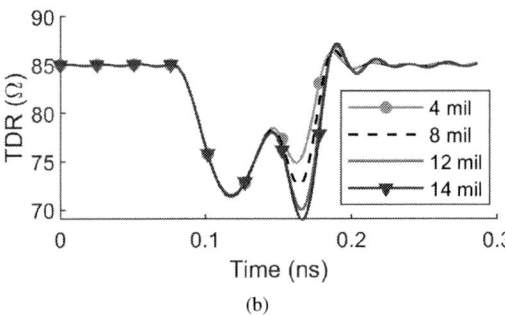

(b)

Fig. 5. TDR for the short and the long via with different stub lengths. (a) Short via. (b) Long via.

and time domain. We can show that while reducing the via stub improves the via impedance, the pad of the via has a comparable influence and unless the pad is further optimized a very short stub will yield no improvement.

REFERENCES

[1] M. Cocchini, J. Fan, B. Archambeault, J. L. Knighten, X. Chang, J. L. Drewniak, Y. Zhang, and S. Connor, "Noise Coupling Between Power/Ground Nets Due To Differential Vias Transitions in a Multilayer PCB," in *2008 IEEE International Symposium on Electromagnetic Compatibility.* IEEE, Aug. 2008.

[2] G.-H. Shiue, C.-L. Yeh, L.-S. Liu, H. Wei, and W.-C. Ku, "Influence and Mitigation of Longest Differential Via Stubs on Transmission Waveform and Eye Diagram in a Thick Multilayered PCB," *IEEE Transactions on Components, Packaging and Manufacturing Technology,* vol. 4, no. 10, p. 1657–1670, Oct. 2014.

[3] C.-L. Yeh, Y.-C. Tsai, C.-M. Hsu, L.-S. Liu, S.-H. Tsai, Y. H. Kao, and G.-H. Shiue, "Influence of Via Stubs with Different Terminations on Time-Domain Transmission Waveform and Eye Diagram in Multilayer PCBs," in *2012 IEEE Electrical Design of Advanced Packaging and Systems Symposium (EDAPS).* IEEE, Dec. 2012.

[4] Y. Zhang, M. Bohra, N. Pham, P. R. Paladhi, W. D. Becker, and D. M. Dreps, "Signal Integrity Characterization of Channels With Asymmetric Via Stubs," in *2020 IEEE 29th Conference on Electrical Performance of Electronic Packaging and Systems (EPEPS).* IEEE, Oct. 2020.

[5] "Ansys Electronics Desktop 2024 R1." [Online]. Available: ansys.com

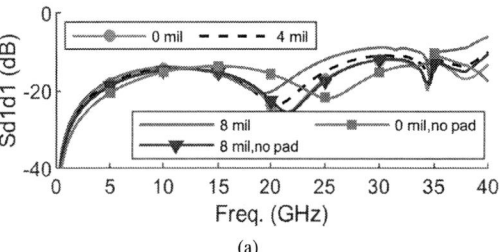

(a)

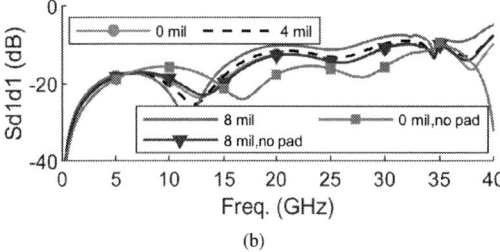

(b)

Fig. 6. Return loss for the short and the long via with different stub lengths. In some cases the via pad was removed. (a) Short via. (b) Long via.

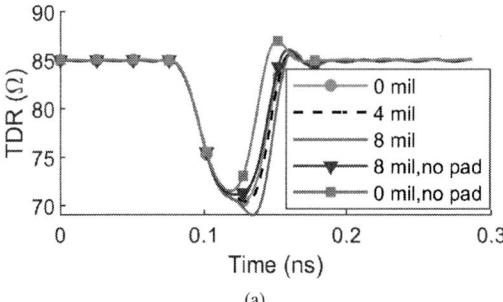

(a)

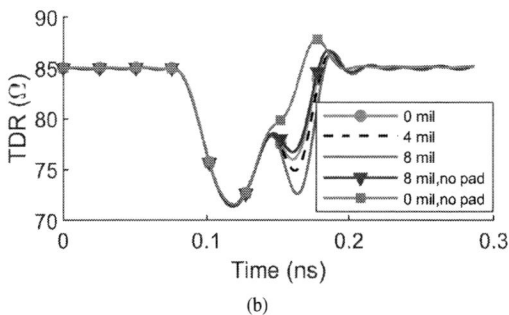

(b)

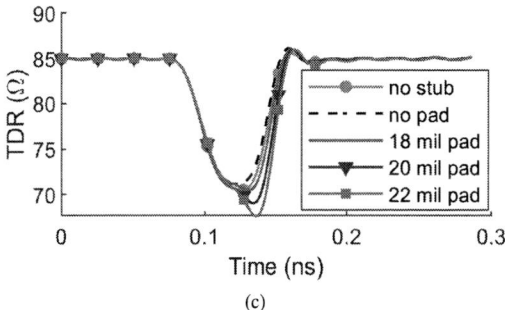

(c)

Fig. 7. TDR for the short and the long via with different stub lengths. In some cases the via pad was removed. (a) Short via. (b) Long via. (c) Additional variations of pad size for short via and an 8 mil stub. The pad size for the model without a stub is 20 mil.

979-8-3503-5124-8/24 $31.00 © 2024 IEEE

Reinforcement Learning Based Automatic Router for Power Delivery Network Prototypes

Felix Yuan
Client Graphics and AI Group
Intel Corporation
felix.yuan@intel.com

Abinash Roy
Client Graphics and AI Group
Intel Corporation
abinash.roy@intel.com

Abstract—Reinforcement learning (RL) is applied to an automated power delivery network design process. Initially, a prototype board is created with multiple power rails. Using two RL agents, a power delivery network (PDN) is created. Firstly, each power rail is routed from the voltage regulator to the integrated circuit (IC) pins. Secondly, the metal width of each rail is expanded to decrease the resistance and inductance of the rail.

Keywords— Automatic Router, Reinforcement Learning, Power Integrity, Power Delivery Network

I. INTRODUCTION

In modern computing, processors draw large amounts of current operating with high clock frequencies. To maintain correct voltage levels, well designed power delivery networks (PDNs) are required. When designing robust PDNs, engineers manually route power rails from the voltage regulator modules (VRM) to the integrated circuit (IC) pins. Because of the repetitive nature and time required for PDN design, research has been done to automate the design process.

In current literature, automatic routing has been explored for PDNs using conventional routing algorithms [1], [2]. Although these algorithms are effective, machine learning (ML) is being explored as a tool to improve automatic routing. ML can yield potential benefits such as decreased routing time and more advanced routing strategies. Currently, reinforcement learning (RL) is already being used to design PDNs, but the routing capability is limited to a single layer [3]. However, depending on rail count, routing congestion, electrical requirement, and product features, power rail routing typically happens in multiple metal layers.

In this paper, an automatic router is developed that uses two RL agents to route power rails from the VRM to the IC pins using graph-based routing on multiple layers. In addition to routing, RL is used to increase metal thickness to optimize the PDN. This method allows contributions to previous works by combining an automatic router architecture similar to [1], [2], while using RL to route PDNs on multiple layers. Therefore, the proposed router is more efficient and intelligent.

II. METHODS

A. Reinforcement Learning

Reinforcement learning is a type of machine learning where an agent learns to perform a task through trial and error.

Generally, the agent interacts with a simulated environment with state s_t. For each state, the agent is given an observation, which the agent uses to select an action a_t. After each action, a reward is given to define the desirability of the action. To make decisions, the agent follows a stochastic policy $\pi_\theta(a_t \mid s_t)$. This policy is then able to be updated based on the rewards given for each action.

In this auto-router, a proximal policy optimization (PPO) algorithm is used to train the RL agents. PPO uses an objective

$$L^{CLIP}(\theta) = \mathrm{E}_t[\min(r_t(\theta)A_t, \mathrm{clip}(r_t(\theta), 1-\varepsilon, 1+\varepsilon)A_t] \quad (1)$$

where $r_t(\theta)$ defines a probability ratio between the previous policy and the current policy [4]. In the objective, the probability ratio is clipped between $1+\varepsilon$ and $1-\varepsilon$, meaning that policy updates are limited to a smaller range. When policy updates are too large, instabilities may be caused since a large update could mean the policy corresponding to the maximum reward is missed. This allows PPO to be used for more stable training compared to other RL algorithms.

B. Architecture of Automatic Router

Fig. 1 presents the architecture of the PDN router. Initially, a PCB is converted into a 3D array that is used as the environment for the RL agents. The array uses integer encoding to represent each component on the PCB, such as the VRM, IC location, and keep out zones.

RL agent I is used to create the routes from each VRM location to their respective IC location. The observation of the environment given is a 5x5x5 grid, centered around the head of the power rail. The direction of the closest IC pin relative to the head of the power rail is also given to the agent for the x-axis, y-axis, and z-axis. As shown in fig. 2, using these observations, the RL agent can choose to move the head of the power rail in the north, east, south, and west directions, in addition to up and down between board layers.

RL agent II is employed to expand each power rail after all rails have successfully been routed from the VRM to the IC pins. When a rail is selected for its physical growth, each rail square is evaluated. The squares adjacent to an empty space or another rail are stored in a list. For each square in the list, the RL agent is given the option of either expanding or not

---

979-8-3503-5124-8/24 $31.00 © 2024 IEEE

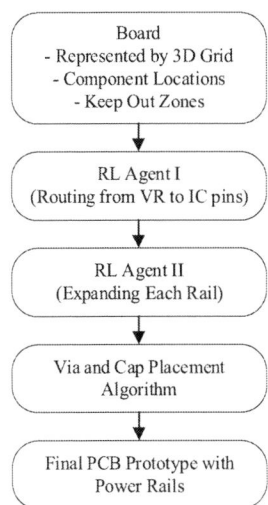

Figure 1: Architecture of the reinforcement learning based automatic router.

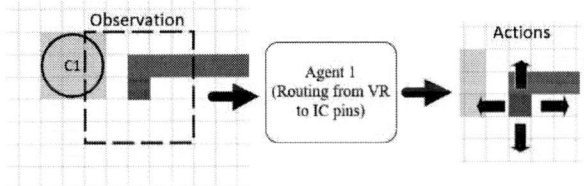

Figure 2: Observation and possible actions of RL agent I.

expanding. For this agent, the observation given is a 7x7 grid centered around the square that is being evaluated for growth. The agent will output a 1 to expand, and 0 to ignore the square. The router will keep expanding until all available space is routed.

Next, a via stitching algorithm is utilized to place vias to connect different layers of the PCB. To do this, the algorithm looks for overlaps of the same rail on different layers and designates those areas for via stitching.

Finally, a capacitor placement algorithm is utilized to place decoupling capacitor ports near the IC pins.

C. Agent I Training

When training agent I, an example board floorplan is created with random locations for 3 VRMs and IC locations. Actions are taken until all rails have finished routing. This can occur if a rail collides with a keep out zone, another rail, or the rail reaches its respective IC pin. When finished, the environment will be reset, meaning a new PCB will be generated.

Possibility	Reward
Rail Reaches IC	+300
Rail collides with keep out zone or another rail	-80
Move in direction of IC	+10
Sparse Bonus	$1*N_{empty}$

Table 1: Possible rewards for agent I after each step.

After each action, a reward is recorded for the PPO algorithm to use when updating the policy. Table 1 shows the possible rewards given to the agent, organized from most significant to least significant. The largest reward is given for the rail reaching the IC pins. A large negative reward is given if the rail collides with a keep out zone or another rail. A small reward is given any time the rail moves closer to the IC pins. Finally, a reward is given for the number of empty squares around the head of the rail. To calculate this value, for each empty square in the $\pm$ x and y directions, a reward of 1 will be given, for a maximum of +8 reward.

D. Agent II Training

During training, the environment used is a 10x10 grid with 3 rails pre-routed from the VRM to the IC pins. After each edge square is chosen to expand, a reward will be given to the PPO RL agent.

To calculate the reward given after each step, the voltage from the VRM to the IC pins are calculated by modelling the board as a resistive grid [5]. For voltage calculation, current draw used is 1 A for each rail. Because of Ohm's Law, a lower voltage potential from the VRM to the IC pins corresponds to a lower resistance. The reward function $r(s)$ at state s_t after each step of the rail growth is

$$r(s) = C_1 \sum_{i=1}^{n_{rails}} (V_i - V'_i) + C_2 \qquad (2)$$

where V and V' are the voltage before and after the step. i is the index of the rail, and C_1 and C_2 are constants that are tuned to incentivize rail growth. The reward function is designed so that all rails are considered when expanding a single rail. Decreasing the resistance of one rail but increasing the resistance of another rail by a larger amount is not rewarded positively.

III. VALIDATION OF AUTOMATIC ROUTER

A prototype PCB in fig. 3 shows the top 3 metal layers to be routed by the proposed auto-router with each layer being

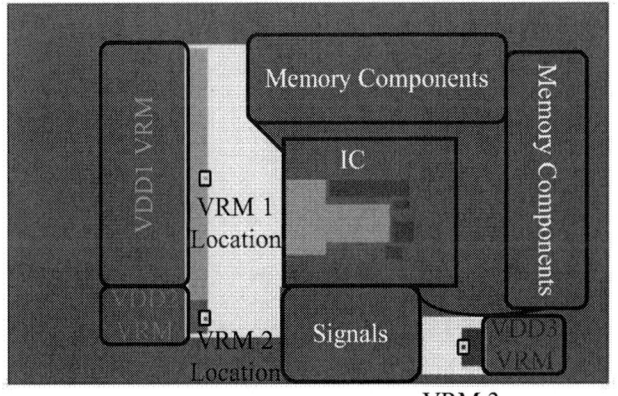

Figure 3: Initial Layout of VRM and IC locations on the top layer.

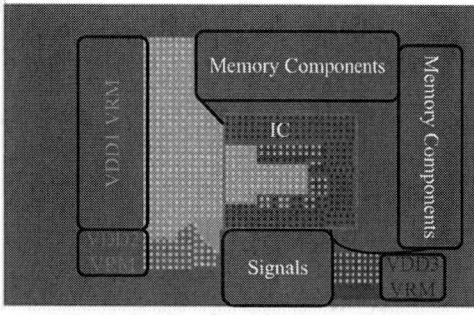

Layer 1

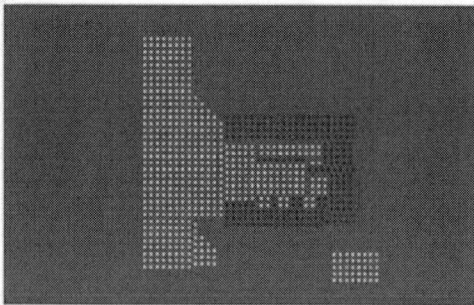

Layer 2

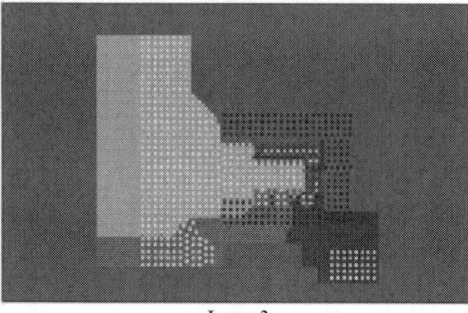

Layer 3

Figure 4: Finalized layout after routing and expanding rails.

Rail	Manual Routing		Automatic Routing	
	DC Resistance (mΩ)	Inductance (pH)	DC Resistance (mΩ)	Inductance (pH)
VDD1	1	1	1.046	1.012
VDD2	12.974	15.411	10.885	14.539
VDD3	9.505	7.438	8.593	7.364

Table 2: DC resistance and loop inductance of each power rail, for manual routed design and automatic routed design. The values are normalized to the values of VDD1 in the manual design.

design, the DC resistance of the automatically routed design is improved in VDD2 and VDD3, however the resistance of VDD1 is increased. The inductance of the automatically routed design is also reduced on VDD2 and VDD3, while an increase is measured in VDD1.

Using this design, the automatic router is shown to be effective in board level designs. Our on-going work is to extend this routing architecture for the application of IC package substrate routing.

IV. Conclusion

Creating PCB layouts can be repetitive and tedious, allowing for the use of automation. As a result, a reinforcement learning based approach to automatic prototyping of power rail routing is presented in this paper. Reinforcement learning is used to find a path from the VRM to the IC pins, as well as to increase the thickness of the rails.

Improvements can be made to this method of automatic board routing. Currently, the router is more effective at routing designs with less power layers. In the future, more training can be implemented to allow routing of many layers, such as on an IC package. Future work will also include techniques requiring advanced planning, such as adding parallel paths on different layers.

V. References

[1] R. Bairamkulov, A. Roy, M. Nagarajan, V. Srinivas, and E. G. Friedman, "SPROUT—Smart Power Routing Tool for Board-Level Exploration and Prototyping," *IEEE Trans. Comput.-Aided Des. Integr. Circuits Syst.*, vol. 41, no. 7, pp. 2263–2275, Jul. 2022.

[2] R. Coutts, A. Roy, M. Nagarajan, V. Srinivas, and P. Penzes, "Automatic Package Router for Power Delivery Network," in *2021 IEEE 30th Conference on Electrical Performance of Electronic Packaging and Systems (EPEPS)*, Austin, TX, USA: IEEE, Oct. 2021, pp. 1–3.

[3] S. Han, O. W. Bhatti, W.-J. Na, and M. Swaminathan, "Reinforcement Learning Applied to the Optimization of Power Delivery Networks with Multiple Voltage Domains," in *2023 IEEE MTT-S International Conference on Numerical Electromagnetic and Multiphysics Modeling and Optimization (NEMO)*, Canada: IEEE, Jun. 2023, pp. 147–150.

[4] J. Schulman, F. Wolski, P. Dhariwal, A. Radford, and O. Klimov, "Proximal Policy Optimization Algorithms," *arXiv*, 2017.

[5] P. Zegarmistrz, S. A. Mitkowski, A. Porębska, and A. M. Dąbrowski, "Nodal analysis of finite square resistive grids and the teaching effectiveness of students' projects," *2nd World Conference on Technology and Engineering Education*, Sep. 2011.

represented with a 150 by 92 grid. The top layer includes 1 IC and 3 VRMs. Memory components and IO signal traces on the board severely limit routing space. Layer 2 is a dedicated ground layer to maintain a solid current return path.

Fig 4. displays the board after the routing and expansion is finished. The power rails are routed sequentially from VDD1 to VDD3. VDD2 and VDD3 must be routed on the 3rd layer to reach their respective IC pins. After all rails are routed, agent II is used to expand each rail, until all available space is utilized for the power rails. Next, the via placement algorithm is utilized to place power and ground vias. Power vias are shown in yellow, while ground vias are dark grey. Capacitor ports are also placed, shown in red.

The automatic router's prototype is compared to a manually routed design for the same board. Using an industry standard E&M solver, the DC resistance and loop inductance of each rail between the VR inductor location and the IC is measured and shown in table 2. Each value is normalized to the VDD1 rail of the manually routed design. Compared to the manually routed

979-8-3503-5124-8/24 $31.00 © 2024 IEEE

Improve CLK Phase Noise Performance by Mitigating Antiresonance Phenomenon of Power Net with a π-Type Filtering Structure

Xinlin Tang
Signal Integrity Department
Chengfang Information Co., Ltd.
Shanghai, CHINA
xinltang6@gmail.com

Shuxiang Li
CHG Signal Integrity Department
CISCO System Inc.
Shanghai, CHINA
shuxili@cisco.com

Tao Fang
Department of Computer and Information Science
University of Macau
Macau, CHINA
taylefang@gmail.com

Yuan Fang
Signal Integrity Department
Chengfang Information Co., Ltd.
Shanghai, CHINA
Fang24311@gmail.com

Abstract— In this article, we studied the phenomenon of test failures in CLK phase noise assessment, identifying the antiresonance in the power supply network of clock chip. We found that the jitter of the CLK signal is strongly correlated with the filtering effect of the power supply network of the clock chip. To improve the jitter performance of CLK signal, we implemented a π-type filtering structure into the existing circuit topology. This modification successfully mitigated the antiresonance peak. In order to minimize jitter as much as possible, we have optimized and compared the π-type filtering structure from simulation and test perspective. Finally, we have provided a general conclusion regarding this issue.

Keywords—Antiresonance, Phase Noise, π-type filtering Structure

I. INTRODUCTION

Clock phase noise analysis is essential in electronic system design, particularly in the fields of communications and signal processing. This phenomenon presents as random phase variations in signals, which can adversely affect system performance. The evaluation of phase noise is critical to ensure the integrity, accuracy, and reliability of clock signals that are key to timing synchronization in digital systems. Additionally, phase noise testing is vital for assessing the quality of oscillators and frequency synthesizers in Radio Frequency (RF) systems, where precise frequency modulation is imperative.

Precise phase noise characterization allows engineers to identify and attenuate noise sources, leading to improve system performance and reduce interference in communication channels. Therefore, a deep understanding and thorough testing of phase noise are essential for the development and optimization of electronic systems.

Digital systems have stringent requirements for clock signal jitter, and oscillators and clock generators are typically characterized by their phase noise performance [1]. Therefore, when the jitter performance of a clock signal fail to meet system requirement, it becomes necessary to analyze phase noise characteristic curves to identify predominant noise sources within particular frequency band. Our research has found that the power supply input to the clock generator can significantly affect the phase noise of the clock signal. If severe antiresonance

occurs in the power supply filtering network, the jitter of the clock signal will noticeably increase [2].

This paper includes 5 sections. Section 2 presents the initial phase noise test results of the CLK signal, which unfortunately failed to meet the chip requirement. Section 3 explores the relationship between antiresonance of power net and the π-type filtering structure, which can improve the phase noise and jitter performance obviously. Section 4 studies the influence of π-type filtering structure optimization to the phase noise. Conclusively, Section 5 summarizes the findings and insights garnered throughout the research.

II. THE FAILURE OF CLK PHASE NOISE TEST

In Fig. 1(a), the initial phase noise test result of RMS jitter of CLK signal is 1.42593psec, which failed to meet the digital receiver requirement, less than 500fsec. Phase noise elements disperse signal power across adjacent frequencies, producing noise sidebands [3]. Failure to adhere to phase noise specifications can heighten receiver noise levels, compromise receiver anti-interference capabilities, and impede digital signal detection. Under normal circumstances, if the jitter of a clock signal does not meet the requirements, then the phase noise performance of that clock signal is also typically poor. Additionally, power supply noise can lead to variations in signal propagation delay, thereby affecting system timing [4]. Furthermore, the phase noise of a CLK signal is influenced by the power supply network of CLK generator. Therefore, the focus of our research is on optimizing the power supply network design for clock generator.

Because power supply noise can introduce jitter, therefore we scrutinized the voltage regulator module(VRM), topology of the power filtering network, depicted in Fig. 2(a). The ferrite bead in this structure is with 0.18ohm DCR and 220ohm impedance at 100MHz(FB_DCR0p18_220ohm). During time domain simulation, an antiresonance point at 385.5KHz was identified, featuring a resonance peak of 22.918dB, as illustrated by the red curve in Fig. 3. It is widely acknowledged that antiresonance points may induce circuit instability, escalate interference or noise, and undermine overall system performance. Thus, we think that the presence of this resonance point leads to the failure of the clock phase noise test.

III. ADDING A π-TYPE FILTERING STRUCTURE TO SOLVE THE PROBLEM

Fig. 4(a) depicts a topology composed of two capacitors. In Fig. 4, C_b is with smaller capacitance than C_a. Between the self-resonant frequencies, the larger capacitor becomes inductive and the smaller capacitor becomes capacitive [5]. From [5],we know in Fig4.(a), the antiresonance peak value is

$$Z_{p1} = \frac{R_c^2 + R_a R_b}{R_a + R_b} \tag{1}$$

where R_a and R_b are the parasitic resistances(ESR), and L_a is the parasitic inductance of the capacitors C_a. $R_c = \sqrt{L_a/C_b}$.

If we add R_u in the middle of C_a and C_b, the peak impedance varies according to R_u. In Fig. 4(b), the antiresonance peak value as.

$$Z_{p2} = \frac{R_c^2[1 - \frac{k^2}{1+\left(\frac{R_u}{R_c}\right)^2}] + [R_a + \frac{R_u k^2}{1+(R_u/R_c)^2}]R_b}{[R_a + \frac{R_u k^2}{1+(R_u/R_c)^2}] + R_b} \tag{2}$$

Here k is the coupling coefficient. M is the mutual inductance between the two loops given by $M = kL_a$.

$$Z_{p2} - Z_{p1} = -\frac{R_C^2 A}{R_u(R_a + A + R_b)} R_b < 0, \tag{3}$$

where $A = \frac{R_u k^2}{1+(R_u/R_c)^2}$.

From Eq. 3, it is evident that adding a resistor between the two capacitors results in a lower peak value at the resonance point compared to having only the two capacitors. Although adding a resistor is positive in suppressing antiresonance, they still can't meet the requirement of digital chip for clock jitter. So a π-type filtering structure is proposed, composed of two capacitors and a resistor, can effectively suppress the anti-resonance phenomenon, thereby improving the stability of the power supply.

We optimized the original topology with a π-type filtering structure, composed of two 10uf capacitors and a 1ohm resistor, as depicted in Fig. 2(b). The AC simulation result is showed by the blue curve in Fig. 3. Based upon this simulation result, we tested the RMS jitter, which is 440.616fsec showed in Fig. 1(b). Evidently, it can meet the chip requirement, jitter less than 500fsec. Based on the analysis above, we know that for a clock generator, if the filtering circuit of the chip's power supply network is not well-designed, the generated clock signal will have greater phase noise. Regardless of whether a resistor is added in series in the power supply network or a π-type structure is incorporated, this phenomenon can be effectively improved.

However, different π-type networks have varying filtering effects. Although the anti-resonance peak value can serve as the most important indicator for judging the filtering effect, when it comes to reducing clock jitter, what we actually care about is the filtering effect within a certain frequency range. For example, the clock jitter tested in this article originates from phase noise between 10KHz and 20MHz. Therefore, in the next section, we have studied the transmission characteristics of the filtering network, hoping that it can provide guidance when optimizing the π-type network.

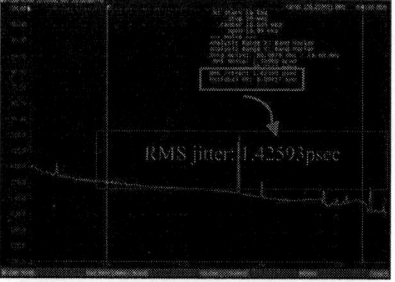

(a) Phase noise test result of initial topology

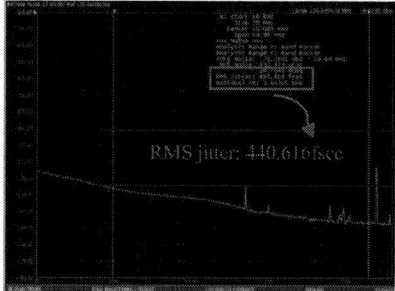

(b) Phase noise test result of adding a π-type filtering structure

Fig. 1. Test result comparison between initial topology and topology with a π-type filtering structure

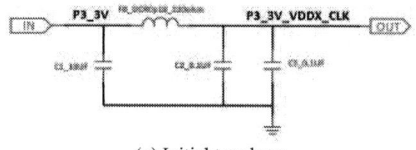

(a) Initial topology

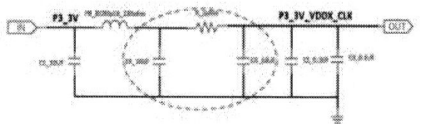

(b) Topology after adding a π-type filtering structure

Fig. 2. Topology of power filtering network

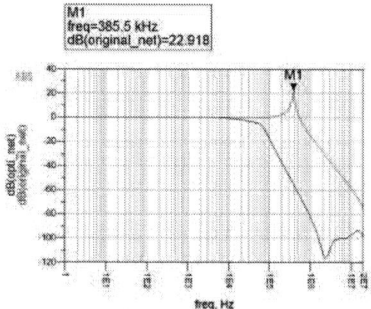

Fig. 3. Power filtering simulation result of initial and π-type topology

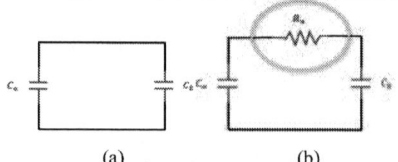

Fig. 4. Topology without resistor and with resistor

IV. GUIDANCE TO IMPROVE THE FILTERING ABILITY

From Eq. 3, we know the antiresonance peak value is not only influenced by resistor but also by capacitor. We next conducted a study about the component selection of two capacitors within the π-type structure, and made a comparison from time domain and frequency domain perspectives.

In Fig. 5, the blue curve is the simulation result of the π-type filtering circuit with two 10uf capacitors and one 1ohm resistor, while the pink curve is the simulation result of the π-type filtering circuit with one 10uf capacitor, one 0.1uf capacitor and one 1ohm resistor. Notably, the pink curve is smoother and with a lower antiresonance peak value compared to the blue curve. However, within the frequency range we are concerned with, the design represented by the pink curve performs better at low frequency(10KHz~100KHz), while the design represented by the blue curve performs better at high frequency(100KHz~20MHz).

Therefore, we compared the results of the frequency domain simulation, as shown in Fig. 6, where the blue curve represents the design with two 10uf capacitors. Looking at the insertion loss, the design with two 10uf capacitors has a lot of filtering effects, but this contradicts the results of the time domain simulation. So we tested the jitter of the π-type filtering circuit with one 0.1uf capacitor, one 10uf capacitor and one 1ohm resistor. The jitter is about 419fsec, showed in Fig. 7, less than the result of Fig. 1(b). Based on the comprehensive analysis of the experiments above, if we use the jitter results of the tests as the basis for judging the filtering effect, then the insertion loss simulation in the frequency domain cannot help us optimize the π-type network.

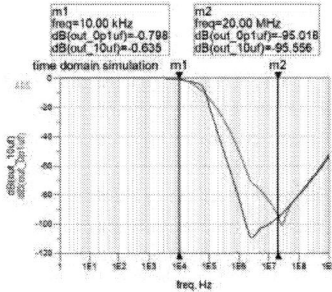

Fig. 5. Comparison between two kinds of π-type filtering structure from time domain perspective

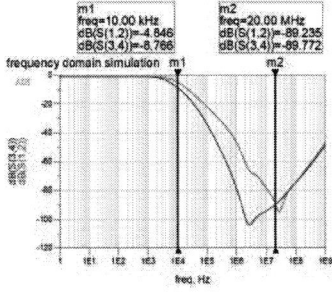

Fig. 6. Comparison between two kinds of π-type filtering structure from frequency domain perspective

Therefore, we can only summarize some patterns from the simulation results in the time domain. Let's go back to Fig. 5, focusing on the frequency range from 10KHz to 20MHz. Although both types of π-type filtering circuit can eliminate antiresonance, the lower frequency range has a greater weight, so the jitter test result is better. Therefore, when using time domain simulation to optimize the π-type filtering circuit, we should focus on the filtering effect in the low-frequency range.

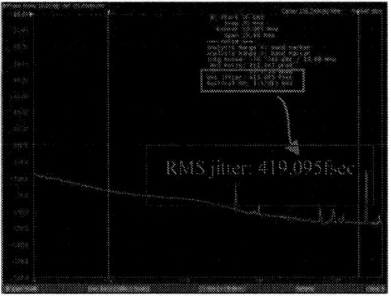

Fig. 7. Phase noise test result of adding a π-type filtering structure, which is with one 10uf capacitor, one 0.1uf capacitor and one 1ohm resistor

V. SUMMARY

This paper, by testing the jitter of the clock signal, has discovered the relationship between clock jitter and the power supply network of the clock chip. Then, through analysis and testing, it compared the effects of adding resistor or π-type filtering structure to the power supply network. Finally, we analyzed how to optimize the π-type structure from the perspectives of time domain and frequency domain simulations. After eliminating the antiresonance peak, the results of the low-frequency segment in the time domain simulation are very important for us to optimize the π-type structure. At the same time, the two capacitors in the π-type structure should be chosen from different models as much as possible, as this is conducive to expanding the filtering range. This paper provides us with a perspective for analyzing clock jitter, which has certain guiding significance for the design of high-speed systems.

REFERENCES

[1] Kester, Walt . "Converting Oscillator Phase Noise to Time Jitter." Tutorial MT-008, Analog Devices ,2005.

[2] Sun, S. , Ren, K. , Ding, W. , Chow, D. , Hoang, T. , & Shumarayev, S. , et al. "PDN-Noise-to-Jitter Transfer in High-Speed Transceivers." DesignCon 2012.

[3] DAISHINKU CORP. (KDS)," Clock jitter and phase noise", White paper, 20190205.

[4] Klokotov, D. , Shi, J. , Wang, Y. , & Com. "Distributed Modeling and Characterization of On-Chip/System Level PDN and Jitter Impact." DesignCon 2018.

[5] Davis, and K. A. . "Effect of a magnetically coupled resistive loop on antiresonance." Electronics Letters 52.13(2016):1162-1164.

[6] Truong, B. D. , Roundy, C. , Andersen, E. , & Roundy, S., "Analysis of Resonance and Anti-Resonance Frequencies in a Wireless Power Transfer System: Analytical Model and Experiments." IEEE Transactions on Circuits and Systems, II. Express briefs 7(2019)

A Tunable Inductor Peaking Technique for Optical Communication Systems

Festim Iseini[1,2,3], Han-Ting Lin[2], Nicola Pelagalli[1], Andrea Malignaggi[1],
Corrado Carta[1,4], Gerhard Kahmen[1,3] and Andreas Weisshaar[2]

[1]IHP GmbH – Leibniz Institute for High Performance Microelectronics, 15236 Frankfurt (Oder), Germany
[2]School of Electrical Engineering and Computer Science, Oregon State University, 97331 Corvallis, USA
[3]Brandenburg Technical University Cottbus, 03046 Cottbus, Germany
[4]Technische Universität Berlin, 10587 Berlin, Germany
(email: iseini@ihp-microelectronics.com)

Abstract—A tunable inductor peaking technique for loss compensation in optical communication systems is presented. The proposed technique is based on creating a tunable damped resonator with a fixed peaking inductor and a high-frequency transistor with low intrinsic capacitance. The tuning technique is first demonstrated with a generic equivalent circuit model and then implemented and evaluated using the high-speed bipolar transistor in 130 nm IHP SG13G2 BiCMOS technology, featuring f_t/f_{max} of 350/450 GHz. Our proposed technique has been simulated in a real case scenario demonstrating the effectiveness of the tunable inductor peaking technique for loss compensation in optical communication systems.

I. Introduction

Data traffic has been rapidly growing in recent years. In particular, with the Internet of Things constantly involving everyone on a daily basis [1], the need of data center infrastructures capable of satisfying the increasing data demand is getting more attention in the scientific community. Fiber-based communication systems are usually preferred to satisfy the broadband specifications that data centers require [2]. Optical transceivers require broadband electronic integrated circuits (EIC) to drive the optical modulators [3]. Electro-optical performance, especially bandwidth (BW), crucially depends on the integration and packaging technology between the EIC and the photonic integrated circuit (PIC) and needs to be taken into account in the design process of the entire system [4]. Due to the fabrication process variations, the additional losses introduced by the interconnections or by the fabrication of the optical modulator itself can be very difficult to control. Tunable peaking (TP) techniques, generally implemented on the EIC (due to its accessibility), can be used as frequency equalizers to shape the overall frequency response and compensate for these losses. The most common techniques used for this purpose are the continuous-time linear equalizers (CTLE) [5], which make use of variable capacitors. This solution presents some limitations: it is bulky and consumes space and additional power. An alternative solution proposed in this work is the implementation of resonator-based tunable inductors. In this case, the range of frequency operation is limited by the resistance and the capacitance

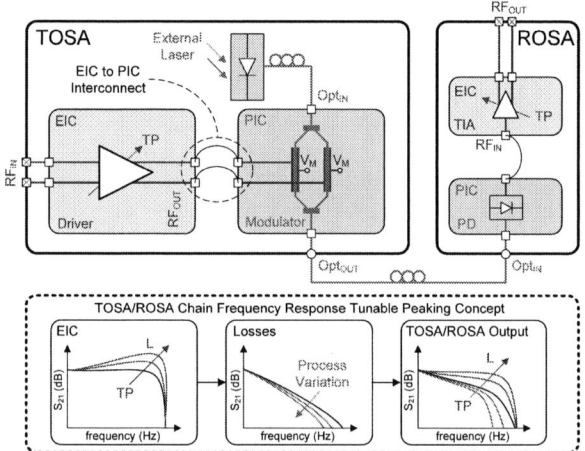

Fig. 1: General block diagram of an optical transmitter and receiver sub-assemblies (TOSA/ROSA) chain in an optical communication system (top) and corresponding sub-block frequency responses (bottom).

introduced by the active device between the inductors generating a resonance. The effective inductance can be controlled by properly biasing the active device. Switch-based variable inductors employing MOS transistors have been implemented, for example, for voltage-controlled oscillators [6]. This paper proposes the application of a transistor as a variable resistor to tune the damping of the resulting resonator and, hence, tune the effective peaking inductance. The proposed technique results in a compact and power efficient TP technique for circuits in optical transmitter and receiver sub-assemblies (TOSA/ROSA) (Fig. 1). This paper investigates the feasibility and implementation of the proposed technique, first with the evaluation of the resonator circuit model and implementation with a transistor, and finally with a real scenario demonstration of a simulated simple TOSA chain.

II. Tunable Inductor Analysis and Evaluation

Fig. 2 shows the basic RLC resonator circuit model used to demonstrate the proposed technique. The analysis can be

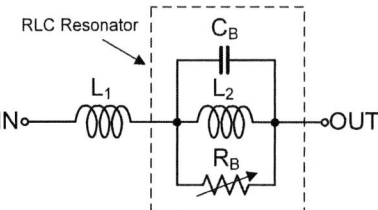

Fig. 2: RLC resonator circuit used for the initial investigation.

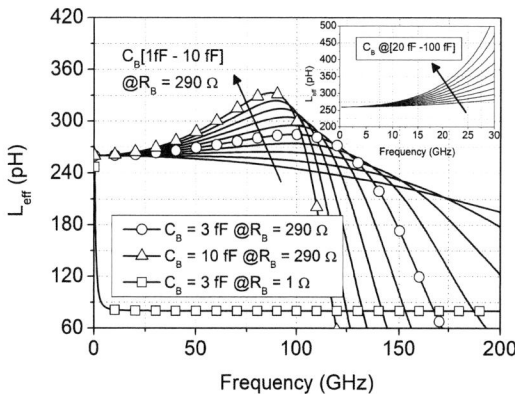

Fig. 4: Effective inductance of the circuit in Fig. 2 for varied capacitance C_B and $R_B = 290\,\Omega$. Inset: capacitance variation from 20 fF to 100 fF.

readily extended to multiple resonators; however, for the purpose of this work, the analysis is limited to a single resonator. The purpose of the tunable inductor is to have different states representing different levels of peaking in a given range of frequency. In contrast to the conventional implementation of tuned resonators, it has been found that the quality factor of the effective inductor is not important for tunable inductor peaking applications. For the initial evaluation, $L_1 = 80\,\text{pH}$ and $L_2 = 180\,\text{pH}$ are chosen accordingly to the following driver demonstration without taking their parasitics into account. L_1 is used as an offset and the analysis focuses here on the RLC resonator. Fig. 3 shows the effective inductance ($L_{\text{eff}} = \text{imag}(Z_{\text{RLC}})/\omega$) as a function of R_B while $C_B = 3\,\text{fF}$. Fig. 4 shows the effects of C_B when $R_B = 290\,\Omega$. The larger C_B, the lower the resonance frequency, underdamping the resonator and limiting the frequency range of operation. This behavior can be compensated by reducing R_B to increase the damping factor ξ of the resonator defined in (1).

$$\xi = \frac{1}{2R_B}\sqrt{\frac{L_2}{C_B}} \qquad (1)$$

In a real case considering the nonidealities of the circuit (Fig. 5), the most critical component is the parasitic capacitance C_{SUB} of the inductor, which lowers the resonance frequency further decreasing performance. This can be avoided by properly designing the inductor and reducing as much as possible its parasitic capacitance. In this evaluation L_1 and

L_2 are uncoupled but can be coupled to improve inductor performance [7].

III. TUNABLE INDUCTOR IMPLEMENTATION

The variable resistance R_B and small capacitance C_B can be realized with a properly sized transistor Q_B (either bipolar or MOS) and by varying the base or gate voltage to create the desired resonance and effective inductance. We have implemented this approach with bipolar transistors in the high-speed IHP SG13G2 technology (see Fig. 5). Bipolar transistors have been preferred over MOS transistors due to their overall lower capacitance, so that a higher resonant frequency and a larger BW can be achieved. The base voltage applied to the npn transistor forward-biases the two p-n junctions [8] creating a variable output collector-emitter resistance. The device dimensions have been chosen for low capacitance and sufficient resistance range to obtain different peaking states at 100 GHz. Fig. 6 compares EM simulations of the layout (using ADS Momentum) and its lumped model given in Fig. 5. For this comparison the values of R_B and C_B in Fig. 5 were extracted from the PDK of Q_B for different values of V_{be}.

IV. DRIVER DEMONSTRATION

The simulation testbench for the driver demonstration is depicted in Fig. 7. The driver consists simply of a common-emitter (CE) cascode amplifier (generally used as the output

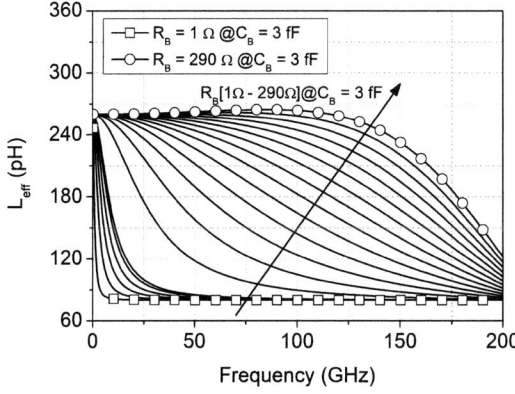

Fig. 3: Effective inductance of the circuit in Fig. 2 for varied resistance R_B and $C_B = 3\,\text{fF}$.

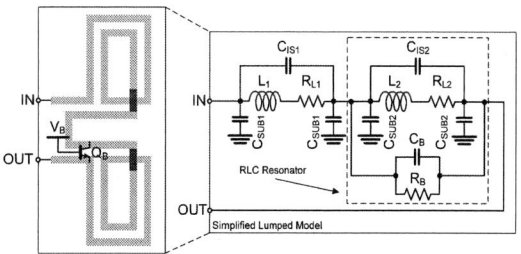

Fig. 5: Layout (left) and corresponding lumped model (right) of the implemented tunable inductor.

979-8-3503-5124-8/24 $31.00 © 2024 IEEE   137

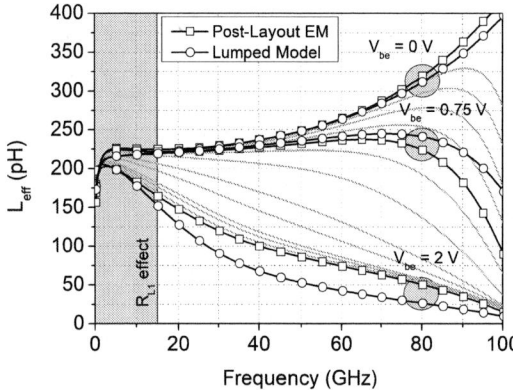

Fig. 6: Comparison between EM simulation and lumped model of the inductor in Fig. 5 with R_B and C_B extracted from the PDK model of the transistor for different values of V_{be}.

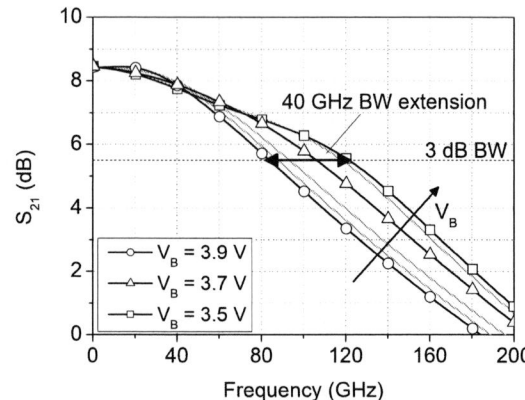

Fig. 8: Small-signal simulation of the driver for different inductor peaking states. Shaded curves represent intermediate states that can be achieved by fine-tuning V_B.

stage of a driver [9]) with a 3 dB BW of 80 GHz. Fig. 8 shows the different states of the driver with the tunable inductor. Different peaking levels can be identified resulting in an equivalent bandwidth extension of up to 40 GHz. To further quantify the improvements with the proposed technique, time-domain simulations with 0.25 ps signal rise/fall times have also been carried out. The resulting eye diagram for 300 Gbps non-return-to-zero (NRZ) signaling is shown in Fig. 9. An improvement in the eye opening of up to 45% can be observed.

V. Conclusion

In this paper, a detailed study and implementation of a tunable inductor peaking technique has been presented. The proposed technique can be effectively used to compensate and control loss variations in optical communication systems. Our proposed technique represents a good alternative to existing techniques with potentially smaller chip footprint and lower power consumption and design complexity.

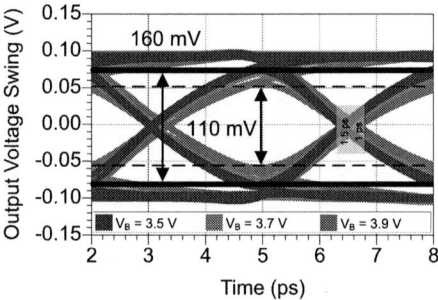

Fig. 9: Time-domain simulation of the driver for different states at 300 Gbps NRZ.

VI. Acknowledgment

This research was supported in part by a DAAD scholarship, IHP, and CDADIC.

References

[1] Cisco Visual Networking Index: Forecast and Trends, 2018–2023.
[2] J. J. Maki, "Cloud Optics–IEEE 802.3 Ethernet, OIF, and MSA Defined Optical Specifications in Data-Center Aligned Form Factors," in *Advanced Photonics 2018, OSA Technical Digest (online) (Optica Publishing Group, 2018)*, paper SpTu3F.1.
[3] E. Sackinger, *Broadband circuits for optical fiber communication.* John Wiley and Sons, 2005.
[4] E. Sentieri, et al. "12.2 A 4-channel 200Gb/s PAM-4 BiCMOS transceiver with silicon photonics front-ends for gigabit Ethernet applications," *2020 IEEE ISSCC*, pp. 210-212, 2020.
[5] Z. Jia, et al. "A 200-Gb/s PAM-4 Feedforward Linear Equalizer with Multiple-Peaking and Fixed Maximum Peaking Frequencies in 130nm SiGe BiCMOS," *2022 IEEE ICTA*, pp. 104-105, 2022.
[6] A. Tanabe, et al. "A novel variable inductor using a bridge circuit and its application to a 5–20 GHz tunable LC-VCO," *IEEE JSSCC*, pp. 883-893, 2011.
[7] H.-T. Lin and A. Weisshaar, "Robust and Efficient Design of On-Chip Compact Delay Units Based on Bridged T-Coil," *2023 IEEE EPEPS*, pp. 1-3, 2023.
[8] R. L. Schmid, et al. "On the analysis and design of low-loss single-pole double-throw W-band switches utilizing saturated SiGe HBTs," *IEEE TMTT*, pp. 2755-2767, 2014.
[9] F. Iseini, et al. "Lumped Ultra-Broadband Linear Driver in 130 nm SiGe SG13G3 Technology," *2022 IEEE BCICTS*, pp. 136-139, 2022.

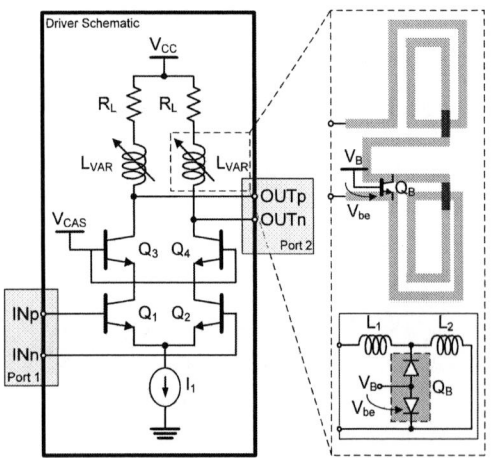

Fig. 7: Simulation testbench for the proposed technique implemented in a driver circuit (left), and layout and equivalent circuit of the variable inductor (right).

979-8-3503-5124-8/24 $31.00 © 2024 IEEE

Tunable True-Time-Delay Unit Based on Bridged T-Coil

Han-Ting Lin[1], Festim Iseini[2] and Andreas Weisshaar[1]

[1]School of EECS, Oregon State University, Corvallis, Oregon 97331, USA

[2]IHP GmbH - Leibniz Institute for High Performance Microelectronics, 15236 Frankfurt (Oder), Germany

(email: linha@oregonstate.edu)

Abstract—This paper presents the design of tunable true-time-delay (TTD) units with both discrete and continuous time delay states. Tuning is achieved by varying the capacitances of a bridged T-coil (BTC) circuit. The effectiveness of the design approach is demonstrated with both switched capacitor banks and varactors. All tunable TTD units have been designed in a Tower Semiconductor 0.18 μm SiGe BiCMOS process. Measurements of a fabricated two-state delay unit (20 ps and 25 ps delay) with capacitance tuning implemented with nfet switches show a tunable range of 25% over an 8 GHz bandwidth with < 1.8 dB insertion loss and > 15.8 dB return loss. Full-wave simulation results for a varactor-tuned design demonstrate a tuning range of 16 ps - 25 ps (±25%) over an 8 GHz bandwidth.

Index Terms—tunable true time delay, on-chip delay unit, bridged-T coil, varactor

I. INTRODUCTION

Phased array antennas have been employed in communication and radar systems for many decades. With the increasing demand of 5G, future 6G, and satellite communications, phased array antenna systems have resurfaced as a popular research topic. True time delay (TTD) units are widely used components in RF systems and diverse signal processing applications. In particular, TTDs are increasingly being employed in phased array antenna systems to eliminate beam squinting effects in wide-band applications. Numerous studies have been done for the design and implementation of tunable on-chip analog TTD units. Some research work has employed active devices, g_m-RC or g_m-C all-pass filters, within the delay lines [1], but the control circuit of the active device increases the size, complexity, and power consumption of the TTD units. The trombone delay topology implements switches for selecting various delay paths to realize tunable time delays [2], [3], [4]. Path selection with fixed TTD units is more easily implemented, and passive delay lines generally provide a wider frequency bandwidth but at the expense of a larger footprint. We introduce a different design approach for tunable TTD based on a capacitively tuned BTC, as illustrated in Fig. 1. By tuning the capacitances of a BTC while constraining the return loss to an acceptable minimum level, we can utilize a single BTC circuit as a tunable TTD unit with multiple time delay states.

In the following, we will first review the BTC design for a lowpass fixed TTD unit, and will then apply the capacitance tuning method to achieve tunable time delay. Following this, we will present simulation and measurement results pertaining

to the tunable TTD units with switched capacitors. Finally, we will demonstrate by full-wave simulation a tunable TTD unit with a 1 ps tuning step implemented by varactors. All TTD units have been designed in the Tower Semi 0.18 μm SiGe BiCMOS process [5], and post layout full-wave electromagnetic (EM) simulations have been done in Cadence EMX.

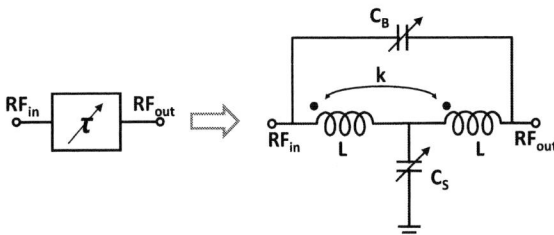

Fig. 1: Tunable TTD unit realized as a capacitively tuned BTC circuit.

II. DESIGN METHODOLOGY

The BTC is a lumped element circuit designed as an all-pass network (APN) with a primary objective of achieving impedance matching across a broad bandwidth. To realize tunable time delay states, the capacitance values of the BTC are varied accordingly. The capacitance tuning range is constrained to maintain the BTC's matching condition up to an acceptable return loss minimum and to stay within a maximum delay variation within the bandwidth. Capacitance tuning is typically implemented in terms of a switched bank of capacitors or with varactors.

A. BTC Design for Lowpass Fixed Time Delay

For an ideal matched time delay unit the scattering parameter S_{21} (transfer function) is an exponential function ($e^{-s\tau}$) of delay τ. A second-order Padé approximation is used to realize a BTC with maximally flat group delay response over a finite bandwidth, as expressed by equation (1) [6].

$$S_{21} = e^{-s\tau} \approx \frac{1 - \frac{\tau}{2}s + \frac{\tau^2}{12}s^2}{1 + \frac{\tau}{2}s + \frac{\tau^2}{12}s^2} \tag{1}$$

The values of the BTC elements can readily be determined using the even-odd mode analysis technique to solve the network depicted in Fig. 1 [7]. The element values are given in terms of the desired fixed delay time τ and system impedance Z_0 as $L = Z_0\tau/3$, $k = 0.5$, $C_B = \tau/(12Z_0)$, and $C_S = \tau/Z_0$.

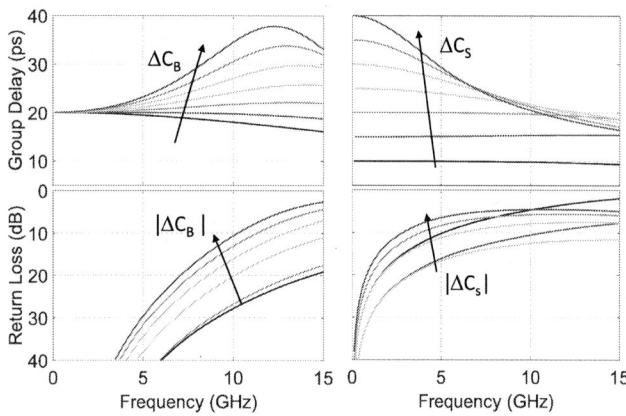

Fig. 2: Ideal circuit simulation of the capacitively tuned BTC in Fig. 1 for a 20 ps time delay design. C_S is changed from 0 fF to 1,200 fF in steps of 200 fF. C_B is changed from 13 fF to 133 fF in steps of 20 fF.

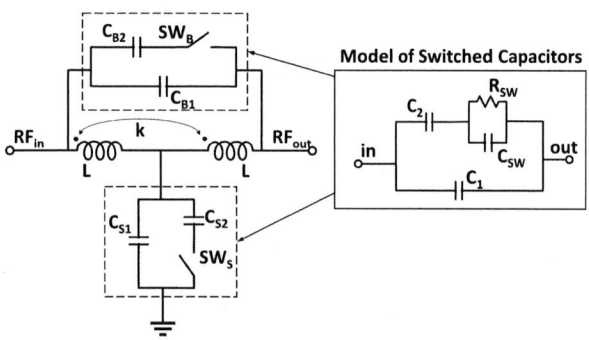

Fig. 3: Tunable TTD unit based on BTC with capacitance.

For a lowpass TTD design with $\tau = 20$ ps and $Z_0 = 50\,\Omega$, the corresponding elements values are $L = 333$ pH, $C_B = 33$ fF, and $C_S = 400$ fF ([8]).

B. Switched Capacitor Bank Tuning

By tuning C_S and C_B of the BTC shown in Fig. 1, various time delay states can be obtained. Figure 2 shows an analysis of the group delay and return loss variations with varying C_S and C_B. This analysis is based on the circuit simulation of the ideal BTC with $C_S = 400$ fF and $C_B = 33$ fF for a 20 ps time delay. It shows that increasing C_S raises the base group delay, but with the trade-off of reducing return loss and bandwidth and increasing time delay variation. On the other hand, varying C_B does not change the base group delay but can be used to expand the time delay bandwidth while maintaining a sufficiently large return loss.

The realization of the capacitively tuned BTC with a switched capacitor bank is illustrated in the schematic shown in Fig. 3. In this design, SW_S and SW_B are both turned off for a group delay of 20 ps and turned on for a group delay of 25 ps. We imposed the constraint of a minimum of 16 dB return loss from DC to 8 GHz. The BTC design is based on the 20 ps fixed TTD unit with effective capacitances $C_S = 400$ fF and $C_B = 33$ fF for the switched off case. We applied the C_S / C_B tuning analysis first to find C_S for the desired base group

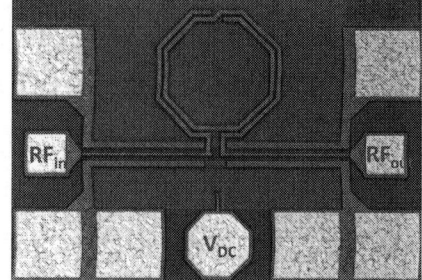

Fig. 4: Die photo of tunable lowpass TTD unit including RF ports and DC control pad.

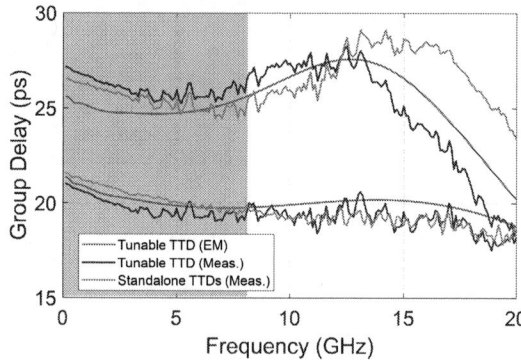

Fig. 5: Measurement and EM simulation results of group delay response for 20 ps and 25 ps designs.

delay, and then to tune C_B to realize the 25 ps TTD within the imposed constraints. The resulting capacitance values for the 25 ps TTD state are $C_S = 600$ fF and $C_B = 73$ fF, respectively.

The on-chip switch is modeled as a parallel RC circuit as shown in Fig. 3. Here, nfet RF switches were chosen primarily because of their low resistance when switched on. Assuming the R_{SW} of an ideal switch (zero when turned on and infinite when turned off), the effective capacitances C_{off} and C_{on} in the switch off and on states, respectively, are given as

$$C_{\text{off}} = (C_2 \parallel C_{\text{SW,off}}) + C_1 \tag{2a}$$
$$C_{\text{on}} = C_1 + C_2. \tag{2b}$$

Capacitances C_1 and C_2 have fixed values and are realized by metal-insulator-metal (MIM) capacitors. $C_{\text{SW,off}}$ is the switch capacitance in the off state. By solving equations (2a) and (2b), the initial values for C_1 and C_2 can be determined. After implementing the layout and the PDK model for the nfet switches, fine-tuning is needed for C_1 and C_2 together with the number of fingers of the nfet switches to compensate for the parasitics in the layout and the switches. Fine-tuning was done by optimization in the Advanced Design System (ADS) [9]. Figure 4 shows a die photo of the fabricated tunable TTD unit. The utilized area measures $170 \times 175\,\mu m^2$, excluding the RF and DC pads. This footprint area is the same as that of a standalone fixed TTD unit with either 20 ps or 25 ps delay.

Figure 5 compares the overall time delay performance measurements of tunable TTD unit with post-layout EM simula-

979-8-3503-5124-8/24 $31.00 © 2024 IEEE

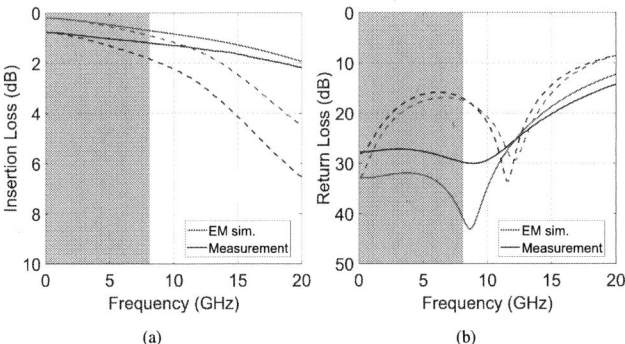

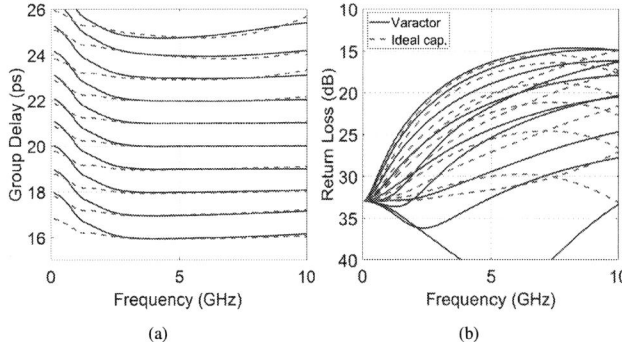

Fig. 6: Measurement and EM simulation results of 20 ps (solid) and 25 ps (dashed) designs. (a) Insertion loss.(b) Return loss.

Fig. 7: (a) Group delay states and (b) return loss comparison between BTC EM simulation with ideal capacitance (red dashed lines) and varactor (blue solid lines) obtained with tuned C_B and C_S values.

tions done in Cadence EMX ([10]), along with measurements from standalone TTD units of 20 ps and 25 ps, respectively. The performance differences between the EM simulation and measurements are insignificant in the group delay response within the effective bandwidth from DC to 8 GHz. The maximum measured insertion loss is 1.8 dB and 1.2 dB at 8 GHz for the switched-off and switched-on cases, respectively, as shown in Fig. 6(a). The return loss results in Fig. 6(b) show that the lowest measured return loss is 15.8 dB at 6 GHz, which closely aligns with the design constraint of 16 dB.

C. Varactor Tuning

To achieve more time delay states with the switched capacitor bank approach, the number of transistor switches and capacitor needs to be increased, which adds further parasitics and increases circuit size. The calculation method for effective capacitance will be more complicated and might not lead to a viable solution. Varactor tuning is a commonly used method for realizing different capacitance values over a continuous range. We have implemented the continuously tuned BTC unit with varactors and done EM simulations of the layout together with the varactor PDK model in Cadence EMX ([10]). The EM simulation results of the layout together with varactors as well as with ideal capacitors are shown in Fig. 7. The ideal capacitance values were obtained from the C_S / C_B tuning analysis. The dimensions of the varactors were optimized to cover the tuning range for the effective capacitances of C_S and C_B. Figure 7(a) shows the group delay states for a tuning range from 16 ps to 25 ps and with a 1 ps tuning step size. The corresponding return loss is shown in Fig. 7(b). The effective varactor capacitance is increased at lower frequency near DC. The excess capacitance near DC adds additional time delay to the group delay performance over frequency. However, despite these challenges, the overall performance of group delay and return loss is good.

III. CONCLUSION

We have presented a design method for capacitively tuned true time delay units based on a BTC. We have implemented a tunable BTC with two delay states in a Tower Semi 0.18 μm SiGe BiCMOS process. The measured results for the tunable BTC with delay states 20 ps and 25 ps showed good performance over an 8 GHz bandwidth with a minimum return loss of 15.8 dB and a maximum insertion loss 1.8 dB. To realize more delay states, we have design a tunable BTC with varactors for ten delay states from 16 ps to 25 ps with a 1 ps step size in a Tower Semi 0.18 μm SiGe BiCMOS process. Full-wave electromagnetic simulation in Cadence EMX demonstrates the viability of this multi or continuous delay-state design approach using varactors. The proposed tunable TTD topologies can be applied in different applications that require fine delay resolution with a compact tunable TTD unit. Furthermore, these topologies can be implemented in existing path switching topologies, including the trombone topology, to offer additional delay states and increased tuning resolution without increasing the footprint of the circuit.

ACKNOWLEDGMENT

This research was supported by the Center for Design of Analog-Digital Integrated Circuits (CDADIC). We thank Tower Semiconductor for providing fabrication support.

REFERENCES

[1] S. K. Garakoui et al., "Compact cascadable g m -c all-pass true time delay cell with reduced delay variation over frequency," IEEE Journal of Solid-State Circuits, vol. 50, no. 3, pp. 693–703, 2015.

[2] S. Park and S. Jeon, "A 15–40 GHz CMOS true-time delay circuit for uwb multi-antenna systems," *IEEE Microw. Wireless Compon. Lett.*, vol. 23, no. 3, pp. 149–151, 2013.

[3] M. H. Ghazizadeh and A. Medi, "Novel trombone topology for wideband true-time-delay implementation," IEEE Transactions on Microwave Theory and Techniques, vol. 68, no. 4, pp. 1542–1552, 2020.

[4] D. Hao et al., "A low insertion loss variation trombone true time delay in GaAs pHEMT monolithic microwave integrated circuit," *IEEE Microw. Wireless Compon. Lett.*, vol. 31, no. 7, pp. 889–892, 2021.

[5] Tower Semiconductor, website: https://towersemi.com/.

[6] G. H. Golub and C. F. Van Loan, Matrix computations. JHU press, 2013.

[7] D. M. Pozar, Microwave engineering. John Wiley & Sons, 2011.

[8] H.-T. Lin and A. Weisshaar, "Robust and efficient design of on-chip compact delay units based on bridged t-coil," in Proc. IEEE 32nd Conf. Electr. Perform. Electron. Packag. Syst. (EPEPS), 2023, pp. 1–3.

[9] *PathWave Advanced Design System*. (2024). Keysight Technologies. [Online]. Available: https://www.keysight.com/us/en/home.html.

[10] *EMX Planar 3D Solver*. (2024). Cadence. [Online]. Available: https://www.cadence.com/en_US/home.html.

979-8-3503-5124-8/24 $31.00 © 2024 IEEE

An Efficient Machine Learning Approach for PSIJ Analysis in a Chain of CMOS Inverters

1st Ahsan Javaid
Dept. of Electronics
Carleton University
Ottawa, Canada
Email: Ahsanjavaid@cmail.carleton.ca

2nd Ramachandra Achar
Dept. of Electronics
Carleton University
Ottawa, Canada
Email: Achar@doe.carleton.ca

3rd Jai Narayan Tripathi
Dept. of Electronics
Indian Institute of Technology Jodhpur
Jodhpur, India
Email: Jai@iitj.ac.in

Abstract—In this paper, an efficient machine learning approach based on the knowledge-based and recurrent neural networks to predict power supply induced jitter in the presence of multiple power supply noises is presented. The proposed approach provides a reasonable accuracy and a significant increase in speed compared to conventional approaches.

Keywords—Recurrent artificial neural network (RANN), power supply induced jitter (PSIJ), power supply noise.

I. INTRODUCTION

In modern high-speed high-density circuits, power supply noise can significantly degrade the system performance by altering the timing behavior of the output from its ideal timing, commonly known as power supply induced jitter (PSIJ). Minimizing PSIJ becomes critical and consequently power integrity modeling and analysis becomes increasingly challenging [1].

A detailed power integrity analysis with large number of SPICE based circuit simulations is required to predict reasonably accurate jitter in the presence of multiple noises which can make the process prohibitively CPU expensive. In order to address the above computational burden, several efficient approaches can be found in the literature for PSIJ estimation [2]-[9]. The estimation of jitter in the presence of supply noise, ground noise and transmission media was presented in [2]. *Kim et al.* [3] presented analytical modeling of PSIJ transfer function for inverter chains in the presence of a single noise source. Efficient semi-analytical method for determining PSIJ for a chain of inverters in the presence of multiple noise sources was proposed in [4]. Statistical analysis for determining jitter in parallel single-ended buffers can be found in [5]. In [6], propagation delay-based method to obtain PSIJ sensitivity profile for a CMOS buffer chain was investigated. The development of knowledge-based neural network (KBNN) for analysis of PSIJ and PSIJ transfer function for a single CMOS module in the presence of supply noise were presented in [7] and [8]. An hybrid approach using knowledge-based and deep belief neural networks was introduced in [9] that predicts jitter in the presence of multiple noise sources.

In this paper, analysis in [7] which was focused on a single inverter is further advanced to predict PSIJ for a chain of CMOS inverters in the presence of multiple noise sources such as: supply noise, data noise and ground noise. For this purpose, the knowledge-based neural network (KBNN) is combined with the recurrent artificial neural network [10] (RANN). First, an accurate PSIJ error vector associated with the first inverter in the CMOS chain is computed using training data generated from two types of models, analytical expressions (such as in [4]) as well as a circuit simulator. Next, the error vectors for the N-cascaded inverters are computed by only employing the fast-to-evaluate analytical model [4]. In this process, training data from a computationally expensive circuit simulator is not required, resulting in a significant speed-up during the data generation process. All the above error vectors are used for training of the RANN. Next, a knowledge-based neural network is developed by combining the outputs of both the RANN and analytical relations to obtain the final PSIJ response. Results from a case study of inverters based on 22 nm CMOS technology [11] demonstrate that the proposed model predicts the PSIJ with reasonable accuracy while providing significant speed-up compared to HSPICE.

II. DEVELOPMENT OF THE PROPOSED K-RANN MODEL

This section presents an efficient estimation of power supply induced jitter using a hybrid neural network approach for a chain of inverters. A chain of N-cascaded CMOS inverters where all the inverters consist of an n-channel (M_n) and p-channel (M_p) transistors is shown in Fig. 1. The input terminals of the chain are connected to power supply (V_{DD}) and PRBS generator $(v_{in}(t))$. On the other hand, output of the chain is connected to a load capacitor (C_L). Also, multiple noise sources such as: power supply noise $(v_s(t))$, input data noise $(v_d(t))$ and ground bounce noise $(v_g(t))$ are injected into the system. Note that the dashed lines in Fig. 1 represent the paths of the injected noises. In the presence of various noise sources, CMOS inverters can suffer from instantaneous power fluctuations in the PDN and the consequent timing jitter. The corresponding jitter at the output of the each stage are represented as $(J_1, J_2, \ldots, J_N)$ (see Fig. 1).

A. The Proposed K-RANN Approach for PSIJ Estimation

In this section, KBNN in [7] is further advanced to predict jitter for a chain of CMOS inverters in the presence of multiple

979-8-3503-5124-8/24 $31.00 © 2024 IEEE

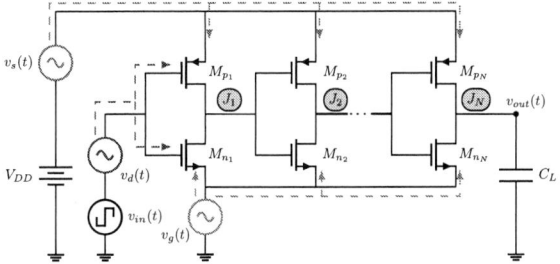

Fig. 1: Inverter chain with multiple noise sources

noise sources. The proposed K-RANN model for the Nth inverter in a chain (using analysis in [7]) can be expressed as, $J_{K_N}(\theta) = \mathcal{E}_N(\theta) + J_{A_N}(\theta)$ where $J_{K_N}(\theta)$, $J_{A_N}(\theta)$ and $\mathcal{E}_N(\theta)$ represent the final PSIJ response, analytical PSIJ response and the RANN-based error response of the Nth inverter, respectively. The error response $\mathcal{E}_N(\theta)$ can be obtained once recurrent neural network is trained. For training purposes, the PSIJ error-based sequential data corresponding to N-cascaded inverters is required. Here, θ represents the input parameters associated with the injected noise sources. For simplicity, in the subsequent sections, dependency of training data on 'θ' is omitted.

B. Recurrent Artificial Neural Network (RANN)

A recurrent artificial neural network [10] is a type of deep learning model that has an ability to generate dynamic sequences. The inputs $(x_1, \ldots, x_n, \ldots, x_N)$ propagate into RANN in a sequence and the corresponding recurrent (or hidden) states $(r_1, \ldots, r_n, \ldots, r_N)$ are updated by processing the information embedded in the previous states $(r_0, \ldots, r_{n-1}, \ldots, r_{N-1})$, respectively. Note that r_0 is a null vector since the algorithm has no information preceding the first element in the sequence. The updated recurrent states for the nth input is given as [10],

$$r_n = \sigma(U\,x_n + R\,r_{n-1} + a) \tag{1}$$

where $\sigma(.)$, U, R, a and r_{n-1} represent the activation function, weight connections, feedback connections, biases and previous recurrent state, respectively. The final updated state (r_N) which contains information of all the previous inputs is used along with weights V and biases b to compute the output of the RANN given as,

$$\mathcal{E}_N = \sigma(V\,r_N + b) \tag{2}$$

C. Data Generation and Training of the RANN

In this section, the PSIJ error vectors related to N-cascaded inverters are generated which form the training data set for the development of the RANN based model. Let the error vector corresponding to the second inverter be defined as,

$$J_{E_2} = J_{E_1} + J_{E_{1,2}} \tag{3}$$

where J_{E_1} is the error vector associated with the first inverter which is the difference [7] between PSIJ responses obtained

from circuit simulator (J_{H_1}) and analytical relations (J_{A_1}) given as, $J_{E_1} = J_{H_1} - J_{A_1}$. J_{E_1} is used to train an artificial neural network (see Section II-D). $J_{E_{1,2}}$ is the error between the PSIJ responses at the output of the first and second inverters, defined as, $J_{E_{1,2}} = J_{H_2} - J_{H_1}$.

Let $\mathcal{E}_1$ represent the error response (after training) from the corresponding ANN ($J_{E_1} \approx \mathcal{E}_1$). Next, J_{E_1} is modeled as a first recurrent state with input $\mathcal{E}_1$. First, multiplying the above expression with weight U yields, $U J_{E_1} \approx U \mathcal{E}_1$. Since $\mathcal{E}_1$ is the first input in the sequence, an initial (or null) state (r_0) with corresponding weight (R) and bias (a) along with an activation function $(\sigma(.))$ are introduced on both sides of $U J_{E_1} \approx U \mathcal{E}_1$. Above expression can be rewritten as,

$$\sigma(U J_{E_1} + R\,r_0 + a) \approx \sigma(U \mathcal{E}_1 + R\,r_0 + a) \tag{4}$$

Next, comparing left side of (4) with (1) represents the current recurrent state where $x_1 = J_{E_1}$. Hence, the first updated recurrent state for the first inverter in the chain is given as,

$$r_1 \approx \sigma(U \mathcal{E}_1 + R\,r_0 + a) \tag{5}$$

The process of obtaining J_{H_2} via HSPICE is more CPU expensive than the process of generating J_{H_1}. Since analytical relations provide good PSIJ approximation, the PSIJ differences between the consecutive inverters in the chain also provide good error responses. Therefore, the error between the first and second inverter is approximated as, $J_{E_{1,2}} \approx \mathcal{E}_{1,2}$ where $\mathcal{E}_{1,2} = J_{A_2} - J_{A_1}$. Here, J_{A_2} represents the PSIJ analytical response for the second inverter. Using above analysis, $J_{E_{1,2}}$ can also be modeled as a second recurrent state (r_2) using $\mathcal{E}_{1,2}$ (second input in the sequence), previous state (r_1) along with the corresponding weights and biases (U, R, a), represented by $r_2 \approx \sigma(U \mathcal{E}_{1,2} + R\,r_1 + a)$.

Similarly, the error related to the Nth inverter is given as, $J_{E_N} = J_{E_{N-1}} + J_{E_{N-1,N}}$ where $J_{E_{N-1}}$ can be modeled as a previous state r_{n-1} (similar to r_1 while updating r_2). Also, $J_{E_{N-1,N}}$ represents the error between the PSIJ responses at the output of inverters $N-1$ and N, defined as, $J_{E_{N-1,N}} \approx \mathcal{E}_{N-1,N}$. Using above analysis, $J_{E_{N-1,N}}$ can also be modeled as a final recurrent state (r_N) while using the previous state (r_{N-1}) with input $\mathcal{E}_{N-1,N}$ (last input in the sequence). Here, $\mathcal{E}_{N-1,N} = J_{A_N} - J_{A_{N-1}}$ where $J_{A_{N-1}}$ and J_{A_N} are the PSIJ analytical responses for inverters $N-1$ and N, respectively. Using (r_N) along with (2), the output of the RANN can be computed. Also, the Back Propagation Through Time (BPTT) algorithm [10] is used to adjust weights and biases.

D. Data Generation of J_{E_1} and Training of the ANN

In this section, details of the PSIJ error vector construction associated with the first inverter in the chain is presented (to obtain the training data set J_{E_1}). First, J_{A_1} is determined using fast-to-evaluate analytical relation [4]. Next, an accurate PSIJ training data (J_{H_1}) is needed which requires a simulation of the entire chain (as shown in Fig. 1) including all the inverters. This direct simulation of N-cascaded inverters can be a computationally expensive process. Therefore, a chain of N-cascaded inverters are replaced with their corresponding

979-8-3503-5124-8/24 $31.00 © 2024 IEEE

total resistors $(R_2, \ldots, R_N)$ and capacitors $(C_2, \ldots, C_N)$ in parallel, depending on the transistors region of operations (saturation, linear or cut off). The above-mentioned total resistors and capacitors associated with each inverter can be obtained using 1-bit simulation from HSPICE, evaluated at the midpoint of the rising edge. Once J_{E_1} is computed, it is used to train an ANN employing Levenberg-Marquardt training algorithm [7]. Note that when $N = 1$, the RANN expression (2) represents ANN-based error response $(\mathcal{E}_1)$.

III. RESULTS AND DISCUSSIONS

In this section, the proposed methodology to predict power supply induced jitter is validated. For this purpose, a chain of 11 cascaded inverters in the presence of supply noise, ground bounce and data noise is considered. All the inverters are powered with a DC supply voltage of 1.2V and input PRBS has a frequency of 125 MHz. The output of a chain is connected to a 100 fF load capacitor. Also, the PSIJ responses are evaluated at midpoint of the rising edge, simulating over 500 bits. All the injected noises are in the form of sine waves (typically generate the higher jitter) with amplitude, frequency, and initial phase of $(75 \text{ mV}, f_n, 21°)$, $(25 \text{ mV}, f_n, 45°)$ and $(15 \text{ mV}, f_n, 55°)$. The frequencies of all the noises (f_n) are varied from 10 MHz to 1 GHz.

First, a 1-bit simulation is employed to extract the resistors $(R_2, \ldots, R_{11})$ and capacitors $(C_2, \ldots, C_{11})$ for 11-cascaded inverters via HSPICE. The sum of all the above resistors and capacitors (including load) are computed to be $216 \, k\Omega$ and $126 \, fF$, respectively. The total time required for 1-bit simulation is 0.15 sec.

Next, all inverters except the first inverter in the chain are replaced with the above-mentioned lumped resistors and capacitors. The training data sets J_{H_1} and J_{A_1} were obtained using HSPICE and analytical relations for 20 data (training) points. Next, the error vector J_{E_1} is constructed which is used along with the Levenberg-Marquardt algorithm to develop an ANN. The data generation and evaluation times for J_{H_1}, J_{A_1} and J_{E_1} are 38.6 sec, 0.109 sec and 0.001 sec, respectively. The training process of ANN required a total of 14 epochs to achieve a training error of 10^{-12} in 0.015 sec.

Next, the analytical expressions [4] are employed to generate the sequential data $(\mathcal{E}_1, \mathcal{E}_{1,2}, \mathcal{E}_{2,3}, \ldots, \mathcal{E}_{10,11})$ via computing the PSIJ error between all the adjacent inverters. The total data generation time for the above generated set of data (evaluated at 50 points) is 2.7 sec. The above set of data is used to train the RANN using the BPTT training algorithm. The training of RANN required a total of 120 epochs to achieve a training error of 10^{-12} in 7.25 sec. Note that retraining of the RANN is required if the size of the chain is altered.

Next, the proposed K-RANN model $(J_{K_{11}})$ is constructed by combining the error-based output of the RANN $(\mathcal{E}_{11})$ with the analytical relation based PSIJ response $(J_{A_{11}})$, both evaluated at 25 equally-spaced points. The final jitter $(J_{K_{11}})$ response is compared with validating data set (PSIJ response using the conventional approach (HSPICE)). The corresponding plots for all the approaches (including analytical model)

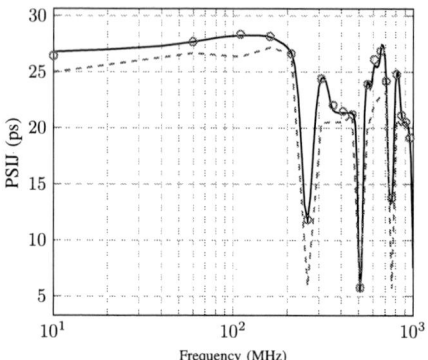

Fig. 2: PSIJ responses using the proposed approach (solid line), HSPICE (circles) and analytical model (dashed line).

are shown in Fig. 2. It can be seen from the above results that the proposed model matches the conventional approach more accurately compared to analytical PSIJ response. The total evaluation time required (excluding data generation and training times) for K-RANN and HSPICE approaches are 0.22 sec and 225 sec, respectively. The proposed approach provides a significant speed up of 1022 with a relative error of approximately 0.95% compared to using HSPICE.

IV. CONCLUSIONS

In this paper, KBNN and RANN were combined to predict jitter induced by multiple noise sources: power supply noise, ground bounce noise and input data noise in CMOS inverter circuits. Validating example demonstrates the accuracy and efficiency of the proposed method compared to HSPICE.

REFERENCES

[1] A. V. Mezhiba and E. G. Friedman, "Power Distributed Network in High Speed Integrated Circuits", Kluwer Academics, 2004.

[2] V. K. Verma, D. Junjariya and J. N. Tripathi, "Modeling Power Supply and Ground-Bounce Induced Jitter for a VDM Circuit driving Long Lines," *IEEE Symposium on Electromagn. Compat.*, pp. 657-661, MI, USA, 2023.

[3] H. Kim, J. Kim, J. Fan and C. Hwang, "Precise Analytical Model of Power Supply Induced Jitter Transfer Function at Inverter Chains," *IEEE Trans. Electromagn. Compat.*, vol. 60, no. 5, pp. 1491- 1499, Oct. 2018.

[4] J. N. Tripathi, P. Arora, H. Shrimali, and R. Achar, "Efficient jitter analysis for a chain of CMOS inverters," *IEEE Trans. Electromagn. Compat.*, vol. 62, no. 1, pp. 229–239, Feb. 2020.

[5] J. Kim, "Statistical analysis for pattern-dependent simultaneous switching outputs (SSO) of parallel single-ended buffers," *IEEE Trans. Circuits Syst. I, Reg. Papers*, vol. 64, no. 1, pp. 156–169, Jan. 2017.

[6] X. J. Wang and T. Kwasniewski, "Propagation Delay-Based Expression of PSIJ Sensitivity for CMOS Buffer Chain", *IEEE Trans. Electromagn. Compat.*, Vol. 58, No. 2, pp. 627-630, April 2016.

[7] A. Javaid, R. Achar and J. N. Tripathi, "Development of KBNNs for Analysis of PSIJ in CMOS Inverter Circuits," in *IEEE Trans. Microw. Theory Tech.*, vol. 71, no. 4, pp. 1428-1438, April 2023.

[8] A. Javaid, R. Achar and J. N. Tripathi, "Efficient Estimation of PSIJ via Jitter Transfer Function and Knowledge-based ANNs," *Proc. IEEE Workshop on Signal and Power Integrity*, pp. 1-3, Aveiro, Portugal, 2023.

[9] A. Javaid, R. Achar and J. N. Tripathi, "Prediction of PSIJ via Deep Belief Network and KBNN," *Proc. IEEE Workshop on SPI*, Portugal, 2024.

[10] J. Mazumdar and R. G. Harley, "Recurrent Neural Network Trained with Back Propagation Through Time Algorithm to Estimate Nonlinear Load Harmonic Currents," in *IEEE Transactions on Industrial Electronics*, Sept. 2008.

[11] PTM, Nanoscale Integration and Modeling (NIMO) Group, 2011, http://ptm.asu.edu/.

979-8-3503-5124-8/24 $31.00 © 2024 IEEE

Modeling Microwave S-parameters using Frequency-scaled Rational Gaussian Process Kernels

Thijs Ullrick
IDLab
Ghent University - imec
Ghent, Belgium
Thijs.Ullrick@UGent.be

Dirk Deschrijver
IDLab
Ghent University - imec
Ghent, Belgium
Dirk.Deschrijver@UGent.be

Wim Bogaerts
Photonics Research Group
Ghent University - imec
Ghent, Belgium
Wim.Bogaerts@UGent.be

Tom Dhaene
IDLab
Ghent University - imec
Ghent, Belgium
Tom.Dhaene@UGent.be

Abstract—**This work presents a machine learning technique to model the complex-valued scattering parameters (S-parameters) of passive microwave devices as a function of frequency and a set of design variables. The proposed Gaussian process (GP) model intricately models the real and imaginary parts of the S-parameters by employing a physics-informed kernel, adept at representing complex holomorphic functions and incorporating the Hermitian symmetry inherent in scattering parameters. Additionally, to extend the kernel's capabilities to higher dimensions beyond standard GP techniques, it is extended with a frequency scaling, enhancing the modeling capacity. The resulting physics-informed frequency-scaled GP model accurately predicts the S-parameter values at desired parameter configurations in the design space. One application example demonstrates the superiority of the new kernel, compared to standard GP kernels.**

Index Terms—**Gaussian processes (GP), kernels, machine learning (ML), Microwave filters, S-parameters**

I. INTRODUCTION

PARAMETRIC macromodeling is indispensable for the characterization of high-frequency electromagnetic (EM) systems, and plays a crucial role in design space exploration, optimization, and sensitivity analysis. Several widely used macromodeling techniques rely on vector fitting (VF) to build a rational function approximation [1]. A major advantage of these rational models is their seamless conversion into state-space form, facilitating integration in SPICE-like solvers for time-domain simulations [2]. These rational models, however, do not provide any uncertainty estimation, limiting their applicability for design optimization purposes.

Recent advancements in macromodeling utilize machine learning techniques such as artificial neural networks (ANN) [3], [4] and support vector machines (SVM) [5] to address the limitations of standard approaches like VF [1] and AAA [6]. However, ANNs, effective for high-dimensional and non-linear functions, require substantial amounts of training data and are prone to overfitting. Conversely, SVMs offer robust regularization but lack probabilistic interpretability.

Stochastic models, such as Gaussian processes (GP), offer a promising alternative due to their data efficiency and posterior

This work was supported by the 'Flemish Research Foundation (FWO-Vlaanderen) under grant G031421N' and 'Onderzoeksprogramma Artificiële Intelligentie (AI) Vlaanderen' programs.

variance estimation, particularly advantageous in computationally expensive optimization scenarios [7]. Standard GPs, however, are typically adopted for modeling smooth functions, while microwave S-parameters often exhibit a dynamic behavior. In fact, this property of the GP derives from the typical covariance function used among the data points, also known as kernels.

While S-parameters may exhibit dynamic behavior, changes in the overall frequency response due to adjustments in the device's geometrical dimensions or dielectric properties are usually smooth or can be represented as a compression/expansion transformation along the frequency axis. In this regard, the present work introduces a novel kernel for modeling parametric S-parameters. It combines the rational Szegő kernel, originally proposed by Bect et al. [8], with standard GP kernels to extend its applicability to a multi-dimensional settings. Moreover, a parametric scaling in the frequency dimension is incorporated in the kernel, significantly enhancing its performance in modeling microwave S-parameters.

II. METHODOLOGY

A. Gaussian Process Modeling

GPs are probabilistic models that define distributions over functions, where the joint distribution of any collection of points on that function follows a multivariate normal distribution. In particular, the GP is data-driven: it doesn't depend on a fixed number of parameters, such as the poles and residues of a rational VF macromodel. Instead, its complexity and capacity to express patterns increase with the volume of training data. This adaptability makes GPs highly data-efficient, facilitating accurate predictions even when the dataset is limited in size.

GPs are defined by a mean function, representing the expected value of the function at each point, and a covariance function, also referred to as the kernel, which captures the correlation between pairs of points. In many cases, prior information on the underlying stochastic process or function, such as the periodicity or smoothness, can be encoded in either the mean function or kernel. By properly incorporating these assumptions, GPs can achieve high accuracy, even when trained on small datasets.

979-8-3503-5124-8/24 $31.00 © 2024 IEEE

Once the kernel is defined, new function values are predicted via Gaussian Process Regression (GPR), also known as Kriging. In essence, GPR entails two primary steps: specifying a prior distribution based on assumptions about the underlying data-generating process, typically achieved through the design of appropriate mean and kernel functions, followed by updating this prior using Bayes' theorem to derive the posterior distribution. The posterior distribution in GPR captures both the predictive mean and the uncertainty associated with each prediction, making it a powerful tool for regression tasks.

B. Rational Szegö Kernel

Few physics-informed kernels have been introduced in the literature for the modeling of microwave S-parameters [8], [9]. For instance, the delayed GP method introduced by Garbuglia et al. [9] demonstrates comparable accuracy to delayed vector fitting for modeling elecrically long interconnects prone to significant cross-talk. However, it fails to leverage the intricate relation of the real and imaginary part of the scattering parameters, and instead, models them as independent variables using separate GPs. To use the data more effectively, a Multi-Output Gaussian Process (MOGP) is adopted in this work. An MOGP is an extension of the standard GP that can simultaneously model multiple related outputs. This is particularly useful in scenarios where outputs are correlated or share common characteristics, allowing for more efficient and accurate predictions compared to modeling each output independently.

Bect et al. [8] recently introduced a novel covariance function, referred to as the rational Szegö kernel, for modeling complex-valued functions. This kernel has been designed to represent a space of complex holomorphic functions and incorporates the Hermitian symmetry inherent in the frequency response of dynamical systems. It effectively captures the correlation between the real and imaginary parts, leading to a fitting procedure that converges significantly faster compared to standard GP kernels. For a detailed discussion of its derivation and properties, readers are referred to [8]. The Szegö kernel is adopted in present work as the covariance function of the MOGP and can be expressed as follows

$$K_{sz}(s_0, s_1) = \begin{bmatrix} \Re(\frac{k+c}{2}) & \Im(\frac{-k+c}{2}) \\ \Im(\frac{k+c}{2}) & \Re(\frac{k-c}{2}) \end{bmatrix} \quad (1)$$

with

$$k(s_0, s_1) = \frac{\sigma^2}{2\alpha + s_0 + s_1^*}$$
$$c(s_0, s_1) = \frac{\sigma^2}{2\alpha + s_0 + s_1} \quad (2)$$

where α and σ^2 are the hyperparameters of the rational kernel and $s = j2\pi f$ is the Laplace variable.

C. Rational Kernel Extension to Higher Dimensions

While the Szegö kernel excels in representing complex-valued functions across frequency, the dynamics associated with the parameterization of these functions can be effectively captured by standard GP kernels. In this work, the Szegö kernel is extended to higher dimensions by combining it with a standard Matérn 5/2 kernel [9], for modeling S-parameter variations with respect to design variables. Leveraging the distinct strengths of each kernel, the covariance function of the MOGP can be expressed as

$$K_{cm}(s_0, \mathbf{x}_0, s_1, \mathbf{x}_1) = K_{sz}(s_0, s_1) \circ K_{mat}(\mathbf{x}_0, \mathbf{x}_1) \quad (3)$$

where $\mathbf{x}_0$ and $\mathbf{x}_1$ are vectors containing the design variables.

D. Frequency-scaled Kernels

Microwave S-parameters often exhibit compression or expansion along the frequency axis. Assuming the frequency response remains unaffected by other transformations or changes, this implies that we can express $S(s, \hat{\mathbf{x}}_0) = S(\gamma s, \hat{\mathbf{x}}_1)$. Here, γ represents the scaling of the frequency response relative to the Laplace variable s as the design variables are tuned from $\hat{\mathbf{x}}_0$ to $\hat{\mathbf{x}}_1$. Consequently, the parameter configurations $(s, \mathbf{x}_0)$ and $(\gamma s, \mathbf{x}_1)$ are highly correlated. However, stationary kernels like the SE kernel, which rely solely on the distance between points, struggle to capture such correlations effectively. To address this, the covariance function (3) is enhanced by incorporating a linear frequency scaling with respect to the design variables, leading to

$$K_{fs} = K_{cm}(s_0(1 + \gamma \cdot \mathbf{x}_0), \mathbf{x}_0, s_1(1 + \gamma \cdot \mathbf{x}_1), \mathbf{x}_1) \quad (4)$$

where γ serves as an additional hyperparameter describing the linear frequency scaling. It is noteworthy that the use of a frequency-scaling coefficient to improve modeling accuracy is inspired by techniques employed in rational modeling [10].

III. APPLICATION EXAMPLE

In this application example, the proposed GPR framework is evaluated for modeling the transmission of a double-folded microstrip band-stop filter, formed by a dielectric substrate between a top metallization and a bottom ground plane. As illustrated in Fig. 1, the geometry of the upper layer consists of two stubs, with identical length and spacing, folded onto the two sides of a transmission line. The S-parameters of the device are simulated for a set of 15 equispaced frequencies within the $[5, 25]$ GHz range using ADS Momentum. The chosen design parameters are the stub length $L \in [1.5, 3.0]$ μm and line-stub spacing $S \in [0.0, 0.3]$ μm.

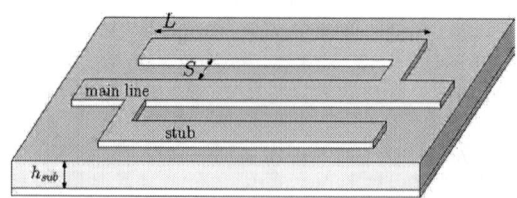

Fig. 1: Double-folded microstrip band-stop filter geometry.

In the following analysis, three kernel functions are compared: two separate GPs for independent modeling of the real and imaginary part using a Matérn kernel K_{mat}, an

979-8-3503-5124-8/24 $31.00 © 2024 IEEE

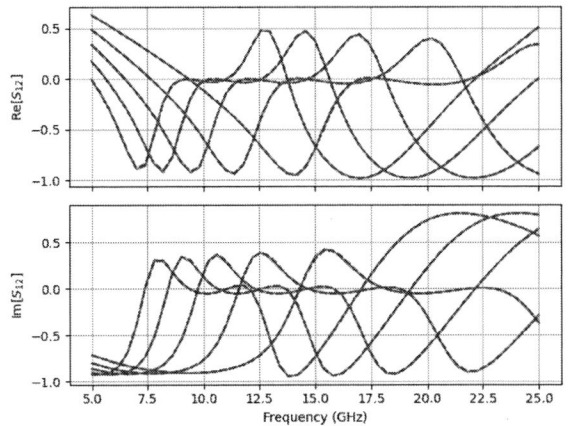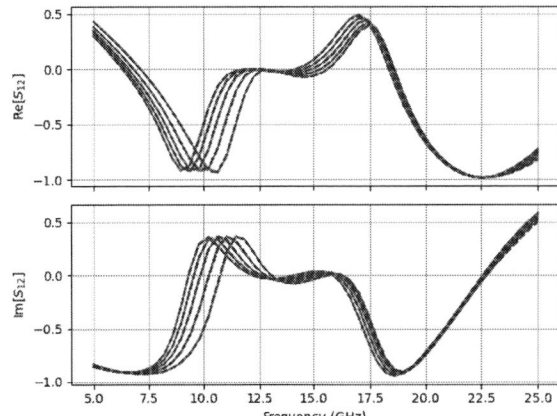

Fig. 2: Comparison of real and imaginary part of S_{12}, as predicted with K_{fs} (blue), and the original S-parameter (red), for varying L (left) and S (right).

TABLE I: Prediction accuracy of the GP models

$N_f \times N_L \times N_S$	RMSE (dB)			MAE (dB)		
	K_{mat}	K_{cm}	K_{fs}	K_{mat}	K_{cm}	K_{fs}
$15 \times 2 \times 2$	-3.7	-4.5	-23.7	2.8	2.0	-13.5
$15 \times 3 \times 3$	-8.9	-8.7	-36.6	-0.8	-0.5	-25.5
$15 \times 4 \times 4$	-11.0	-12.5	-40.9	-3.6	-3.0	-31.3
$15 \times 5 \times 5$	-12.3	-16.4	-41.4	-4.6	-6.0	-32.5

MOGP using the composite Szegö Matérn kernel K_{cm}, and an MOGP using the frequency-scaled enhanced version K_{fs}. The implementation of the GP is done in Python using the GPyTorch library. It is worth noting that the implementation employs real-valued kernels and inputs, where the outputs of the MOGP correspond to the real and imaginary parts, respectively.

The training data is generated by simulating the frequency response of the double folded stub filter on a uniform N×N grid within the 2D design space. The hyperparameters of each model are then selected by minimizing the marginal log-likelihood. Once the models are fit to the data, their accuracy is quantified in terms of the root-mean-squared error (RMSE) and maximum absolute error (MAE), which are evaluated on a set of 400 (f, L, s) samples chosen randomly according to an Latin hypercube design. Both metrics are reported in Table I for each kernel.

The new kernel yields highly accurate predictions despite the significant variability in the S-parameters. Indeed, this is demonstrated in Fig. 2, which plots S_{12} for varying L and S respectively. The predicted real and imaginary parts (blue) accurately match the validation samples (red) computed via EM simulation. In particular, the model effectively captures the compression and expansion observed in the S-parameters.

IV. CONCLUSION

The rational kernel used in this work enables detailed modeling of the real and imaginary parts of highly dynamic S-parameters. Additionally, extending the kernel to higher dimensions, particularly by incorporating a linear frequency scaling, has shown superior performance in modeling parametric microwave S-parameters compared to standard kernels, which lack sufficient accuracy.

REFERENCES

[1] S. Grivet-Talocia and B. Gustavsen, *Passive macromodeling*. Wiley series in microwave and optical engineering, Hoboken, New Jersey: John Wiley & Sons, Inc, 2016.

[2] Y. Ye, T. Ullrick, W. Bogaerts, T. Dhaene, and D. Spina, "SPICE-Compatible Equivalent Circuit Models for Accurate Time-Domain Simulations of Passive Photonic Integrated Circuits," *Journal of Lightwave Technology*, vol. 40, pp. 7856–7868, Dec. 2022.

[3] F. Feng, W. Na, J. Jin, W. Zhang, and Q.-J. Zhang, "ANNs for Fast Parameterized EM Modeling: The State of the Art in Machine Learning for Design Automation of Passive Microwave Structures," *IEEE Microwave Magazine*, vol. 22, pp. 37–50, Oct. 2021.

[4] M. Schierholz, I. Erdin, J. Balachandran, C. Yang, and C. Schuster, "Parametric S-Parameters for PCB based Power Delivery Network Design Using Machine Learning," in *2022 IEEE 26th Workshop on Signal and Power Integrity (SPI)*, (Siegen, Germany), pp. 1–4, IEEE, May 2022.

[5] R. Trinchero, M. Larbi, H. M. Torun, F. G. Canavero, and M. Swaminathan, "Machine Learning and Uncertainty Quantification for Surrogate Models of Integrated Devices With a Large Number of Parameters," *IEEE Access*, vol. 7, pp. 4056–4066, 2019.

[6] Y. Nakatsukasa, O. Sète, and L. N. Trefethen, "The AAA Algorithm for Rational Approximation," *SIAM Journal on Scientific Computing*, vol. 40, pp. A1494–A1522, Jan. 2018. Publisher: Society for Industrial and Applied Mathematics.

[7] C. E. Rasmussen and C. K. I. Williams, *Gaussian processes for machine learning*. Adaptive computation and machine learning, Cambridge, Mass: MIT Press, 2006. OCLC: ocm61285753.

[8] J. Bect, N. Georg, U. Römer, and S. Schöps, "Rational kernel-based interpolation for complex-valued frequency response functions," Sept. 2023. arXiv:2307.13484 [cs, math].

[9] F. Garbuglia, T. Reuschel, C. Schuster, D. Deschrijver, T. Dhaene, and D. Spina, "Modeling Electrically Long Interconnects Using Physics-Informed Delayed Gaussian Processes," *IEEE Transactions on Electromagnetic Compatibility*, vol. 65, pp. 1715–1723, Dec. 2023.

[10] T. Ullrick, D. Spina, W. Bogaerts, and T. Dhaene, "Wideband parametric baseband macromodeling of linear and passive photonic circuits via complex vector fitting," *Scientific Reports*, vol. 13, p. 15407, Sept. 2023. Number: 1 Publisher: Nature Publishing Group.

PSIJ based Optimal PDN Design for Cost-Effective SSD using Reinforcement Learning

Taein Shin[1], Seonguk Choi[1], Jungmin Ahn[1], Junghyun Lee[1], Keunwoo Kim[1], Haeseok Suh[1], Hyunah Park[1], Haeyeon Kim[1], Hyunjun An[1], Jinwook Song[2], and Joungho Kim[1]

[1]School of Electrical Engineering, Korea Advanced Institute of Science and Technology (KAIST), Daejeon, South Korea
[2]Solution Development Team (Memory Business), Samsung Electronics, Hwaseong, South Korea
E-mail: taeinshin@kaist.ac.kr

Abstract— **In this paper, we first propose power supply noise induced jitter (PSIJ) based optimal power distribution network (PDN) design using reinforcement learning (RL) to achieve cost-effective solid state drive (SSD). Compared to the traditional PDN impedance design, designing based on PSIJ, which includes current noise and timing information, allows for more precise PDN optimization. The proposed method allows for the development of an optimal PDN design that meets the target PSIJ with minimal decoupling capacitor (decap) area usage. In order to achieve a cost-effective SSD board design, the RL reward includes both PSIJ and the decap area. This enables the RL agent to learn to satisfy the target PSIJ using minimal resources from decap library. To derive PSIJ, the process includes adding decaps to the z-parameters of the SSD board's PDN plane to extract PDN impedance and modeling PSIJ in the frequency domain through jitter sensitivity. Finally, the successful optimization is demonstrated through performance comparison with a representative genetic algorithm (GA).**

Keywords—Decoupling capacitor, Power Distribution Network, Power Supply Noise Induced Jitter, Reinforcement Learning, Solid State Drive

I. INTRODUCTION

The speed and bandwidth of chip-package and interconnect systems have significantly increased, making timing jitter a crucial factor in overall performance. Power supply-induced jitter (PSIJ) has become a notable component of total jitter. Unlike other common jitter factors, PSIJ associated with the power domain is increasingly significant due to the power supply package technology not scaling down in line with chip technology and the effect of the clock path lengthening as chip technology decreases [1]. Therefore, this PSIJ has become a key factor in Power Integrity (PI) design.

The conventional power distribution network (PDN) design method involves calculating the target impedance for each frequency band and then designing based on that. Previous researches on PDN design through de-cap optimization also rely entirely on target impedance [2]-[4]. However, this PDN impedance is essentially defined by allowing only a certain ratio of AC voltage relative to the DC voltage, making PDN design based on this approach overly optimistic. A slightly advanced approach is to determine the target impedance more precisely based on the current noise spectrum, but this method also has limitations [4]. In the case of PDN design based on PSIJ, more precise design is possible because it includes not only current noise information but also time-based information.

We introduce reinforcement learning (RL) to design the PDN based on PSIJ. RL stores weights learned from vast

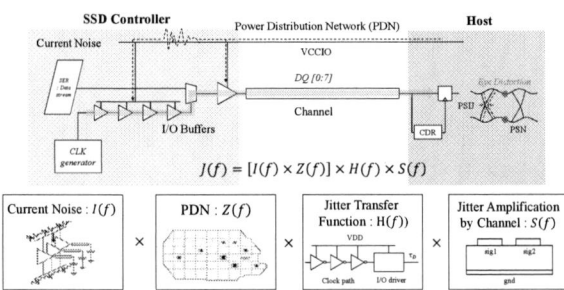

Fig. 1. Schematic of SSD I/O Interface for PSIJ including current noise, PDN impedance, jitter transfer function, and jitter amplification.

amounts of data in neural networks, making it highly proficient in non-linear regression and offering significant advantages in optimization. RL offers not only higher performance compared to traditional random search (RS) or genetic algorithms (GA) but also significantly reduces time consumption due to its reusability [5].

Accordingly, in this paper, we first propose PSIJ based optimal PDN design using RL to achieve cost-effective SSD. As shown in Fig. 1, PSIJ is determined by current noise, PDN impedance, jitter transfer function and channel in SSD I/O Interface. Our design target is PDN with decaps. Section 2 provides an explanation of the RL framework used for optimizing PDN based on PSIJ. In Section 3, the modeling process to derive PSIJ and the setup of the SSD board level PDN are briefly described. Finally, Section 4 validates the performance by comparing the proposed method with GA.

II. PROPOSAL OF PSIJ BASED RL METHOD USING ENCODER-DECODER NETWORK MODEL

RL consists of two main components: an agent responsible for learning and taking actions based on that learning, and an environment that evaluates the actions taken by the agent. Fig. 2 depicts a schematic diagram of the proposed method. First, we will examine the Markov decision process (MDP) parameters that the RL aims to solve. The state (*s*) refers to the situation of the given problem, which includes the port locations and the candidates for decaps of the target SSD board PDN. The action (*a*) corresponds to the design parameters we aim to optimize, including the total number of decaps, as well as the size and location of each decap. Therefore, RL methods train the policy (*π*) network to maximize the reward (*r*) by finding optimal actions (*a*) for a given state (*s*). Finally, the reward serves as the criterion for evaluating how well the agent's actions are derived. In this paper, if the target PSIJ is not satisfied, the reward is set to 0; if it is satisfied, the decap area consumed is subtracted from a certain constant value.

979-8-3503-5124-8/24 $31.00 © 2024 IEEE

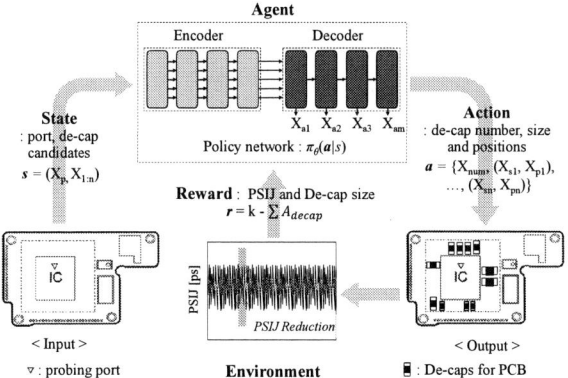

Fig. 2. Overall schematic of the proposed PSIJ based RL framework.

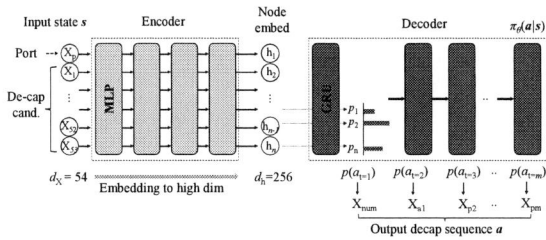

Fig. 3. Policy network configuration with encoder-decoder as agent.

$$Reward = 0 : target\ PSIJ < output\ PSIJ$$

$$Reward = k - \sum A_{decaps} : target\ PSIJ > output\ PSIJ \quad (1)$$

This setup enables the agent to learn to use the minimal decap area while achieving the target PSIJ.

Next, we use a policy network-based encoder-decoder model for agent as shown in Fig. 3. The encoder consists of a series of multilayer perceptron (MLP) networks that embed the state (*s*) value, while the decoder uses this information to sequentially derive action values based on a gated recurrent unit (GRU) network. Through this sequential decoder configuration, previous information is used to derive subsequent information, leading to improved performance [6]. In this setup, each decoder (one GRU network) sequentially selects nodes (actions *a*) with the highest probability distribution function (PDF), determining the total number of decaps, the size of the n-*th* decap, and the location of the n-*th* decap. Furthermore, for the iterative learning of RL, the algorithm used is REINFORCE based on policy gradient [6]. Based on this proposed RL framework, it becomes possible to solve the problem of meeting the target PSIJ while using a minimal size of decaps.

III. MODELING OF PSIJ AND DECAP ASSIGNMENT FOR SSD BOARD LEVEL PDN

To account for changes in PSIJ based on decap assignment, fast and accurate modeling is necessary for optimal PDN design. First, the PDN impedance should be determined based on the decap assignment. The target PDN adopted the 0.75 CVDD net of the SSD board, which corresponds to the 6-*th* layer out of a total of 12 layers. Consequently, each decap added includes an additional via with an inductance of 0.2 nH for bottom routing. In this net, there are a total of 53 candidate locations for decaps, and the problem is set to place up to a maximum of 8 EA decaps. The RL agent considers both the

TABLE I. ELECTRICAL CHARACTERISTICS AND SIZES ON DECAP CANDIDATES

| | Capacitance | ESR [mΩ] | ESL [nH] | Std. | Physical Size | |
					Area	Thickness
1	47 uF	4.8	0.3	1608	1.28 mm²	0.8 mm
2	22 uF	4.5	0.2			
3	10 uF	5.7	0.3			
4	4.7 uF	5.9	0.2			
5	2.2 uF	8.6	0.2	1005	0.5 mm²	0.5 mm
6	1 uF	11.7	0.2			
7	200 nF	17.2	0.1	0603	0.18 mm²	0.3 mm
8	100 nF	31.2	0.1			

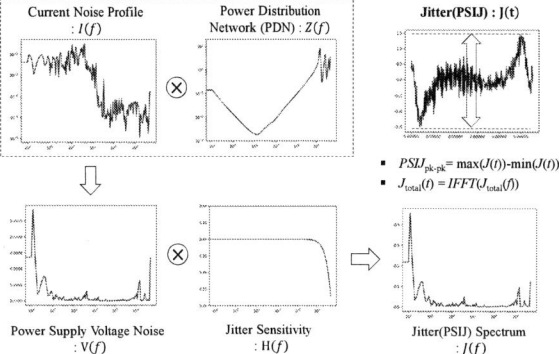

Fig. 4. Overall process for PSIJ modeling for SSD I/O Interface including SSD Board level PDN.

electrical characteristics and physical size from the 8 types of decap libraries in Table I to find the optimal combination. Each decap Z parameter including ESR and ESL is cascaded with the board PDN Z parameter according to the size and location to derive the PDN impedance.

Fig. 4 illustrates the PSIJ modeling process in the frequency domain. By multiplying the frequency profiles of PSIJ in each domain, the jitter spectrum is derived [7]. First, the power supply voltage noise $V(f)$ is obtained by multiplying the current noise profile $I(f)$ by the board-level PDN impedance $Z(f)$ based on decap assignment. Then, by multiplying this result by the jitter sensitivity of the SSD controller's clock path $H(f)$, the jitter spectrum can be determined. Finally, applying the inverse Fourier transform (*IFFT*) to the jitter spectrum yields the peak-to-peak PSIJ $J(t)$ in the time domain, which is a key criterion for PDN optimization.

$$PSIJ_{pk-pk} = max(PSIJ(t)) - min(PSIJ(t)) \quad (2)$$

Since the focus is currently on the board-level PDN, the impact of jitter amplification caused by channels, which is more significant at relatively high frequencies, is disregarded. Accordingly, it can be confirmed that low-frequency PSIJ is a significant component as shown in Fig. 4.

IV. VERIFICATION OF THE PROPOSED METHOD

In this chapter, the optimization results and performance comparison are discussed. Fig. 5(a) depicts how the proposed RL converges in reward as training steps increase at target PSIJ 2.5 ps case, showing how decap assignments change at each training step. The RL agent begins to converge around the 300-*th* training step and reaches complete convergence around the 1000-*th* step. This indicates successful learning has occurred. Accordingly, at the 100th step, having only one decap resulted in a reward of 0, failing to meet the target PSIJ.

979-8-3503-5124-8/24 $31.00 © 2024 IEEE   149

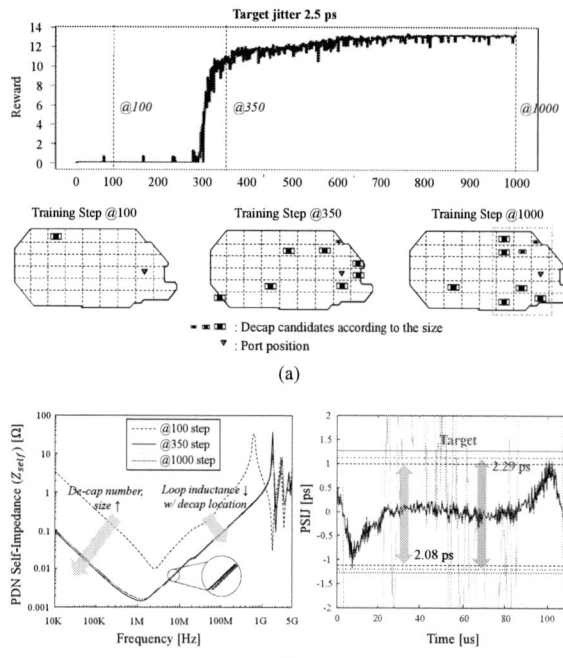

(a)

(b)

Fig. 5. Results of the optimization: performance improvement depending on training steps (a) Reward convergence and decap design results (b) PDN impedance and time domain PSIJ.

Starting from around the 350-*th* step where convergence begins, the number of decaps reaches the maximum of 8 EA, satisfying the target PSIJ. In the final convergence phase (1000-*th*), decaps are positioned closer to the ports, and their sizes are also optimized further.

This can be more closely observed in Fig. 5(b) through the PDN impedance profile. Initially, due to insufficient decaps, the impedance values are high throughout the entire bandwidth. However, around the 350th step, as the number and size of decaps increase, the impedance significantly decreases. Below 1 MHz, the impedance decreases based on the number and size of decaps. Beyond 1 MHz, the impedance is influenced by loop inductance, which are determined by the location of decaps and ESL. Accordingly, by the 1000-*th* step, reduction of the size of decaps (reducing penalties) slightly increases the impedance below 1 MHz while optimizing their location to lower the impedance above 1 MHz. As a result, based on the time domain PSIJ graph, it can be confirmed that the PSIJ is rather increased from 2.08 ps to 2.29 ps. This signifies the optimization direction aimed at satisfying the target of 2.5 ps while utilizing the minimum decap area.

Finally, Table II compares the optimal design results between RL and GA for various target PSIJ 2.5 ps, 3.0 ps, and 3.5 ps. GA was trained with a population size of 50 and 100 generations, total 5000 iterations. The proposed RL results are based on 10 inferences. In all cases, the proposed RL demonstrates superior performance compared to GA as the reward of 13.28, 16.48, and 18.40. Even when considering RL's learning over 1000 iterations versus GA's 5000 iterations, the RL outperforms. Fig. 6 compares the total area of decaps corresponding to optimization penalties required to meet the target PSIJ. As shown to RL achieving higher rewards, it is observed to have lower penalty values. This means that the

TABLE II. PSIJ OPTIMIZATION RESULTS BY THE PROPOSED METHOD

Target PSIJ	Method	Total Reward	PSIJ [ps]	Decap Area [mm^2]
2.5 ps	GA {50×100}	12.08	2.29	8.96
	Proposed {10}	13.28	2.29	8.36
3.0 ps	GA {50×100}	15.84	2.48	7.08
	Proposed {10}	16.48	2.87	6.76
3.5 ps	GA {50×100}	16.84	2.77	6.58
	Proposed {10}	18.40	3.21	5.80

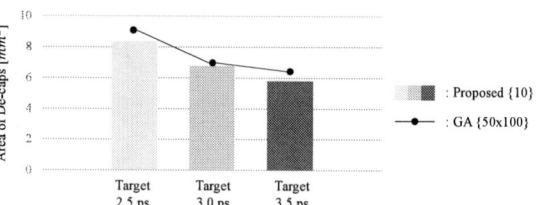

Fig. 6. Comparison of the optimization penalty (area of decaps)

proposed RL minimizes the use of decaps more effectively compared to GA while satisfying the same target PSIJ.

V. CONCLUSION

This paper demonstrates the optimization of PDN decap design solely based on PSIJ. To achieve this, PSIJ based RL method for optimal PDN design of cost-effective SSD is proposed. The proposed RL-based method, encompassing the RL framework from decap assignment to PSIJ modeling, has demonstrated superior performance compared to conventional optimization methods. This method has proved the potential for precise PI design that minimizes decap usage, using PSIJ as the metric.

ACKNOWLEDGMENT

This work was supported by Samsung Electronics Co., Ltd (IO201207-07813-01). We would like to acknowledge the technical support from ANSYS Korea. This research was supported by National R&D Program through the National Research Foundation of Korea (NRF) funded by Ministry of Science and ICT (NO. 2022M3I7A4072293).

REFERENCES

[1] D. Oh, "System level jitter characterization of high speed I/O systems," *2012 IEEE International Symposium on Electromagnetic Compatibility*, Pittsburgh, PA, USA, 2012, pp. 173-178.

[2] H. Park et al., "Transformer Network-Based Reinforcement Learning Method for Power Distribution Network (PDN) Optimization of High Bandwidth Memory (HBM)," *in IEEE Transactions on Microwave Theory and Techniques*, vol. 70, no. 11, pp. 4772-4786, Nov. 2022.

[3] L. Zhang, L. Jiang, J. Juang, Z. Yang, E. -P. Li and C. Hwang, "Decoupling Optimization for Complex PDN Structures Using Deep Reinforcement Learning," *in IEEE Transactions on Microwave Theory and Techniques*, vol. 71, no. 9, pp. 3773-3783, Sept. 2023.

[4] J. Song et al., "Novel Target-Impedance Extraction Method-Based Optimal PDN Design for High-Performance SSD Using Deep Reinforcement Learning," *in IEEE Transactions on Signal and Power Integrity*, vol. 2, pp. 1-12, 2023.

[5] R. S. Sutton and A. G. Barto, *Reinforcement Learning: An Introduction*. Cambridge, MA, USA: MIT Press, 1998.

[6] S. Choi et al., "Deep Reinforcement Learning-Based Optimal and Fast Hybrid Equalizer Design Method for High-Bandwidth Memory (HBM) Module," *in IEEE Transactions on Components, Packaging and Manufacturing Technology*, vol. 13, no. 11, pp. 1804-1816, Nov. 2023

[7] T. Shin *et al.*, "Modeling and Analysis of System-Level Power Supply Noise Induced Jitter (PSIJ) for 4 Gbps High Bandwidth Memory (HBM) I/O Interface," *2021 IEEE Electrical Design of Advanced Packaging and Systems (EDAPS)*, 2021.

979-8-3503-5124-8/24 $31.00 © 2024 IEEE

A Study on How Capacitance of Power Filtering Circuit Influences the Antiresonance Frequency

Shuxiang Li
CHG Signal Integrity Department
CISCO System Inc.
Shanghai, CHINA
shuxili@cisco.com

Xinlin Tang
Signal Integrity Department
Chengfang Information Co., Ltd.
Shanghai, CHINA
xinltang6@gmail.com

Tao Fang
Department of Computer and Information Science
University of Macau
Macau, CHINA
taylefang@gmail.com

Yuan Fang
Signal Integrity Department
Chengfang Information Co., Ltd.
Shanghai, CHINA
Fang24311@gmail.com

Greg Fu
CHG Signal Integrity Department
CISCO System Inc.
Shanghai, CHINA
guanfu@cisco.com

Stephen Scearce
Enterprise Routing, Wireless & SMB Hardware
CISCO System Inc.
North Carolina, USA
sscearce@cisco.com

Abstract—This article is dedicated to solving the problem of how to help engineers to optimize the power filtering circuit design. Through a thorough investigation, we discovered a significant relationship between the total capacitance in the circuit and the occurrence of antiresonance point. When the ferrite bead or inductor is constant, an increase in the total circuit capacitance leads to a proportional decrease in the antiresonance frequency. This inverse relationship is mathematically expressed in Eq. (4), which illustrates the ratio between the increase in capacitance and the corresponding decrease in antiresonance frequency. The finding contributes to a deeper understanding to the PDN (power distribution network) design and offers insights for optimizing circuit design to minimize the impact of antiresonance phenomenon.

Keywords—Antiresonance Frequency, Optimization, Capacitance of power filtering circuit

I. INTRODUCTION

In the realm of electronic design, the relentless pursuit of energy efficiency has prompted a continuous reduction in power voltages, underscoring the pivotal role of PDN in the design process. The design of PDN is now more critical than ever, as it directly impacts the minimization of power consumption and the effective management of noise introduced by power supplies. However, the antiresonance phenomenon of power net at the board level poses significant risks, potentially triggering a cascade of detrimental effects that may disrupt the performance of electronic systems. To address this challenge, extensive researches have been conducted to optimize die-level PDN to ensure compatibility with package and board-level PDN designs [1-3], thereby avoiding undesired antiresonance. It has been established that antiresonance peaks are intricately linked to parallel-connected capacitors with distinct self-resonant frequencies [4].

Circuits are particularly susceptible to frequency distortion near antiresonance frequency, leading to signal distortion and the generation of unwanted noise [5]. Such distortion can compromise the quality and bandwidth of signal transmission.

Further research has revealed that the peak frequency of power supply noise, as determined by Fast Fourier Transform (FFT) analysis, are in close agreement with the measured antiresonance peak frequency [6]. This finding underscores the critical need to eliminate antiresonance peak at the board level.

The purpose of this paper is to investigate the relationship between capacitance and antiresonance frequency. We explored whether antiresonance would change when the total capacitance is held constant but individual capacitor values differ, while keeping the ferrite bead as same. For instance, we compared a circuit composed of ten 0.1uf capacitors with a total capacitance of 1uf to a circuit with a single 1uf capacitor to determine if there is a significant difference in their antiresonance frequencies. Additionally, we examined how increasing capacitance shifts antiresonance peak to lower frequency [6]. All additional capacitors we added realize their effects on the antiresonance frequency.

The paper is organized as four parts. Section 2 presents the network model with antiresonance characteristic that we utilized in our subsequent study. Section 3 delves into the relationship between antiresonance and capacitance. Section 4 summarizes the findings of our study in a mathematical formula. Through this research, we aim to contribute new insights into PDN design, optimize circuit performance, and mitigate the adverse effects of antiresonance.

II. THE FILTERING CIRCUIT WITH ANTIRESONANCE

We analyzed a filtering circuit of power supply network, as depicted in Fig. 1. In this topology, the ferrite bead is with 0.18ohm DC resistance (DCR) and an 220ohm impedance at 100MHz(FB_DCR0p18_220ohm). During the time domain simulation, we identified an antiresonance point at 386.4KHz with a peak of 22.943dB, as illustrated by the red curve in Fig. 2.

979-8-3503-5124-8/24 $31.00 © 2024 IEEE

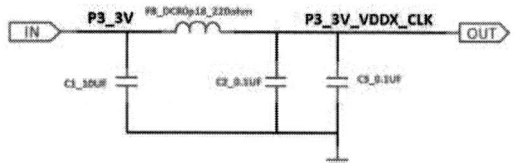

Fig. 1. Topology of the filtering circuit

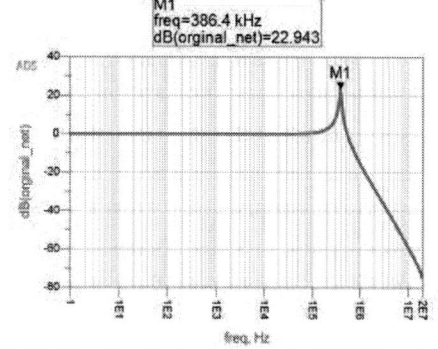

Fig. 2. Time domain simulation result of initial topology

III. THE RELATIONSHIP BETWEEN CAPACITANCE AND ANTIRESONCE

A. Research on the total capacitance in different combinations and antiresonance

To further study how antiresonance frequency changes when capacitors are combined in different ways. We take the initial topology in Fig. 1 as an example. We found that the 10uf capacitor in front of the ferrite bead does not affect the antiresonance frequency and peak value. Keeping the ferrite bead and other conditions to be constant, we only change the number and value of capacitors.

As Table 1 showed, we set C_{total} (the total capacitance of capacitors) = 0.2uf as initial status. It is composed of two 0.1uf capacitors, so the C_{num} (number of each capacitor) = 2. The simulated A_{freq} (antiresonance frequency) is 385.5KHz. Taking the initial state as a reference, we simulated 8 different cases with varying capacitances. Different cases are with different A_{freq}, so we get the parameter F_{ratio} (current antiresonance frequency ÷ initial antiresonance frequency).

TABLE I. The relationship between antiresonance frequency and capacitance

Case	C (uf)	C_{num}	C_{total} (uf)	A_{freq} (KHz)	F_{ratio}	C_{ratio}
Initial	0.1	2	0.2	385.50	\	\
No.1	0.1	3	0.3	315.50	0.818	1.5
No.2	0.47	1	0.47	256.50	0.665	2.35
No.3	0.22	3	0.66	210.90	0.547	3.3
No.4	1	1	1	205.10	0.532	5
No.5	2.2	1	2.2	134.30	0.348	11
No.6	3.3	1	3.3	112.50	0.292	16.5
No.7	4.7	1	4.7	88.72	0.230	23.5
No.8	47	2	94	12.42	0.032	470

From Table 1, we can see that the larger the capacitance, the lower the antiresonance frequency. In order to study the impact

of different capacitor combinations on the resonance frequency, we simulated various capacitor configurations. We had depicted the relationship between F_{ratio} and C_{total} in Fig. 3. We found that as long as the C_{total} remains constant, the antiresonance frequency is almost unchanged with different combinations of capacitors.

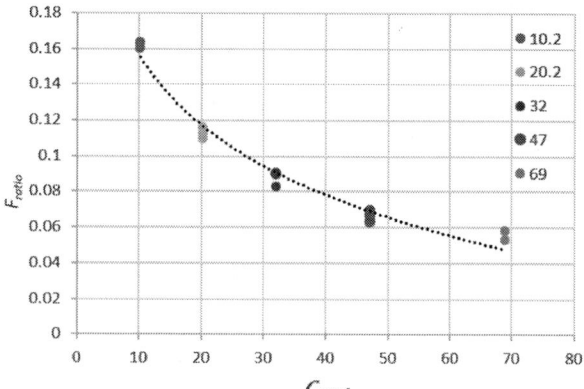

Fig. 3. The relationship between C_{total} and F_{ratio}

This allows filter simulations to concentrate on studying the ferrite bead and its connections. Moreover, for simplicity and efficiency in obtaining quick results, combinations of capacitors with equal total capacitance are preferable over complex combinations involving numerous smaller capacitance values. This finding streamlines the simulation process for those involved in time domain simulations, facilitating getting results.

B. Concrete study on the impact of increased capacitance to the antiresonance point

In circuit theory, the resonance point usually refers to the resonance frequency, with the formula [7],

$$A_{freq} = \frac{1}{2\pi\sqrt{LC}} \tag{1}$$

where L is the value of ferrite bead (unit: Henry), C is the value of capacitance (unit: Farad).

When we add a capacitor C_{add} in the initial equation,

$$A_{add} = \frac{1}{2\pi\sqrt{L(C+C_{add})}} \tag{2}$$

So,

$$\frac{C+C_{add}}{C} = \left(\frac{A_{add}}{A_{freq}}\right)^{-2} \tag{3}$$

From Eq. (3), we know $\frac{C+C_{add}}{C}$ and $\frac{A_{add}}{A_{freq}}$ have an exponential relationship during ideal circumstances. To delve deeper into the relationship between the F_{ratio} and C_{add}, we had derived a formula from the data in Table 1 and Fig. 3. Subsequently, we plotted the capacitance ratio, C_{ratio} (current C_{total} ÷ initial C_{total}) against F_{ratio} in Fig. 4.

It's evident from the plot that there exists a strong positive exponential relationship between the two variables. We got an exponential relationship, expressed as follows,

$$C_{ratio} = F_{ratio}^{-0.45} - 0.02 \tag{4}$$

979-8-3503-5124-8/24 $31.00 © 2024 IEEE

This relationship provides valuable insights into how adjustments in capacitance influence the antiresonance frequency, aiding in the optimization of circuit designs for enhanced performance. We verify the reliability of the formula by simultaneously increasing the total capacitance to 80 times as its initial value , employing three different ferrite beads with distinct initial total capacitance values.

For instance, in Fig. 1, we replaced the FB_DCR0p18_220ohm with a ferrite bead featuring 0.1ohm DCR and 220ohm impedance at 100MHz . Subsequently, we adjusted the initial C_{total} from 0.2uf to 0.1uf. We increased C_{total} by 80 times to 8uf, represented by blue star in Fig. 4, getting F_{ratio}=0.125(the initial A_{freq} is 616.6KHz, when the initial C_{total} is increased by 80 times to 8uf, the A_{freq} is changing to 76.92KHz, so F_{ratio} =0.125=76.92KHz/616.6KHz).

Similarly, we repeated this process for a ferrite bead with 0.15ohm DCR and 330ohm impedance at 100MHz, the initial C_{total} is still 0.2uf. For another ferrite bead with 470ohm impedance at 100 MHz and 0.25ohm DCR, the initial C_{total} is 1uf. When the C_{total} is increased by 80 times, getting F_{ratio} =0.128=30.2KHz/236KHz(illustrated by the green triangle in Fig. 4, A_{freq} is changing from 236KHz to 30.2KHz) and F_{ratio}=0.109 =8.92KHz/81.85KHz(illustrated by the red circle in Fig. 4, $A_{freq}$is changing from 81.85KHz to 8.92KHz), respectively.

Upon analysis, we observed that the F_{ratio} obtained by increasing the C_{total} by 80 times under the three different conditions had errors controlled within ±10%. This demonstrates the usability and generality of Eq. (4) in accurately predicting how increasing capacitance can lower the antiresonance frequency to the desired value during subsequent evaluations and debugging, obviating the need for simulations. This approach streamlines the design process and enhances efficiency.

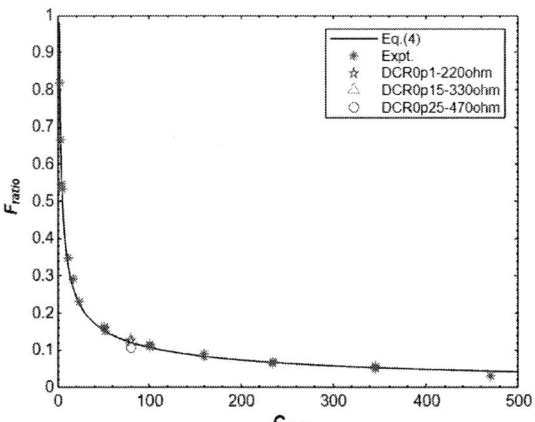

Fig. 4. The relationship between C_{ratio} and F_{ratio}

IV. Summary

In this study, we investigate the influence to antiresonance frequency after changing capacitance. Our findings reveal that this antiresonance frequency shows no correlation with the capacitor preceding the ferrite bead. Moreover, by keeping the ferrite bead constant and fixing the total capacitance, we observe fluctuations in the position and peak of the antiresonance point within ±10%. This discovery significantly simplifies the simulation process, particularly in circuits with numerous small capacitors, facilitating quicker acquisition of simulation results. Our investigation further reveals a consistent decrease in the antiresonance frequency as capacitances are added to the initial circuit. This prompted a deeper exploration to the relationship between the C_{ratio} and F_{ratio}, leading to the derivation of Eq. (4).

To validate the robustness of Eq. (4), experiments were conducted across three different circuit topologies. Notably, our results showed that increasing the C_{total} by 80 times resulted in F_{ratio} = 0.125, 0.128, and 0.109, respectively. These values exhibited errors within ±10% compared with those predicted by Eq. (4), affirming its reliability and applicability. Eq. (4) enables engineers to efficiently determine the necessary capacitance adjustments to achieve a desired antiresonance frequency. This helps to make informed decisions during circuit optimization and tuning process, enhancing overall design precision and performance.

Reference

[1] P. Larsson, "Resonance and Damping in CMOS Circuits with On-Chip Decoupling Capacitance," IEEE Trans. on Circuits and Systems-1 vol.45, no.8, pp.849-858, Aug. 1998.

[2] R. Kobayashi, et al., "Effects of Critically Damped Total PDN Impedance in Chip-Package-Board CoDesign," Proc. of IEEE EMC Symposium, Pittsburgh 2012.

[3] W. Kim, "Estimation of Simultaneous Switching Noise From Frequency-Domain Impedance Response of Resonant Power Distribution Networks," IEEE Trans. on CPMT, vol.1 no.9, pp. 1359-1367, Sept. 2011.

[4] C. R. Paul, "Effectiveness of multiple decoupling capacitors," IEEE transactions on electromagnetic compatibility, vol. 34, no. 2, pp. 130–133, 1992.

[5] Ichimura, W. , Kiyoshige, S. , Terasaki, M. , Kobayashi, R. , & Sudo, T. , "Anti-resonance peak frequency control by variable on-die capacitance," 2013 9th International Workshop on Electromagnetic Compatibility of Integrated Circuits (EMC Compo). IEEE, 2013.

[6] W. Ichimura et al., "Anti-resonance peak frequency control by variable on-die capacitance," 2013 9th International Workshop on Electromagnetic Compatibility of Integrated Circuits (EMC Compo), Nara, Japan, 2013, pp. 171-174, doi: 10.1109/EMC Compo. 2013. 6735195.

[7] Sundeep, S. , Wang, J. B. , Griffo, A. , & Alvarez-Gonzalez, F., "Anti-resonance Phenomenon and Peak Voltage Stress within PWM Inverter Fed Stator Winding," IEEE Transactions on Industrial Electronics, 2021, 68(12): 11826-11836.

Simulation method for Quasi-static solver to effectively model parasitic components between Package and PCB

Silvia Simone[1], Fabio Pareschi[1], Davide Lena[2], and Gianluca Setti[3]

[1]DET, Politecnico of Torino, corso Duca degli Abruzzi 24, 10129 Torino, Italy.
[2]STMicroelectronics s.r.l., Torino, Italy.
[3]CEMSE, King Abdullah University of Science and Technology (KAUST), Saudi Arabia.
Email: {silvia_simone, fabio.pareschi}@polito.it, davide.lena@st.com, gianluca.setti@kaust.edu.sa.

Abstract—A methodology to simulate Package and PCB with a quasi-static solver is developed. The widespread cascade method, where Package and PCB are first simulated standalone to reduce the required computational effort and then recombined, is improved to consider electromagnetic interactions between the two systems. The proposed approach provides results very similar to that achieved with the full system simulation, but with a computational cost that is very similar to that of the standard cascade method.

Index Terms—quasi-static solver, electromagnetic simulation, Q3D, Package, Printed Circuit Board (PCB), parasitic components, power integrity.

I. INTRODUCTION

Nowadays, considering the effect of parasitic components in the early design phase is fundamental for a thorough analysis that guarantees compliance with the electromagnetic compatibility standard regulation, but also to reduce the risk of iteration of the production process, ensuring that all constraints are met. In particular for power integrity analysis, knowing the values of parasitic elements in a very accurate way is essential for design considerations. For that reason, electromagnetic simulation of Printed Circuit board (PCB) and Package (PKG) of the device under test must be performed.

For design of converters and power electronic equipment a quasi-static tool is adequate when the structure is not electrically large, therefore when the dimension of the structure is approximately lower than $\lambda/10$, where λ is the signal wavelength, to precisely model the parasitic components for the co-simulation with the electrical circuit [1]. However, performing a complete simulation of PCB and PKG can be time and memory consuming. Especially if the system consists of components of different dimensions subdividing it into sections with similar size can be useful.

The current industry adopted method is to model PCB and PKG separately due to the limited computational resources [2]. Some investigations on the difference between considering the full model and the cascade of the separate objects are also considered [3], instead a technique to extract only the value of interconnections is developed [4]. For high frequencies applications, mainly for signal integrity analysis,

some techniques have also been studied [5]–[8]. However, all aforementioned methodologies involve full wave solver, or results are compared with a hybrid solver, cascading the different analyzed elements [9].

In this work the quasi-static solver of Ansys, Q3D, is used. It can be successfully employed when the frequencies involved and the dimensions of the device satisfy the explained solver validity condition. A typical application is a DC/DC converter with a switching frequency in the range of kHz-MHz, where lumped elements of parasitic components can be used for power integrity analysis. With the adopted tool, a methodology to reduce the computational effort of electromagnetic simulations is explained, obtaining more accurate values respect to simply cascading the two separately simulated elements, that is the easiest methodology to adopt but which neglects the electromagnetic contribution due to the interaction between the two systems. Instead, with the proposed solution the accuracy of the obtained results is comparable to the simulation of the full system, nevertheless maintaining a significant reduction in the employed resources. In Section II the method used for analyzing the PCB and PKG with a quasi-static solver is explained, and it is applied to a practical case showing the results in Section III. Finally, the conclusion is drawn.

II. ADOPTED METHODOLOGY

When considering a structure consisting of PCB and PKG, most accurate results are obtained with the simulation of the whole system (i.e., PCB + PKG) represented in Figure 1.

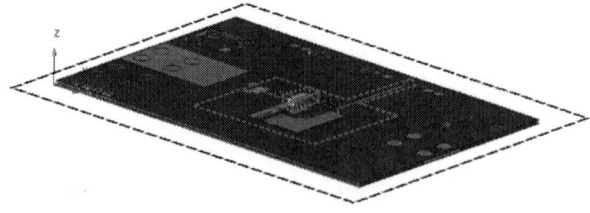

Fig. 1. Complete model of PCB highlighted in dark blue, signal traces in green and PKG in light blue, used for the simulation of whole system to obtain parasitic components considering the interaction between the structures.

979-8-3503-5124-8/24 $31.00 © 2024 IEEE

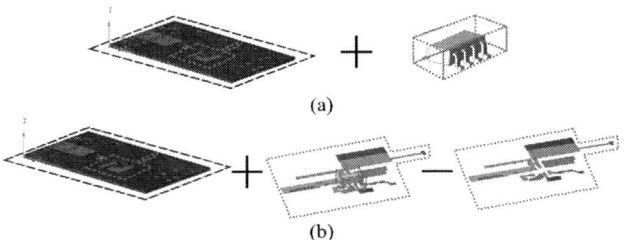

(a)

(b)

Fig. 2. Cascade method of T-parameters (a): simulation of PCB and simulation of PKG; proposed method (b): simulation of PCB, simulation of PKG+signal traces connected to it, simulation of signal traces only.

However, due to the limited computational resources, the most widespread technique consists in simulating the PCB and the PKG separately, as represented in Figure 2(a), and subsequently cascading the two matrices of S-parameters obtained from the electromagnetic simulations, conventionally using ABCD or T-parameters.

In this paper, we propose first to simulate the PCB alone, but instead of cascading results with that obtained from the simulation of the PKG alone, a simulation of the system composed of the PKG plus the PCB traces connected to it is performed and results are combined. Since in this way the PCB traces are present twice, to eliminate the contribution a simulation of only the traces is performed and subtracted from the previous results. This procedure is schematized in Figure 2(b).

To be able to correctly add and subtract the different contributions of the S-parameters obtained as a result of the individual electromagnetic simulations, the following considerations are taken into account. Q3D simulator defines *nets* that indicate the electrical connections of conductors [10], in each net a *sink* and multiple *sources* can be defined, they represent terminals connecting points where currents can go in or out. For each net the inductance and resistance between each source and the sink can be obtained converting the S-parameters into Z-matrix and closing the corresponding sink terminal to GND, as in Figure 3(a).

For each net, the direct values of parasitic inductance and resistance between a source and sink defined in that same net (self-inductance and self-resistance) are obtained considering the imaginary part $\text{Im}(\cdot)$ and real part $\text{Re}(\cdot)$ of the specific element of the Z-matrix of the modified system as:

$$L_{i,i} = \text{Im}\left(\frac{Z_{mod_{i,i}}}{2\pi f}\right), \quad R_{i,i} = \text{Re}(Z_{mod_{i,i}}) \quad (1)$$

whereas for the parasitic components defined between different sources on the same net (mutual-inductance and mutual-resistance) we have:

$$L_{i,j} = \text{Im}\left(\frac{Z_{mod_{i,j}}}{2\pi f}\right), \quad R_{i,j} = \text{Re}(Z_{mod_{i,j}}) \quad (2)$$

The S-parameters obtained for the three simulations in Figure 2(b) are converted into Z-parameters considering for each net the sink terminal connected to GND. The obtained parameters of the PCB are summed to that of the PKG+traces simulation, while the matrix obtained from the traces simulation is subtracted. With the final matrix obtained by using equations (1) and (2) it is possible to compute the corresponding parasitic values for each net.

Conversely, it is also possible to consider the parasitic resistances and inductances between different sources on different nets (mutual-inductance and mutual-resistance). In this case, the matrix is computed as represented in Figure 3(b). All the sinks of the nets are now connected to GND and, in the same way as before but starting from different matrices, the parameters obtained from the PCB simulation are added to the parameters of the PKG+traces simulation and subtracted to the traces one. In this case we are interested in the mutual parasitic resistance and inductance between terminals in different nets only, so that equations (2) need to be used.

Reducing the Z-matrix as explained allows to isolate only the quantities of interest, therefore the new matrix of each system can be directly added and subtracted, and subsequently the mathematical operations to derive L and R can be performed.

III. OBTAINED RESULTS

The explained method is applied to the simulation of a model of PCB and PKG with the quasi static-solver Q3D. Results are compared to the case of the simulation of the complete system which takes into account all the interactions between PCB and PKG, and to the most widespread method, i.e., the case of cascading PCB and PKG considering transmission parameters. Terms of comparison are the simulation time of each block according to the methodology adopted, and the accuracy computed by means of the maximum relative error between the resistance (or inductance) computed with the approximated method and with the full simulation, and averaged for the different considered cases. Data are collected in Table I. It is possible to observe that the simulation time of the proposed method is similar with that of the cascade one, and significantly reduced compared to the whole system simulation, decreasing the resource effort required by the electromagnetic simulation. Furthermore, with the proposed

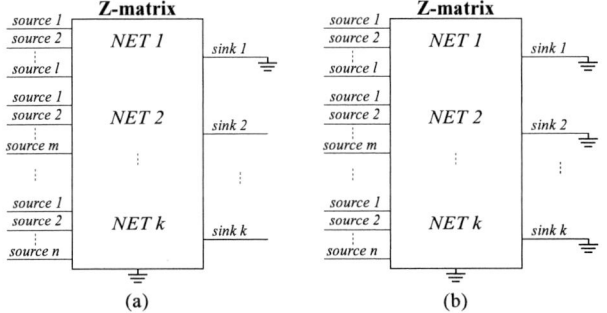

Fig. 3. Z-matrix converted from S-matrix obtained from electromagnetic simulation. (a): closing alternately the sink of each net to GND to compute the parasitic values of that net, (b): closing sink of every net to GND to compute the parasitic values between different nets.

979-8-3503-5124-8/24 $31.00 © 2024 IEEE

TABLE I
PERFORMANCE COMPARISON BETWEEN THE FULL SYSTEM SIMULATION,
THE STANDARD CASCADE APPROACH AND THE PROPOSED APPROACH.

Methodology	Accuracy	Simulation	Time
Full system	100%	PCB+PKG	32h5min
Cascade	74%	PCB	18h18min
		PKG	6min
Proposed	90%	PCB	18h18min
		PKG+Traces	14min
		Traces	10min

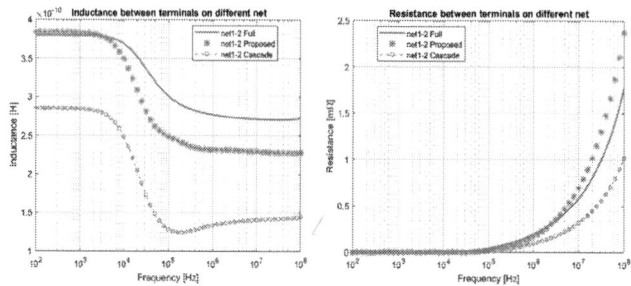

Fig. 5. Comparison between (mutual) parasitic inductance and (mutual) resistance between two sources on different nets (from source1 on net1 to source1 on net2) obtained starting from Z-matrix in Figure 3(b) using the three different methods explained.

ACKNOWLEDGMENT

This publication is part of the project PNRR-NGEU which has received funding from the MUR - DM 352 / 2022.

REFERENCES

[1] S. Simone, F. Pareschi, G. Setti, and D. Lena, "Equivalent skin effect model for time-domain analysis starting from electromagnetic simulator values," in *2024 25th International Conference on Thermal, Mechanical and Multi-Physics Simulation and Experiments in Microelectronics and Microsystems (EuroSimE)*, Apr. 2024, pp. 1–6, doi:10.1109/EuroSimE60745.2024.10491514.

[2] Z. Chen, "A general co-design approach to multi-level package modeling based on individual single-level package full-wave s-parameter modeling including signal and power/ground ports," in *2012 IEEE 62nd Electronic Components and Technology Conference*, May 2012, pp. 1687–1694, doi:10.1109/ECTC.2012.6249066.

[3] J. Zhang, A. C. Scogna, T. Vincent, H. Shi, and H. Liu, "System level simulation solutions for high-speed channels - package and PCB interface," in *2013 IEEE International Symposium on Electromagnetic Compatibility*, Aug. 2013, pp. 593–598, doi:10.1109/ISEMC.2013.6670481.

[4] Y.-C. Chang, S. S. H. Hsu, D.-C. Chang, J.-H. Lee, S.-G. Lin, and Y.-Z. Juang, "A de-embedding method for extracting s-parameters of vertical interconnect in advanced packaging," in *2011 IEEE 20th Conference on Electrical Performance of Electronic Packaging and Systems*, Oct. 2011, pp. 219–222, doi:10.1109/EPEPS.2011.6100231.

[5] L. Teoh, C. Chiang, A. C. Scogna, K. Khrone, and H. Lee, "De-embedding method and segmentation approach using full wave solver for high speed channels in PCB/package co-design," in *2014 IEEE Electrical Design of Advanced Packaging & Systems Symposium (EDAPS)*. IEEE, Dec. 2012, pp. 33–36, doi:10.1109/EDAPS.2014.7030808.

[6] H. Dsilva, A. Jain, S. J, and A. Kumar, "Effective segmentation approach of package-to-PCB modeling using full-wave EM field solver," in *2020 IEEE Electrical Design of Advanced Packaging and Systems (EDAPS)*, Dec. 2020, pp. 1–3, doi:10.1109/EDAPS50281.2020.9312900.

[7] A. C. Scogna, C. Chiang, L. Lau, L. Teoh, and H. Lee, "Modeling methodologies for multi level PCB-package co-simulation & co-design," in *2013 IEEE 22nd Conference on Electrical Performance of Electronic Packaging and Systems*, Oct. 2013, pp. 57–57, doi:10.1109/EPEPS.2013.6703466.

[8] Z. Chen, "A lumped/discrete port de-embedding method by port connection error-cancelling network in full-wave electromagnetic modeling of 3d integration and packaging with vertical interconnects," in *2013 IEEE 63rd Electronic Components and Technology Conference*, May 2013, pp. 1980–1987, doi:10.1109/ECTC.2013.6575850.

[9] A. C. Scogna, C. T. Chiang, K. Krohne, L. K. Teoh, and H.-Y. Lee, "Signal integrity analysis for high speed channels in pcb/package co-design interface: 3d full wave vs. 2d/hybrid approach & full model vs. segmentation approach," in *2013 IEEE 15th Electronics Packaging Technology Conference (EPTC 2013)*, Dec. 2013, pp. 585–588, doi:10.1109/EPTC.2013.6745787.

[10] "Q3d extractor help," 2024.

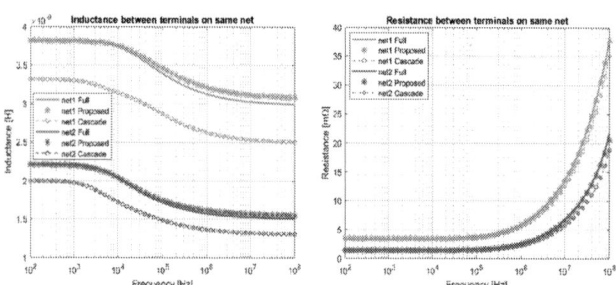

Fig. 4. Comparison between (self) parasitic inductance and (self) resistance on the same net (computed for source1 on net1 and for source1 on net2) starting from Z-matrix in Figure 3(a) using the three different methods explained.

approach, the parasitic effect between PCB and PKG is not completely neglected, as it happens in the case of the cascade approach, but it is possible to obtain an accuracy comparable to the full method. The values of parasitic inductances and resistances are computed for the case of the whole simulation, the cascade, and the proposed one starting from the electromagnetic simulation results. Parasitic inductance and resistance between sink and source in the same net, considering two nets, are represented in Figure 4. Parasitic inductance and resistance between sources on the two different nets previously considered are depicted in Figure 5. From the comparison it is possible to observe that with the proposed methodology the worst accuracy achieved at low frequencies is 98%, with the cascade method instead the value is 75%. Considering the entire frequency range the lowest accuracy for the presented method is about 70%, while for the other case is about 45%.

IV. CONCLUSION

A new methodology to model PCB and PKG in order to consider the interaction between the two elements in a quasi-static simulator is described, reducing the computational time by approximately half the time required for the full system simulation. Furthermore, better results are obtained compared to the most common method used that is the cascade of the two elements, where the interaction between PCB and PKG is neglected. The results obtained show comparable values respect to the complete simulation in the entire range of frequencies valid for the quasi-static simulator, and in particular completely overlapping at low frequencies, maintaining the computational cost of the cascade method.

979-8-3503-5124-8/24 $31.00 © 2024 IEEE

Analysis of Echo and Crosstalk Cancellation in Simultaneous Bidirectional Transceivers for Dense Die-to-Die Interconnects

Tong Liu, *Student Member, IEEE,* Taeyang Sim, and Samuel Palermo, *Senior Member, IEEE*
Analog and Mixed-Signal Center, Texas A&M University, College Station, TX, USA

Abstract—The impact of echo and crosstalk cancellation in die-to-die interconnect links that employ simultaneous bidirectional (SBD) signaling is analyzed in the context of a 24-wire transceiver operating over a four-layer interposer. Simulation results show that these techniques allow for a 4-rank transceiver system with $40\mu m$ bump pitch to occupy only $240\mu m$ die edge width and signal up to 128Gb/s per data channel wire over a very tight $4\mu m$ signal-to-signal pitch, resulting in a 46.9Tb/s/mm edge density. The proposed crosstalk cancellation circuitry improves the inbound signal voltage margin at 32GHz by 22% at a BER=10^{-12}.

Index Terms—Crosstalk cancellation, die-to-die interconnects, echo cancellation, simultaneous bidirectional (SBD) signaling, ultra-short-reach (USR).

I. INTRODUCTION

DIE-TO-DIE interconnects are in increasingly high demand in cloud computing and artificial intelligence applications. In order to support this, die-to-die interconnect roadmaps have increased edge density metrics [1]–[4]. For example, recent IEEE HIR and UCIe-A standards have edge bandwidth requirements of 4Tb/s/mm and 10.536Tb/s/mm with $40\mu m$ and $45\mu m$ bump pitch, respectively, based on 2D or 2.5D architectures [5], [6]. Further increases in edge density necessitates either increasing the data rate per wire or reducing the signal-to-signal pitch to accommodate more data lanes. As the data rate per wire increases, channel loss will be higher, necessitating that the system cancel more inter-symbol interference (ISI) caused by the insertion loss. Utilizing bidirectional signaling [7], [8] is a choice to scale data rate while keeping the same bandwidth. However, utilizing bidirectional signaling in die-to-die interface requires hybrid circuitry compatible with single-ended signaling system, and echo cancellation circuits to reduce reflections at the channel. As the signal-to-signal pitch of the data lanes decreases, a major issue that needs to be addressed is crosstalk. As a result, additional crosstalk cancellation mechanisms need to be considered when the signal pitch scales down.

This work presents an SBD die-to-die transceiver front-end capable of supporting data transmission rates up to 128 Gb/s/wire. The proposed transceiver is based on an inverter-based front-end design [7] with active high-pass cancellation circuits for echo and crosstalk cancellation [9]. The overview of the transceiver front-end architecture is described in Section II. Section III discusses the interposer channel design and the

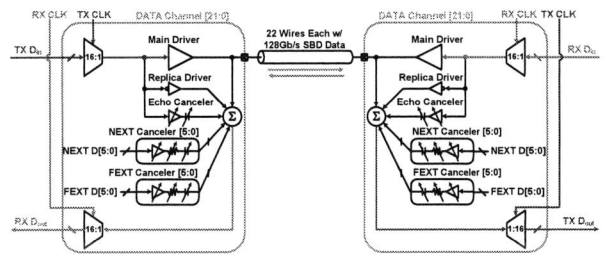

Fig. 1: SBD transceiver block diagram with echo and crosstalk cancellation.

front-end circuit model used for simulation. Simulation results are presented in Section IV. Finally, Section V concludes the paper.

II. TRANSCEIVER ARCHITECTURE

Fig. 1 shows the proposed die-to-die interconnect transceiver front-end architecture that consists of totally 22 single-ended data wires that transmit bidirectional data streams between dies. There will be 2 unidirectional forwarded-clock channels transmitting the TX and RX clock. On the TX side, multiple signals are added at the hybrid input to cancel the outbound signal, reflections and crosstalks of RX data. A replica driver with inverted TX input bits at the hybrid input cancels the outbound signal from the TX. An active high-pass path featuring a tunable driver and capacitors mitigates reflections between the TX driver and the channel. Additionally, two sets of six-segment high-pass filters are utilized for cancelling crosstalk. For NEXT cancellation, the input data comes from the aggressor TX side input, while for FEXT cancellation, the input data is from the aggressor TX side hybrid output, which is the aggressor RX side input data.

III. SBD SYSTEM MODEL

Fig. 2(a) shows bump map of current design with a $40\mu m$ bump pitch. These signals are routed through an interposer with one bump layer for redistribution layer (RDL) and four signal layers [10]. Operating the transceivers at 128Gb/s/wire in a four-layer interposer system allows for a 46.9Tb/s/mm edge density considering only SBD data wires. Fig. 2(b) shows the proposed layer stack. The signal-to-shield pitch is $2\mu m$ and

979-8-3503-5124-8/24 $31.00 © 2024 IEEE   157

signal-to-signal pitch is $4\mu m$. Table I shows the dimensions and material properties of the interposer.

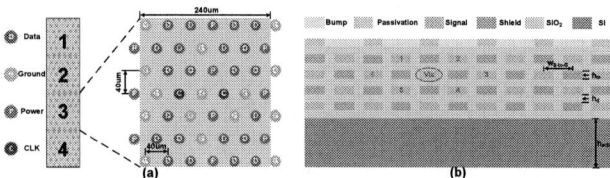

Fig. 2: (a) Proposed 4-rank SBD transceiver bump map. (b) 4-layer interposer layer stack.

TABLE I
DIMENSIONS AND MATERIAL PROPERTIES OF SILICON INTERPOSER CHANNEL

Parameters	Symbol	Value
Signal-to-Shield pitch	$w_{S\text{-}to\text{-}G}$	$2\mu m$
Thickness of metal	h_m	$1\mu m$
Thickness of SiO$_2$	h_d	$1\mu m$
Thickness of Si substrate	h_{sub}	$100\mu m$
Relative permittivity of SiO$_2$	ϵ_{SiO_2}	4.1
Loss tangent of SiO$_2$	$tan\delta_{SiO_2}$	0.001
Conductivity of Cu	σ_{Cu}	$5.8 \times 10^7 \sigma/m$
Conductivity of Si substrate	$\sigma_{Silicon}$	$20\ \sigma/m$

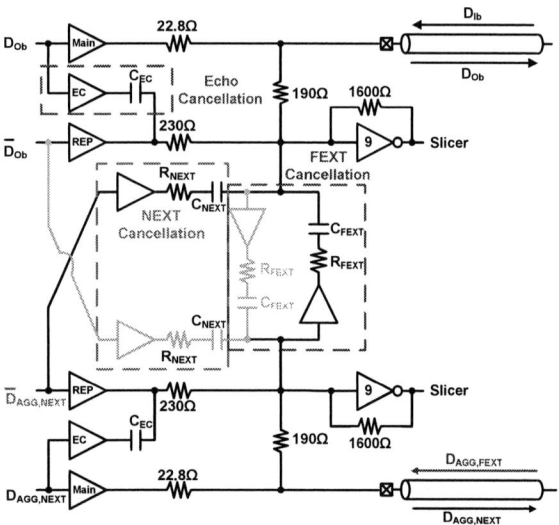

Fig. 3: Circuit model for SBD transceiver system simulation.

To determine the number of aggressors for cancellation, an EM simulation involving all 96 wires of the four-layer interposer was conducted to find which channels contribute most to the crosstalk. Based on the simulation results, it was found that six adjacent channels can cover the majority of the crosstalk, which are labeled in Fig. 2(b).

Fig. 3 shows the SBD transceiver front-end block diagrams for simulation. This model is a simplified version that considers only one victim and one aggressor, which have the same power-sum near-end crosstalk (PSNEXT) and far-end crosstalk (PSFEXT) as six aggressors to reduce the

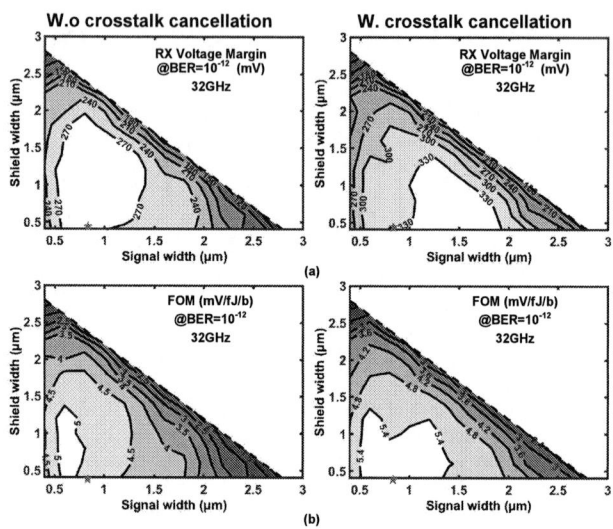

Fig. 4: 128Gb/s simulation results: (a) TIA output voltage margin at BER=10^{-12} and (b) FOM.

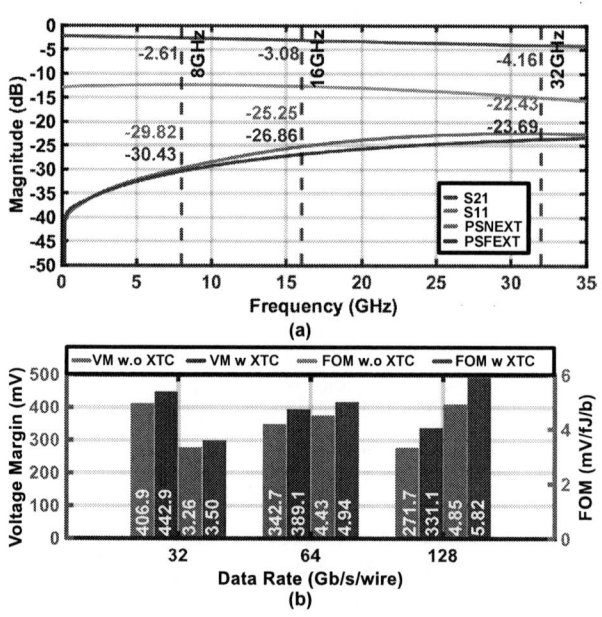

Fig. 5: (a) Selected channel S-parameters. (b) Simulated BER=10^{-12} voltage margin and FOM without and with crosstalk cancellation at different data rates.

complexity of adapting resistor and capacitor values. The main and replica drivers have equivalent resistance of 22.7Ω and 137.6Ω respectively for impedance matching.

IV. SIMULATION RESULTS

Fig. 4(a) shows the TIA output voltage margin at a BER of 10^{-12} both without and with crosstalk cancellation, for different channel signal and shield widths. In the statistical simulation here, a 2mV estimated TIA output noise is assumed. Fig. 4(b) shows the corresponding figure-of-merit

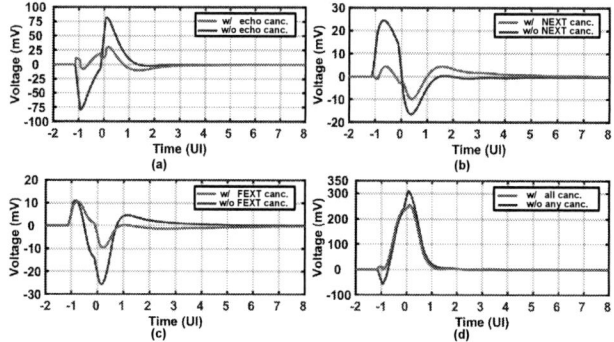

Fig. 6: Simulated TIA output: (a) Outbound-signal pulse response with and without echo cancellation. (b) PSNEXT aggressor pulse response with and without NEXT cancellation. (c) PSFEXT aggressor pulse response with and without FEXT cancellation. (d) Total pulse response with and without both echo and crosstalk cancellation.

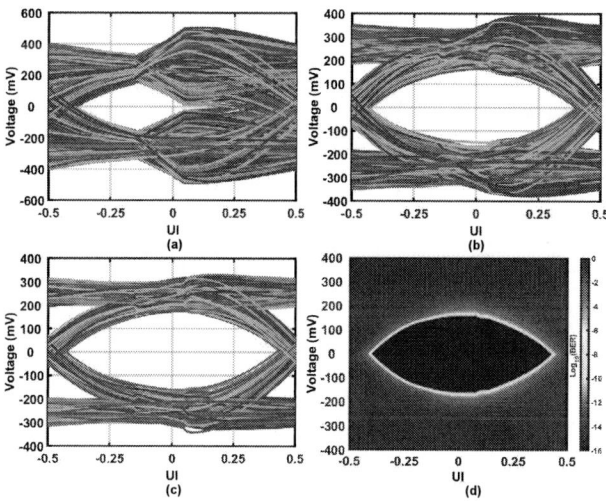

Fig. 7: Transient-simulated 64Gb/s (128Gb/s SBD) TIA output eye diagrams (a) without echo and crosstalk cancellation, (b) with only echo cancellation, and (c) with both echo and crosstalk cancellation. (d) Statistical 64Gb/s eye diagram with echo and crosstalk cancellation.

(FOM), which is defined by the ratio of the voltage margin to the line energy. The line energy here includes main and replica drivers, echo and crosstalk cancellation circuits. Based on the results, a channel of $0.8\mu m$ signal width and $0.4\mu m$ shield width is selected considering the margin and FOM, it also gives 40Ω characteristic impedance at 32GHz. Fig. 5(a) shows the s-parameters, PSNEXT, and PSFEXT responses of the selected channel. At 32GHz, realative to considering the total 95 aggressors, a six aggressor approximation is within 0.5dB and 2.2dB for PSNEXT and PSFEXT, respectively.

Fig. 5(b) shows the voltage margin and FOM at various SBD data rates, with the relative improvement in voltage margin and FOM with crosstalk cancellation increasing at higher data rates due to higher insertion loss and crosstalk.

The voltage margin improves from 271.7mV to 331.1mV at the highest 128Gb/s/wire. While including the crosstalk cancellation circuitry does have a slight power penalty, the system is overall more efficient with the FOM increasing from 4.85mV/fJ/b to 5.82mV/fJ/b. It is anticipated that the area penalty will be minor, as the crosstalk cancellation incorporates both inter-channel routing parasitics as a baseline resistance value and additional tunable resistor and capacitor elements.

Fig. 6 shows the TIA output pulse responses at 64Gb/s unidirectional operation, with and without the corresponding cancellation circuitry activated and the time UI=0 being the nominal sampling time. Fig. 6(a)-(c) are derived by applying input pulse signals only at D_{Ob}, $D_{AGG,NEXT}$, and $D_{AGG,FEXT}$ respectively, while connecting the other data input nodes to 0.5VDD. Estimated inter-channel wiring parasitics, along with tunable RC values, are considered in the crosstalk cancellation settings. Fig. 6(d) is derived by applying input pulse signals at D_{Ob}, $D_{AGG,NEXT}$, $D_{AGG,FEXT}$, and D_{Ib} to show the total received inbound signal considering outbound non-idealities with and without cancellation circuitry activated. From the simulation results, the maximum echo pulse amplitude suppression is 75mV, while NEXT and FEXT pulses are decreased by 20mV and 15mV, respectively.

Fig. 7(a)-(c) show transient eye diagrams with different cancellation configurations. The input signal at each side is an 800mV 64Gb/s PRBS31 pattern with different seeds. The eye height with crosstalk cancellation activated is improved from 311mV to 333mV. Fig. 7(d) shows the corresponding statistical eye under the same noise estimation as before, with the shape matching the transient simulation results.

V. CONCLUSION

This paper presents a dense die-to-die SBD transceiver front-end architecture with echo cancellation that operates at a max data rates of 128Gb/s/wire and achieves a 46.9Tb/s/mm edge density. Additional high-pass filter-based crosstalk cancellation allows for a tight $4\mu m$ signal-to-signal pitch and a 22% improvement in inbound signal voltage margin with low power overhead.

ACKNOWLEDGMENT

This work was supported by the Semiconductor Research Corporation (SRC) Texas Analog Center of Excellence (TxACE) under Grant 3160.045.

REFERENCES

[1] B. Dehlaghi et al. JSSC, vol. 51, no. 11, pp. 2690–2701, 2016.
[2] M.-S. Lin et al. JSSC, vol. 55, no. 4, pp. 956–966, 2020.
[3] Y.-Y. Hsu et al. Symposium on VLSI Technology, pp. 1–2, 2021.
[4] K. Seong et al. ISSCC, pp. 114–116, 2023.
[5] "Heterogeneous integration roadmap." [Online]. Available: https://eps.ieee.org/images/files/HIR_2024.
[6] "Universal chiplet interconnect express." [Online]. Available: https://www.uciexpress.org/specifications.
[7] Y. Nishi et al. JSSC, vol. 58, no. 4, pp. 1062–1073, 2023.
[8] Y.-H. Fan et al. JSSC, vol. 55, no. 2, pp. 439–451, 2020.
[9] B. Min et al. Analog Integrated Circuits and Signal Processing, vol. 88, pp. 233–243, 2016.
[10] K. Cho et al. Transactions on CPMT, vol. 8, no. 9, pp. 1658–1671, 2018.

Transformer Based Channel Identification

Priyank Kashyap*, Yeujiang Wen*†, Yongjin Choi*, Chris Cheng*, Paul D. Franzon†

*Hewlett Packard Enterprise, Storage Division, USA

†ECE Dept., North Carolina State University, Raleigh, NC, USA

Email*:{priyank.kashyap, yongjin.choi, chris.cheng}@hpe.com

Email†:{wyuejia, paulf}@ncsu.edu

Abstract—Once an electrical system completes assembly, high-speed channel characterization is complex as access points are limited, and probing by a network analyzer is not a possibility. Further active channels may now have different types of I/O from NRZ to PAM4 levels, adding additional complications. This work describes a method for predicting a channel's behavior and implicitly learning the underlying system parameters in two stages. The first stage uses an autoregressive transformer structure to predict the channel's behavior as a single-bit response (SBR). The second stage uses the model's latent space and unsupervised dimensionality reduction to determine the underlying system parameters. The results show that the model can predict the SBR for completely unseen channels for both NRZ and PAM-4 despite the model training using only NRZ waveforms. Further, unsupervised dimensionality reduction enables the determination of system parameters, specifically the length of the overall transmission line.

Index Terms—Channel-behavior, High-speed link, Transformer

I. INTRODUCTION

High-speed channels enable data transmission from a transmitter to a receiver. However, these channels contain many components, such as vias, transmission lines, and connectors, complicating their design. Additionally, once a channel is fully assembled, it is difficult to probe the link to ensure it matches the designed system. There are limited probe points, with most probing limited to the channel's input and outputs. Compounding this is that the channel's performance is a significant bottleneck in the overall performance of the end-to-end link. Therefore, ensuring that the assembled system matches the intended one is critical. To that end, this problem resembles one where one determines the channel parameters given some desired performance or the inverse problem.

With the advancements in machine learning, there has been significant progress in solving the inverse problem for channel design. Trinchero et al. [1] use LS-SVM to determine 4 trace properties for 8 desired sets of eye characteristics. Roy et al. [2] use a lifelong approach to solve the inverse problem using a deep neural network (DNN). Both works aim to determine the underlying channel's geometric properties based on its impact on an eye-opening at the receiver's end, limiting their scope to only solving the inverse problem.

A different class of methods looks at generating both forward and inverse solutions using a single model. Ambasana et al. [3] propose using invertible neural networks (INNs), a class of networks that can jointly train on the forward and inverse problem, to predict the channel and receiver design

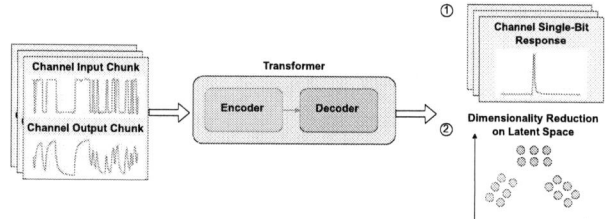

Fig. 1: Overview of the proposed approach. ① The trained transformer model predicts the channel's single-bit response. ② Using the latent space of the encoder from the transformer and dimensionality reduction techniques, the approach identifies the relevant system parameters.

parameters. Ma et al. [4] use two networks back-to-back, where the first network is trainable and performs the inverse task, and the second is a pre-trained network for the forward problem. These works improve the model's utility as it serves both the forward and inverse problems. Notably, in both of these methods, the models attempt to recover the inverse solution by using a latent space directly. The training of an INN relies on a latent space to generate the inverse solution. In contrast, the two-network solution is an auto-encoder where the latent variables are the inverse solution.

With that in mind, in this paper, we aim to determine the inverse solution implicitly. The paper proposes an autoregressive transformer to predict the channel's behavior as a single-bit response (SBR) using a PRBS sequence at the channel's input and output. The approach then leverages the trained latent space between the encoder and decoder within the transformer and applies unsupervised dimensionality reduction on the training set. Then, during inference, the dimensionality reduction maps the test sample latent space to the lower dimensional manifold to determine the underlying system parameters. Such an approach has multiple advantages, primarily because the model's output is interpretable based on the underlying physical parameters. Further, the results show that though the model trains on one modulation scheme, it can predict a channel's SBR for a different one, namely, train on NRZ and predict PAM-4.

II. PROPOSED APPROACH

Fig. 1 shows the proposed approach, whereby the transformer learns to predict a channel's SBR. Then, we can determine different channel parameters using the dimensionality

979-8-3503-5124-8/24 $31.00 © 2024 IEEE

reduction on the latent space. The rest of the section describes the model architecture for predicting the channel's SBR and discusses uniform manifold approximation and projection (UMAP), the dimensionality reduction technique we employ.

A. Transformer

Transformer models are the backbone of recent advances in natural language processing and have been adopted in different domains in SI/PI. We make certain assumptions to leverage transformers to model a high-speed channel. The work assumes that a finite number of distinct voltages is possible due to the scope resolution when measuring waveforms. Equipped with this understanding, we treat each voltage level as a distinct word in a dictionary, enabling their direct use with transformer models.

The transformer used in this work is as proposed by Vaswani et al. [5], with modifications that allow the model to work with multivariate time series. The model contains positional encoding, which enables the model to learn the temporal dependency between different time steps, thereby predicting a single-bit response with post-cursor intersymbol interference (ISI). Further, the models use multi-headed attention, which enables them to focus on different parts of the input sequence to determine the appropriate SBR. The model trains using teacher forcing, with a causal mask on the SBR, which enables it to predict the voltage at the next time step.

B. Unsupervised Dimensionality Reduction

As the model architecture comprises an encoder-decoder structure, we hypothesize that the latent space between the two models can distinguish the underlying physical parameters. The latent space for a single input tends to contain many values, necessitating dimensionality reduction techniques. Though principal component analysis (PCA) would enable dimensionality reduction, given that the principal components are linear,it is challenging to use for a feature-rich space. Further, t-distributed stochastic neighbor embedding (t-SNE) is non-deterministic; thus, one has to fit a new reduced space whenever new data is presented.

An alternative is UMAP, which uses manifold learning to reduce dimensionality. Like t-SNE, UMAP is also nonlinear, but it has the advantage of being reusable by mapping unseen data to an existing manifold [6]. Such a technique gives insight into the transformer model's prediction abilities by tying the latent space to known physical parameters.

III. DATA COLLECTION AND PREPROCESSING

This section describes the data collection used to train the transformer models.

The data collection considers a wide range of differential channels traversing a motherboard, backplane, and back to the motherboard, as shown in Fig. 2. The motherboard is an 8-layer PCB with an FR4 dielectric constant of 2.85, whereas the backplane is a 13-layer PCB with an FR4 dielectric constant of 2.82. For each stackup, we design a segment, as shown in Fig. 2, with two microstrips, a stripline, and vias to connect

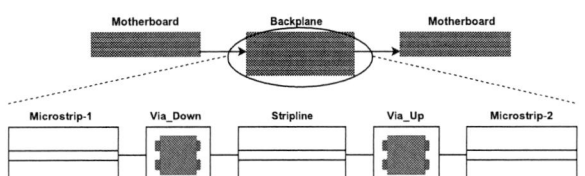

Fig. 2: End-to-end channel along the different stackup. Zoomed in view of the fixed segment that contains two microstrips, two vias and a stripline. Each stack up contains a segment for a total of 5 physical parameters.

TABLE I: Design parameters for each segment

	Motherboard (mils)	Backplane (mils)
Via Stub	[4, 20, 30]	[1.5, 20, 40]
Stripline	[1000, 10000]	[1000, 10000]
Microstrip-1	1000	2000
Microstrip-2	1000	2000

the two transmission lines. To create different channels, we select a motherboard via (for both segments), a backplane via, and the lengths of the three striplines. TABLE I shows the parameter ranges for each of the structures discussed previously. In addition to the physical channel parameters, we vary the edge rate and bit period. We use Latin hypercube over these 7 parameters to generate 300 and 100 S-parameters as training and testing channels. We use HSpice to convert the S-parameters to .rfm format to ensure the link is causal and passive. With the channel as a .rfm model, we first simulate the link using HSpice and collect waveforms on both ends for a PRBS-15 bitstream. Then, we collect the SBR for each channel. HSpice collects each waveform at a resolution of 1 ps for 1000 ns.

After collecting the time series at both ends of the link, the preprocessing first aligns the waveforms at both ends of the channels, eliminating any delay. After alignment, the preprocessing downsamples the channel input-output pair at the same scale of 10 ps interval. After downsampling, the preprocessing digitizes the channel input and output separately, representing the two domains. Similarly, the preprocessing downsamples at 10 ps resolution and digitizes the channel's SBR. For the results shown in the following section, the post-digitization gives us 340 levels for the channel's input and output waveforms, whereas the SBR has 256 levels.

IV. EXPERIMENTAL RESULTS

The transformer model was trained on a GeForce RTX 3090-TI GPU for 100 iterations, with a batch size of 32. The training uses an Adam optimizer with a custom-scheduled learning rate. Further, the model training uses early stopping on the validation set to prevent overfitting. If the validation loss does not decrease for 10 training iterations, the training stops, and the best validation loss model is loaded. Each training iteration takes 22 seconds, and the model converges within 25 training iterations.

979-8-3503-5124-8/24 $31.00 © 2024 IEEE

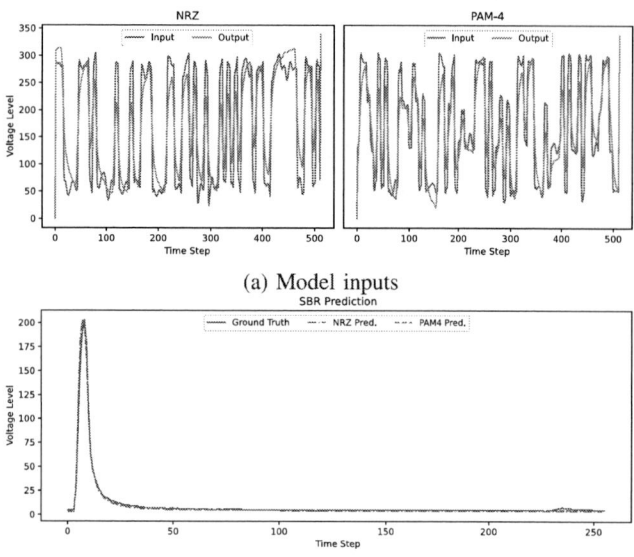

(a) Model inputs

(b) SBR comparison

Fig. 3: Test case for a channel using NRZ and PAM-4 modulation. The generated SBR and the true SBR have minor errors and minor shift the results.

As the model predicts the output autoregressively for any channel input-output pair, the model makes 256 predictions for the pair to get the complete SBR. The entire process of generating the 256-time step SBR takes 27 seconds, whereas the main cursor generation, which has 50-time steps, takes 5 seconds.

Fig. 3a shows the inputs to the model for an unseen channel for both NRZ and PAM-4 modulation. The ends of each waveform are specialized tokens reserved for <START> and <END>, which enable the model's input to be up to 512 time steps. In the case of the results shown here, the channel waveforms are 512 time steps. Fig. 3b shows the results for the same channel with both modulations. It is evident from the figure that the generated SBR is accurate and shows a high correlation with the simulated ground truth, regardless of the modulation. Further, minor errors are present in the waveform towards the end of the SBR or approximately 60 UIs after the cursor. The root-mean-squared error (RMSE) for the SBRs in Fig. 3b are 2.55 and 1.78 for NRZ and PAM-4, respectively. Across the test, the RMSE for NRZ and PAM-4 are 3.14 and 3.08, respectively.

We then look at the UMAP space for the latent variables that the encoder in the transformer generates. Fig. 4 shows a 2-component UMAP on the latent space, with each channel input-output waveform chunk as a single point in the plot. We then visualize each sample with the color indicating the channel length for the training set. Then, to determine the length of an unknown channel's PRBS chunk, we map its latent space into UMAP space. Fig. 4 shows these test set samples with a red outline. As is evident from Fig. 4, the latent space is continuous, with the channel length increasing as one goes along the UMAP-1 dimension. Further, for the test cases, the channel's length roughly corresponds to the correct

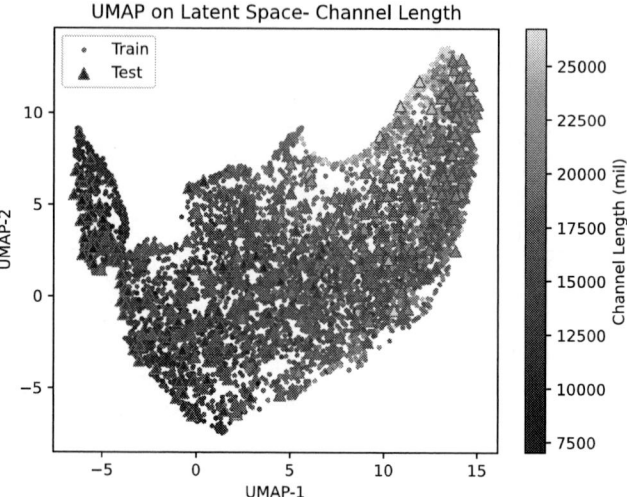

Fig. 4: UMAP results on the latent space with the color of the sample indicating the length of the channel. The outlined samples show the test samples overlapping training samples with similar lengths.

training set length, enabling us to identify it. We visualize the total channel length as we found it to be the dominant parameter for the overall loss; however, one can change it to other design parameters.

V. CONCLUSION

This paper presents a method to determine the channel's behavior using transformers. The model represents voltages as words and entire waveforms as sentences, thereby leveraging transformer models' ability to learn temporal information. The proposed method can recover the channel's SBR with an RMSE of 3.14. Further, the model performs equally well on PAM-4 waveforms despite not training on them. The paper then performs dimensionality reduction on the model's latent space using UMAP to create a nonlinear mapping to a lower-dimensional manifold to accurately identify the channel parameters, such as the channel length.

REFERENCES

[1] R. Trinchero, M. A. Dolatsara, K. Roy, et al., "Design of high-speed links via a machine learning surrogate model for the inverse problem," in *Electrical Design of Advanced Packaging and Systems (EDAPS)*, 2019, pp. 1–3.

[2] K. Roy, M. A. Dolatsara, H. M. Torun, et al., "Inverse design of transmission lines with deep learning," in *IEEE 28th Conference on Electrical Performance of Electronic Packaging and Systems (EPEPS)*, 2019, pp. 1–3.

[3] N. Ambasana, O. W. Bhatti, M. A. Dolatsara, et al., "Invertible neural networks for high-speed channel design & parameter distribution estimation," in *2021 IEEE 30th Conference on Electrical Performance of Electronic Packaging and Systems (EPEPS)*, 2021, pp. 1–3.

[4] H. Ma, E.-P. Li, Y. Wang, et al., "Channel inverse design using tandem neural network," in *IEEE 26th Workshop on Signal and Power Integrity (SPI)*, 2022, pp. 1–3.

[5] A. Vaswani, N. Shazeer, N. Parmar, et al., "Attention is all you need," *Advances in Neural Information Processing Systems*, vol. 30, 2017.

[6] L. McInnes, J. Healy, and J. Melville, "UMAP: Uniform manifold approximation and projection for dimension reduction," *arXiv preprint arXiv:1802.03426*, 2018.

979-8-3503-5124-8/24 $31.00 © 2024 IEEE

Design and Analysis of Ultra High Bandwidth (UHB) Interconnection-based GPU-Ring for the AI Superchip Module

Jungmin Ahn*, Seonguk Choi*, Taein Shin*, Junghyun Lee*, Jiwon Yoon*, Keunwoo Kim*, Keeyoung Son*, Haeseok Suh*, Taesoo Kim*, Hyunah Park*, Hyunjun An*, Jinwook Song**, and Joungho Kim*

*School of Electrical Engineering, Korea Advanced Institute of Science and Technology (KAIST), Daejeon, Republic of Korea
**Solution development Team, Samsung Electronics, Hwaseong, Republic of Korea
Email: jungminahn@kaist.ac.kr

Abstract— In this paper, we propose an ultra-high bandwidth (UHB) graphic processing unit (GPU)-ring structure for next-generation superchip modules. The interconnection via switches in multi-GPU system involves high latency and power consumption, which undermines the performance of artificial intelligence (AI). To address this issue, UHB GPU-ring architecture connects 4 GPU-high bandwidth memory (HBM) modules with short interconnection and high data rate on the same computing node. Proposed UHB GPU-ring architecture integrates the GPU-HBM modules directly, which enables to function 4 GPU-HBM modules as 1 module, decreasing the non-uniform memory access (NUMA) effect. To ensure the signal integrity of the proposed UHB GPU-ring, we designed the hardware system with the continuous time linear equalizer (CTLE) and decision feedback equalizer (DFE) at the receiver referring to Peripheral Component Interconnect Express (PCIe) 5.0 system. As a result, we verified 32TB/s bandwidth and 0.6 pJ/bit between GPU-HBM modules, which enhances bandwidth by 17.7 times and energy consumption by 2.5 times. The result showed that the UHB GPU-ring can be a promising solution for the next generation superchip module.

Keywords— Artificial intelligence, equalizer, graphic processing unit, high bandwidth memory, high speed interconnection

I. INTRODUCTION

Demand for high performance computing has been surged because of the remarkable improvement in artificial intelligence (AI), especially large language model (LLM). LLMs require extensive parallel computation and substantial data storage to function effectively. This trend is rapidly advancing the development of systems built with GPUs and high bandwidth memory (HBM). However, a single GPU-HBM module is insufficient to meet the computational and memory demands of these complex models, necessitating the use of multi-GPU systems. Therefore, demand for multi-GPU system-based supercomputing architecture has been increased. To efficiently implement multi-GPU configurations, various interconnection topologies, like NVswitch and NVLink are being developed [1].

However, multi-GPU system induces non-uniform memory access (NUMA) effect, which is timing difference between local and remote memory access [2]. This can lead to uneven memory distribution, where some GPUs might have more local memory access while others have to rely more on remote memory. This imbalance can be the bottleneck of LLM computation, reducing overall system performance. To overcome this problem, multi-

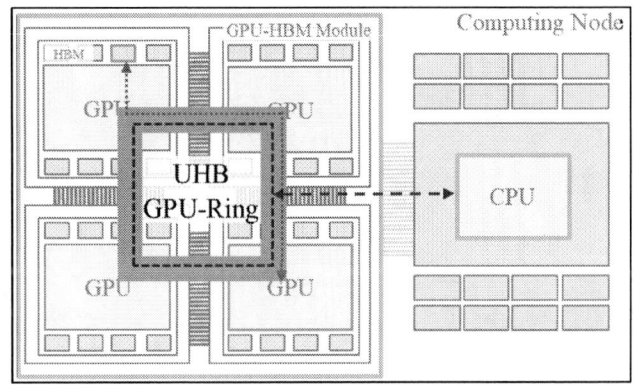

Fig. 1. Proposed UHB GPU-ring architecture connects 4 GPU-HBM modules on the same computing node. The GPU and HBM located in different modules communicate via the GPU-ring denoted by the red line. The GPU-ring enables the CPU to treat four modules as a single module denoted by the blue line.

GPU system should minimize the latency between the GPU-HBM modules [2]. Therefore, various methods are being explored to optimize interconnection topologies. Currently, each GPU-HBM modules are connected individually via switches [3]. Bandwidth of GPUs interconnection has been increased with the progression of generations, enhancing the data transfer rates and interconnection methodology. Nonetheless, the bandwidth of GPU-links still falls short of meeting the demands of LLMs with large parameter size.

In this paper, we propose an ultra-high bandwidth (UHB) interconnect for GPU-ring architecture for AI superchip module. By connecting 4 GPU-HBM modules with the proposed UHB, multi-GPU operates as a high-performance single GPU-HBM module as shown in Fig. 1. To verify the hardware system of the proposed architecture, we conducted channel simulation considering the channel and equalizer characteristics. To ensure signal integrity (SI) for the target bandwidth, continuous time linear equalizer (CTLE) and decision feedback equalizer (DFE) are utilized at the receiver. Based on the results, we demonstrate the UHB interconnection enhance the bandwidth between GPU-HBM modules and expect to decrease the NUMA effect.

II. PROPOSED UHB GPU-RING ARCHITECTURE

Proposed system has 4 GPU-HBM modules on the same computing node and each GPU-HBM module is interconnected with the proposed ring architecture. Proposed architecture

979-8-3503-5124-8/24 $31.00 © 2024 IEEE

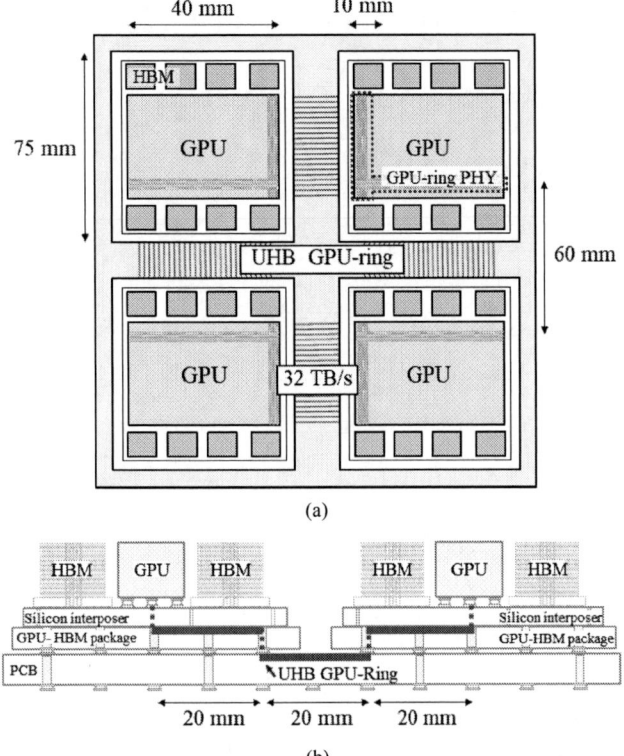

(a)

(b)

Fig. 2. (a) Floorplan of UHB GPU-ring for multi-GPU computing is described. (b) The full system channel of the proposed architecture is presented with the cross-sectional view.

connects the modules through the package and PCB channels. This interconnection scheme allows the 4 modules to function as a single unit. 4 GPU-HBM modules are connected to a computing processing unit (CPU), allowing the CPU to operate by treating these 4 modules as a single superchip module. GPU-HBM modules are connected within a single node, resulting in short interconnection length and low energy consumption.

To minimize the NUMA effect by hardware design solution, the bandwidth between modules must meet at least the sum of memory bandwidth of the 4 GPU-HBM modules. Based on HBM3, each HBM has a bandwidth of 819 GB/s, thus each GPU-HBM module has 6.5 TB/s memory bandwidth [4]. Assuming each HBM has a bandwidth of 1 TB/s, the system is designed to achieve 32 TB/s (32 HBM for the proposed architecture) bandwidth. To achieve the target bandwidth and decrease the NUMA effect, signal should be transferred at 128 Gb/s with the proposed channel structure.

Full system configuration and physical dimension of the proposed system is illustrated in Fig. 2. All physical dimensions are based on modern GPU and HBM packages [4]. The total length and transition to connect the GPU-HBM module is determined based on the designed dimensions shown in Fig. 2. To connect the GPU-HBM modules, a minimum of 60 mm is required considering the physical dimension shown in Fig. 2 (a). In Fig. 2 (b), the dimension of package and PCB channel and

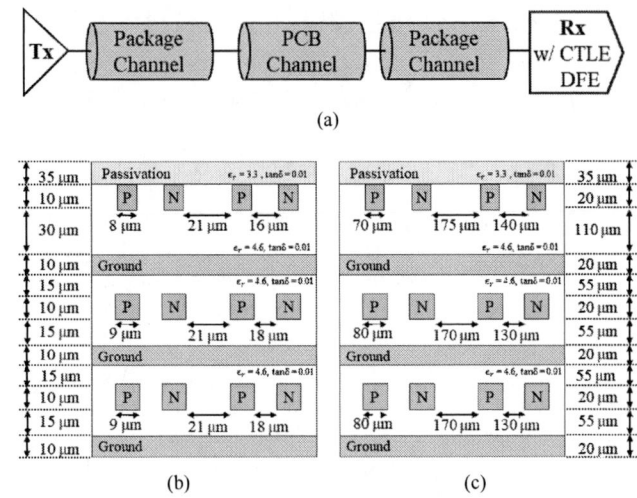

(a)

(b)   (c)

Fig. 3. (a) Full system of proposed UHB GPU-ring architecture. The stack-up and physical dimension of designed (b) package channel and (c) PCB channel with differential signal. (Proposed architecture has 1-layer micro-strip channel and 4-layer strip channel. For the simplicity, only 3 layers are shown.)

the transition of the channels are shown. Full channel system and dimension for each channel is shown in Fig. 3. Both package and PCB channels are designed with the differential channel, differential channel has higher resilience to crosstalk than the single-ended channel. Dimension of the proposed channel structure and the characteristic of the stack-up is shown in Fig. 3 (b) and (c). The channel design and analysis are utilized by the electromagnetic (EM) simulation and advanced design system (ADS). Considering that the interconnection length between GPUs in the conventional supercomputer rack ranges from 1m to 3m, this structure is expected to be more energy efficient with lower latency [1].

III. ANALYSIS OF UHB-GPU-RING ARCHITECTURE

To verify the proposed UHB interconnection between GPU-HBM modules, an analysis is performed through package and PCB channels. The effect of the interposer channel is negligible and therefor omitted in this paper. First, channel design for the proposed architecture considering the characteristic of the interconnection and physical dimension is conducted. Second, to achieve the targeted bandwidth (32 TB/s) through the designed channel, simulations are conducted to verify the proposed channel architecture at hardware-level for the high-performance computing.

In this paper, we analyze the micro-strip channel with differential signaling scheme for the proposed channel. Because, for the proposed stack-ups, micro-strip line has higher insertion loss than the strip-line. As shown in Fig. 4, differential channel insertion loss at 64 GHz, which is the Nyquist frequency for the 128 Gb/s, is -18.654 dB. As the proposed channel has larger attenuation level than -15 dB, which needs equalization to guarantee the SI, we utilized CTLE and DFE [5].

The simulation set-up and characteristics of the channels are shown in Table 1. CTLE is configured with 2 zeros and 3 poles and DFE is configured with 3 taps. Parameter values for each

979-8-3503-5124-8/24 $31.00 © 2024 IEEE   164

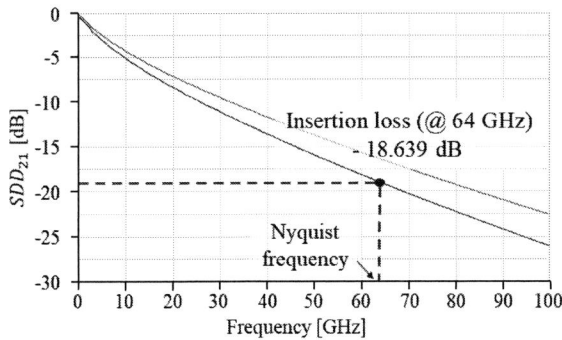

Fig 4. Differential insertion loss of strip line(green) and micro-strip line(gray) is shown.

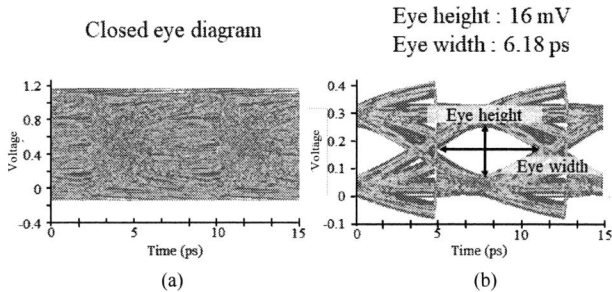

Fig 5. Eye diagram results of 128 Gb/s signal for the proposed architecture is presented. (a) Eye diagram results without CTLE and DFE, (b) with proposed CTLE and DFE design is shown.

TABLE I.   DESIGN PARAMETER AND CHARACTERISTIC OF THE PROPOSED UHB GPU-RING ARCHITECTURE

	Package channel	PCB channel
Channel Length	20 mm	20 mm
Number of channels	200	200
Differential insertion loss	0.382 dB/mm	0.17 dB/mm
Energy efficiency	0.6 pJ/bit	
Data rate	128 Gb/s	
Nyquist frequency	64 GHz	

TABLE II.   CTLE AND DFE DESIGN PARAMETER OF THE UHB GPU-RING ARCHITECTURE

CTLE parameter	Parameter value	CTLE parameter	Parameter value	DFE parameter	Parameter value
Zero 1	8 GHz	Pole 2	40 GHz	c_1	-0.043
Zero 2	20 GHz	Pole 3	55 GHz	c_2	0.001
Pole 1	15 GHz	DC gain	5 dB	c_3	0.016

$$H(s) = \frac{A_{DC}\omega_{p1}\omega_{p2}\omega_{p3}}{\omega_{z1}\omega_{z2}} \frac{(s+\omega_{z1})(s+\omega_{z2})}{(s+\omega_{p1})(s+\omega_{p2})(s+\omega_{p3})} \quad (1)$$

$$y[n] = x[n] + c_1 \cdot y[n-1] - c_2 \cdot y[n-2] + c_3 \cdot y[n-3] \quad (2)$$

CTLE and DFE are listed in Table 2. The transfer function, which incorporates the zero and pole frequencies, is obtained through Eq. (1). All zero and pole frequencies are tuned to amplify the signal at the Nyquist frequency. DFE tap values are adaptively chosen and the equation is shown in Eq. (2). By using the equalizer, eye width has opened to 6.18 ps (0.39 UI), and the eye height has reached 0.16 V, shown in Fig 5. Based on the results, proposed channel is capable to use referring to the PCIe 5.0 specification [6]. By the proposed UHB GPU-ring, bandwidth is improved from 1.8 TB/s to 32 TB/s [1]. Therefore, the proposed architecture can be a promising solution for the supercomputing module to have less NUMA effect.

IV. CONCLUSION

In this paper, we proposed the UHB GPU-ring architecture for future supercomputing module by enhancing the bandwidth between GPU-HBM modules. The proposed architecture is designed at the PCB and package levels to shorten the distance and integrate more I/O channels between the modules. To verify the proposed channel, we designed and analyzed the micro-strip channel, which is the worst case of the proposed channel architecture. To enhance the signal integrity and achieve the target bandwidth, we utilize CTLE and DFE at the receiver. The proposed architecture successfully increased the bandwidth to 32 TB/s, which is increased 17.7 times, and energy efficiency increased 2.5 times, respectively [1]. GPU-ring structure, with its short interconnection lengths, would provide the power-efficient and high bandwidth interconnection for the super-computing node. This result enables multiple GPU-HBM modules to be treated as a single GPU-HBM, thereby minimizing the NUMA effect. The proposed architecture can be utilized as a solution for next-generation supercomputers with complex AI models.

ACKNOWLEDGMENT

We would like to acknowledge the technical support from ANSYS Korea. This work was supported by Samsung Electronics Co., Ltd (IO201207-07813-01). It is the result of research conducted in cooperation with the Solution Development Team of the Memory Division

REFERENCES

[1] "A. Li et al., "Evaluating Modern GPU Interconnect: PCIe, NVLink, NV-SLI, NVSwitch and GPUDirect," in IEEE Transactions on Parallel and Distributed Systems, vol. 31, no. 1, pp. 94-110, 1 Jan. 2020, doi: 10.1109/TPDS.2019.2928289.

[2] J. Meredith, P. Roth, K. Spafford and J. Vetter, "Performance Implications of Non-uniform Device Topologies in Scalable Heterogeneous Architectures," in IEEE Micro, vol. 31, no. 5, pp. 66-75, Sept.-Oct. 2011, doi: 10.1109/MM.2011.79

[3] "NVSWITCH : The World's Highest-Bandwidth On-Node Switch", NVIDIA, 2018

[4] Memory, JEDEC Standard High Bandwidth. "DRAM Specification." Standard JESD235A (2015)

[5] "AN 835: PAM4 Signaling Fundamentals", Intel, 2019

[6] PCI-SIG, "PCI Express® Base Specification Revision 5.0, Version 1.0", May 22, 2019

Recent Advances in Signal Integrity Simulation and Analysis of Interposers

Jonatan Aronsson
CEMWorks Inc.
Winnipeg, Canada
aronsson@cemworks.com

Feng Ling
Xpeedic
Bellevue, USA
feng.ling@xpeedic.com

Abstract—This paper presents recent advances for analyzing signal integrity in interposer-based interconnect systems. It proposes a BEM solver methodology that reduces the computational resources by over 40 times compared to FEM.

Index Terms—Interconnect, Signal Integrity, Boundary Element Method

I. INTRODUCTION

In the rapidly evolving landscape of electronic packaging and system integration, the shift towards 3D integrated circuits (ICs) and the adoption of interposers is a significant industry trend. Interposers provide shorter interconnect lengths, enhancing signal speed, reducing latency and minimizing parasitic effects, which are important for maintaining signal integrity for high-speed communication. As electronic systems push towards higher throughput and denser configurations, traditional signal integrity analysis methodologies struggle with maintaining accuracy and efficiency. Full-wave electromagnetic (EM) simulators enable the accurate prediction of signal integrity across these complex structures. There are several numerical methods available, each offers unique benefits, including the Boundary Element Method (BEM), the Finite Element Method (FEM), and the Finite-Difference Time-Domain (FDTD) method. This work introduces a BEM-based methodology that offers a more efficient solution compared to the traditional FEM approach. While FDTD is another alternative, it often presents challenges in maintaining accuracy due to stability and dispersion errors.

This paper is organized as follows: Section II details the methodology of the BEM solver; Section III presents numerical results validating the solver against a FEM solver for a canonical interconnect and demonstrates the solver on a real-world High-Memory Bandwidth (HBM) routing.

II. FORMULATION

The discretized mixed-potential form of the electric field integral equation (EFIE) can be written in matrix form as

$$ik_0 \left(\mathbf{V} + \mathbf{S} - \frac{1}{k_0^2} \mathbf{D}^T \cdot \mathbf{P} \cdot \mathbf{D} \right) \mathbf{J} = -\eta_0^{-1} \mathbf{b} \quad (1)$$

where $\mathbf{J}$ is the unknown electric current, $\mathbf{S}$ is the matrix form of the surface impedance operator, which can either be in a global form [1] or approximated as a local operator when the skin depth is much smaller than the cross-sectional dimensions of the conductor, $\mathbf{D}$ is the matrix form of the divergence operator [2] and $\mathbf{I}$ is the identity matrix. The other matrices elements are defined as

$$[\mathbf{V}]_{mn} = \int_S d\mathbf{r} \, \mathbf{\Lambda}_m(\mathbf{r}) \cdot \int_{S'} d\mathbf{r}' \, \mathbf{G}^A(\mathbf{r}, \mathbf{r}') \mathbf{\Lambda}_n(\mathbf{r}') \quad (2)$$

$$[\mathbf{P}]_{mn} = \int_S d\mathbf{r} \, \nabla \mathbf{\Lambda}_m(\mathbf{r}) \int_{S'} d\mathbf{r}' \, g^\phi(\mathbf{r}, \mathbf{r}') \nabla \mathbf{\Lambda}_n(\mathbf{r}') \quad (3)$$

$$b_m = \int_S d\mathbf{r} \, \mathbf{E}_{inc}(\mathbf{r}) \cdot \mathbf{\Lambda}_m(\mathbf{r}) \quad (4)$$

and $\mathbf{\Lambda}$ is the surface basis function, $\mathbf{E}_{inc}$ is the incident electric field due to the ports, $\mathbf{G}^A$ is the Green's Function for the magnetic vector potential and g^ϕ is the Green's Function for the electric scalar potential. The magnitude of the terms in (1) scale with different frequency dependency which cause a numerical breakdown with the finite-precision arithmetic at low frequencies. Several remedies have been proposed, in this work we split the unknowns into current and charge components and use the augmented version of EFIE (A-EFIE) [2] which can be written in matrix form as

$$\begin{bmatrix} \mathbf{V} + \mathbf{S} & \mathbf{D} \\ \mathbf{D}^T & k_0^2 \mathbf{I} \end{bmatrix} \cdot \begin{bmatrix} ik_0 \mathbf{J} \\ c_0 \rho \end{bmatrix} = \begin{bmatrix} -\eta_0^{-1} \mathbf{b} \\ 0 \end{bmatrix} \quad (5)$$

where ρ is the unknown charge. This formulation is directly adaptable for the simulation of interposers by assuming the dielectrics in the stackup are of infinite extent and utilizing the Layered Media Green's Function (LMGF) in equations (2) and (3). Alternatively, the analysis can be extended through a multi-region BEM formulation that includes both electric and magnetic currents at the dielectric interfaces. The A-EFIE can be extended to solve the multi-region BEM by adding unknowns for magnetic currents and charges as elaborated in [3], [4]. Utilizing LMGF simplifies the problem by reducing the number of unknowns; however, it introduces approximations by assuming the infinite extent, and the evaluations of the LMGF are computationally expensive. In contrast, the multi-region approach can leverage the closed-form free-space Green's Function, which enables highly efficient generation of compressed *H*-matrix representations [5], [6]. Therefore, despite the increase in the number of unknowns, the multi-region BEM may yield faster computational solutions.

979-8-3503-5124-8/24 $31.00 © 2024 IEEE

III. NUMERICAL RESULTS

This section illustrates the application of the A-EFIE BEM described in the previous section to two practical interconnect benchmarks, comparing its performance with the industry-standard FEM solutions. The first case involves a canonical interposer interconnect, while the second evaluates a real-world HBM interconnect.

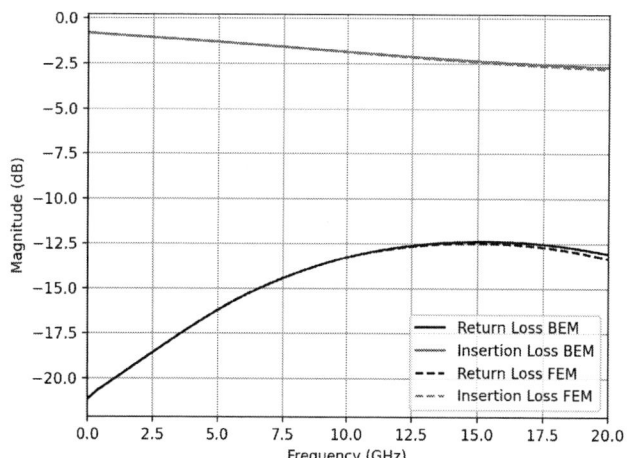

Fig. 2. S-parameters for a single trace of the interposer interconnect, comparing the BEM and FEM solutions.

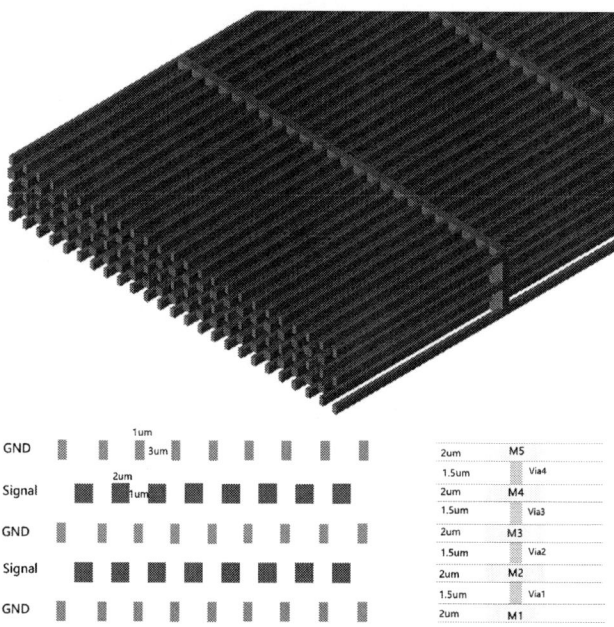

Fig. 1. The top image shows a 3D representation of the interposer interconnect benchmark with alternating signal (red) and ground (green) traces. The bottom image shows a cross-sectional view and stackup diagram of the interconnect layers, including the thickness, width and spacing. The length of the interconnect is 2,000 μm.

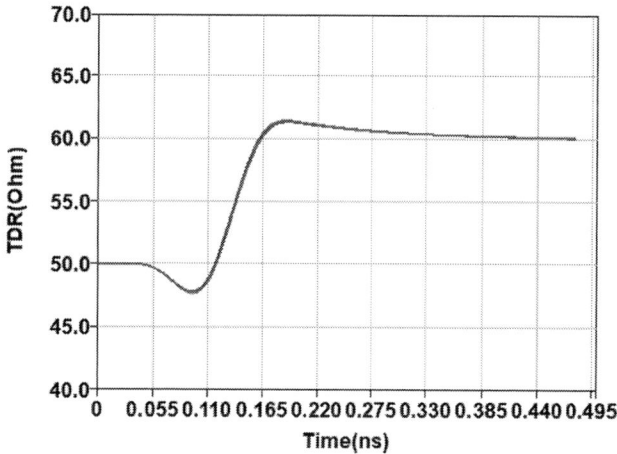

Fig. 3. TDR results for a single trace of the interposer interconnect using the BEM (red line) and FEM (green line).

A. Interposer Interconnect

The first benchmark is a canonical interconnect from an interposer design. Fig. 1 shows a 3d view of the 48 signal nets and a cross-sectional view of the dimensions and stackup. Utilizing the proposed BEM solver, this simulation required 11 minutes and 15 GB of memory on a computer with four Intel Xeon Gold 6252 processors, 96 cores and 1 TB of memory) a significant reduction compared to the 1,265 minutes and 655 GB required by the commercial FEM solver. Utilizing the proposed BEM solver, this simulation required 11 minutes and 15 GB of memory, a significant reduction compared to the 1,265 minutes and 655 GB required by the commercial FEM solver. The simulations were conducted on a computer with Intel Xeon Gold 6252 processors, featuring 96 cores and 1TB of memory. Fig. 2, 3 and 4 display the return and insertion loss for a single trace, Time Domain Reflectometry (TDR), the Power Sum Near End Crosstalk (PSNEXT), and the Power Sum Far End Crosstalk (PSFEXT), respectively. The results demonstrate that the BEM closely matches the FEM solver across all metrics, verifying the accuracy of the proposed solver.

B. HBM Interconnect

The second simulation focuses on HBM routing on a Chip-on-Wafer-on-Substrate (CoWoS-S) platform, shown in Fig. 5. The model used 736 thousand basis functions for the electric currents and convergence with ten frequency samples. The solver simulated five frequencies in parallel, took 291 minutes and utilized 273 GB of cumulative memory. These simulations were also performed on a computer equipped with Intel Xeon Gold 6252 processors, featuring 96 cores and 1TB of memory. The return and insertion loss for all traces is presented in Fig. 6. Additionally, Fig. 7 displays the TDR for all ports, incorporating a 50 ps port extension.

IV. CONCLUSION

This paper demonstrated the BEM's efficiency in analyzing complex interconnects like interposers and HBM routing. Compared to the FEM, BEM significantly reduced computational time and memory usage by approximately 100 and

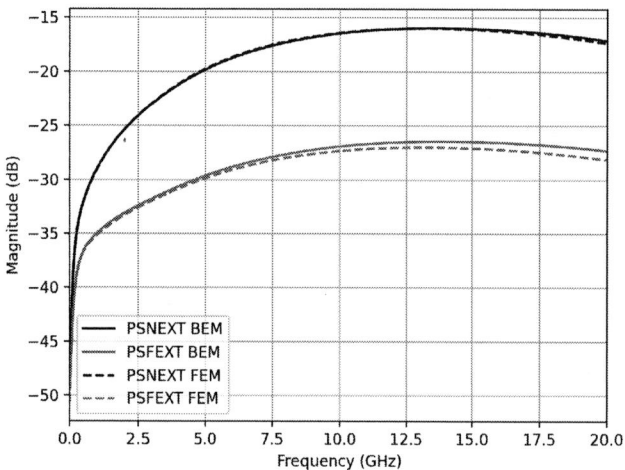

Fig. 4. PSNEXT (black) and PSFEXT (red) from 0 to 20 GHz for the interposer interconnect. The BEM solution is represented by solid lines and the FEM solution is the dashed lines.

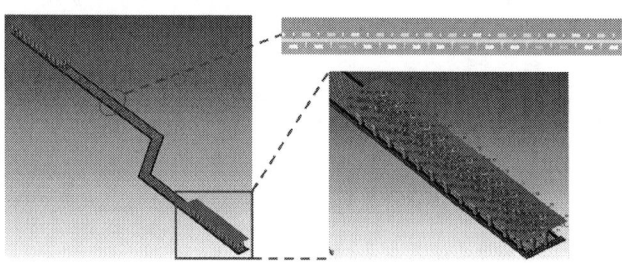

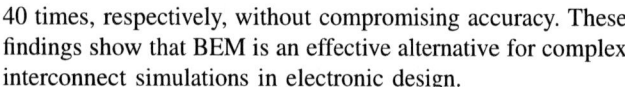

Fig. 5. Layout of an HBM interconnect using CoWoS-S technology, featuring 48 signal nets. The left panel illustrates the interconnect assembly, while the right panel offers a detailed view of the connection terminals.

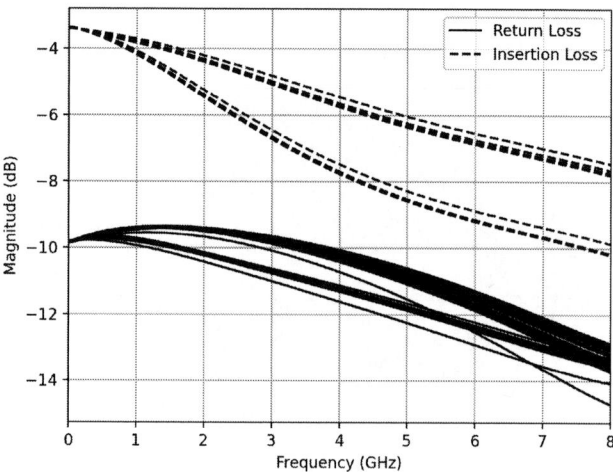

Fig. 6. Return Loss (solid lines) and Insertion Loss (dashed lines) for all the HBM nets.

40 times, respectively, without compromising accuracy. These findings show that BEM is an effective alternative for complex interconnect simulations in electronic design.

REFERENCES

[1] Z.-G. Qian, W. C. Chew, and R. Suaya, "Generalized impedance boundary condition for conductor modeling in surface integral equation," IEEE Trans. Microw. Theory Techn., vol. 55, no. 11, pp. 2354–2364, Nov. 2007.

[2] Z. G. Qian and W. C. Chew, "An augmented electric field integral equation for high-speed interconnect analysis," Micro. Opt. Technol. Lett., vol. 50, no. 10, pp. 2658–2662, Oct. 2008.

[3] M. Taskinen and P. Ylä-Oijala, "Current and charge integral equation formulation," IEEE Trans. Antennas Propag., vol. 54, pp. 58–67, Jan. 2006.

[4] T. Xia et al., "An Enhanced Augmented Electric-Field Integral Equation Formulation for Dielectric Objects," in IEEE Trans. Antennas Propag., vol. 64, no. 6, pp. 2339-2347, June 2016

[5] J. Aronsson and V. Okhmatovski, "Vectorial Low-Frequency MLFMA for the Combined Field Integral Equation," in IEEE Antennas and Wireless Propagation Letters, vol. 10, pp. 532-535, 2011

[6] S. Ambikasaran and E. Darve, The Inverse Fast Multipole Method, preprint, arXiv:1407.1572, 2014.

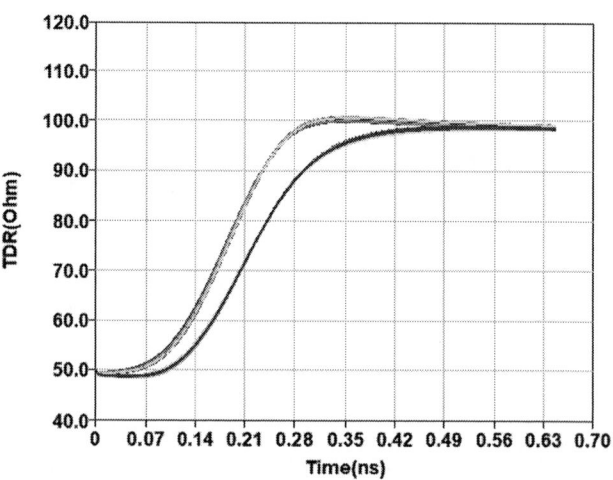

Fig. 7. TDR for all the HBM nets with a 50ps port extension.

979-8-3503-5124-8/24 $31.00 © 2024 IEEE

Multiphysics-Informed ML-Assisted Chiplet Floorplanning for Heterogeneous Integration

Vinicius C. Do Nascimento[1], Seunghyun Hwang[1], Michael J. Smith[2],
Qiang Qiu[1], Cheng-Kok Koh[1], Ganesh Subbarayan[2], Dan Jiao[1]

[1]*Elmore Family School of Electrical and Computer Engineering, [2]School of Mechanical Engineering*
Purdue University, West Lafayette, IN 47907, USA

Abstract—This paper presents an efficient and accurate method to perform multiphysics-informed floorplan and placement of heterogeneously integrated chiplets. Traditional multiphysics simulations, often impractical in optimization due to high computational cost, are replaced by a high-fidelity and efficient generative model via image-based machine learning (ML). Utilizing the ML model for fast performance assessment, we further accelerate the physical design by developing a dynamic rank-revealing algorithm for solving the underlying large-scale constrained optimization problem. Application to chiplet floorplanning and comparison with prevailing methods have demonstrated the superior performance of the proposed work.

Index Terms—System in Package (SiP), Physical Design, Multiphysics, Machine Learning

I. INTRODUCTION

The floorplan of chiplets in heterogeneously integrated systems in package (SiP) must consider multiphysics performance while minimizing the wirelength, which has not been well studied. Analyzing the multiphysics performance of a 3D SiP remains time-consuming, even with today's most efficient simulation tools. Furthermore, the underlying optimization problem is difficult to solve since it is not only non-convex and nonlinear but also involves a high-dimensional design space with a large number of constraints.

In this work, we propose a dynamic rank-revealing optimization method that extends the concepts proposed in [1] to address the convoluted nonlinear and non-convex optimization problem. Using this method, the chiplets' shape and coordinates are optimized directly without any floorplan representation. Furthermore, instead of evaluating solution quality with costly multiphysics simulations across a SiP during optimization, we guide the optimization process with an efficient conditional image generative model [2]. This model is trained via image-based machine learning from the results of rigorous heat transfer simulations. It is capable of performing real-time precision prediction of 3-D package thermal maps. This ML model is also integrated with prevailing optimization methods used for floorplanning, such as the Simulated Annealing (SA) optimization with Corner Block List (CBL) representation. The proposed method has shown a superior capability in identifying global optima in high-dimensional spaces while meeting a multitude of multiphysics constraints.

This work was supported by Rapid-HI (Heterogeneous Integration) Design Institute (an Elmore ECE Emerging Frontiers Center) and an NSF Future of Semiconductors (FuSe) grant under award No. 10002201.

II. MULTIPHYSICS-CONSTRAINED CHIPLET FLOORPLANNING

The chiplet floorplanning problem may be realized by making the minimization of the netlist wirelength W the objective composed with the legalization function, L, subject to multiphysics constraints,

$$\max_{\mathbf{x} \in X} \quad W(L(\mathbf{x}))^{-1}$$
$$\text{s.t.} \quad \mathbf{g}(\mathbf{x}) \leq \mathbf{0} \tag{1}$$

where optimization variables $\mathbf{x}$ define the chiplets' position and shape by containing the coordinates of each lower-left corner and aspect ratio

$$\mathbf{x} = \begin{bmatrix} x_1 & \cdots & x_{n_{cpt}} & y_1 & \cdots & y_{n_{cpt}} & \alpha_1 & \cdots & \alpha_{n_{cpt}} \end{bmatrix}. \tag{2}$$

The wirelength is estimated based on the Half-Perimeter Wirelength metric (HPWL). The inequality constraints $g_j(\mathbf{x}) \leq g_{j_{\max}}$ are cast in each term c_j as follows

$$c_j(\mathbf{x}) = \begin{cases} 1, & \text{if } g_j(\mathbf{x}) \leq g_{j_{\max}} \\ 1/(1 + g_j(\mathbf{x}) - g_{j_{\max}}), & \text{otherwise} \end{cases} \tag{3}$$

Then, we set $f_c(\mathbf{x})$ as the objective function with embedded constraints

$$f_c(\mathbf{x}) = \begin{cases} f(\mathbf{x}), & \text{if } \sum_j^{N_c} c_j = N_c \\ \epsilon \left(\sum_j^{N_c} c_j - N_c \right)/N_c, & \text{otherwise} \end{cases} \tag{4}$$

where N_c is the number of imposed constraints, and ϵ is a small number. In this framework, we implement any number of arbitrary hard constraints, such as fixed outline $x_i + \Delta x_i < x_{outline}$, peak temperature for each chiplet $T_{i_{pk}} < T_{i_{max}}$, and positional boundaries, for instance, i^{th} chiplet constrained to the North boundary $\text{chiplet}_i \subseteq \text{N}$.

A. Legalization

Floorplan legalization algorithms move blocks to remove overlap given an arbitrary arrangement. They are typically designed to remove only a small percentage of remaining overlapping after the general optimization. The legalization step is often realized by setting it up as a convex optimization problem using linear or quadratic programming. Since the legalization is in the optimization loop in (1), we set up a fast and straightforward legalization strategy that does not burden the objective function evaluation. Algorithm 1 describes our

979-8-3503-5124-8/24 $31.00 © 2024 IEEE

Algorithm 1 Spread Overlapping Blocks

Require: blocks coordinates
1: $\mathbf{x}_{events} \leftarrow$ sorted blocks' x-coord. with ID and left label
2: $i_0 \leftarrow 0$ ▷ sweep line cursor at 1st event
3: **for** $iteration \leftarrow 0$ **to** n_{max} **do** ▷ $O(n_{overlaps})$
4:    $valid \leftarrow$ **true** ▷ assume no overlapping
5:    $I \leftarrow I()$ ▷ empty interval tree with y-intervals
6:    **for** $i \leftarrow i_0$ **to** $|\mathbf{x}_{events}|$ **do** ▷ $O(n_{blocks})$
7:       **if** ISLEFT(x_i) **then** ▷ i^{th} event is b_i left corner
8:          **for each** $b_j \in I.\text{overlap}(b_i)$ **do** ▷ $O(n_{y,o})$
9:             **if** $(b_{j,x_u} + b_{i,\Delta x} < x_o)$ **and**
10:                 $(b_{j,y_u} + b_{i,\Delta y} > y_o)$ **then**
11:                $b_{i,x_l} \leftarrow b_{j,x_u}$ ▷ b_i right of b_j
12:             **else if** $(b_{j,x_u} + b_{i,\Delta x} > x_o)$ **and**
13:                 $(b_{j,y_u} + b_{i,\Delta y} < y_o)$ **then**
14:                $b_{i,y_l} \leftarrow b_{j,y_u}$ ▷ b_i above b_j
15:             **else**
16:                **if** $b_{j,x_u} - b_{i,x_l} < b_{j,y_u} - b_{i,y_l}$ **then**
17:                   $b_{i,x_l} \leftarrow b_{j,x_u}$ ▷ b_i right of b_j
18:                **else**
19:                   $b_{i,y_l} \leftarrow b_{j,y_u}$ ▷ b_i above b_j
20:             $valid \leftarrow$ **false**
21:             **break**
22:          $I.\text{insert}(b_i)$ ▷ $O(\log k)$ for k nodes in I
23:       **else** ▷ i^{th} event is b_i right corner
24:          $I.\text{remove}(b_i)$ ▷ $O(\log k)$ for k nodes in I
25:    **if** $\neg valid$ **then** ▷ moved a block
26:       $i_0 \leftarrow$ index of b_{j,x_l} for $\min_j b_{j,x_l} \in I$
27:       swap $\mathbf{x}_{events}$ due to moved block
28:       **break** ▷ back cursor to leftmost block in I
29:    **if** $valid$ **then** ▷ valid layout, stop legalization
30:       **break**

Algorithm 2 Dynamic Rank-Revealing Optimization

Require: $\mathbf{x}_{ref}, \mathbf{x}_{st_{1\dots n}}$
1: $f_{opt} \leftarrow \infty$
2: $\mathbf{x}_{opt} \leftarrow \mathbf{x}_{ref}$
3: $\mathbf{x}_{fl_{1\dots n}} \leftarrow$ RANDOMINITIALIZATION($n_{fl}, \mathbf{x}_{st_{1\dots n}}$)
4: **for** $iteration \leftarrow 1$ **to** max iterations **do**
5:    $\mathbf{S}_0 \leftarrow []$ ▷ initialize empty matrix
6:    $\mathbf{x}_{dyn_{1\dots n}} \leftarrow \mathbf{x}_{st_{1\dots n}} \cup \mathbf{x}_{fl_{1\dots n}} \cup \mathbf{x}_{gr_{1\dots n}}$
7:    $\mathbf{x}_{gs_{1\dots n}} \leftarrow$ GAUSSIANSAMPLES($\mathbf{x}_{dyn_{1\dots n}}$)
8:    $\mathbf{x}_{dyn_{1\dots n}} \leftarrow \mathbf{x}_{dyn_{1\dots n}} \cup \mathbf{x}_{gs_{1\dots n}}$
9:    **for** $i \leftarrow 1$ **to** n_{ref} **do**
10:       $\mathbf{x}_{seeds_{1\dots m}}$ from $\mathbf{x}_{dyn_{1\dots n}}, \mathbf{x}_{ref_i}$ ▷ $m := \sum_i^n N_{s_i}$
11:       $\mathbf{S}_0(\mathbf{x}_{ref}) \leftarrow f(\mathbf{x}_{seeds_{1\dots m}})$ ▷ [1]
12:       $\mathbf{S}_0 \leftarrow [\mathbf{S}_0; \mathbf{S}_0(\mathbf{x}_{ref})]$
13:       $\mathbf{S}_0' \leftarrow$ offset each row $N_r - 1$ times ▷ [1]
14:       $\mathbf{x}_{rank_{1\dots k}} \leftarrow$ RANK POINTS($\mathbf{S}_0'$) ▷ Alg. in [1]
15:       Evaluate $f(\mathbf{x}_{rank_{1\dots k}})$
16:       $\mathbf{x}_{gr_{1\dots n}} \leftarrow$ GRADIENTSTEPS($\mathbf{x}_{seeds}, f(\mathbf{x}_{seeds})$)
17:       Update f_{opt} and $\mathbf{x}_{opt}$ from evaluated values
18:       Update $\mathbf{x}_{ref_{cand}}$ from $\mathbf{x}_{seeds_{1\dots m}}, \mathbf{x}_{rank_{1\dots k}}$
19:       Update $\mathbf{x}_{fl_{1\dots n}}$ scores
20:       $\mathbf{x}_{ref} \leftarrow$ NEWREFERENCE($\mathbf{x}_{ref_{cand}}$)
21: **return** $\mathbf{x}_{opt}$ ▷ return optimal

---

sweep line algorithm using interval trees to detect overlapping chiplets and push them over to the right or upwards depending on their position in the floorplan.

Note that Algorithm 1 can lead to blocks outside the fixed outline. If any block violates the maximum x_o, the blocks are compacted in the horizontal direction, and likewise for the vertical direction.

B. Optimization Method Assisted by ML Model

Algorithm 2 depicts the dynamic rank-revealing optimization method developed in this work, which is an expansion of [1]. An initial reference point and static samples are required for each one of the n optimization variables as in [1]. A set of randomly initialized floating samples is updated at each iteration according to the best function value provided by the variable's sample across all iterations, penalized against the clustering of the points. The sample candidates come from Gaussian sampling around the existing samples for that iteration and gradient steps taken around each reference point. Since the seed matrix is populated from m cross samples around a reference point (called seed points $\mathbf{x}_{seeds_{1\dots m}}$ where $m := \sum_i^n N_{s_i}$ for N_{s_i} samples for the ith variable), a

numerical gradient is readily available, indicating if the current reference is a local optimal. From this gradient, a new sample for each variable is created from the point selected from a gradient step toward the direction of optimality for each reference point. In this floorplanning work, it is observed that taking a user-defined number ($3 \cdot i$) of steps contributes positively to the quality of the optimizer. Each iteration of the optimization starts with an empty seed matrix. An inner loop appends rows to the seed matrix built from n_{ref} adaptively selected references. The rank-revealing points are selected and evaluated inside the adaptive referencing inner loop. The dynamic samples are updated for the newly evaluated points, and a new reference is selected. The reference is selected from a short list of the best-evaluated points penalized against crowding.

Fig. 1 depicts the major components implementing the proposed chiplet floorplanning method according to Algorithm 2. It starts with a random reference point for a layout. Seed points are generated, and their corresponding legalized layouts are created parallelly on the CPU. The valid layouts are selected for thermal evaluation using the ML model of [2]. It is important to note that memory bandwidth and inference overhead are computational bottlenecks when data is transferred back and forth from CPU to GPU. However, these can be mitigated by batching the simulations for the machine-learning model. For the chiplet floorplanning solved in this work, a batch size of around 15 is found to provide the best performance of around 215 ms per layout. The simulations' inputs are assembled into a single batched tensor for inference on the GPU. Unique input tensors are filtered for thermal evaluation

979-8-3503-5124-8/24 $31.00 © 2024 IEEE

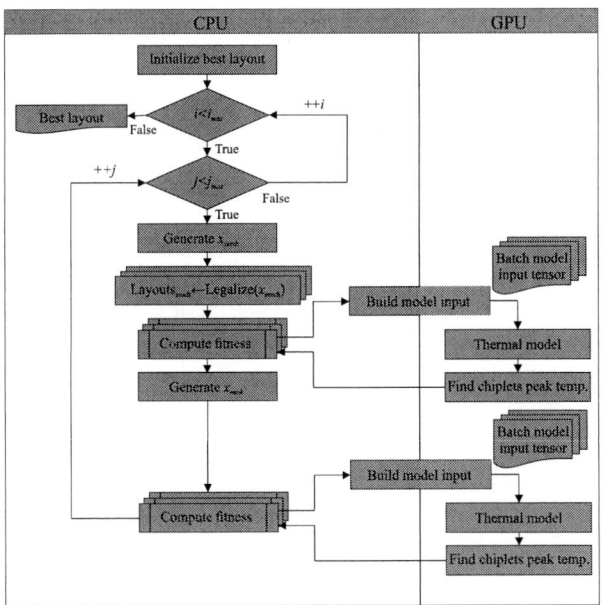

Fig. 1. Rank revealing chiplet floorplanning framework aided by multiphysics ML model.

in the same batch to avoid redundant simulations. Building the input tensors takes an average of 7.7 ms per candidate solution, and finding the chiplets' peak temperature takes an average of 1.2 ms per layout thanks to the fast ML model. From the resulting seed matrix, the rank-revealing algorithm selects points for evaluation in the same parallel and batched manner.

III. RESULTS

We considered a working example of chiplets in package floorplanning from the classic GSRC n10 benchmark, and we placed the newly defined chiplets on top of an interposer. The interposer has a fixed area of 500×500. Extending 50 units beyond the interposer is a mold compound filling the space between the chiplets and above them. Each chiplet in this example has a different peak temperature allowance $(90, 100, 80, 80, 80, 100, 80, 120, 100, 80°C)$ corresponding to the function, materials, and technology node used for each. The objective of floorplanning is to minimize the wirelength while satisfying thermal constraints.

We perform the task using the proposed thermal-aware method with dynamic rank-revealing optimization guided by the multiphysics ML model generated using [2]. And we compare the method against the approach without the thermal constraints, and the SA-CBL thermal-aware optimization. The Simulated Annealing is set up for 2,000 optimization iterations with 125 iterations per temperature (250,000 total), of which a valid layout is found 44,283 times for thermal simulation, running for 341.83 minutes. We run the thermal-aware optimization with the proposed method for the same wall-time.

The non-thermal-aware runs for 88.2 seconds. Fig. 2 depicts the optimal design results for the three validation methods. In the first column, we plot the temperature fields on the surface

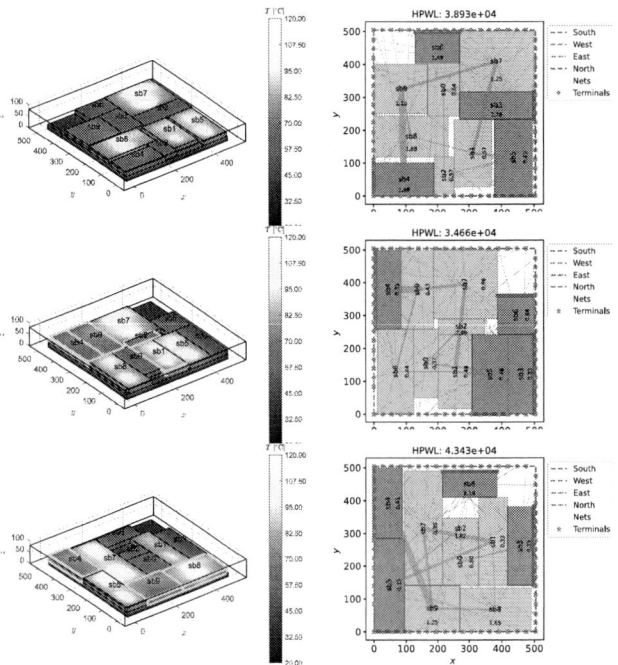

Fig. 2. Chiplet floorplanning results. (Upper) Proposed method. (Middle) Non thermal-aware method. (Lower) SA-CBL approach.

of the chiplets and interposer (the mold compound of the package is made transparent). The second column shows a top view of the chiplets, terminals, and connecting nets. The rows of the figure depict the results for the thermal-aware, the non-thermal-aware, and the SA-CBL optimization, respectively. The resultant wirelength estimated through the HPWL for each method is 3.893e+04, 3.466e+04, and 4.343e+04, respectively as depicted in the title of the second column figures. Note that the non-thermal aware optimization yields the shortest wirelength since it is not constrained by temperature. However, this unawareness results in four chiplets with peak temperature violation, highlighted in green boxes in the first column of Fig. 2. The SA-CBL approach is shown to have three temperature violations, whereas the proposed method meets all thermal constraints while achieving a much shorter wirelength.

REFERENCES

[1] D. Jiao and V. C. do Nascimento, "Fast rank-revealing method for solving large global optimization problems and its applications," *IEEE Trans. Microwave Theory and Techniques*, 2024. [Online]. Available: https://ieeexplore.ieee.org/document/10431823

[2] M. J. Smith, S. Hwang, V. C. Do Nascimento, Q. Qiu, C.-K. Koh, G. Subbarayan, and D. Jiao, "Real-time precision prediction of 3-d package thermal maps via image-to-image translation," in *2023 IEEE 32nd Conference on Electrical Performance of Electronic Packaging and Systems (EPEPS)*, 2023, pp. 1–3.

Accuracy Study of the Differential Surface Admittance Operator for Lossy Metal Characterization

M. Huynen[1], V. Okhmatovksi[2], D. De Zutter[1], and D. Vande Ginste[1]

[1] quest, IDLab, Department of Information Technology, Ghent University/imec, Ghent, Belgium
[2] Department of Electrical and Computer Engineering, University of Manitoba, Winnipeg, Canada

Abstract—The analytic, complete solution of a boundary integral formulation consisting of the electric field integral equation (EFIE) and the differential surface admittance (DSA) operator is presented for lossy conductors. Through a Galerkin Method of Moments with two complete sets of orthogonal vector spherical harmonics as basis functions, the complete system of equations is solved, including a closed-expression for the DSA elements via a generalized Fourier series. A comparison with the Mie series solution shows the range and accuracy of the DSA-EFIE formulation in dealing with lossy materials, and paves the way for a future study into the operator's fundamental properties, leading to improved tools for interconnect modeling that alleviate the identified dense-mesh and low-frequency breakdown.

Index Terms—method of moments, single-source, differential-surface admittance.

I. INTRODUCTION

In the context of boundary integral equations (BIEs) for interconnect modeling, a couple of challenges have led to a plethora of solutions and approaches as the problem is inherently difficult to tackle. The inclusion of good conductors with a finite conductivity is one such example. The proposed techniques range from simple but limiting approximations, such as the Leontivich boundary condition, to full-out integral operators that have to tackle the numerical integration of a Green's function that is quickly declining yet displaying highly oscillatory behavior, which may prove challenging and costly [1]. A different approach is the differential surface admittance (DSA) operator [2], which is a single-source BIE that places a fictitious current density on the surface while preserving the fields outside of the replaced volume. The operator relation between the current and the electric field on the boundary is expressed as the difference between two Poincaré-Steklov (PS) operators, which can be constructed through the volume's eigenfunctions [2], [3], by means of conventional operators including the medium's Green's function [4] or via the Fokas method [5]. The first and latter avenue provide a fast alternative as the PS operators are constructed in non-traditional ways.

The technique has been continuously developed since its inception for 2-D cross-sections [2], including the augmentation of more shapes such as circles [6], triangles and polygons [7], arbitrary polygons [5] and cross-sections [4], the inclusion

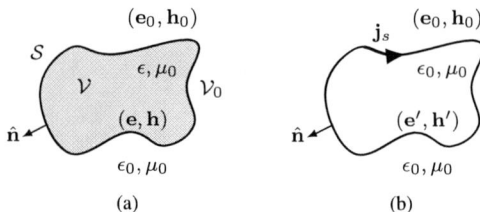

Fig. 1: Illustration of the equivalence principle with (a) the original situation and (b) the single-source equivalence.

of magnetic materials [8] and the extension to 3-D objects such as cylinders and cuboids [3], [9]. All the publications and applications have established the DSA technique as an accurate, competitive and diversely applicable BIE flavor. Nevertheless, given its relatively recent development, some fundamental questions on the eigenmode-based approach remain unanswered; in the context of interconnects in particular, the low-frequency and dense-mesh breakdown need to be understood fully to alleviate these adverse phenomena.

Consequently, we take the first steps to investigate these matters by examining scattering at a lossy sphere, inspired by the approach taken in [10], [11], where a Galerkin Method of Moments, entirely relying on vector spherical harmonics as test and basis functions, is employed to reach a fully analytical solution for all appropriate integral operators. In this paper, we achieve closed expressions for the DSA operator as well, including a closed sum for the summation over the eigenmodes, like it was found for a cuboid with entire domain basis function [9], which vastly improves the accuracy and computational complexity of the method. A comparison to the standard Mie series solution shows the versatility and applicability of this technique and provides the preparation for a fully fledged spectral analysis in future publications, which opens up the way for improved BIE formulations for interconnect characterization, as the root cause of low-frequency and dense-mesh breakdown can be identified and circumvented.

II. THE DIFFERENTIAL SURFACE ADMITTANCE OPERATOR

Consider a homogeneous dielectric volume $\mathcal{V}$ with (complex) relative permittivity $\epsilon = \epsilon_r \cdot \epsilon_0 + \sigma/(j\omega)$ and permeability μ_0 as represented in Figure 1a. Following the single-

979-8-3503-5124-8/24 $31.00 © 2024 IEEE

source equivalence theorem [12], this material can be swapped out for the background medium, for convenience taken as free space, by placing a surface current density $\mathbf{j}_s$ on the boundary $\mathcal{S}$ (see Figure 1b) while preserving field distribution outside of $\mathcal{V}$. Without a material discontinuity, these fields can now be computed through the standard electric field integral equation (EFIE).

The imposed unknown surface current density $\mathbf{j}_s$ that acts as a source of the fields in the EFIE is obtained via the DSA operator. Here, we concentrate on the 3-D expansion of its original form involving the sum over all the eigenmodes of $\mathcal{V}$ [3]:

$$\mathbf{j}_s(\mathbf{r}) = \mathcal{Y} \circ \mathbf{e}_0^t \qquad\qquad \mathbf{r} \in \mathcal{S}, \qquad (1)$$

$$= -\eta \sum_\nu \left[\frac{\mathcal{K}_\nu}{\mathcal{N}_\nu^2} \int_\mathcal{S} (\hat{\mathbf{n}} \times \mathbf{h}_\nu^*(\mathbf{r}')) \cdot \mathbf{e}_0^t(\mathbf{r}')\, \mathrm{d}\mathbf{r}' \right] (\hat{\mathbf{n}} \times \mathbf{h}_\nu(\mathbf{r})),$$

with $\mathcal{Y}$ the DSA operator, k_0 and k the wavenumber of the background medium and the volume's medium, respectively, η the contrast parameter $(k^2 - k_0^2)/(j\omega\mu_0)$, $\mathcal{K}_\nu$ a shorthand for $k_\nu^2/[(k_\nu^2-k^2)(k_\nu^2-k_0^2)]$, and $\mathbf{h}_\nu$ the magnetic eigenmode of the PEC cavity with the same shape of $\mathcal{V}$ filled by the background medium, with k_ν its wavenumber and $\mathcal{N}_\nu$ its normalization constant.

III. GALERKIN MoM ON A SPHERE

A. Expansion into basis functions

To investigate a lossy metal sphere, we follow the same Helmholtz decomposition procedure in vector spherical harmonics as applied in [11] to expand the unknown surface quantities $\mathbf{j}_s$ and $\mathbf{e}_0^t$ as

$$\mathbf{j}_s = \sum_{n'=0}^N \sum_{m'=-n'}^{n'} \alpha_{n'm'}^{(1)} \mathbf{u}_{n'm'}^{(1)} + \alpha_{n'm'}^{(2)} \mathbf{u}_{n'm'}^{(2)}, \qquad (2)$$

$$\mathbf{e}_0^t = \sum_{n''=0}^N \sum_{m''=-n''}^{n''} \beta_{n''m''}^{(1)} \mathbf{u}_{n''m''}^{(1)} + \beta_{n''m''}^{(2)} \mathbf{u}_{n''m''}^{(2)}, \qquad (3)$$

with $\mathbf{u}_{nm}^{(1)}$ and $\mathbf{u}_{nm}^{(2)}$ two sets of orthogonal, complete sets of vector basis functions defined on $\mathcal{S}$, the surface of a sphere with radius a, as

$$\mathbf{u}_{nm}^{(1)}(\theta, \phi) = \sqrt{d_{nm}} \nabla^t Y_n^m(\theta, \phi), \qquad (4)$$

$$\mathbf{u}_{nm}^{(2)}(\theta, \phi) = \sqrt{d_{nm}} \hat{\mathbf{r}} \times \nabla^t Y_n^m(\theta, \phi). \qquad (5)$$

More details on the definition and normalization choice of the scalar spherical harmonics Y_n^m and their normalization constant d_{nm} can be found in [11].

The spherical magnetic eigenmodes $\mathbf{h}_\nu$ are typically characterized into two distinct classes, i.e., transversal magnetic (TM$_r$) and transversal electric (TE$_r$) eigenmodes, with corresponding definitions:

$$\mathbf{h}_{nms}^{\mathrm{TM}} = \nabla \times [r j_n(k_{ns} r) Y_n^m(\theta, \phi) \hat{\mathbf{r}}] = \mathbf{p}^{(1)} \qquad (6)$$

$$\mathbf{h}_{nms}^{\mathrm{TE}} = \frac{1}{\kappa_{ns}} \nabla \times \nabla \times [r j_n(\kappa_{ns} r) Y_n^m(\theta, \phi) \hat{\mathbf{r}}] = \mathbf{q}^{(1)}, \qquad (7)$$

where $j_n(x)$ is the spherical Bessel function, $k_{ns} = x_{ns}/a$ with x_{ns} the roots of $[x j_n(x)]' = 0$, and $\kappa_{ns} = y_{ns}/a$ with y_{ns} the roots of $j_n(x) = 0$.

B. Galerkin discretization

The discretization strategies for the EFIE [10] operator and the one for the DSA operator profit from the same orthogonality properties of the spherical harmonics Y_n^m since both the surface $\mathbf{u}^{(i)}$ and volume $\mathbf{p}^{(1)}/\mathbf{q}^{(1)}$ vector harmonics are composed out of them. Hence, when we introduce the expansions (2) and (3) into (1) and focus first on the TM eigenmodes (6), the orthogonality results in a single expansion term of $\mathbf{j}_s$ corresponding to its counterpart in $\mathbf{e}_0^t$:

$$\alpha_{nm}^{(1)} = \mathcal{Y}_n^{1,1} \beta_{nm}^{(1)}, \qquad (8)$$

with $k_{ns} = x_{ns}/a$ in

$$\mathcal{Y}_n^{1,1} = \frac{-1}{j\omega\mu_0 a} \sum_{s=1}^\infty \frac{2 k_{ns}^2 (k^2 - k_0^2)}{(k_{ns}^2 - k^2)(k_{ns}^2 - k_0^2)\left[1 - \frac{n(n+1)}{x_{ns}^2}\right]}. \qquad (9)$$

Analogously for the TE eigenmodes (7), a relation is deduced for the second set of expansion coefficients:

$$\alpha_{nm}^{(2)} = \mathcal{Y}_n^{2,2} \beta_{nm}^{(2)}, \qquad (10)$$

with $\kappa_{ns} = y_{ns}/a$ in

$$\mathcal{Y}_n^{2,2} = \frac{-1}{j\omega\mu_0 a} \sum_{s=1}^\infty \frac{2 \kappa_{ns}^2 (k^2 - k_0^2)}{(\kappa_{ns}^2 - k^2)(\kappa_{ns}^2 - k_0^2)}. \qquad (11)$$

C. Generalized Fourier series

Evaluating (9) and (11) with an acceptably high level of accuracy proves difficult as the summation is slow to converge and requires the roots x_{ns} and y_{ns}, which need to be computed numerically and accurately as well. Hence, the need for a closed sum inspired by its equivalent for the entire domain basis approach of the 3-D DSA for cuboids [9], which is found here by applying the theory of the generalized Fourier series. For the current spherical geometry, it states that a square-integrable function on the interval $[0, a]$ can be expanded as

$$f(r) = \sum_{s=1}^\infty c_s j_n(k_{ns} r) = \sum_{s=1}^\infty \frac{\langle f, j_n(k_{ns} r)\rangle}{||j_n(k_{ns} r)||^2} j_n(k_{ns} r), \quad (12)$$

with $\langle f, g \rangle = \int_0^a f(r) g(r) r^2\, \mathrm{d}r$. When we apply this expansion to the following functions:

$$f(r) = \frac{(ka)^2 j_n(kr)}{[ka j_n(ka)]'}, \qquad g(r) = \frac{ka j_n'(kr)}{j_n(ka)}, \qquad (13)$$

and their counterparts with $k = k_0$, their generalized Fourier expansion coefficients coincide with the sums in (9) and (11), respectively. As such, these sums are replaced by the following closed expressions:

$$\mathcal{Y}_n^{1,1} = \frac{-1}{j\omega\mu_0 a} \left[\frac{(ka)^2 j_n(ka)}{[ka j_n(ka)]'} - \frac{(k_0 a)^2 j_n(k_0 a)}{[k_0 a j_n(k_0 a)]'} \right]. \qquad (14)$$

$$\mathcal{Y}_n^{1,1} = \frac{1}{j\omega\mu_0 a} \left[\frac{(ka) j_n'(ka)}{j_n(ka)} - \frac{(k_0 a) j_n'(k_0 a)}{j_n(k_0 a)} \right]. \qquad (15)$$

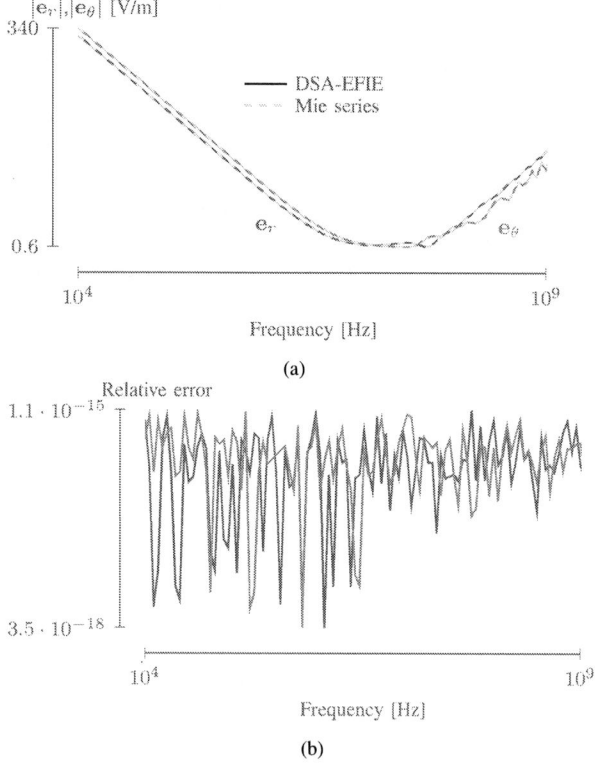

(a)

(b)

Fig. 2: Electric field components (a) and their relative error (b) over a broad frequency range for scattering at a copper sphere with radius $1\,\text{m}$ by an unit radial electric dipole at $r = 10\,\text{m}$ from the origin. (a) The field components $\mathbf{e}_r$ (red) and $\mathbf{e}_\theta$ (green) are recorded at $(r, \theta) = (2, \pi/4)$. The Mie series solution is also provided (dashed, gray).

Combining these algebraic expressions with the two sets for the EFIE operator [11], leads to a diagonal set of matrix equations that is trivially solved and leads to closed expressions for the expansion coefficients of $\mathbf{j}_s$ and $\mathbf{e}_0^t$.

IV. NUMERICAL RESULTS

To validate the analytic results, we consider scattering at a copper ($\sigma = 5.8 \cdot 10^7\,\text{S/m}$) sphere with radius $1\,\text{m}$ from $10\,\text{kHz}$ up to $1\,\text{GHz}$. The source is a dipole located on the z-axis at a distance of $10\,\text{m}$ from the origin with unit dipole moment oriented along the z-axis. As an example, we compare the radial and tangential field component at $r = 2\,\text{m}$ at an angle of $\theta = \pi/4$ with the Mie series solution. All methods' solutions are computed with the built-in routines of Wolfram Mathematica 12 with 50 terms each.

Both components are shown in Figure 2a and show an excellent agreement. This is confirmed in Figure 2b, where the relative error between the DSA-EFIE and the reference result is reported. The maximum relative error over the entire frequency range is just $1.1 \cdot 10^{-15}$ while the average for both components is $3.7 \cdot 10^{-16}$, confirming machine precision accuracy. In other words, the DSA captures the current crowding very well over a broad frequency range thanks to the analytical

expressions for the infinite summations, proving the accuracy and reliability of the method for handling finite conductors, hence confirming its appositeness for interconnect modeling.

V. CONCLUSION

An analytical solution for scattering at a lossy metal sphere through the differential surface admittance operator was presented. By means of a Method of Moments procedure with vector spherical harmonics as basis and test functions, closed expressions for the differential surface admittance operator dealing with the inside problem were obtained. Moreover, a closed formula was derived for the infinite series involving spherical Bessel zeroes, which greatly improves the accuracy and convergence of the analytical solution. Through comparing the results with the Mie series, it is shown that a rigorous solution has indeed been found that will form the basis for a spectral analysis in future work, whose insight into inherent frequency-dependent properties of the analytically solved DSA-EFIE system will lead to improved discretized BIE formulations for interconnect characterization.

REFERENCES

[1] J. Peeters, I. Bogaert, and D. De Zutter, "Calculation of MoM interaction integrals in highly conductive media," *IEEE Trans. Antennas Propag.*, vol. 60, no. 2, pp. 930–940, Feb. 2012.

[2] D. De Zutter and L. Knockaert, "Skin effect modeling based on a differential surface admittance operator," *IEEE Trans. Microw. Theory Techn.*, vol. 53, no. 8, pp. 2526–2538, Aug. 2005.

[3] M. Huynen, M. Gossye, D. De Zutter, and D. Vande Ginste, "A 3-D differential surface admittance operator for lossy dipole antenna analysis," *IEEE Antennas Wireless Propag. Lett.*, vol. 16, 2017.

[4] U. R. Patel and P. Triverio, "Skin effect modeling in conductors of arbitrary shape through a surface admittance operator and the contour integral method," *IEEE Trans. Microw. Theory Techn.*, vol. 64, no. 9, pp. 2708–2717, Sep. 2016.

[5] D. Bosman, M. Huynen, D. De Zutter, and D. Vande Ginste, "Construction of the differential surface admittance operator with an extended fokas method for electromagnetic scattering at polygonal objects with arbitrary material parameters," *Computers & Mathematics with Applications*, vol. 128, pp. 44–54, Dec. 2022.

[6] U. R. Patel, B. Gustavsen, and P. Triverio, "An equivalent surface current approach for the computation of the series impedance of power cables with inclusion of skin and proximity effects," *IEEE Trans. Power Del.*, vol. 28, no. 4, pp. 2474–2482, Oct. 2013.

[7] T. Demeester and D. De Zutter, "Construction of the Dirichlet to Neumann boundary operator for triangles and applications in the analysis of polygonal conductors," *IEEE Trans. Microw. Theory Techn.*, vol. 58, no. 1, pp. 116–127, Jan. 2010.

[8] D. Bosman, M. Huynen, D. De Zutter, H. Rogier, and D. Vande Ginste, "A 2-D differential surface admittance operator for combined magnetic and dielectric contrast," *Computers & Mathematics with Applications*, vol. 102, pp. 175–186, Nov. 2021.

[9] M. Huynen, K. Y. Kapusuz, X. Sun, G. Van der Plas, E. Beyne, D. De Zutter, and D. Vande Ginste, "Entire domain basis function expansion of the differential surface admittance for efficient broadband characterization of lossy interconnects," *IEEE Trans. Microw. Theory Techn.*, vol. 68, no. 4, pp. 1217–1233, 2020.

[10] G. Hsiao and R. Kleinman, "Mathematical foundations for error estimation in numerical solutions of integral equations in electromagnetics," *IEEE Trans. Antennas Propag.*, vol. 45, no. 3, pp. 316–328, 1997.

[11] O. Goni and V. I. Okhmatovski, "Analytic solution of surface–volume–surface electric field integral equation on dielectric sphere and analysis of its spectral properties," *IEEE Trans. Antennas Propag.*, vol. 69, no. 12, pp. 8479–8493, 2021.

[12] E. Martini, G. Carli, and S. Maci, "An equivalence theorem based on the use of electric currents radiating in free space," *IEEE Antennas Wireless Propag. Lett.*, vol. 7, pp. 421–424, 2008.

979-8-3503-5124-8/24 $31.00 © 2024 IEEE

Compact Fiber Weave Model for Full Wave Solvers

Stefan de Araujo
University of Texas at Austin
dearaujo2@utexas.edu

Daniel de Araujo
Siemens
Daniel.deAraujo@siemens.com

Bhyrav Mutnury
AMD
bhyrav.mutnury@amd.com

Abstract—**In modern, high-speed PCBs, the fiber weave effect can introduce differential in-pair skew leading to eye closure. Using 3D solvers, skew was computed by constructing simplified 4-cell models directly from datasheets.**

Index Terms—fiber weave effect, simulation, in-pair skew

I. INTRODUCTION

The fiber weave effect is a well documented phenomenon arising from the pattern in the underlying fiberglass bundles embedded within the resin in printed circuit boards (PCBs). This pattern creates differing electrical properties within the layer due to the inhomogeneous structure of the fiberglass and resin which have different dielectric constants. During the manufacturing process, the exact positioning of the fiber bundles is not controlled, so the effect must be accounted for during the design process for both single and multiple layer alignments. This paper presents a minimal model to estimate best, worst, and average skew for an arbitrary design.

II. BACKGROUND

As signal transmission frequencies increase, the delay caused by the fiber weave effect begins to become a limiting factor. When a differential pair runs parallel to these fiber bundles, the relative permittivity (κ) of the material underneath each trace may not be the same. This leads to different propagation delays and introduces differential skew. This differential skew causes mode conversion and differential signal loss. This is especially evident with looser weave such as the 1080 shown in Fig. 1.

One trace may align with a fiber bundle and another with a resin-rich area. The propagation delay due to this bad alignment can lead to eye closure, increased crosstalk, and increased Electromagnetic Interference (EMI) [1].

The weave of the fiberglass fabric consists of two types of fiber bundles, the warp and weft/fill. During the fabric making process, the warp bundles are held tight while the fill are passed in between. As these are two separate bundles, the amount of threads and spacing between bundles may differ between the warp and fill.

There are many different methodologies to model the impact of this effect with various levels of simplifications. Some techniques utilize a full 3D solver with the fiber geometry within the model [2] [3] whereas others reduce the problem to 2 dimensions [4] [5]. While having the full three-dimensional geometry of the fiber bundles may lead to accurate results,

Fig. 1. Microscopic view of 1080 (left) and 1086 weaves (right). Source: [6].

the computational and mesh complexity increases. However, a two-dimensional model is much less flexible to complex trace routing such as non-orthogonal angles.

One method is to explicitly model the fiber bundles within each section with different dielectric constants based on an equivalent volume of glass [3]. The geometry is constructed directly from the weave specification and split into multiple layers to be used in a volumetric solver. This method can account for both warp and fill having different effective dielectric constant as well as resin, however, this incurs a much higher meshing cost as a single dielectric layer may require three to seven layer separations when used in a layered solver.

This paper proposes a fiber weave four-cell equivalent dielectric model - an expressive yet efficient modeling methodology in order to model differential skew caused by the fiber weave effect.

III. MODELING STRATEGY

The end goal of the model is to find the equivalent κ for the four different sections of the weave (intersect, warp, fill, and resin, highlighted in Fig. 2) and model a printed circuit board accurately for the fiber weave effect. The intersect is the part of the weave which both the warp and fill bundles overlap, and it is the section with the largest density of fiberglass. The warp and fill regions consist only of their corresponding fiberglass bundles, and the resin area has no fiberglass bundles. In order to model the fiber weave effect, the weave itself must be modeled first. To ensure that the model is accurate, the parameters are taken from a datasheet directly. For these models, the 1080 and 1086 weaves are used with specifications coming from Isola [6] and AGY [7].

The model is constructed from the fiber count, fiber thickness, weave width, and weave spacing. Using the AGY ECD450-1/0 fiberglass yarn specification, there are approximately 204 fibers in the yarn, the diameter ranges 0.2 to 0.249 mils (5.08 to 6.32 micron), approximated as 0.236 mils (6

979-8-3503-5124-8/24 $31.00 © 2024 IEEE

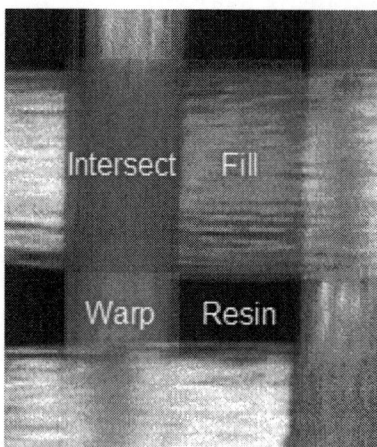

Fig. 2. Labeled weave sections. Image from [6].

Weave	Thickness	Width	Spacing
1086 Warp	1.44	10.80	16.60
1086 Fill	1.00	14.70	17.10
1080 Warp	1.60	8.20	17.00
1080 Fill	1.10	12.10	22.40

TABLE I
ISOLA WEAVE SPACING AND SIZE SPECIFICATION.

micron) for the stripline cases to model a strong skew [7]. For the microstrip tests, the minimum, maximum and mean values were modeled to capture the range and expected skew. The radius of the fibers (r) is used to compute the cross-sectional area of an individual fiber. Then, using the filament count (c), the total area of a bundle can be calculated. Averaging this area over the width of the warp or fill (w) gives an approximation of the equivalent fiberglass thickness (t_{eq}) for a given point on the warp or fill as shown in Fig. 3. Equation 1 shows the calculation for this equivalent volume. The weave specification is included in Table I.

$$t_{eq} = \frac{\pi c r^2}{w} \qquad (1)$$

The dielectric constant was calculated as a linear interpolation across the materials for a given point with t_{tot} being the thickness of the substrate. The dielectric constant of resin (κ_r) and fiberglass (κ_g) used are 3 and 6 respectively. Since the resin rich areas have no fiberglass, the κ was 3. For the warp

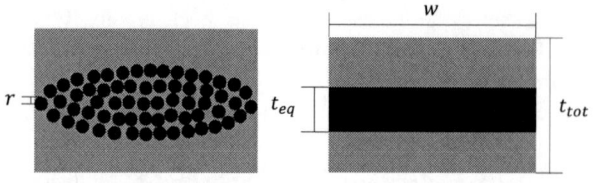

Fig. 3. Weave represented as individual fibers in a bundle (left). Equivalent thickness cell based on bundle volume (right).

Weave	Intersect κ	Fill κ	Warp κ	Resin κ
1080	4.826	3.737	4.088	3.00
1086	4.433	3.607	3.826	3.00

TABLE II
κ VALUES FOR 1080 AND 1086 WEAVES.

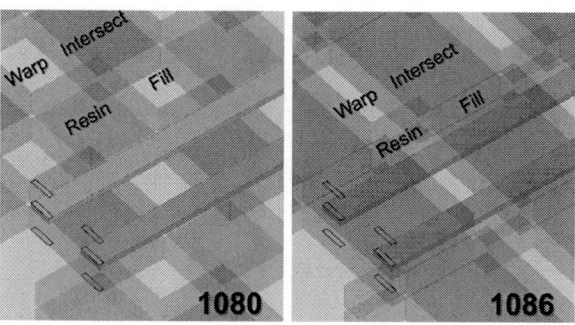

Fig. 4. 1080 (left) and 1086 (right) model in HyperLynx 3D Explorer.

and fill areas, the ratio of resin to fiberglass was computed and the κ was the weighted average between the two materials as shown in Equation 2. The intersection used the combined amount of fiberglass from warp and fill (adding t_{eq} of each). The calculated κ are found in table II

$$\kappa = \frac{t_{eq}}{t_{tot}} \kappa_g + \frac{t_{tot} - t_{eq}}{t_{tot}} \kappa_r \qquad (2)$$

Given the different dielectric constants for each region, a microstrip and stripline model was constructed in HyperLynx Full-wave Solver to study the effect (Fig. 4). Using parametric sweeps in HyperLynx 3D Explorer, the position of the traces to the weave was offset to model different alignments, test cases were created, solved, and automatically measured to generate the results. A method similar to [8] was used to extract in-pair skew. Since we assumed no specific data rate, we weighted the skew across frequency equally.

IV. RESULTS

As seen in Table III, the model demonstrates the fiber weave effect in both the 1080 and 1086 weaves. Fig. 5 shows the skew across different offsets where 0 offset is a balanced offset with minimal skew. The model behaves as expected with the half period having the worst skew.

The surface and contour plots in Fig. 6 demonstrate the effects of the multiple ply alignment. The mean (μ) of having the 2 plys aligned (equivalent to one thick layer) is approximately the same as independent alignment (two smaller layers with independent fabric positioning), but the standard deviation (σ) is smaller with independent ply alignment. These results correlate well with other works.

In [4], the effect was modeled by cascading smaller simple models based on simplified cross sections of the transmission lines. This model used the 106 weave which has approximately half as much fiberglass as the 1080 weave. The study found

979-8-3503-5124-8/24 $31.00 © 2024 IEEE

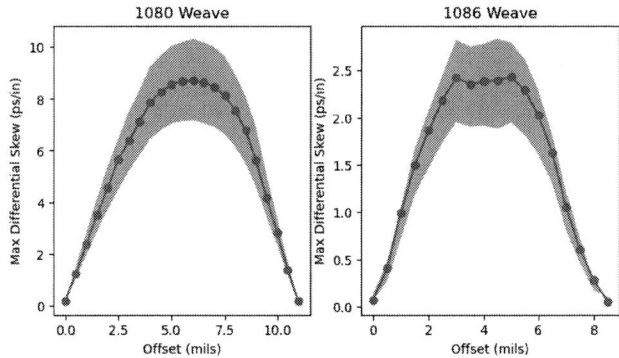

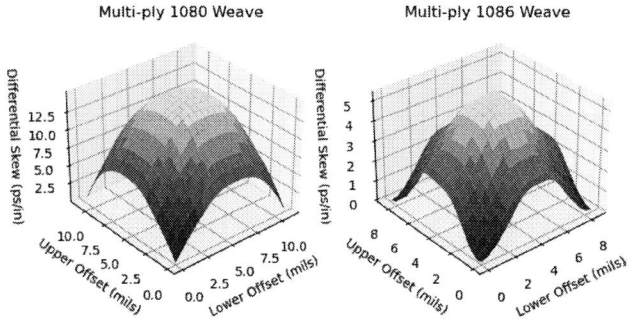

Fig. 5. Mean microstrip skew with highlighted error from filament diameter range.

Weave Type	μ	σ	Max Skew
1080 Stripline	9.775	4.949	14.912
1080 Microstrip	6.336	3.294	10.016
1086 Stripline	2.892	2.001	5.478
1086 Microstrip	1.847	1.059	3.068

TABLE III
SKEW SIMULATION RESULTS FOR $2r = 6\mu$M.

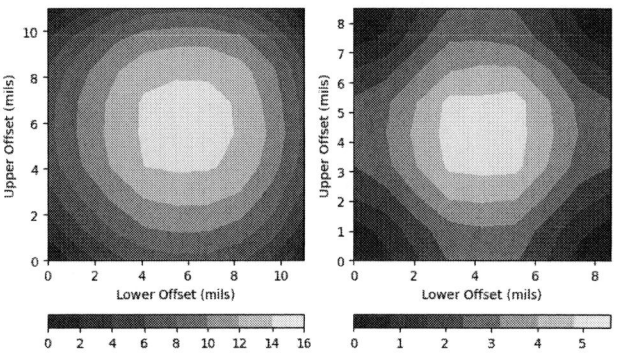

Fig. 6. Stripline Skew vs Offset surfaces: 1080 (left) and 1086 (right) vs. bottom ply offset (x) and top ply offset (y).

a maximum skew of about 2.07 ps/in for a 4.1 in microstrip. The cascaded model in [4] split the model into cases, one where the trace is parallel to the fibers (case 1) and the case where the trace runs perpendicular to the fibers (case 2). This model does not account for the still uneven spread of glass fibers when simulating case 2. Since there is twice as much fiberglass and [4] did not account for the fiber weave effect in about half the cases, the resulting 9.45 ps/in skew aligns with the results of the study.

For [3], the maximum skew calculated in the 1080 weave microstrip was approximately 6 ps/in. This model used κ values of 3.5 and 6 for the resin and glass fibers respectively. After accounting for the difference in κ values, their results fall within the lower range of the 1080 microstrip range seen in Fig. 5.

V. CONCLUSION

As long as printed circuit boards continue to have periodically varying dielectric constants of unknown offsets, the fiber weave effect will continue to skew differential pairs. We can, however, model the impact of this effect to minimize the impact of this effect. This paper proposed and tested a four cell equivalent dielectric modeling strategy which derives the structure of the model directly from datasheets and can be applied to any weave type. This model utilizes simplified geometry to avoid excess mesh complexity while still allowing for arbitrary routing. By applying a tiled grid of regions with different dielectric constants, the model can accurately recreate an arbitrary weave. This gridded region is simple to apply to a complex PCB and can be used to test best, worst, mean, and

the probability of a feasible design for a given weave. The results show that this model has a similar level of accuracy to other techniques while being flexible and computationally efficient.

REFERENCES

[1] J. Loyer, R. Kunze, and X. Ye, "Fiber weave effect: Practical impact analysis and mitigation strategies," *DesignCon*, 2007.

[2] N. Ghassemi, W. Han, and H. Tournier, "Study of fiberweave effect through simulation and measurement on performance of differential stripline at high frequency," in *2017 IEEE International Symposium on Electromagnetic Compatibility & Signal/Power Integrity (EMCSI)*, pp. 193–197, 2017.

[3] A. Manukovsky, Y. Shlepnev, and S. Mordooch, "Quantification of delay and skew uncertainty due to fiber weave effect in pcb interconnects," in *2023 IEEE 32nd Conference on Electrical Performance of Electronic Packaging and Systems (EPEPS)*, pp. 1–3, 2023.

[4] T. Zhang, X. Chen, J. E. Schutt-Ainé, and A. C. Cangellaris, "Statistical analysis of fiber weave effect over differential microstrips on printed circuit boards," in *2014 IEEE 18th Workshop on Signal and Power Integrity (SPI)*, pp. 1–4, 2014.

[5] C.-H. Wang, M.-T. Lu, J.-R. Huang, C.-S. Chen, and R.-B. Wu, "Modeling and mitigating fiber weave effect using layer equivalent model and monte carlo method," in *2022 IEEE 72nd Electronic Components and Technology Conference (ECTC)*, pp. 1851–1857, 2022.

[6] "Understanding glass fabric." https://www.isola-group.com/wp-content/uploads/Isola-Glass-Fabric-04_2022.pdf.

[7] "Agy technical product guide." https://www.agy.com/wp-content/uploads/2014/03/AGY_Technical_Product_Guide-Revised.pdf.

[8] S.-J. Moon, J. Li, X. Zhang, C.-P. Kao, B. Lee, H. Dsilva, and J.-R. Guo, "Intra-pair skew metric, eips (effective intra-pair skew)," in *2021 IEEE 25th Workshop on Signal and Power Integrity (SPI)*, pp. 1–4, 2021.

979-8-3503-5124-8/24 $31.00 © 2024 IEEE

Megtron 6 and 8 Characterization Methodology

Stefan de Araujo
University of Texas at Austin
dearaujo2@utexas.edu

Daniel de Araujo
Siemens
Daniel.deAraujo@siemens.com

Roger Delbue, Ryan Keegan
Teledyne LeCroy Inc
(roger.delbue, ryan.keegan)@teledyne.com

Abstract—**Laminate characteristics play an important role in design and simulation. We establish a methodology to electrically characterize two types of laminates (Megtron 6 and Megtron 8) using two methods: measurement-based model and de-embedding.**

Index Terms—**laminate, characterization, optimization**

I. Introduction

Electrical material characterization is a critical aspect of printed circuit board (PCB) design. Electrical properties such as dielectric constant, loss tangent and surface roughness can affect the robustness of a design and whether it satisfies specification requirements. Extracting these values from manufactured samples help validate the nominal data sheet values as well as provide insight to improve the modeling assumptions.

Relying on manufacturer's data sheets provides a good starting point for many of these values. These are often close to the properties of the final products, but, due to manufacturing tolerances and changes during the board manufacturing process, the final values may also differ.

By utilizing lab measurements taken from manufactured boards, the electrical properties of these boards can be estimated. These measurements, however, include additional loss and artifacts due to transitions between the VNA cable, connectors, and the board. There are different approaches to handle these probing artifacts as well as different ways to evaluate the correctness of the material parameters in both time and frequency domain. This methodology automates the manual process of material characterization across multiple metrics in frequency, impedance, and time.

II. Motivation

When high performance materials are required to meet stringent applications such as the PCIe Gen 6 Interposer (Fig. 1) a good characterization methodology is needed [1].

The process of estimating the dielectric material properties for transmission line modeling is often manual and difficult [2]. Automating this process can save time by efficiently searching the design space, especially when multiple metrics for accuracy are required. Some methods such as [3] require that a board of a specific construction is made (such as a board with two striplines with different trace widths) in order to be characterized.

In this work, we will use two approaches to extract the material properties: de-embedding with a 2x Thru that removes

Fig. 1. PCIE Gen 6 Interposer

the launch effects and a measurement based model (MBM) approach that explicitly accounts for them for both a single-ended stripline board using Megtron 6 [4] and a differential microstrip board using Megtron 8.

III. Methodology

Estimating transmission line characteristics from a single measurement is error prone since the vector network analyzer (VNA) is calibrated to the end of its cables. The transition from the cables to connector to board to transmission line needs to be accounted for. The methodology can use MBM or de-embedded reference data to guide the estimation of dielectric constant, loss tangent and surface roughness.

A. MBM reference

With the MBM approach, the launch/transition is explicitly modeled along with the transmission line section we wish to characterize. By having multiple lengths and keeping the launch consistent, we can separate the effects of each. For Megtron 8 differential microstrip the lengths of interest were 40mm and 167mm. For Megtron 6 single-ended stripline, the lengths were 1in and 2in.

B. De-embedded reference

For the de-embedded approach, the shorter length measurement, often referred as the 2x Thru is split into two halves and its effects "removed" through de-embedding on the longer length. With the connector/board/transition effects removed, the remaining length can be used directly.

For the Megtron 6 data, the de-embedding technique was the In-Situ De-embedding (ISD) [5] using the 1in measurement to

979-8-3503-5124-8/24 $31.00 © 2024 IEEE

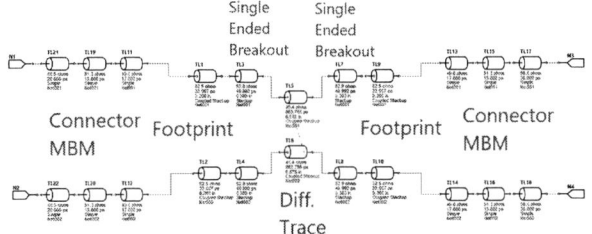

Fig. 2. SI/PI Linesim MBM Schematic

Case	DK	Df	Cannonball Rough
Meg6 DeEmb	3.52	0.0095	$1.15 \cdot 10^{-6}$
Meg6 MBM	3.29	0.0097	$0.56 \cdot 10^{-6}$
Meg6 Manual	3.40	0.0046	N/A
Meg8 DeEmb	3.15	0.0024	$1.92 \cdot 10^{-6}$
Meg8 MBM	3.12	0.0031	$1.47 \cdot 10^{-6}$
Meg8 Manual	3.10	0.0015	$2.00 \cdot 10^{-6}$

TABLE I
EXTRACTED PARAMETERS

de-embed the 2in data, and the Megtron 8 data used the 2x Thru de-embedding technique in [6] to de-embed the launch, transitions, and 40mm from the 167mm.

C. Optimization

The parameter extraction process involves correlating a simulation to reference measurements. Within the simulation, the dielectric constant, loss tangent, and surface roughness (Cannonball model) were the parameters of interest; all three were optimized together to minimize the error. The Time Domain Reflectrometry (TDR) impedance, group delay, and insertion loss were the metrics used.

The objective function was a weighted sum of Mean Absolute Errors (MAE, $\mathcal{E}$) where each metric was scaled to have a similar magnitude to allow all three to be optimized simultaneously as seen in (1). F and T are the sets of frequencies and time steps which the parameters are evaluated, and $\mathcal{Z}_t$, $\mathcal{D}_f$, and $\mathcal{L}_f$ are the TDR impedance, group delay, and insertion loss at a specific time step/frequency respectively.

$$
\begin{aligned}
\mathcal{E} = {} & \frac{w_Z}{|T|} \sum_{t \in T} \left| \mathcal{Z}_t \Big|_{\text{sim}} - \mathcal{Z}_t \Big|_{\text{actual}} \right| + \frac{w_D}{|F|} \sum_{f \in F} \left| \mathcal{D}_f \Big|_{\text{sim}} - \mathcal{D}_f \Big|_{\text{actual}} \right| \\
& + \frac{w_L}{|F|} \sum_{f \in F} \left| \mathcal{L}_f \Big|_{\text{sim}} - \mathcal{L}_f \Big|_{\text{actual}} \right|
\end{aligned}
\tag{1}
$$

In this work, we used $w_Z = 3$, $w_D = 10^{11}$ and $w_L = 10$.

First, a schematic model was created in HyperLynx Signal Integrity and Power Integrity (SI/PI) [7] to extract the s-parameters and then calculate the error between the simulated and measured insertion losses, impedances, and delays.

To find the best fit parameters for the model, we used HyperLynx Design Space Explorer (HL-DSE) to optimize parameters using the figure of metric in (1). Given that we are minimizing the error in impedance, group delay, and insertion loss across multiple lengths where the laminate parameters can affect multiple outputs, the search of the design space is very difficult to do manually. We used the HL-DSE SHERPA search algorithm with 500 evaluations for each study.

To improve the balance between the accuracy of each metric, the weighting of each needs to adjusted so that one does not dominate over the others.

D. Experiments

The initial MBM optimization struggled to converge and this was primarily due to the modeling of the launch using lossless transmission lines. While this would capture the impedance and delay effects, the loss of the connector and transitions would be then captured in the transmission line loss which would be excessive for the shorter length and insufficient for the longer one. Once lossy elements were included that could be independently controlled from the transmission line parameters, the transition loss could separated and closure on both lengths achieved.

For the de-embedded cases, the error in (1) can be used directly, however, for the multi-length MBM cases, we found it best to scale the error proportional to the relative length ratio. As part of the optimization, geometric features were given a narrow range of variability of 5% from the nominal/measured values to account for manufacturing inconsistencies.

IV. RESULTS

Good correlation was achieved for the MBM model as seen on Figures 3, 5 and for the de-embedded approach shown in Figures 4, and 6.

The de-embedded optimization outperforms optimizing over the "raw" data in the error metrics (seen by comparing figures 3, 4 and 5, 6) as there are fewer unknowns and parameters to optimize. The removal of the launch and transition artifacts is helpful. However, this assumes there is consistency in the fabrication so that the launch on the shorter measurement is representative of the others. Variations will result in errors being introduced. In Figure 3 we can see the impedance asymmetry in the TDR plot. When de-embedded from the longer length, this translates into artifacts that can be see in Figure 4. When comparing the TDR of the Megtron 6 vs. 8 boards, the former had much better impedance control/symmetry consistency (MAE 1in: 0.07, 2in: 0.10) vs. the latter (MAE 40mm: 0.44, 167mm: 1.05). Optimizing below these values is essentially fitting noise.

Using the measured stack-up values of the Megtron 6 board instead of the nominal values had better parameter convergence. No cross-section data was available for the Megtron 8 test boards at this time.

V. CONCLUSION

The presented characterization methodology utilizes advanced optimization tools to efficiently search the different parameters across multiple metrics and test cases simultaneously. This was demonstrated with both single-ended and differential test cases with stripline and microstrip transmission lines. This process can be adapted to arbitrary sets of parameters, and the targets set for any desired metrics. Both unmodified lab

979-8-3503-5124-8/24 $31.00 © 2024 IEEE

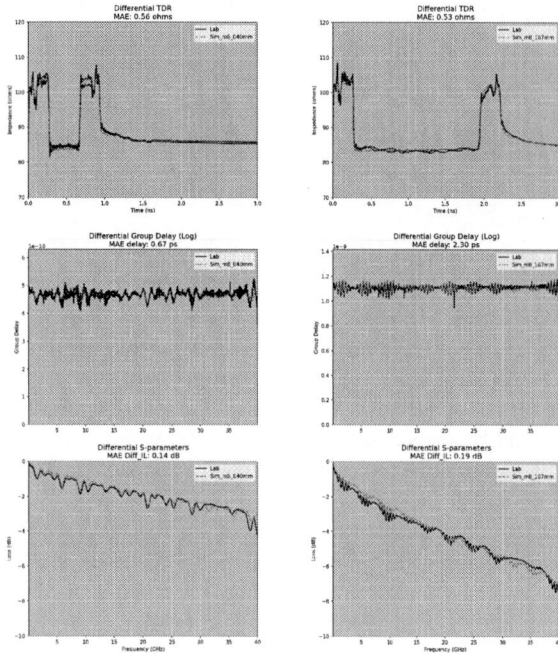

Fig. 3. Megtron 8 best simulation and measurement

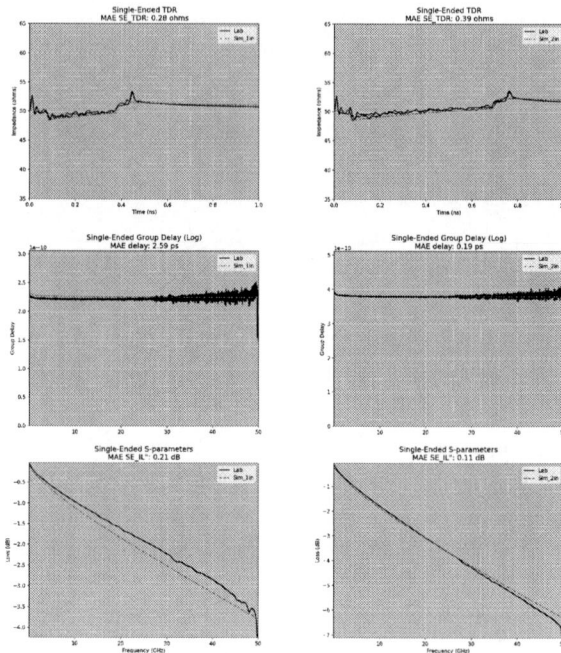

Fig. 5. Megtron 6 best simulation and measurement

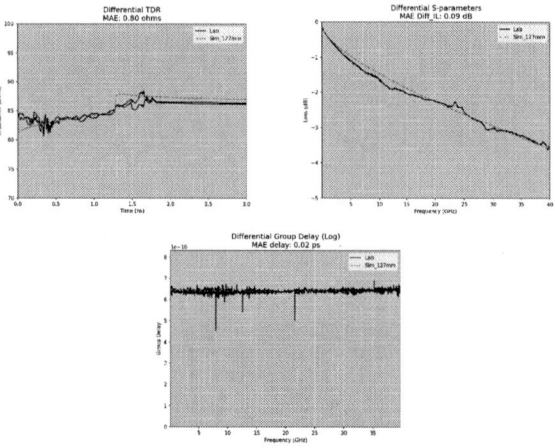

Fig. 4. Megtron 8 de-embedded best simulation and measurement

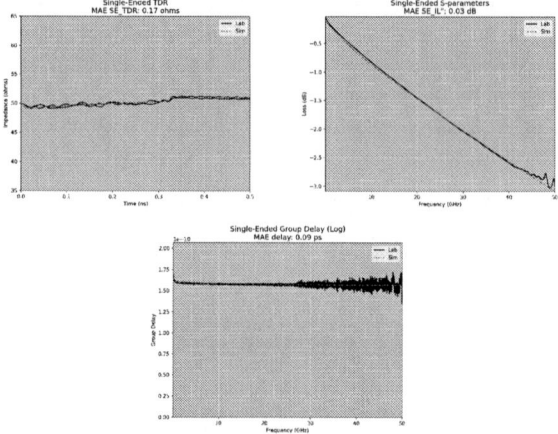

Fig. 6. Megtron 6 de-embedded best simulation and measurement

measurements and de-embedded measurements for a Megtron 6 and Megtron 8 board were used to evaluate this methodology and demonstrated good results. A simple test of the measurement quality consists of comparing the TDR metric reversing the propagation direction. This establishes a self-consistency and quality value below which further optimization is futile.

REFERENCES

[1] R. Delbue, R. Keegan, D. de Araujo, and R. Wolff, "Simulation and measurement correlation of a transparent pcie gen6 design," in *2024 DesignCon Proceedings*, 2024.

[2] K. Hu, "Pcb parameter extraction for signal integrity modeling," in *2022 IEEE International Symposium on Electromagnetic Compatibility and Signal/Power Integrity (EMCSI)*, pp. 93–96, 2022.

[3] L. Hua, B. Chen, S. Jin, M. Koledintseva, J. Lim, K. Qiu, R. Brooks, J. Zhang, K. Shringarpure, and J. Fan, "Characterization of pcb dielectric properties using two striplines on the same board," in *2014 IEEE International Symposium on Electromagnetic Compatibility (EMC)*, pp. 809–814, 2014.

[4] F. Guo, K. Aygün, W. D. Becker, S. G. Talocia, J. A. Hejase, W.-W. Wong, T. Zhou, H. Barnes, Z. Peng, A. Pelger, M. Sahouli, J. Schutt-Aine, F. Ling, E. Griese, P. R. Paladhi, R. Sharma, N. Pham, T.-M. Winkel, E. Fledell, M. J. Hill, B. Silva, K. Hu, J. Aronsson, C. Liu, Y. Jeong, and A. E. Yilmaz, "The ieee eps packaging benchmark suite," in *2021 IEEE 30th Conference on Electrical Performance of Electronic Packaging and Systems (EPEPS)*, pp. 1–4, 2021.

[5] "In situ de-embedding (isd) user's guide," *https://ataitec.com/products/isd/*, 2024.

[6] "Maui studio," *https://www.teledynelecroy.com/support/user/*, 2023.

[7] "Hyperlynx product family," *https://eda.sw.siemens.com/en-US/pcb/hyperlynx/*, 2024.

979-8-3503-5124-8/24 $31.00 © 2024 IEEE

Full-Wave Analysis for Ground Via Placement with Layered Media Integral Equations

Alireza Niazi and Vladimir Okhmatovski

University of Manitoba, Winnipeg, MB, Canada, http://www.umanitoba.ca

Abstract—As signaling rates exceed 40 Gbps, detailed electromagnetic analysis of the Ground Return Vias (GRVs) becomes essential. Incorrect GRV placement, whether too far apart or poorly arranged in a ball-grid array, can cause a signal via's insertion loss (IL) to increase substantially. Historically, signal integrity (SI) engineers have placed GRVs near signal layer transitions following conventional practices, often without a precise understanding of the optimal quantity and spacing required. This paper proposes accurate full-wave electromagnetic analysis framework for via placement optimization. The method solves the layered medium Mixed Potential Integral Equation (MPIE) coupled with the Surface-Volume-Surface Electric Field Integral Equation (SVS-EFIE), making it suitable for analyzing composite 3D metal/dielectric structures within layered substrates encountered in GRV placement problems.

Index Terms—PCB, Ground Return Via, Electromagnetic modeling, Signal integrity

I. INTRODUCTION

This paper presents a computational framework suitable for accurate full-wave analysis of coupling between the signal and ground vias found in electronic packages. Analyzing the interaction between signal vias and the ground return vias (GRVs) requires capturing frequency-dependent impedance in the return path. We aim at developing capability to model effect of multiple GRVs on return loss and insertion loss, particularly near resonant frequencies. Accuracy level in such analysis depends on handling the metal of the signal and ground vias as impenetrable material with simple surface impedance model or as penetrable material accounting for redistribution of the current in the vias due to proximity and skin-effect.

Commonly used frameworks for electromagnetic analysis include the finite element method (FEM) and layered media integral equations with layered media Green's functions (LMIE). LMIE formulations confine field quantities to metal surfaces and/or their volumes eliminating the need for distretization of the layered substrate dielectrics. This leads to smaller matrix equations compared to FEM that fully discretize dielectric substrates and metal components. The reduction in the size of the matrix equations in solution of LMIEs comes at the expense of the matrices in such equations being dense as opposed to the sparse matrix equations in FEM, which may or may not offer computational advantage depending on a problem at hand.

Our LMIE formulation couples the layered medium MPIEs for metal regions with SVS-EFIEs for dielectric regions. It is termed SVS-S-EFIE [4] and allows for full-wave electromagnetic analysis of general 3D metal/dielectric composite

structures embedded into a planar layered media of infinite extent. Electric field Green's functions are computed using a mixed-potential approach developed by Michalski and Zhang [3]. It enables RWG Method of Moments (MoM) discretization of the surface currents on metal regions considered impenetrable with loss handled through surface impedance in MPIE and piece-wise discretization of the electric field in the metal regions considered to be penetrable [2]. If finite dielectric volume regions are present in the model (e.g. air gap enforcing substrate truncation effect [1]), volume polarization currents in these regions are also discretized using tetrahedral meshes with piece-wise constant basis functions for approximation of polarization current components within each tetrahedron [4].

In this work we consider a model consisting of one signal via and varying numbers of ground return vias (GRVs) from 1 to 8. Initial findings regarding the computed S-parameters in models with infinite dielectric layers are outlined in this paper and compared with results obtained using a commercial electromagnetic analysis tool. Advancements in model sophistication for analysis of GRVs will be presented at the conference.

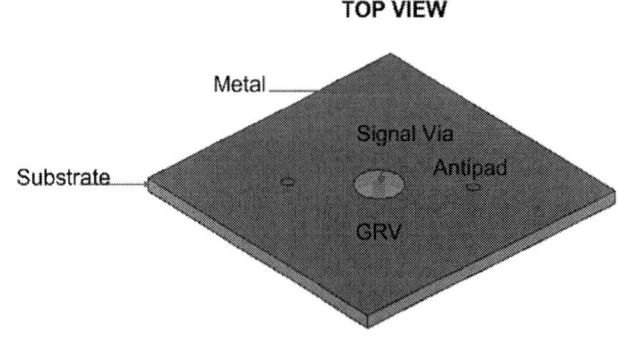

Fig. 1. Signal via and GRVs in layered substrate (courtesy of MathWorks).

II. 3D COMPOSITE OBJECTS FORMULATION OF SVS-S-EFIE IN LAYERED MEDIUM

To illustrate the formulation of SVS-S-EFIE for analyzing general interconnects in designs, we consider the example of a via in a printed circuit board (PCB), as depicted in Figures 1 and 2. The model consists of a homogeneous, non-magnetic dielectric region, $R1$, with a volume of V_{R1} collectively denoting penetrable dielectric regions such as non-ideal penetrable metals, air gap in the substrate to emulate

979-8-3503-5124-8/24 $31.00 © 2024 IEEE

its truncation, etc. Additionally, there is a circuit region, $R2$, which includes metallization and via layers enclosed by the surface ∂V_{R2}. The volume is made up of sub-volumes $\partial V_{L_2,R2}$ and surface is made up of sub-surfaces $\partial V_{L_2,R1}$, which are confined to the layer of the substrate (layer 2). The vias are located within the substrate layer 2, as shown in Figure 1 and described in Table I.

Using the surface equivalence principle (SEP) and the volume equivalence principle (VEP), the electric field in layered medium can be expressed, as follows:

$$
\boldsymbol{E}^p(\boldsymbol{r}) = \boldsymbol{E}_p^{\text{inc}}(\boldsymbol{r}) + \iota\omega\mu_0 \int\limits_{\partial V_{L2,R2}} ds' \overline{\boldsymbol{G}}_{eb}^{p,2}(\boldsymbol{r},\boldsymbol{r}') \cdot \mathcal{J}_{R2}(\boldsymbol{r}') +
$$

$$
k_0^2(\epsilon_{R1} - \epsilon_{L2}) \int\limits_{V_{L2,R1}} \overline{\boldsymbol{G}}_{eb}^{p,2}(\boldsymbol{r},\ \boldsymbol{r}') \cdot \boldsymbol{E}_{R1}(\boldsymbol{r}') dv', \quad (1)
$$

where observation point $\boldsymbol{r}$ can reside in any layer, i.e. $\boldsymbol{r} \in L_p, p = 1,2,3$; $\overline{\boldsymbol{G}}_{eb}^{pp'}(\boldsymbol{r},\boldsymbol{r}')$ is the dyadic Green's function for the electric field in a multilayered medium with $\boldsymbol{r}$ in layer p and $\boldsymbol{r}'$ in layer p' [4], $\iota = \sqrt{-1}$, ω is the angular frequency, ϵ_0 and μ_0 are the permittivity and permeability of free space, respectively, and $k_0 = \omega\sqrt{\epsilon_0\mu_0}$ is the wavenumber in free space. $\boldsymbol{E}_{R1}$ denotes the electric field within the volume $V_{L2,R1}$ of region $R1$, which is assumed to be fully embedded in the second layer of the substrate. In Equation (1), $\epsilon_{R1} = \epsilon_{R1} + \frac{\sigma_{R1}}{i\omega\epsilon_0}$ represents the complex relative permittivity of the finite dielectric region $R1$. The equivalent surface current density on the metal and via layer forming region $R2$ is assumed to be fully residing in layer of the substrate (layer 2) and is indicated by $\mathcal{J}_{R2}$. Details on single-source representation of the electric field $\boldsymbol{E}_{R1}$ using fictitious surface current $\boldsymbol{J}$ and constraining of the observation point in (1) to the surfaces ∂V_{R1} and ∂V_{R2} resulting in a system of two coupled integral equations can be found in [4]. The Method of Moments (MoM) reduces the coupled LMIEs to the system of linear algebraic equations with respect to the vectors of unknown coefficients $[I]$ in the expansion of the fictitious surface current $\boldsymbol{J}$ in region $R1$ using RWG functions, and the vectors of unknown coefficients $[\mathcal{I}]$ in the expansion of the current $\mathcal{J}$ in the metal region $R2$ using RWG functions as shown in [4] also.

III. NUMERICAL RESULTS

To examine the GRVs return and insertion loss in a packages, we study a single signal via with 1 to 8 GRVs surrounding it. The GRVs geometry is depicted in Figures 1 and 2, which also include cross-sectional views of the metallization and via layers in the package. Figure 2 shows a one-layer dielectric substrate that supports the metal and via layers. GRVs connect top ground plane which is taken to be finite and the bottom ground plane which is assumed to be infinite and handled by layered media Green's function. The substrate features a relative permittivity of 3.15, a loss tangent of 0.002, and a thickness of 0.22 mm. The vias, represented as finite metal structures within the substrate, are also shown in Figure 2, with their dimensions detailed in Table I.

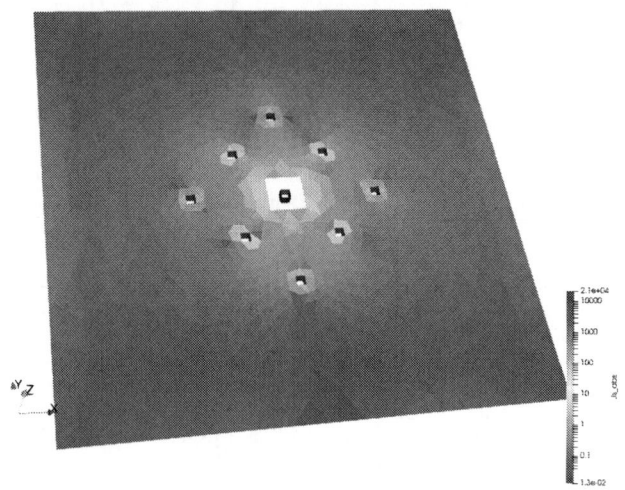

Fig. 2. Magnitude of the surface electric current $\mathcal{J}$, on PEC top ground plane and vias obtained with RWG MoM solution of MPIE at 100 MHz. Eight GRVs connect top finite ground plane to infinite bottom ground plane while infinite dielectric substrate is assumed.

In this preliminary study we conduct simulations using our StratUM3D academic solver, which applies the Method of Moments (MoM) solution of MPIE [4], and the FEKO commercial solver [5] with PEC metal model. In the first scenario, the model incorporates 1 GRV, with metal surfaces discretized into 1,500 triangles. In the second scenario, 4 GRVs are discretized using 1,650 first-order triangles.

TABLE I
MATERIAL, ANALYSIS, AND GEOMETRY PARAMETERS

Material Parameters	
Number of Metal Layers	3 (finite ground/via/infinite ground)
Number of Dielectric Layers	3 (air-substrate-air)
Relative Permittivity	3.15
Dielectric Loss Tangent	0.002
Substrate Thickness	220 micron
Metal	Lossless
Ground Plane Thickness	35.56 micron
Analysis Information	
Minimum Frequency	100 MHz
Maximum Frequency	60 GHz
Frequency Step	100 MHz
Reference Impedance	50 Ohm
Geometry Information	
Via Diameter	190 micron
Antipad Diameter	780 micron

The S-parameters, computed using both the StratUM3D solver and the FEKO commercial solver, are illustrated in Figure 3 for scenarios involving 1 to 8 GRVs. Although discernible impacts from the placement of GRVs are apparent at specific frequencies such as 40GHz. Future studies will include effect on GRVs return and insertion loss from metal loss handled through Leontovich impedance boundary condition in the MPIE and substrate truncation effects emulated through insertion of air gap into infinite layer of the substrate [1].

979-8-3503-5124-8/24 $31.00 © 2024 IEEE

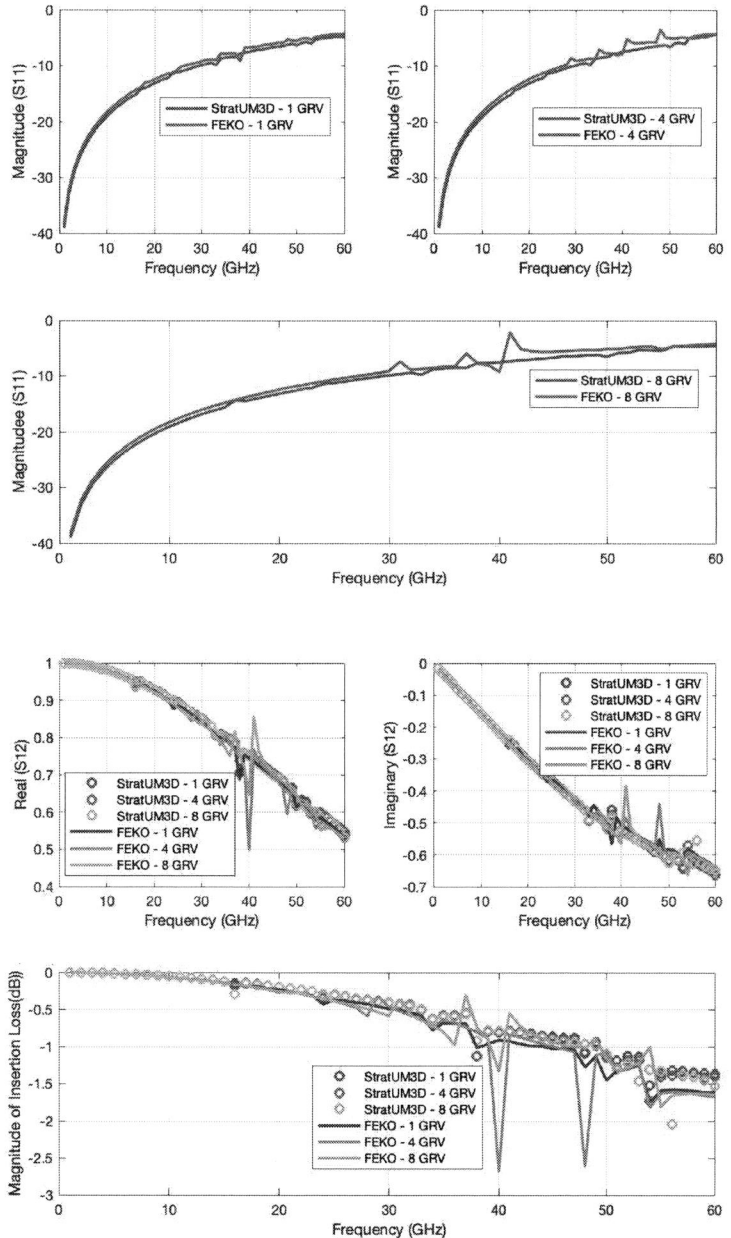

Fig. 3. S-Paremeters computed in StratUM3D and its comparison against FEKO in the frequency range from 100 MHz to 60 GHz.

IV. CONCLUSION

The paper describes a computational framework for conducting full-wave analysis of Ground Return Vias (GRVs) placement with layered media integral equations. It has been observed that GRVs interactions lead to the formation of resonance when two or more GRVs are present. Within the frequency range corresponding to the resonant response, the return path may degrade abruptly, resulting in significantly heightened transmission path discontinuities and insertion loss.

REFERENCES

[1] A. Niazi, et.al., "Full-Wave Analysis of Interconnects in Finite Substrates with Layered Media Formulation of SVS-EFIE for 3D Composite Metal-Dielectric Structures," 2023 IEEE EPEPS, Milpitas, USA.

[2] A. Aljamal, et.al., "Delta-Gap Source Excitation Model in SVS-EFIE for 3-D Interconnect Characterization," 2019 IEEE EPEPS, Montreal, CA.

[3] K. A. Michalski and D. Zheng, "Electromagnetic scattering and radiation by surfaces of arbitrary shape in layered media. I. Theory," IEEE Trans. Antennas Propag., vol. 38, no. 3, pp. 335-344, March 1990.

[4] Okhmatovski, V.I. Zheng, S. (2024). Theory and Computation of Electromagnetic Fields in Layered Media, IEEE Press/Wiley, 2024.

[5] "FEKO User's Manual," Altair Engineering, 2022

979-8-3503-5124-8/24 $31.00 © 2024 IEEE

Tree-Based Boosting for Efficient Estimation of S-Parameters for Package Electrical Analysis

Doğanay Özese[*], Mustafa Gökçe Baydoğan[*], Ahmet Cemal Durgun[†], and Kemal Aygün[‡]

[*]Department of Industrial Engineering, Boğaziçi University, İstanbul, Türkiye
[†]Department of Electrical and Electronics Engineering, Middle East Technical University, Ankara, Türkiye
[‡]Assembly and Test Technology Development, Intel Corporation, Chandler, AZ 85226 USA
doganay.ozese@std.bogazici.edu.tr, mustafa.baydogan@bogazici.edu.tr, acdurgun@metu.edu.tr, kemal.aygun@intel.com.tr

Abstract—We propose a gradient boosted tree surrogate model for S-parameter prediction in high frequency structures with limited training data. Compared to data-hungry neural networks, our approach achieves reasonable accuracy and trains significantly faster.

Index Terms—surrogate model, gradient boosted trees, second level interconnect

I. INTRODUCTION

In recent years, the adoption of surrogate models (SMs) in electromagnetic (EM) design and optimization has seen significant growth due to their enhanced optimization efficiency [1]. These models are computationally efficient and are developed using advanced statistical learning techniques, effectively replacing the need for resource-intensive EM simulations in the optimization workflow [2], [3].

SMs are trained with some known (i.e. labeled) data and generalized for an unknown solution space. However, the involvement of 3D EM simulations in the loop to obtain labeled data significantly increases the overall computational complexity, especially when the number of design choices and parameters increases [2]. Moreover, balancing data requirements and model complexity is a critical consideration in the development of SMs.

As dynamic systems are highly nonlinear and generate multiple outputs, generally at multiple frequency levels, this makes neural network (NN)-based learners (i.e. deep NN architectures) a natural choice as a SM. NNs, while powerful, often suffer from being excessively data-hungry because of their large number of parameters. This high demand for data is a significant drawback in scenarios where data is limited, as deep NN architectures may not perform well without sufficient data [4]. This limitation conflicts with the goal of rapid design space exploration and optimization, since a substantial number of experiments are needed to train a robust model. Furthermore, the complexity of hyper-parameter tuning and the necessity for extensive sampling further complicate the use of NNs. For quicker and more efficient optimization, we require models that are not only fast learners but also less dependent on vast amounts of data.

This work was supported in part by Semiconductor Research Corporation (SRC) under the contract 2023-PKG-3195.

We introduce a methodology to characterize the input space for predicting high-speed I/O interconnect S-parameters over a wide frequency range using gradient boosted trees (GBT). Gradient boosting constructs an ensemble predictor through gradient descent within a functional space, iteratively combining weaker models (base predictors) to form a strong predictor [5]. Unlike deep NNs, GBT are less data-hungry and incorporate regularization techniques, such as shrinkage and tree pruning, which prevent overfitting and enhance generalization. The iterative learning mechanism allows for continuous model improvement with feedback which reduces the need for extensive hyper-parameter tuning. These characteristics make GBT ideal for fast design space exploration and effective optimization, particularly in data-limited scenarios.

Generally, GBT is preferred as the state-of-the-art approach in single output situations [6]. However, our application requires the prediction of S-parameters over a wide frequency range. To reduce the problem complexity, we apply principal component analysis (PCA) on the outputs to reduce the target dimensions, where we simply apply singular value decomposition on S-parameters to represent the real and imaginary parts in lower dimensional space. On the reduced target space, we still have multiple outputs which necessitates the use of a multi-target implementation of GBT. Here, Catboost [5] is used to build a model on the design parameters to infer the transformed target space.

Our experiments with low, moderate, and high training size settings reveal the robustness of the proposed strategy. The GBT model consistently outperforms its NN-based counterpart in test accuracy and training efficiency for smaller training sizes, achieving lower NMSE and MSE values with significantly shorter training durations. However, with larger training sizes, the NN-based approach offers more precise predictions.

II. METHODOLOGY

The methodology begins with an input space $X_{N \times P}$, characterized by P design features and N instances, representing the geometric parameters of the design. Correspondingly, the output space $Y_{N \times M \times F}$ comprises complex S-parameters, consisting of N instances, M parameter indices, and F frequency points, capturing both real and imaginary components for each S-parameter. To manage this high-dimensional output effectively, PCA is employed as a dimensionality reduction

979-8-3503-5124-8/24 $31.00 © 2024 IEEE

technique, where the output response is parameterized as a linear combination of the eigenvectors corresponding to the largest singular values as given in (1). Here, S_{mn}, p_k, and a_{mn}^k are the S-parameter entry, the eigenvectors, and the associated coefficients (i.e. loadings), respectively. Utilizing PCA, the number of points to be predicted is significantly reduced from $M \times F$ to a more manageable number of principal components, i.e. C, where C is dynamically determined by PCA to explain a specified ratio of the variance.

$$S_{mn}(f) = \sum_{k=1}^{C} a_{mn}^k p_k(f) \tag{1}$$

Once the dataset is transformed into this reduced-dimensional space using the training data, GBT is trained on $X_{N \times P}$ to predict $Y_{N \times C}$ coefficients. GBT is constructed through a nonparametric gradient descent strategy, es expressed in (2), starting with a simple base prediction. The method sequentially adds tree-based models ($\hat{S}_k$) that are trained on the derivative of the loss of the previous models ($L(S, \hat{S}_{k-1})$) where L is the loss function. This approach corrects errors iteratively, where the learning rate(γ_k) is a crucial parameter in this process, as it controls the contribution of each new model, ensuring that adjustments to the predictions are made gradually. Regularization techniques such as shrinkage and tree pruning are incorporated to prevent overfitting and improve generalization. Although there are many hyper-parameters of the recent GBT implementations, it is known that they are robust to the settings of the most of them [6]. The most important hyper-parameters are the number of trees, the learning rate and the tree depth to control the learning process. It is also known that they are robust if these parameters are set within a reasonable range [6]. GBT facilitate efficient training, outperforming complex NNs in terms of speed.

$$\hat{S}_k = \hat{S}_{k-1} - \gamma_k \sum_{i=1}^{N} \nabla_{\hat{S}_{k-1}} L(S, \hat{S}_{k-1}) \tag{2}$$

During the prediction phase, PCA coefficients are inferred by the trained GBT. The predictions are subsequently subjected to an inverse transformation, converting them back to the original output space to retrieve the S-parameter predictions. This process is briefly illustrated in Fig. 1.

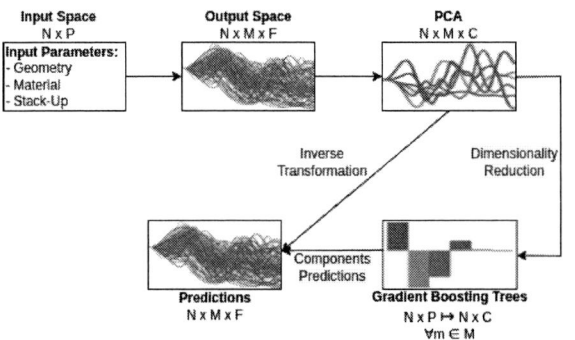

Fig. 1. Flowchart of the Training Process

TABLE I
CONTROL PARAMETERS OF THE PTH STRUCTURE

Parameter		Min	Max	Sample
PTH Pitch (μm)	v_p	300	1200	1110
Core Thickness (μm)	h_{core}	100	1200	512
PTH Diam. (μm)	d_{PTH}	100	250	130
PTH Pad Diam. (μm)	$d_{PTH-pad}$	110	500	234
BU Layer Thickness (μm)	h_{BU}	20	35	21
μvia Diam. (μm)	$d_{\mu via}$	30	70	48
BU Cu Thickness (μm)	$t_{c,BU}$	10	20	19
Core Cu Thickness (μm)	$t_{c,CORE}$	11	40	32
μvia Pad Diam. (μm)	$d_{\mu via-pad}$	31	140	86
μvia Top Antipad Rad. (μm)	r_{a,BU_t}	100	500	220
μvia Bot. Antipad Rad. (μm)	r_{a,BU_b}	100	500	103
PTH Top Antipad Rad. (μm)	r_{a,PTH_t}	50	500	422
PTH Bot. Antipad Rad. (μm)	r_{a,PTH_b}	50	500	129

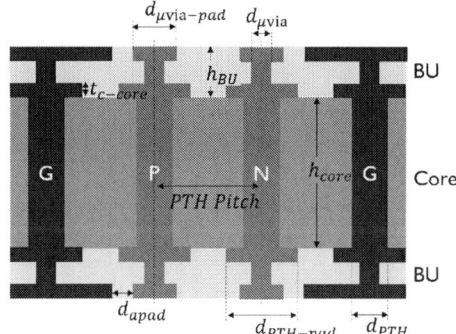

Fig. 2. PTH geometry.

III. NUMERICAL RESULTS

Our learning approach is evaluated on an application focusing on the design of a second level interconnect of a package [7], where the task is to predict the complex S-parameter matrix of a differential pair. The design of the plated through hole (PTH) geometry is depicted in Fig. 2. Parameter bounds of the samples are given in Table I. In order to assess the predictive performance, 680 samples based on Latin Hypercube Sampling (LHS) are determined. These are then simulated using full-wave EM solver to generate their corresponding S-parameters between 0.1-100 GHz at 100 MHz frequency steps. Given that the PTH structure is partially symmetric and reciprocal, the training target data, which are the output channels of the model, are identified as the real and imaginary components of the frequency responses for S_{11}, S_{12}, S_{13}, S_{14}, S_{33}, and S_{34}, resulting in a total of 12,000 output dimensions.

To illustrate the predictive performance of the models, 180 out of 680 samples are left out as test samples. We also select $N \in \{20, 50, 100, 500\}$ to compare the approaches with alternative training sizes. As a benchmark, a spectral transposed convolutional NN with causality and passivity enforcement layers (S-TCNN + CEL + PEL, TCNN in short) [7] is used. The hyperparameters associated with TCNN adhere to the specifications presented in [7] except the cases where $N \in \{20, 50, 100\}$ for which the number of epochs is set to 150 to avoid overfitting. For our proposal, the components

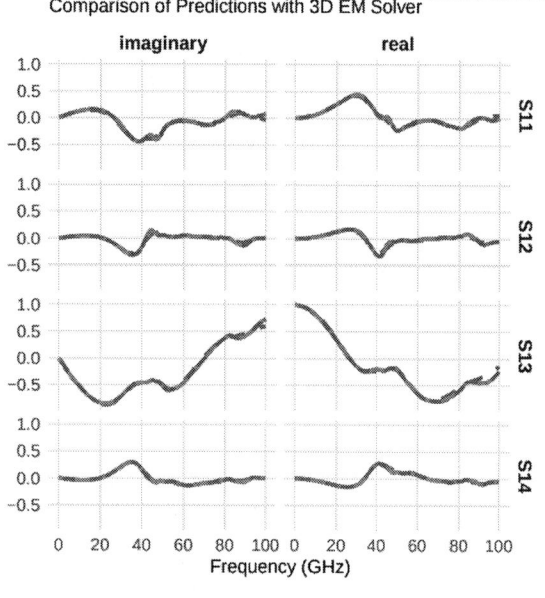

Fig. 3. Comparison of predicted real and imaginary parts of S-Parameters with 3D EM simulation.

behavior. These results suggest that GBT on reduced spaces is a suitable choice for fast design space exploration and optimization in data-limited scenarios.

TABLE II
COMPARISON TCNN AND CATBOOST AT VARIOUS TRAINING SIZES.

Training Size	Method	NMSE	MSE	Training Dur. (sec)
20	PCA	0.3877	0.0049	0.13
	TCNN	0.8214	0.0301	421.19
	GBT	0.6796	0.0261	3.21
50	PCA	0.1115	0.0011	0.19
	TCNN	0.5627	0.0187	432.81
	GBT	0.4547	0.0161	11.25
100	PCA	0.0688	0.0007	0.24
	TCNN	0.3281	0.0108	486.49
	GBT	0.2025	0.0082	25.45
500	PCA	0.0422	0.0004	0.71
	TCNN	0.0624	0.0017	12224.13
	GBT	0.0805	0.0032	71.20

IV. CONCLUSION

This study demonstrates the effectiveness of using GBT for predicting high-speed I/O interconnect S-parameters over a wide frequency range. By integrating PCA to reduce target dimensions, the proposed methodology effectively addresses the challenges associated with data-hungry deep NNs and the high computational complexity of 3D EM simulations. Our experimental results indicate that GBT models offer superior test accuracy and training efficiency, particularly in data-limited scenarios. Although NNs provide more precise predictions with larger training sizes, the GBT approach remains a robust and scalable alternative for rapid design space exploration and optimization. This highlights the potential of GBT models as a valuable tool in the EM design and optimization process, especially when data availability is a constraint. Future work may explore further enhancements in the GBT framework and the integration of additional regularization techniques to further improve model performance.

REFERENCES

[1] F. Zeng, B. Jianguo, Y. Yu, L. Yang, X. Zi, "Optimum Design Of Permanent Magnet Synchronous Generator Based On Maxpro Sampling and Kriging Surrogate Model", IEEJ Trans Elec Electron Eng, vol. 15, no. 2, p. 278-290, 2019.

[2] L. Zhang, X. Huang, J. He, X. Cen, Y. Liu, "Parameter Optimization Study Of Gas Hydrate Reservoir Development Based On a Surrogate Model Assisted Particle Swarm Algorithm", Geofluids, vol. 2022, p. 1-12, 2022.

[3] F. Passos, R. Gonzalez-Echevarria, E. Roca, R. Castro-Lopez, F. Fernández, "A Two-step Surrogate Modeling Strategy For Single-objective and Multi-objective Optimization Of Radiofrequency Circuits", Soft Comput, vol. 23, no. 13, p. 4911-4925, 2018.

[4] A. Adadi, A survey on data-efficient algorithms in big data era. Journal of Big Data, 8(1), 24, 2021.

[5] L. Prokhorenkova, G. Gusev, A. Vorobev, A. V. Dorogush, A. Gulin, CatBoost: unbiased boosting with categorical features. Advances in neural information processing systems, 31, 2018.

[6] A. S. Anghel, N. Papandreou, T. Parnell, A. De Palma, H. Pozidis, Benchmarking and Optimization of Gradient Boosting Decision Tree Algorithms. In Annual Conference on Neural Information Processing Systems, 2018.

[7] H. M. Torun, A. C. Durgun, K. Aygün and M. Swaminathan, "Causal and Passive Parameterization of S-Parameters Using Neural Networks," in IEEE Transactions on Microwave Theory and Techniques, vol. 68, no. 10, pp. 4290-4304, 2020.

explaining 99.5% of the variance in the outputs are selected (i.e. $C = 20$ when $N = 20$ and $C = 40$ for others). Here, Catboost implementation of GBT [5] is utilized and we use the default parameters except the number of trees which is set to 10,000 trees. Performance of target transformation with PCA is also reported. As GBT is trained on the transformed space, its error rate is bounded by PCA reconstruction error. Therefore we also report the information lost by the transformation. The chosen metrics to evaluate the numerical accuracy of both models are the normalized mean squared error (NMSE) across each frequency response in the test set, defined as in (3) together with the Mean Squared Error (MSE). We also consider training duration to illustrate the computational requirements.

$$NMSE = \frac{1}{NM} \sum_{i=1}^{M} \sum_{j=1}^{N} \left(\frac{\sum_{f=1}^{F} \left(S_{ij} - \hat{S}_{ij} \right)^2}{\sum_{f=1}^{F} \left(S_{ij} - \bar{S}_{ij} \right)^2} \right) \quad (3)$$

Table II shows that PCA reconstruction error in terms of NMSE and MSE decreases with increasing training sizes. These errors serve as the lower error bound GBT would achieve if the predictions were perfect. GBT consistently outperforms TCNN in terms of both test accuracy and training efficiency across various training sizes. It achieves lower NMSE and MSE values with significantly shorter training durations compared to TCNN. As the training size increases, both models improve, but GBT remains more efficient and scalable. Fig. 3 demonstrates a sample result, for design parameters listed in Table I, of a 3D EM solver alongside the predictions of TCNN and GBT when $N = 500$. Although TCNN offers more precise predictions compared to the proposed strategy, GBT also provides reasonable prediction

Causal RL Prediction of Fine-Pitch Interconnects Using Neural Networks

Hasan Said Ünal*, and Ahmet Cemal Durgun*

*Department of Electrical and Electronics Engineering, Middle East Technical University, Ankara, Türkiye

acdurgun@metu.edu.tr

Abstract—**In this study, we compare physics-aware neural networks for modeling fine-pitch interconnects. Results show a 5-fold reduction in test loss when imposing DC resistance through analytical equations and preserving the causality relation between resistance and inductance.**

Index Terms—**fine pitch interconnect, causal RLGC parameters, neural network**

I. INTRODUCTION

Fine pitch interconnects are the key building blocks for heterogeneous integration facilitating a high I/O bandwidth density across chiplets. These interconnects can be realized using silicon [1], glass [2], or organic interposers [3], or Embedded Multi-die Interconnect Bridge (EMIB) [4], all comprising micro-bump breakout regions and horizontal transmission lines under 10 μm pitch. Optimizing these communication channels involves numerous design parameters, constraints, and performance metrics, necessitating rigorous optimization with full-wave electromagnetic simulations. While accurate, 3D full-wave analysis is computationally intensive because of small feature sizes and wide frequency ranges. Alternatively, wide-band tabular RLGC modeling offers a faster method, with verified accuracy for horizontal transmission analysis.

Recently, machine learning (ML)-based modeling of high-speed channels for fast design space exploration and optimization applications has become popular. In particular, RLGC parameters prediction using neural networks (NNs) at a single frequency for microstrip [5] and striplines with temperature-dependent conductivities [6] have been reported. In [7], a Gaussian process model was utilized to predict frequency-dependent RLGC parameters, where the frequency is considered as an input parameter. In [8], frequency dependency was handled by separately predicting the DC and AC components of R and G parameters. [9] proposed a training set optimization by maintaining the width/spacing ratio of the lines. In these applications, the ML models were developed using input and output relations and considering the system of interest as a black-box without utilizing the underlying physics.

This study builds on the previous work to improve the prediction accuracy by incorporating known physical infor-

This study has been supported by the 2232 International Fellowship for Outstanding Researchers Program of TÜBİTAK (Project No: 118C198). The views, opinions, and/or findings expressed are those of the authors and should not be interpreted as representing the official views of policies of TÜBİTAK.

mation into the model. We develop and compare two physics-aware NN models with black-box models. One imposes hard constraints on the R parameter and uses the Kramers-Kroning relations to enforce causality, while the other employs physics-based piecewise functional fit models. Comparisons show significant prediction accuracy improvements over black-box methods when trained with additional information.

II. INTERCONNECT MODELING

A typical fine-pitch interconnect comprises several parallel-routed signal and ground traces whose ratio is determined by the bandwidth density and crosstalk requirements. Fig. 1 shows the 2D model of a fine pitch interconnect with a 2:1 signal-to-ground ratio. The design parameters are the trace width (tw), ground width (gw), trace-to-trace spacing (tts), trace-to-ground spacing (tgs), metal thickness (mt), dielectric thickness (dt), dielectric constant, and the metal conductivity.

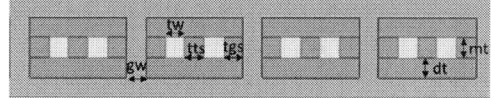

Fig. 1. Fine pitch interconnect geometry.

The dielectric loss tangent is assumed to be zero as the dielectric losses in these interconnects are generally negligible because of short channel lengths. Consequently, the conductance is zero and the capacitance is frequency-independent, not posing any challenges for NN modeling. Hence, we focus on the R and L terms, particularly the self terms. Although the R and L parameters were predicted separately in many black-box ML applications, they are related through the well-known Kramers-Kronig relations. For a causal interconnect, with a total serial impedance given by (1), the frequency-dependent $R_\omega(\omega)$ and $L_\omega(\omega)$ are related through the Hilbert transform relation expressed in (2) [10]. Here, R_0 and L_{ext} are the DC resistance and external inductance of the line.

$$Z(j\omega) = R_0 + R_\omega(\omega) + j\omega L_{ext} + j\omega L_\omega(\omega) \qquad (1)$$

$$\omega L_\omega(\omega) = \frac{1}{\pi} \int_0^\infty \frac{dR_\omega(\omega')}{d\omega'} \ln\left|\frac{\omega' + \omega}{\omega' - \omega}\right| d\omega' \qquad (2)$$

The frequency-dependent behavior arises from the skin effect, which pushes the currents to the conductor surface

979-8-3503-5124-8/24 $31.00 © 2024 IEEE

as the frequency increases. For larger cross-section traces, this effect is significant at lower frequencies. Therefore, the frequency dependency of the resistance is often represented by $\sqrt{f}$. However, this is not valid for fine-pitch interconnects as the current density inside the signal traces can still be high at high frequencies. To account for this issue, [11] proposes a piecewise function, obtained by curve fitting to the measurement data. In this model, the spectrum is divided into low-, middle-, and high-frequency (LF, MF, and HF) regions. In the LF region, the total resistance is the sum of the DC resistances of the signal trace and the ground plane. In the HF region, the resistance follows the $\sqrt{f}$ trend. In the MF region, when the skin effect affects the ground plane but not the signal trace, the resistance is given by (3), where R_{S0} is the DC resistance of the signal line and $k_1, k_2,$ and k_3 are coefficients determined by the current density distribution. The corresponding inductance terms, satisfying the causality, are obtained by dividing the frequency-dependent resistance by $2\pi f$ and adding the external inductance (L_{ext}), as in (4).

$$R_{MF} = R_{S0} + \frac{f}{k_1 f + k_2 \sqrt{f} + k_3} \tag{3}$$

$$L_{MF} = L_{S0} + \frac{1}{2\pi(k_1 f + k_2 \sqrt{f} + k_3)} + L_{ext} \tag{4}$$

III. NEURAL NETWORK ARCHITECTURES

We build four different NN models to test the impact of additional physics-based information on the prediction accuracy. The first model is a black-box NN (NN-BB), comprising linear layers and blindly minimizing the geometrical distance between the predicted and actual results.

The second model utilizes the analytic DC resistances of the signal line and the ground plane. Expectedly, an accurate NN model should predict R_0 at DC, while matching the overall response. This can be achieved by including the distance between the predicted and analytical DC resistances to the loss function, as a soft constraint. A more effective way is referred to as the hard-constrained NN (NN-HC), where $R_\omega(\omega)$ is predicted by the NN, and the analytical resistance value is later added to the predicted response.

The third model is based on the curve fitting method of [11] and is named as piecewise function fit NN (NN-PFF), which has 7 outputs listed in Table I. The transition frequency point between the HF and MF regions is denoted as f_{HF}. Because of the different scales of the fitting parameters k_1, k_2, and k_3, we impose constraints by using sigmoid activation function in the output layer k_1, k_2, k_3, and f_{HF}. The remaining outputs do not have an activation function.

The last model, named Hilbert NN (NN-H) builds on NN-HC by utilizing the Hilbert transform relation between R and L. NN-H has two branches as depicted in Fig. 2. The first branch predicts L_0 and L_{ext}. The second branch predicts an extrapolated resistance [12] and generates a Hilbert transform pair by implementing (2) [10]. The DC inductance L_0 is transferred to the Hilbert layer for improved accuracy. Then R_0 and L_{ext} are added to the predicted resistance and inductance.

All models employ Adam optimizer with a learning rate of 0.07 and L^1 loss. The hidden layers have tanh() activation function and batch normalization. Other critical hyperparameters are listed in Table I.

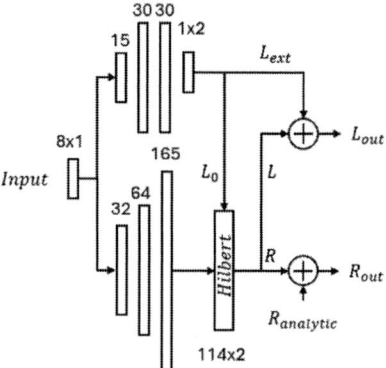

Fig. 2. Hilbert NN (NN-H) model architecture.

IV. NUMERICAL RESULTS

The interconnect model was simulated using Ansys Q2D, collecting RLGC parameters at 114 points from 0 to 20 GHz, with a logarithmic scale for frequencies below 1 GHz. Using the Latin Hypercube Sampling algorithm, 819 samples were generated based on the input parameters and their ranges. The dataset was split into 655 training samples and 164 test samples. Table II lists the input parameters and their ranges.

Table I lists the MAE of all implemented NN models. Expectedly, feeding the NN with physics-based information considerably improves the prediction accuracy. The most significant contribution comes from imposing the hard constraint on the DC resistance. NN-H has the highest accuracy as utilizing the Hilbert transform and training the NN with R and L together, slightly improves the performance. NN-H achieved a test error of 0.0664 Ω/mm and 0.0021 nH/mm for the R and L terms, corresponding to 0.64% and 0.67% mean absolute percentage errors, respectively. Although NN-PFF preserves the causality relation between R and L, the piecewise function model does not completely represent the frequency-dependent behavior of the interconnect. Notably, all models performed similarly for the inductance prediction.

Figs. 3 and 4 compare the predicted per unit length R and L profiles with the simulation results for a sample geometry with design parameters given in Table II. The corresponding MAE errors of NN-H for the R and L terms are 0.0685 Ω/mm and 0.0016 nH/mm, respectively.

V. CONCLUSION

We implemented several physics-aware NN models for causal RL parameter prediction of fine pitch interconnects. The results showed that the prediction accuracy significantly improved when the NN was trained with physics-based information. However, inadequate physical models may lead

979-8-3503-5124-8/24 $31.00 © 2024 IEEE

TABLE I
NEURAL NETWORK ARCHITECTURES AND PERFORMANCE

| | Input | Hidden Layer Nodes | | | | | | Results | | |
		Layer-1	Layer-2	Layer-3	Layer-4	Layer-5	Layer-6	Outputs	MAE R	MAE L
NN-BB	8	32	64(p=0.1)	114	-	-	-	$R_0 + R_w(w)$	0.3472	0.0021
		32	64(p=0.1)	114	-	-	-	$L_w(w)$		
NN-HC	8	32	64(p=0.1)	114	-	-	-	$R_w(w)$	0.0777	0.0025
		32	64(p=0.1)	114	-	-	-	$L_w(w)$		
NN-PFF	8	15	30	30	30	30	7	k_1, k_2, k_3, f_{HF}	0.1407	0.0048
								L_0, L_{S0}, L_{ext}		
NN-H	8	32	64(p=0.1)	165	-	-	-	$R_w(w)$	0.0664	0.0021
		15	30	30	2	-	-	L_0, L_{ext}		

TABLE II
FINE PITCH INTERCONNECT DESIGN PARAMETERS AND THEIR RANGES

Parameter	Range	Sample
Trace width (μm)	0.2–5.0	0.93
Ground width (μm)	0.2–5.0	2.95
Trace to trace spacing (μm)	0.2–5.0	2.90
Trace to ground spacing(μm)	0.2–5.0	3.71
Metal thickness (μm)	0.2–5.0	4.05
Dielectric thickness (μm)	0.2–5.0	3.86
Dielectric constant	2.0–5.0	4.25
Metal conductivity (S/m)	$0.1 - 6.0 \times 10^7$	3.90×10^7

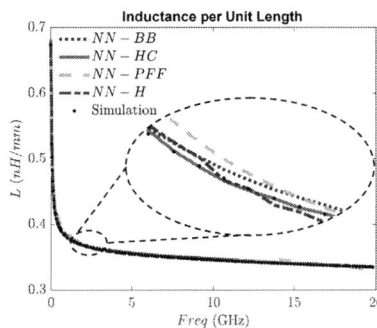

Fig. 4. Comparison of inductance predictions for a sample geometry.

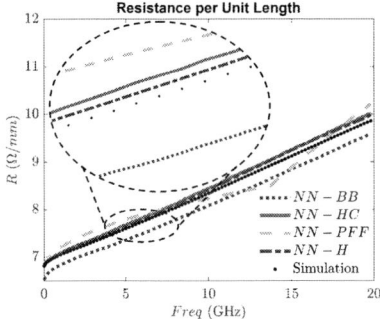

Fig. 3. Comparison of resistance predictions for a sample geometry.

to inferior performance. We also demonstrated that imposing causality relations between the R and L terms also improved the performance. We observed a 5-fold reduction in the test loss for R predictions. The results once more verified that NNs are powerful tools for high-frequency system modeling when they are combined with the underlying physics of the system.

REFERENCES

[1] C. Narayan and S. Purushothaman, "Thin film transfer process for low cost mcm's," in *Proceedings of 15th IEEE/CHMT International Electronic Manufacturing Technology Symposium*, 1993, pp. 373–380.

[2] X. Cui, D. Bhatt, F. Khoshnaw, D. A. Hutt, and P. P. Conway, "Glass as a substrate for high density electrical interconnect," in *2008 10th Electronics Packaging Technology Conference*, 2008, pp. 12–17.

[3] K. Oi, S. Otake, N. Shimizu, S. Watanabe, Y. Kunimoto, T. Kurihara, T. Koyama, M. Tanaka, L. Aryasomayajula, and Z. Kutlu, "Development of new 2.5d package with novel integrated organic interposer substrate with ultra-fine wiring and high density bumps," in *2014 IEEE 64th Electronic Components and Technology Conference (ECTC)*, 2014, pp. 348–353.

[4] R. Mahajan, Z. Qian, R. S. Viswanath, S. Srinivasan, K. Aygün, W.-L. Jen, S. Sharan, and A. Dhall, "Embedded multidie interconnect bridge—a localized, high-density multichip packaging interconnect," *IEEE Transactions on Components, Packaging and Manufacturing Technology*, vol. 9, no. 10, pp. 1952–1962, 2019.

[5] S. Newberry and A. Zadehgol, "A deep neural network modeling methodology for extraction of rlgc parameters in μ-wave and mm-wave transmission lines," in *2022 IEEE International Symposium on Electromagnetic Compatibility & Signal/Power Integrity (EMCSI)*, 2022, pp. 74–79.

[6] Q. Xiao, M. Tang, Z. Liu, and J. Mao, "Distributed parameter modeling for coupled striplines based on artificial neural network," in *2022 IEEE 10th Asia-Pacific Conference on Antennas and Propagation (APCAP)*, 2022, pp. 1–2.

[7] H. M. Torun, M. Larbi, and M. Swaminathan, "A bayesian framework for optimizing interconnects in high-speed channels," in *2018 IEEE MTT-S International Conference on Numerical Electromagnetic and Multiphysics Modeling and Optimization (NEMO)*, 2018, pp. 1–4.

[8] H. Kim, C. Sui, K. Cai, B. Sen, and J. Fan, "Fast and precise high-speed channel modeling and optimization technique based on machine learning," *IEEE Transactions on Electromagnetic Compatibility*, vol. 60, no. 6, pp. 2049–2052, 2018.

[9] B. Pu, H. Kim, X.-D. Cai, B. Sen, C. Sui, and J. Fan, "Training set optimization in an artificial neural network constructed for high bandwidth interconnects design," *IEEE Transactions on Microwave Theory and Techniques*, vol. 70, no. 6, pp. 2955–2964, 2022.

[10] S. Grivet-Talocia, H.-M. Huang, A. Ruehli, F. Canavero, and I. Elfadel, "Transient analysis of lossy transmission lines: an efficient approach based on the method of characteristics," *IEEE Transactions on Advanced Packaging*, vol. 27, no. 1, pp. 45–56, 2004.

[11] D. M. Cortés-Hernández, R. Torres-Torres, M. Linares-Aranda, and O. González-Díaz, "Piecewise physical modeling of series resistance and inductance of on-chip interconnects," *Solid-State Electronics*, vol. 120, pp. 1–5, 2016. [Online]. Available: https://www.sciencedirect.com/science/article/pii/S0038110116000368

[12] H. M. Torun, A. C. Durgun, K. Aygün, and M. Swaminathan, "Causal and passive parameterization of s-parameters using neural networks," *IEEE Transactions on Microwave Theory and Techniques*, vol. 68, no. 10, pp. 4290–4304, 2020.

979-8-3503-5124-8/24 $31.00 © 2024 IEEE

A robust optimization approach for High Bandwidth Memory interposer using Machine Learning

1st Anandajith Jinesh
Analog Design - Signal Integrity
Advanced Micro Devices, Inc.
Markham ON, Canada
anand.jinesh@amd.com

2nd Xuan Chen
Analog Design - Signal Integrity
Advanced Micro Devices, Inc.
Markham ON, Canada
xuan.chen@amd.com

Abstract—**Conventional optimization approaches are computationally inefficient and insufficient for time-sensitive optimization exercises. This paper presents a machine learning (ML) based approach to optimize the interposer structure of High Bandwidth Memory (HBM) by optimizing channel parameters.**
Index Terms—**HBM,machine learning, signal integrity**

I. INTRODUCTION

The technological landscape around the globe has been reshaped by the advancement of ML. These ML algorithms have transformed the world of data-processing, image recognition and beyond. However, as the complexity of these algorithms increase, their computational and data bandwidth has surged.

One of the promising solutions to the challenge is the HBM technology. HBM is a high-performance RAM interface for 3D-stacked DRAM. By facilitating faster data transfer rates and higher bandwidth, HBM enables ML models to operate more efficiently and effectively, particularly in scenarios requiring the processing of large volumes of data in real time. As the technology and bandwidth evolve, designers are called upon to determine the most optimal design. This calls for the need to implement rapid optimization approaches.

Existing studies have profusely discussed methods of improving the power integrity (PI) through the reduction of power supply noise induced jitter (PSIJ) and enhanced capacitor placement. Some of the previous work in the field include [1]–[3]. However, there are only limited studies done on interposer optimization for signal integrity (SI) requirements.

[4] discussed the strategies to improve the SI performance of the interposer by comparing eye-diagrams of models generated through electromagnetic (EM) simulation results. However, this approach is heavily reliant on performing EM simulations. Due to the computational resources required for EM extractions, this approach does not have the capability to evaluate a large number of interposer designs in short periods. [5] discusses the implementation of a convolutional neural network (CNN) in the optimization of HBM, but does not include SI or interposer considerations.

To address this challenge, this paper introduces ML techniques to expedite the design cycle. By leveraging ML algorithms, designers can efficiently explore the vast design space, identify potential trade-offs, and converge on optimal

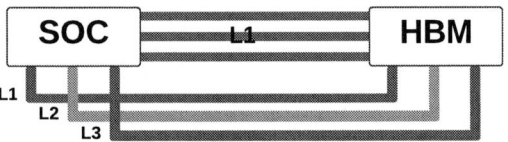

Fig. 1. Interposer Structure.

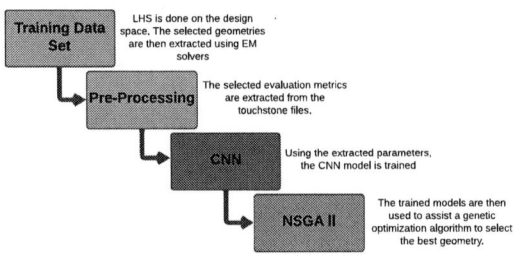

Fig. 2. High Level Algorithm Explanation.

solutions in a fraction of the time compared to traditional design methodologies.

This paper conducted an optimization study under a fixed pitch requirement using both the conventional and ML-based approach. The results and the time consumed by each approach was compared against one another.

II. INTERPOSER STRUCTURE

The interposer in HBM is a silicon die that sits between the HBM memory stacks and the host processor die. It functions as a physical and electrical interface, facilitating the vertical integration of memory dies through through-silicon vias (TSVs) and micro-bumps.

The design of the interposer is critical because it contains the routing for all the input/output signals between the stacked memory dies and the processor. These signals include not only data but also power, clock, and various control signals. By routing these connections through the silicon interposer, manufacturers can achieve a high density of interconnections with reduced parasitic capacitance and inductance, resulting in lower power consumption and higher data transfer speeds.

979-8-3503-5124-8/24 $31.00 © 2024 IEEE

III. Conventional Optimization Approach

Optimization of interposers for HBM structures was conventionally done by performing EM extraction for all possible trace width (TW) and ground width (GW) combinations within the given pitch. The extracted touchstone models are then used to generate eye-diagrams. The best geometry is identified by comparing the eye-diagrams against one another. This approach can ensure a comprehensive evaluation, but it has significant disadvantages. Performing EM extraction can be computationally heavy and can lead to large extraction times. As the pitch increases, the number of combinations will continue to increase. Hence, this approach is not a sustainable practice.

IV. Optimization Algorithm

A. Evaluation Metrics

S-Parameter metrics such as insertion loss (IL), return Loss (RL), far-end crosstalk (FEXT) and near-end crosstalk (NEXT) are crucial in evaluating high-performance communication systems, such as HBM. These parameters are essential for assessing the signal integrity and overall performance of the system. These metrics provides insight into different aspects of the transmission path's performance, reflecting both the effectiveness of the interposer design in HBM and the integrity of signal transmission at high frequencies.

B. Training Data Set

The target for selecting the sampling data points was to effectively capture the the entire design space using minimal number of sampling points. In order to achieve this, a combined approach including a grid search and a quasi-random sampling approach was implemented. For the grid search, the maxima, minima and median points of the dataset was captured. For the quasi-random sampling approach, Latin hypercube sampling (LHS) was implemented.

LHS is a statistical sampling methods to generate data-points from multidimensional distributions. LHS helps in attaining an evenly distributed and comprehensive sampling of the design space. This is achieved by dividing the cumulative distribution function of each variable into equally probably intervals, these intervals are then sampled exactly once. This improves the robustness and efficiency of the model training process [6] . Using both these sampling approaches, twenty sampling point were selected, sixteen points from LHS and four points from grid search. Using these points, interposer models were build and EM extractions were complete. Fig.5, denotes the sampling points which were extracted.

C. Convolutional Neural Network (CNN)

Convolutional neural networks (CNNs) are a class of deep neural networks that have revolutionized the field of computer vision and beyond. For this study, a CNN model was adapted to perform as a regressor to predict the chosen evaluation metrics based on a given interposer geometry.

Convolutional layers for feature extraction, pooling layers to reduce spatial dimensions, and fully connected layers for

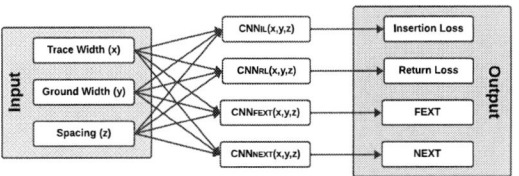

Fig. 3. CNN Block Diagram.

interpretation make up the CNN architecture for numerical regression. The output layer has a single neuron with a linear activation function to predict continuous values, optimized using mean squared error (MSE) loss.

As seen in Fig.3, the processed dataset from the previous section was then used as a training set for the CNN model. The model was trained to predict the IL, RL, FEXT, and NEXT based on the inputs of TW, GW, and Spacing. The four blocks in Fig.3 refer to the four functions that were built to predict the corresponding metrics . To evaluate the CNN models, a comparison was done between the extracted results from the conventional approach, and the output predicted by the CNN model. Fig.4 shows the R-score and the MSE of the CNN model in predicting IL, RL, FEXT, and NEXT at Nyquist frequency.

D. Non-Dominated Sorting Genetic Algorithm II (NSGA II)

NSGA II is a widely used algorithm in the field of multi-objective optimization. Developed by [7], NSGA-II algorithm has a non-dominated sorting approach to classify the population into different fronts based of the level of domination. This method can effectively categorize the solutions based on level of domination.

The functions generated using the CNN was then used as the minimization objectives for the model, while using design rule check (DRC) and design space constraints. The DRC function from Eq.8 was built to use the input geometries to verify the design rules.

$$1\mu m < x < 7\mu m \tag{1}$$

$$1\mu m < y < 7\mu m \tag{2}$$

$$z = (Pitch - x - y)/2 \tag{3}$$

$$min \ f_1(x,y,z) = |CNN_{IL}(x,y,z)| \tag{4}$$

$$min \ f_2(x,y,z) = CNN_{RL}(x,y,z) \tag{5}$$

$$min \ f_3(x,y,z) = CNN_{FEXT}(x,y,z) \tag{6}$$

$$min \ f_4(x,y,z) = CNN_{NEXT}(x,y,z) \tag{7}$$

$$DRC(x,y,z) <= 0 \tag{8}$$

The NSGA algorithm was allowed to search the design space, the points that were evaluated by the algorithms can be see in Fig.5. The model was used to sweep through five hundred points, using random float sampling and ten off-springs per generation. At each evaluated point, the CNN model was called to predict the selected evaluation metrics. After completion of the sweep, the best point was determined using normalized weights.

979-8-3503-5124-8/24 $31.00 © 2024 IEEE

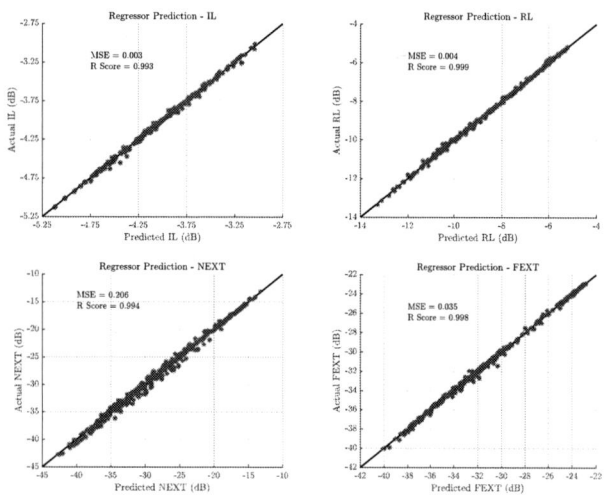

Fig. 4. (a) IL Prediction (b) RL Prediction (c) NEXT Prediction (d) FEXT Prediction.

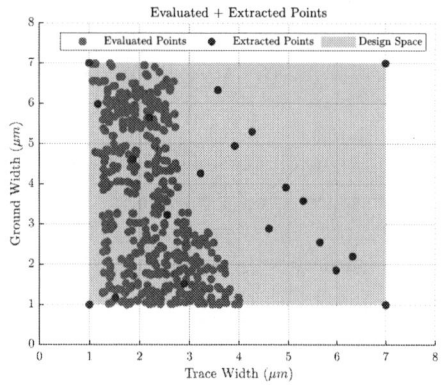

Fig. 5. Design Space.

V. RESULTS

Fig.5 shows a representation of the design space that was evaluated for the study, 1um to 7um for both TW and GW. LHS was performed on the design space to determine the data points that need to be extracted in order to capture the entire design space. Using the extracted points, CNN models were trained to predict IL,RL,FEXT and NEXT. These models were then used as minimization functions for the NSGA algorithm. The NSGA algorithm evaluated various points in the design space and converged on the best geometry. Fig.5 depicts the points that were evaluated by the NSGA algorithm. THE NSGA model thoroughly searched the Low TW and Large GW region of the design space, which meets expectations.

In order to verify the geometry predicted by the ML-based algorithm, the geometry was then EM extracted and compared with the best geometry from the conventional approach. The results of this comparison is presented in Table.I. The ML based approach has a 86.11% reduction in total extraction time. The optimized geometry also has also shown a 46.15%

increase in Eye Width (EW) with crosstalk (xtk) and a 30.28% increase in Eye Height (EH) with xtk.

TABLE I
RESULT SUMMARY

Optimization Method	Conventional	ML Based	Delta
Number of Extractions	144	20	-86.11 %
Extraction Time	864 hrs	120 hrs	-86.11 %
IL	-5.77 dB	-4.27 dB	-26 %
RL	-9.53 dB	-10.21 dB	7.14 %
FEXT	-36.86 dB	-40.52 dB	9.93 %
NEXT	-39.12 dB	-43.69 dB	11.68 %
EW without xtk	0.43 UI	0.61 UI	41.86 %
EW with xtk	0.39 UI	0.57 UI	46.15 %
EH without xtk	168.13 mV	228.20 mV	35.73 %
EH with xtk	156.64 mV	204.07 mV	30.28 %

VI. CONCLUSION

This paper proposed the implementation of ML algorithms for the optimization of HBM interposer. The conventional optimization approach has a proven track record to improve system performance. However, the process is time consuming and computationally inefficient, often requiring substantial computational power and time without guaranteeing the discovery of the optimal solution within a practical time-frame. The ML based approach proposed in this paper can be used to achieve better results with a significant reduction in processing time. In the future, the ML model should be capable of analyzing more parameters to enhance the robustness of the converged geometry.

REFERENCES

[1] T. Shin, H. Park, D. Lho, K. Kim, B. Sim, S. Kim, J. Kim, S. Choi, J. Yoon, J. Song, S. Chun, and J. Kim, "Si/pi co-design of 12.8 gbps hbm i/o interface using bayesian optimization for psij reduction," in *2023 IEEE Symposium on Electromagnetic Compatibility and Signal/Power Integrity (EMC+SIPI)*, 2023, pp. 662–667.

[2] H. Park, H. Kim, H. Kim, J. Park, S. Choi, J. Kim, K. Son, H. Suh, T. Kim, J. Ahn, and J. Kim, "Versatile genetic algorithm-bayesian optimization(ga-bo) bi-level optimization for decoupling capacitor placement," in *2023 IEEE 32nd Conference on Electrical Performance of Electronic Packaging and Systems (EPEPS)*, 2023, pp. 1–3.

[3] H. Park, T. Shin, S. Kim, D. Lho, B. Sim, J. Song, K. Kong, and J. Kim, "Scalable transformer network-based reinforcement learning method for psij optimization in hbm," in *2022 IEEE 31st Conference on Electrical Performance of Electronic Packaging and Systems (EPEPS)*, 2022, pp. 1–3.

[4] K. Cho, H. Lee, H. Kim, S. Choi, Y. Kim, J. Lim, J. Kim, H. Kim, Y. Kim, and Y. Kim, "Design optimization of high bandwidth memory (hbm) interposer considering signal integrity," in *2015 IEEE Electrical Design of Advanced Packaging and Systems Symposium (EDAPS)*, 2015, pp. 15–18.

[5] M. Zhu, Y. Zhuo, C. Wang, W. Chen, and Y. Xie, "Performance evaluation and optimization of hbm-enabled gpu for data-intensive applications," in *Design, Automation and Test in Europe Conference and Exhibition (DATE), 2017*, 2017, pp. 1245–1248.

[6] H. He, Q. Zhou, T. Sun, H. Cheng, and Q. Zuo, "Reliability evaluation based on modified latin hypercube sampling and minimum load-cutting method," in *2015 5th International Conference on Electric Utility Deregulation and Restructuring and Power Technologies (DRPT)*, 2015, pp. 465–471.

[7] K. Deb, A. Pratap, S. Agarwal, and T. Meyarivan, "A fast and elitist multi-objective genetic algorithm: Nsga-ii," *IEEE Transactions on Evolutionary Computation*, vol. 6, no. 2, pp. 182–197, 2002.

Hand-drawn Circuit Schematic Digitization and Netlisting using Machine Learning with Emphasis on Signal Integrity Applications

1st Anuj Mathur
Dept. of Electronics
Carleton University
Ottawa, Canada
Anujmathur@cmail.carleton.ca

2nd Ramachandra Achar
Dept. of Electronics
Carleton University
Ottawa, Canada
Achar@doe.carleton.ca

Abstract—**There is an emerging keen interest in the design and academic communities to automatically generate a simulatable netlist from hand-drawn schematics. This paper presents a novel machine learning-based circuit digitization and netlisting methodology.**

Index Terms—**netlist generation, hand-drawn schematic, computer aided design**

I. INTRODUCTION

Hand-written schematics are often used by designers for development of concepts during the design process. Transferring the hand-drawn schematics to schematic tools is often the initial step in the circuit design process, followed by netlist extraction and simulation. This step can be avoided by automatically generating the netlist through digitization of hand drawn circuit schematics. With larger systems, the digitization process can become a time consuming and laborious task while suffering from inaccuracies.

A netlist, which is used to describe a circuit's electrical connectivity, is comprised of component, text and nodal information. Therefore, the digitization process from a circuit schematic requires electrical component detection and classification, along with text recognition. In recent years, researchers have explored various Machine Learning (ML) models for the recognition of components in the image.

In [1], the authors use the YOLOv5 object detection and classification model [2] for detecting resistors, capacitors, inductors, diodes, and DC sources. The authors use Hough transforms for horizontal and vertical lines, junctions and corner detection to generate a netlist. Additionally, the authors in [3] propose a full end-to-end circuit digitization and netlisting process for resistors, inductors, capacitors, voltage sources, digits, decimals, and unit symbols. By writing the corresponding nets and text to the components, the authors simulated passive linear circuits using the Lcapy Python package [4]. Furthermore, in [5], the authors developed a digitization process of detecting the components, nodes, and creating an end-to-end schematic graphing solution for hand-drawn circuits. The authors detect the circuit components

through a Faster R-CNN model [6], and propose the use of a U-Net [7] for binary segmentation of the foreground circuit.

While some research and development exists, the netlist generation process is often omitted or ignored in the represented works, and further work is required to expand the netlisting process to handle circuit variation and different scenarios. The accuracy of the digitization process is limited by the data used to train models, and the algorithms presented to relate the circuit elements to a proper netlist. In the existing research, the component classes are often limited, or the netlisting process is limited to include simple cases. With smaller component class representation and neglecting orientation and polarity information, a schematic may not be properly realized, or provide meaningful aide to the developer.

This paper presents a digitization and netlisting methodology of handwritten circuit schematics containing passive and active electronic components with an emphasis on signal and power integrity applications. Specifically, this paper provides a methodology for developing a netlist description of a circuit schematic, while accounting for the component accuracy and orientation through a two-stage classification system described in [8]. Compared to existing methods, the proposed methodology accounts for various diverse and additional components, orientation, text detection, junctions, crossovers, terminals, component polarity, power-rails and dependant sources in the netlist generation process. Validating examples are provided to demonstrate applications in the signal integrity field.

II. DEVELOPMENT OF THE PROPOSED DIGITIZATION AND NETLISTING METHODOLOGY

For digitizing circuit schematics, the component and text information must be identified. Object detection and classification models allow for finding possible bounding boxes to identify component locations in the image.

A. Component and Orientation Detection

A two-stage detection and classification ML system was employed on component bounding boxes to get accurate component and orientation information [8]. For component

979-8-3503-5124-8/24 $31.00 © 2024 IEEE

detection, the CGHD electronic hand-drawn schematic dataset [10] was used to train a YOLOv5 object detection model [2]. The CGHD dataset provides annotations on various components (passive and active electronic elements, Integrated Chips, etc.), junctions (corners, joints, crossovers) and text data [11]. The ResNet-50 model developed in [8] was used to classify passive and active electronic components such as resistors, capacitors, inductors, diodes, independent and dependant sources, transistors, and OP-AMP components in the four cardinal orientations [8]. For their applications towards signal integrity, the ResNet-50 model [12] was further trained to include the classification of transmission lines.

B. Text Recognition

For text recognition, each character in the detected text bounding box was resized to $28 \times 28 \times 1$ and characterized through a Convolutional Neural Network (CNN) model. Merged letters (as some upper and lower case English letters are the same shape), and numerical digits trained on the publicly available EMNIST [13] dataset were characterized. Furthermore, the dataset was expanded to characterize the Ω, μ Greek, and the $+$, $-$, . mathematical symbols which commonly appear in schematics. After detecting the component, text and junction information from the schematic image, the text and component elements are removed to only display the wire nets. Furthermore, the image is binarized to have a white background and black nets.

C. Contour Objects

The *findContours* method from Python's Open-CV library was used to identify the separate nets as contour objects. Contour objects are curves presented in the image through the combination of same pixel values. Performing contour analysis as opposed to Hough transform allows for wire nets to be curved or nonlinear and allows the user to draw wire nets freely [14]. Contours are detected randomly from the image and each one is given a unique node number.

D. Junctions, Crossovers and Terminals

As the user may draw circuits with some net junctions accidentally disconnected, any junction bounding boxes present in the circuit are also converted to black. This allows for the junction bounding box to bridge two nets into a continuous contour. To add, crossovers are also a prevalent part of circuit schematics, and need to be addressed for proper netlist generation. In the circuit schematic, crossovers will share the same contour net. Therefore, all crossover bounding boxes are turned white to be removed from the image and an algorithm is used to identify the corresponding wires in a crossover and assign groups of nets together.

For each crossover, the intersection between the bounding box of the crossover and the nodes are found. Also, in the circuit, it may be possible for multiple crossovers to define the same node. With the current implementation, while the nets of each crossover are associated with each other, with multiple crossovers, the same net may have multiple corresponding numbers. Therefore, a dictionary is developed to store and merge all shared nets to be a single net.

Additionally, in circuit schematics, terminals are used for identifying maps between different elements. They can be used to identify the inputs and outputs of the circuit, or can be used to map a sub-system for modular schematic systems. Therefore, terminal orientation are found to identify the input, output, and inout nets from the schematic image properly.

E. Polarity of the Components

The polarity and orientation information of electronic components is critical for the correct netlist generation from a schematic diagram. The lack of, or incorrect orientation information can greatly limit the netlist applications and provide incorrect simulation results. In this work, the output of the secondary classification model provides orientation information to separately generate a netlist descriptor for each component orientation case.

Furthermore, the intersection coordinate points are found for each contour net and component bounding box. Based on the component type, orientation, and the relative intersection coordinate positions, the component polarity in the four cardinal directions is identified. For components with three or more nets connected (e.g. transistors, OP-AMPs, etc.), the relative positions of the intersection points allows for distinction between each connected net and allows for the component to be structured properly. For example, Fig. 1 shows two possible orientations of NPN BJTs.

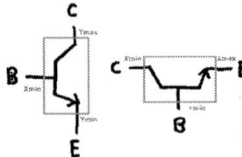

Fig. 1. Netlist structuring for two possible NPN BJT component orientations

F. Power Rails and Dependant Sources

To add, in a circuit schematic image, it is possible for multiple grounds to exist. For every component identified as ground, the associated nets are set to 0. Similarly, for multiple VDD and VSS components, the components are recognized as DC Voltage Sources from the corresponding net to ground.

The circuit schematic may also include dependant voltage and current sources. To find the controlling voltages in the schematic, the text and nets associated with terminal, and junction elements are first found. Additionally, if the associated text of a component contains a V_n character, the associated controlling voltage difference nodes are found through the component nets. For each dependant source, the corresponding texts are identified as the controlling nodes along with the dependant source's corresponding nets to develop the netlist. For current controlling sources, a voltage source component of 0V and corresponding I_n text is required in the schematic.

To generate the netlist description, for each component detected, the intersection with the contour nets is found to assign

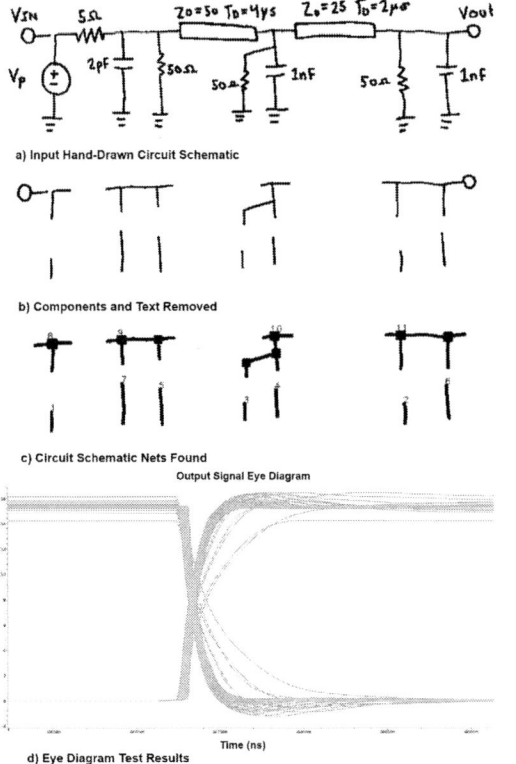

a) Input Hand-Drawn Circuit Schematic

b) Components and Text Removed

c) Circuit Schematic Nets Found

d) Eye Diagram Test Results

```
R1  9  8  5  Ohm;
C1  10  0  1nF;
C2  9  7  2pF;
T1  9  0  10  0  Z0=50  TD=4us;
R2  11  0  50  Ohm;
R3  9  0  50;
T2  10  0  11  0  Z0=50  TD=2us;
R4  10  0  50  Ohm;
V1  8  0  Vp;
C3  11  0  1nF
```

Fig. 2. Eye diagram simulation results using the netlist generated via the proposed circuit digitization process

each component with its associated nets. This approach allows for a netlist description to be developed irrespective of the net's physical geometry. Conversely, while it is also possible to assign each net to multiple component, the proposed approach resembles the current netlisting structure, where a component, and its associated net connections are listed. Additionally, each component is assigned an instance name (e.g. R_n for Resistors, C_m for Capacitors, etc.). Additional details of the proposed methodology are not given due to the lack of space.

III. RESULTS AND EXAMPLES

The proposed methodology allows for users to automatically generate netlists for simulation and modeling of circuit schematics. Generated netlists from this methodology can be simulated on any SPICE simulator such as HSPICE. Fig. 2 a) presents a lossless transmission line model hand-drawn schematic. As seen in Fig. 2 b), the components, and text elements are first detected, and removed from the circuit schematic image. It can be noted that the leftover image contains the wire nets. Further, it can be observed that the V_{in} terminal is disconnected from the input node, whereas it is the same net. By making the junction elements black, the adjacent nets to the disconnect are merged together to a single contour net as seen in Fig. 2 c). It is also noted that the nets found through contour analysis are numbered randomly. The netlist of the circuit schematic is described as:

Even though the nets are ordered randomly, the connections remain electrically correct. Furthermore, the components are passed to the secondary classification model to retrieve the orientation information, and the components are instantiated as netlist elements. Fig. 2 d) shows the corresponding eye-diagram transition as simulated on HSPICE. It is verified that the identical results are offered if a direct schematic is entered and a corresponding netlist is simulated using HSPICE.

IV. CONCLUSION

With increasing AI models and applications, keen interest is emerging in the design community for digitizing and netlisting the hand-drawn circuit schematics. This paper explores digitizing and presented a ML based methodology for netlisting by accounting for component accuracy, orientation, junction elements, and dependant sources. The future work involves handling various unclear hand drawings and text inputs.

REFERENCES

[1] R. Rachala and M. R. Panicker, "Hand-Drawn Electrical Circuit Recognition Using Object Detection and Node Recognition," SN Computer Science, vol. 3, 2022. DOI: 10.1007/s42979-022-01159-0.

[2] Ultralytics, "YOLOv5: A state-of-the-art real-time object detection system," 2021. [Online]. Available: https://docs.ultralytics.com.

[3] W. Uzair, D. Chai, and A. Rassau, "Automated Netlist Generation from Offline Hand-Drawn Circuit Diagrams," 2023. DOI: 10.1109/DICTA60407.2023.00057.

[4] Hayes M. "Lcapy: symbolic linear circuit analysis with Python." 2022. PeerJ Computer Science 8:e875. https://doi.org/10.7717/peerj-cs.875

[5] J. Bayer, A. Roy A. Dengel. "Instance Segmentation Based Graph Extraction for Handwritten Circuit Diagram Images". 2023. 10.48550/arXiv.2301.03155.

[6] S. Ren, K. He, R. Girshick, and J. Sun, "Faster R-CNN: Towards Real-Time Object Detection with Region Proposal Networks," 2016. arXiv:1506.01497

[7] O. Ronneberger, P. Fischer, T. Brox: "U-net: Convolutional networks for biomedical image segmentation". In: Medical Image Computing and Computer-Assisted Intervention–MICCAI 2015: 18th International Conference, Munich, Germany, October 5-9, 2015, Proceedings, Part III 18. pp. 234–241. Springer (2015)

[8] A. Mathur, R. Achar, "Recognition of Electronic Component Orientations from Hand-Drawn Circuit Schematics through a Two Stage Machine Learning System," 2024. DOI: 979-8-3503-6175-9/24

[9] Synopsis. "PrimeSim HSPICE" [Online]. Available: https://www.synopsys.com/implementation-and-signoff/ams-simulation/primesim-hspice.html

[10] F. Thoma, J. Bayer, Y. Li, and A. Dengel, "A Public Ground-Truth Dataset for Handwritten Circuit Diagram Images," 2021. DOI: 10.1007/978-3-030-86198-82.

[11] J. Bayer, "CGHD1152: Circuit Graphs Hand Drawn 1152," Kaggle, 2021. [Online]. Available: https://www.kaggle.com/datasets/johannesbayer/cghd1152.

[12] K. He, X. Zhang, S. Ren, and J. Sun, "Deep Residual Learning for Image Recognition," 2015. arXiv:1512.03385.

[13] G. Cohen, S. Afshar, J. Tapson A. van Schaik., "EMNIST: an extension of MNIST to handwritten letters." 2017. [Online] Available: http://arxiv.org/abs/1702.05373

[14] Itseez, "The OpenCV Reference Manual" 2014. [Online] Available: http://opencv.org/

979-8-3503-5124-8/24 $31.00 © 2024 IEEE

Analysis and Modeling of Controlled Silicon Substrate Roughness for Silver-Based Backside Metallization in Power Electronics Packaging

Mohamed Lamine Faycal Bellaredj
Department of Electrical Engineering
University of Moncton
Moncton, Canada
mohamed.lamine.faycal.bellaredj@umo
ncton.ca

Goran Miskovic
Silicon Austria Labs
Villach, Austria
goran.miskovic@silicon-austria.com

Lukas Vojkuvka
SAL MicroFab
Silicon Austria Labs
Villach, Austria
lukas.vojkuvka@silicon-austria.com

Abstract—In this paper, the analysis of a controlled structuration approach of the silicon (Si) substrate surface roughness through standard acidic wet chemical etching is proposed for the first time, for silver-based backside metallization (BSM) in power electronics packaging applications. Periodically spaced circular openings with diameters and separation distances from 1 um to 5 um were patterned using maskless laser lithography and wet etched in a Si substrate with a standard acidic HNP (HF:HNO₃:H₃PO₄) mixture for 30s. The etched cavities were characterized by scanning electron microscopy (SEM). The extracted etching parameters from SEM observations were used to implement simple analytical models for the estimation of the average arithmetic surface roughness R_a and the normalized etching depth (with respect to the opening's diameter after etching) as a function of the circular openings' dimensions, separation distance and the corresponding underetch. Roughness values ranging from 165 nm to 555 nm were estimated depending on the design specifications. A good correlation was observed between the experimental and theoretical values of the normalized etching depth. The introduced approach allows a rapid estimation of the surface roughness after etching without the need for photoresist removal for profilometry or AFM measurements, which makes it suitable for both rapid prototyping as well as for additional etching cycles after SEM if needed.

Keywords— Analysis, Modeling, Backside metallization), Packaging, roughness, wet chemical etching, HNP, SEM

I. INTRODUCTION

One of the most important parameters in power electronics packaging that guarantees adequate adhesion, heat conduction, ohmic contact, and long-term stability is the backside metallization (BSM) [1-2]. The TiNi(V)Ag stack, which has titanium (Ti) as the adhesion layer, nickel or nickel vanadium (Ni or NiV) as the soldering layer, and silver (Ag) as the protective layer, is the most frequently employed BSM stack on silicon [3–4], generally deposited using either sputtering or evaporation [5–6]. A convenient BSM adhesion interface to the Si substrate's surface requires a customized backside surface for TiNi(V)Ag stack deposition that takes dopant, resistivity, and Si type into consideration [7]. Prior to BSM deposition, the back surface of the Si substrate is chemically, or plasma etched to eliminate native oxide and contaminants. Because of the BSM's multilayer structure, an annealing step is needed either during or after the deposition process to improve the material's adherence to the Si substrate. This can be achieved at the BSM/Si interface by stress adjustment or by forming silicide alloys [8–12]. Sputtering is being employed more and more in BSM deposition because it allows for high vacuum conditions during in situ etching and annealing phases in a sputtering cluster tool [5–6]. Another way to enhance the adherence of BSM to Si is to intentionally roughen the substrate's surface on the backside of the Si substrate using a wet chemical etching process employed to relieve stress following backside grinding [7-8,13-16]. Although isotropic acidic etching of silicon has been extensively studied [13–16], there is a dearth of information regarding the relationship between the chemical etching of Si substrate surface and the BSM. In [8], a failure analysis was carried out to find why TiNiAg BSM peeled off the Si substrate during die removal from adhesive foil following dicing. The cause was found to be a low surface roughness due to over-polishing the wafer after back grinding. The impact of various Si substrate chemical etching techniques and TiNiAg BSM metallization schemes on the mechanical properties and surface roughness of thinned Si wafers were examined in [7,13]. By reducing the Si damage from the backgrinding process and making minor oxidizer modifications in the chemical bath, the wafer and die fracture strengths were increased. In [17-18], we investigated for the first time the effects of uncontrolled Si substrate surface roughness on a sputtered TiNiVAg BSM stack on top of the Si substrate in terms of layer conformability, uniformity, and morphology. Significant morphological changes with defects in the Ag layer and at the Ag/NiV interface were observed in the TiNiVAg BSM stack. To solve or limit the effects of the uncontrolled substrate roughening process on the TiNiVAg BSM, a controlled structuration of the Si substrate surface through a controlled induced surface roughness can be a solution. Extensive work has been undertaken in Si surface structuration for wettability tuning [19], optical applications [20], corrosion resistance and friction/wear reduction [21]. However, and to the best of our knowledge, we couldn't find any research related to controlled Si surface structuration for silver based BSM. In this paper, a periodic structuration of the Si substrate surface for a controlled induced surface roughness is investigated for the first time using wet etching. A simple analytical approach is proposed to estimate the surface roughness based on SEM analysis after wet etching without the need to remove the photoresist. Such an approach speeds up the analysis for rapid prototyping and allows also additional etching cycles if required for a deeper analysis of the etching process.

979-8-3503-5124-8/24 $31.00 © 2024 IEEE

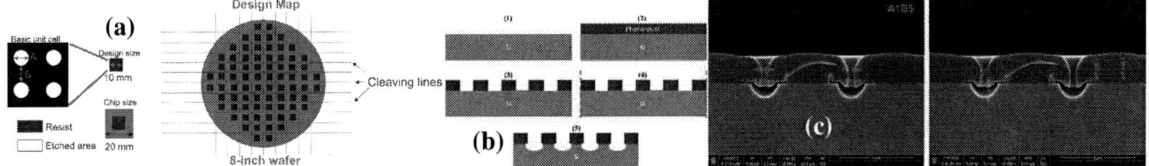

Fig. 1. (a) Design for the wet chemical etching process of the Si wafer where the basic unit cell is made of circular openings with a diameter A separated by a distance B (b) Process flow of the wet etching process (1) starting Si substrate (2) photoresist spin coating (3) photoresist patterning followed by plasma descum (4) wafer dicing for test coupons (5) wet etching process of the Si coupons (c) Cross-sectional SEM pictures of the etched A1B5 coupon etched for 30s.

II. EXPERIMENTAL WORK

A simple design based on circular openings with a diameter A separated by a distance B (both varied from 1 um to 5 um) was used for the active control of the Si substrate surface roughness through wet chemical etching. Square matrixes of 10 mm x 10 mm size were made from a basic unit cell and arrayed with a 10 mm separation distance all over an 8-inch Si wafer to allow a safe dicing of the chips (Fig.1.(a)). The fabrication process flow is shown in Fig.1.(b). The starting substrates were 8-inch polished silicon (Si) wafers with native silicon oxide (SiO_2). First, hexamethyldisilizane (HMDS) adhesion promoter was deposited on the Si surfaces using the vapor primer of the resist track system from Süss MicroTec. A 2-um thick AZ 1518 (MicroChemicals) photoresist was then spincoated ontop of the Si substrate and patterned using the DWL 66+ laser writer (Heidelberg instruments). The exposed photoresist was developed in the AZ 726 MIF developer for 1 min and hard baked at 90 C for 5 min. A descum process using O_2 plasma (Tetra 30 from Diener electronic) was applied to the patterned photoresist for 5 min to get rid of photoresist residues after the development. The patterned Si wafer was then diced into coupons using a mechanical blade. The Si coupons were etched in an acidic solution of fluoridric acid (HF), nitric acid (HNO_3) and phosphoric acid (H_3PO_4) with molar ratios of $HF:HNO_3:H_3PO_4=1:20:20$ at 25C for 30s and characterized using the scios 2 dualBeam (Thermo Fisher Scientific) SEM.

III. RESULTS

The cross-sectional observations of the etched Si coupons allowed the determination of the critical dimensions (CD) before and after the etching as well as the etching depth and underetch as can be seen in Fig.1.(c). These extracted data are summarized in Table.1 where A_f is the opening diameter after etching (critical dimension), B_f the separation distance after etching, T is the design period after etching, A_i is the smallest initial opening diameter (critical dimension) before etching, U_c is the etching undercut and E_D is the etch depth. The differences between the design and the experimental values of AB are due to non-contact laser lithography and its effect on relatively thick photoresists. The surface layer of the photoresist is directly exposed to the laser beam and gets a higher exposure dose than the bulk layer of the photoresist. Consequently, the photoresist development rate is the highest at its surface and decreases (proportionally to the exposure dose) farther from it, which explains the obtained photoresist profiles after etching. Based on the SEM characterization, a simple analytical approach was proposed for the estimation of the surface roughness after etching without the need for the removal of the photoresist

for AFM or profilometry measurements. The proposed approach allows both a rapid prototyping and a more thorough investigation of the surface structuration process through multiple etching cycles. Fig. 2 shows the theoretical 2D profile of the wet etched Si surface based on elliptical etched cavities. Applying the same formalism as in [22], the elliptical cavity equation can be expressed for $-A_f/2 < x < A_f/2$ as:

$$\frac{x^2}{a^2} + \frac{(y-b+E_D)^2}{b^2} = 1 \quad (1)$$

Where a and b are the major and minor semi-axis of the elliptical cavity.
The mean value of the etched y(x) profile can be deduced on a period $T = \frac{A_f}{2} - \left(-B_f - \frac{A_f}{2}\right) = A_f + B_f$ as:

$$y_m = \frac{1}{T}\int_{-\frac{B_f}{2}}^{\frac{A_f}{2}} y dx = \frac{1}{T}\int_{-\frac{A_f}{2}}^{\frac{A_f}{2}} y dx \quad (2)$$

Extracting y from (1) and putting $S(x)=\int y dx$:

$$S(x) = (b - E_D)x - \frac{b}{2a}x\sqrt{a^2 - x^2} - \frac{ab}{2}\arcsin\left(\frac{x}{a}\right) \quad (3)$$

y_m can be expressed using S(x)

$$y_m = \frac{1}{T}\int_{-\frac{A_f}{2}}^{\frac{A_f}{2}} y dx = \frac{2}{T}\int_0^{\frac{A_f}{2}} y dx = \frac{2}{T}S\left(\frac{A_f}{2}\right) \quad (4)$$

If x_m is the x value when $y=y_m$, it can be deduced from (1):

$$x_m = \frac{a}{b}\sqrt{2(b-E_D)y_m - y_m^2 + 2bE_D - E_D^2} \quad (5)$$

The arithmetic average surface roughness R_a can be expressed as:

$$R_a = \frac{1}{T}\int_{-\frac{B_f}{2}}^{\frac{A_f}{2}}|y - y_m|dx = \frac{1}{T}\int_{-\frac{A_f}{2}}^{\frac{A_f}{2}}|y - y_m|dx$$

$$= \frac{2}{T}\int_0^{\frac{A_f}{2}}|y - y_m|dx \quad (6)$$

$$= \frac{2}{T}\left[\int_0^{x_m}(y_m - y)\,dx + \int_{x_m}^{\frac{A_f}{2}}(y - y_m)\,dx\right]$$

$$R_a = \frac{2}{T}[x_m y_m - S(x_m)] \quad (7)$$

The a, b, E_D and U_c values were deduced from the measured SEM cross-sections and used as inputs to compute R_a using MATLAB. The estimated R_a values for the considered designs are given in Table. The normalized etching depth (E_D/A_f) variations as a function of the normalized final separation distance after etching B_f, undercut U_c and initial opening diameter A_i were also investigated (Fig. 3). The normalized etching depth was found to be proportional to the normalized B_f and inversely proportional to the normalized A_i. Increasing the normalized B_f while reducing the A_i values for small cavities will thus increase the normalized etching depth. On the other hand, large cavities with high A_i and small B_f values will result in a small, normalized etching depth. The normalized etching depth was found to be proportional to

TABLE I. Estimated R_a based on (7)

Design	A_i (nm)	A_f (nm)	B_f (nm)	T (nm) = $A_f + B_f$	A_i (nm)	U_C (nm)	E_D (nm)	R_a (nm)
A1B5	400	1848	4230	6078	400	724	850	322
A2B1	1966	2346	610	2956	1966	190	480	165
A2B5	1617	4195	2838	7033	1617	1289	1376	555
A5B1	3837	4585	1350	5935	3837	1350	1110	391
A5B2	5210	6969	110	7079	5210	879,5	1280	306

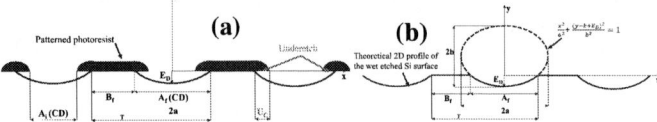

Fig. 2. Theoretical 2D profile of the etched Si surface assuming elliptical cavities (a) with photoresist (b) without photoresist

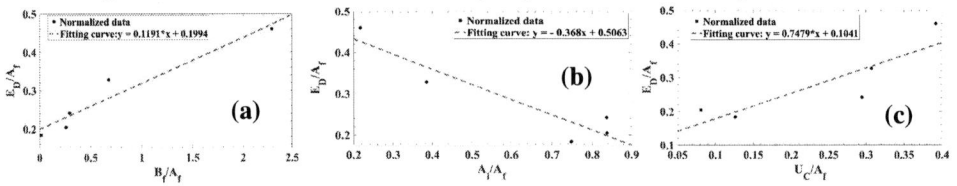

Fig. 3. Normalized etching depth E_D as a function of normalized (a) B_f (b) A_i (c) undercut U_C

the normalized undercut U_C due to the isotropic wet etching process. A good overall correlation was observed between the experimental and modeling results for the normalized etching depth.

IV. CONCLUSION

In this work, we presented the first analysis of a simple structuration approach of a Si substrate to achieve a controlled surface roughness for silver-based BSM in power electronics packaging. Arrays of circular openings periodically spaced were fabricated in a Si wafer using laser lithography, standard acidic wet chemical etching and characterized using SEM microscopy. Simple analytical models were then developed to estimate both the average arithmetic surface roughness and the normalized etching depth. The presented approach enables a quick post-etching surface roughness assessment without the need for photoresist removal for profilometry or AFM measurements, making it appropriate for rapid prototyping and, if necessary, additional etching cycles following SEM.

REFERENCES

[1] P-I. Lee, et al, "Atomic Migration of Cu on the Surface of Si/Ti/Ni/Cu/Ag Thin Films", Journal of Electronic Materials, 51:3624–3636, 2022.

[2] L. C. Wee, et al, "Develop a Time Efficient Method to Enhance the FIB Process on BSM Analysis", STFA 2022, ASM International, 196-200, 2022.

[3] P-C. Wu, et al, "Sputtering of Ag (111) nanotwinned films on Si (100) wafers for backside metallization of power devices", J Mater Sci: Mater Electron (2021) 32:7319-7329.

[4] J. Choi and S. J. An, "Backside Metallization of Ag–Sn–Ag Multilayer Thin Films and Die Attach for Semiconductor Applications", Journal of Electronic Materials, 49, 4265-4271, 2020.

[5] M. Ciacchi, H. Eder, and H. Hirscher, "Evaporation vs. Sputtering of metal layers on the Backside of Silicon wafers", The 17th Annual SEMI/IEEE ASMC 2006 Conference. IEEE, 2006.

[6] A. Rastogi, N. Morin, C. Jones, S. Burgess, R. Trowell, C. Widdicks, I. Moncrieff, M. Ehmann, and S. Schmidbauer, "Productivity challenges in PVD processing in 300mm pilot lines for power semiconductors", IEEE 26th Annual SEMI Advanced Semiconductor Manufacturing Conference (ASMC), pp. 204-208, 2015.

[7] S. Drews, "Single Wafer Surface Conditioning Improving Back-side Metal Adhesion", Advanced packaging, 16. 16-18, 2007.

[8] Y. Wang, et al, "Investigation of Metal Peeling Failure of TiNiAg Metal Thin Films on Silicon Wafers", Advanced Materials Research, vol. 915, pp. 847-850. Trans Tech Publications Ltd, 2014.

[9] D. Resnik, et al, "Investigation of interface properties of TiNiAg thin films on Si substrate", Vacuum 82 (2008) 162–165.

[10] D. Resnik, et al, " Influence of mechanical stress on adhesion properties of DC magnetron sputtered Ti/NiV/Ag layers on n+ Si substrate", Microelectronic Engineering 85 (2008) 1603–1607.

[11] R. Ricciari et al, "Auger electron spectroscopy characterization of Ti/NiV/Ag multilayer back-metal for monitoring of Ni migration on Ag surface", Microelectronics Reliability 55.9-10, 1617-1621, 2015

[12] F-J. Yeh, T-C. Chiu, and K-L. Lin, "The interfacial interaction of Ti/Ni/Ag/Au Multilayer under thermal cycling test", IEEE 14th International Conference on Electronic Materials and Packaging (EMAP), 1-5, 2012.

[13] Zi. M. Chang, and H. M. Ler, "Effect of Wafer Back Metal Thickness and Surface Roughness towards Backend Assembly Processes", IEEE 39th International Electronics Manufacturing Technology Conference (IEMT), 2022.

[14] P. Pal, et al. "High speed silicon wet anisotropic etching for applications in bulk micromachining: a review", Micro and Nano Systems Letters 9, 1-59, 2021.

[15] M. R. Marks, Z. Hassan, and K. Y. Cheong, "Ultrathin Wafer Pre-Assembly and Assembly Process Technologies: A Review", Critical Reviews in Solid State and Materials Sciences, 40:251–290, 2015.

[16] R. Watanabe, et al, "Evaluation of a new acid solution for texturization of multicrystalline silicon solar cells", International Journal of Photoenergy, 2013.

[17] G. Mišković, and M. L. F. Bellaredj, "Effect of Silicon Substrate Roughness on Silver-Based Backside Metallization for Power Electronics Packaging", IEEE 46th International Spring Seminar on Electronics Technology (ISSE), 2023.

[18] P. Malagò, G. Mišković, and M. L. F. BELLAREDJ, "Calculation of the resistance of silver-based backside metallization stack for die-attach packaging applications", IEEE 46th International Spring Seminar on Electronics Technology (ISSE), 2023.

[19] C. G. J. Prakash, and R. Prasanth, "Approaches to design a surface with tunable wettability: a review on surface properties", Journal of Materials Science, 56, 108-135.

[20] O. Poncelet, et al, " Hemispherical cavities on silicon substrates: an overview of microfabrication techniques", Materials Research Express, 2018, vol. 5, no 4, p. 045702.

[21] N. A. H. M. Mahayuddin, et al, " Surface texturing method and roughness effect on the substrate performance: A short review ", Jurnal Tribologi, 2020, vol. 27, p. 8-18.

[22] J. Qu, and A. J. Shih, "Analytical surface roughness parameters of a theoretical profile consisting of elliptical arcs", science and technology, 7(2), 281-294.

Worst-Case Voltage Droop Using Peak Distortion Analysis

Mohamed Sahouli, Isaac Ali, David ReinaMendivil, Gerry Talbot

Advanced Micro Devices, Inc

Email: {Mohamed.Sahouli;Isaac.Ali;David.Reina;Gerry.Talbot}@amd.com

Abstract—**This paper introduces a method for calculating the worst-case voltage droops experienced by a power distribution network. Through peak distortion analysis, each port in a multiport model is stimulated to produce patterns that will result in a large voltage swing at a selected observation point. This approach enables power integrity engineers to anticipate worst-case scenarios at the beginning of the design process and initiate the optimization of power delivery. The methodology is demonstrated using a real system as an example.**

I. INTRODUCTION

In the fast-evolving field of microchip design, it's crucial to ensure power integrity from the early stages of the design process. This is especially important with the introduction of artificial intelligence (AI) chips, which require higher performance and efficiency. Power integrity involves managing the power delivery network (PDN) to ensure a stable and adequate power supply to all chip components [1]. Conducting early power integrity analysis helps in identifying potential issues that could impact the chip's functionality, performance, and reliability.

AI chips handle many calculations simultaneously and process large amounts of data quickly. However, this uses a lot of power which can cause problems [2], [3]. The complex nature of AI algorithms means that how the chip is designed can lead to uneven power distribution and drops in voltage across the chip. If these issues aren't addressed early on, they can be more difficult and expensive to fix later.

Performing power integrity analysis early in the design cycle offers several benefits. It allows designers to identify and mitigate power-related issues before they become critical, ensuring a more robust and reliable chip. Early analysis can also lead to optimized power delivery networks, which improve the overall efficiency and performance of the chip. Furthermore, addressing power integrity issues early can reduce the need for costly redesigns and shorten the overall development time.

Advanced techniques such as peak distortion analysis (PDA) can further enhance the effectiveness of early power integrity analysis. PDA helps compute the worst-case scenario load current stimulus, providing crucial insights into the PDN's performance under various conditions. By guiding important PDN parameters such as capacitor locations, quantity, and impedance targets, PDA can significantly improve the reliability and efficiency of the power delivery network.

This paper presents the PDA for computing the worst-case load stimulus. This analysis can be used in the early design stages to guide important PDN parameters such as capacitor locations, quantity, and impedance targets. The rest of the paper is organized as follows: a section describing the general theory behind PDA and the methodology used for PI analysis. Finally, the proposed method is illustrated using a numerical example from an actual system.

II. PEAK DISTORTION ANALYSIS

In signal integrity applications, peak distortion is well understood [4]–[7], particularly in contexts where the unit interval and symbol levels are precisely defined. This methodology can be extended to power integrity scenarios, albeit with additional complexities in determining the unit pulse. Given that different integrated circuits (ICs) with varying power consumption profiles may utilize the same power rail, analyzing each IC's behavior to ascertain the minimum pulse duration at a specific average power level alongside the idle power consumption is crucial. The latter typically stems from leakage currents and other active components independent of data transmission activities.

Upon establishing the parameters for the unit pulse, which represents the IC in question, this pulse is introduced into PDN in amperes. The resultant effects of this pulse are observable across various nodes within the PDN, with responses recorded at predetermined nodes of interest. These responses, extending over several unit pulse durations, resemble inter-symbol interference phenomena observed in signal integrity fields.

Maintaining the circuit's voltage above a critical minimum threshold is imperative to prevent power delivery failures. Equally, avoiding frequent excursions beyond a specified maximum voltage threshold is essential to ensure long-term device reliability.

Through peak distortion analysis of the unit pulse response, it's possible to devise a sequence of pulses that, when applied, either minimizes or maximizes the voltage level to the most extreme permissible values. This analysis enables the identification of IP activity patterns that lead to such worst-case scenarios. Notably, the specific activity patterns inducing these extreme conditions will differ across various observation nodes within the PDN. Consequently, it's necessary to calculate pairs of activity patterns for each node to achieve the worst-case minimum and maximum voltage levels. For instance, ten distinct activity patterns can be delineated if there are five observation nodes. Assuming each node also serves as a

979-8-3503-5124-8/24 $31.00 © 2024 IEEE

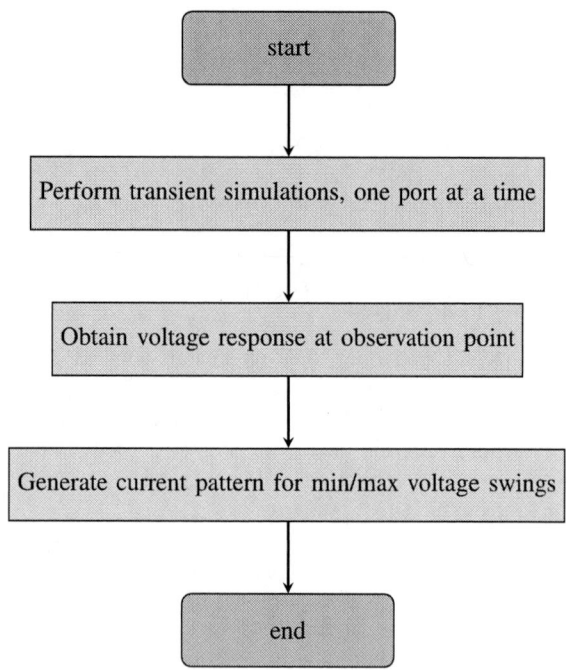

Fig. 1. Proposed PDA Algorithm.

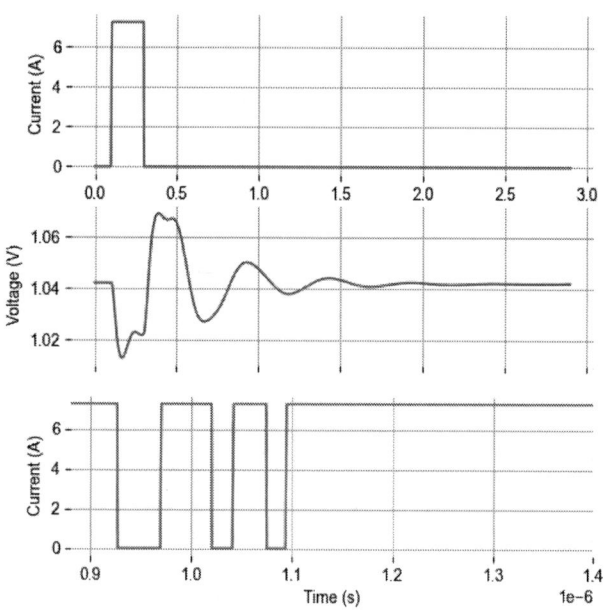

Fig. 2. Top: Signal used to generate voltage responses in the first step; Middle: Resulting voltage at an observation point; Bottom: generated current pattern from the observed voltage response.

connection point for another IC, the number of activity patterns requiring calculation escalates to 100.

Once the required patterns from multiple nodes to a single observation node are known, they can be aligned in time to give combined worst-case excursion values from the ideal.

This approach underscores the importance of a detailed understanding of IP behavior and the PDN's response to different power profiles, facilitating the optimization of power integrity across complex electronic systems. Next, a methodology for using the principle of PDA to compute current patterns that will generate large voltage swings is presented.

III. PROPOSED PDA METHODOLOGY

The previous sections highlighted the importance of designing a PDN to handle modern chips' increased power demand. However, it's often impossible to predict the worst supply voltage noise caused by switching current activity in the integrated circuit (IC), which makes designing robust delivery networks challenging. To tackle this issue, a PDA-based analysis is proposed. This method works well with models that have multiple ports at the die side. Using the superposition principle, the algorithm produces a set of stimuli arranged so that a specific observation port experiences a significant voltage swing.

Given a set of circuits connected to the PDN network at N ports, the PDA methodology systematically automates the characterization of the PDN. It synthesizes patterns to stress power delivery at each port, resulting in minimum and maximum voltage swings at the target ports.

The proposed method involves two main phases. The first phase calculates the responses at each port by conducting a series of transient simulations, with one port of the PDN being excited at a time. The signal used to excite the ports varies and depends on the specific application, often requiring input from the designers of the IP being observed. However, a pulse is generally sufficient. For an N-port system, each transient simulation will yield N responses, resulting in a total of $N \times N$ responses once the first phase is completed.

After completing all the transient simulations, the results are combined. This involves gathering the voltage responses at a specific observation point caused by the excitation of each port. The resulting waveform from each port at the observation point can be used to create a corresponding load stimulus.

This process is illustrated in Fig. 2. The top panel shows the pulse used to excite a specific port, the middle panel presents the response generated at the observation point, and the bottom panel displays an example current pattern resulting from the observed voltage response.

The second phase of the algorithm computes the patterns that will produce the worst-case Vmin and Vmax responses. For this to happen, each port will be launching two target patterns toward the target port (one for each Vmin and Vmax scenario); thus, the complex waveform at each port will have a $2 \times N$ distinct pattern. Note that the Vmin and Vmax patterns are time-separated to prevent the inherent memory of the PDN from causing pattern-to-pattern interference. In addition, because these are SPICE simulations, any non-linear or memory effects due to the circuits will be included in the responses.

Pattern alignment is not part of PDA; however, the analysis algorithm knows which patterns from the N ports target the same port. Knowing this information, the algorithm will

979-8-3503-5124-8/24 $31.00 © 2024 IEEE

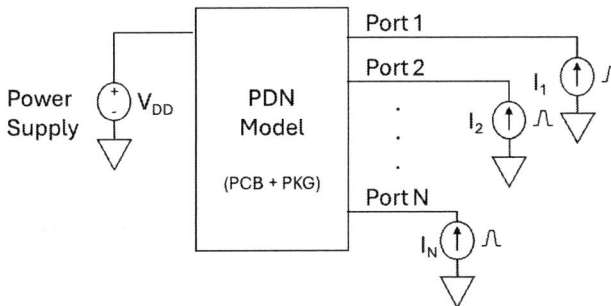

Fig. 3. Representation of layout of the Graphics Card model.

align the patterns to target the same port in time. When the simulation runs, each port will simultaneously launch patterns toward the target due to the algorithm's alignment of contributors. In addition, the algorithm can align patterns to Vmin and Vmax. This will fine-tune the relative delays between the patterns targeting the same port.

To illustrate the PDA algorithm, the proposed method is used on a graphics card model that combines PCB, package, and on-die models (Fig. 3 shows a representation of the layout under study). The die is represented as a multi-port grid, and the PDA is employed to calculate the worst-case stimulus pattern that will produce the worst-case voltage swing at a specific observation point.

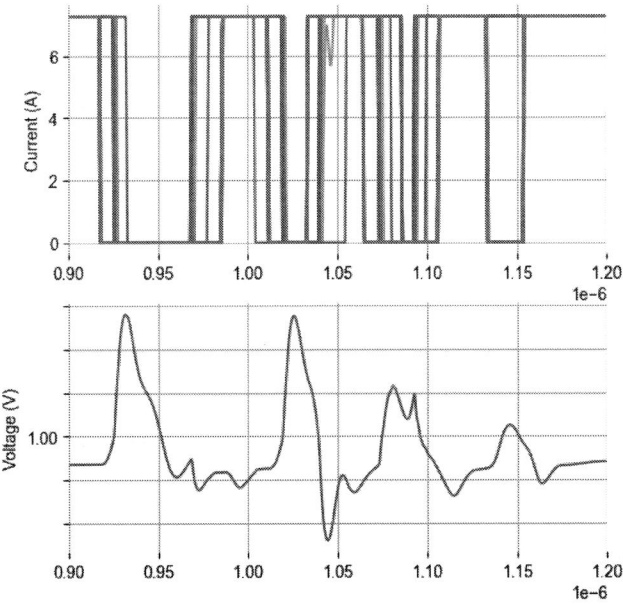

Fig. 4. Sample worst-case voltage and current pattern.

Fig. 4 shows the calculated pattern and the resulting voltage swing caused by the calculated PDA switching activity. The top panel shows the computed current stimulus at each port of the multigrid model, and the bottom panel shows the resulting voltage response at the target port from the input current waveforms from each grid port.

IV. Conclusion

This paper presented a method for generating current patterns to create a worst-case voltage swing at a specific PDN model port. Observing the voltage response to a pulse signal creates worst-case current stimuli leading to large voltage swings. This method allows for early optimization of power delivery in the design cycle and provides essential insight for PI engineers into the network.

Acknowledgment

The authors thank Mark Frankovich, Silqun Leung, and Tawfik RahalArabi for reviewing this work.

References

[1] E. Bogatin, *Signal and Power Integrity - Simplified*, 3rd ed. USA: Prentice Hall PTR, 2018.
[2] M. Hamblen. Power-hungry ai chips face a reckoning, as chipmakers promise 'efficiency'. [Online]. Available: https://www.fierceelectronics.com/ai/power-hungry-ai-chips-face-reckoning-chipmakers-promise-efficiency
[3] C. Hetzner. Ai could gobble up a quarter of all electricity in the u.s. by 2030 if it doesn't break its energy addiction, says arm holdings exec. [Online]. Available: https://fortune.com/2024/04/16/ai-chatgpt-sora-large-language-models-arm-chips-semiconductors-electricity/
[4] R. Shi, W. Yu, Y. Zhu, C.-K. Cheng, and E. S. Kuh, "Efficient and accurate eye diagram prediction for high speed signaling," in *2008 IEEE/ACM International Conference on Computer-Aided Design*, 2008, pp. 655–661.
[5] M. Shimanouchi, M. P. Li, and D. Chow, "Designcon 2011 worst-case patterns for high-speed simulation and measurement," 2011. [Online]. Available: https://api.semanticscholar.org/CorpusID:13311853
[6] W. T. Beyene, "Peak distortion analysis of nonlinear links," in *2013 IEEE 22nd Conference on Electrical Performance of Electronic Packaging and Systems*, 2013, pp. 169–172.
[7] J. Kim, S. Nam, S. Moon, T. Kim, S. You, C. Jo, and Y. Lee, "Effective interface simulation approach based on peak distortion analysis for ucie ips," in *2024 IEEE 74th Electronic Components and Technology Conference (ECTC)*. IEEE, 2024, pp. 1104–1107.

Equalization Techniques for Time Domain Signaling

Shakib Mahmood
Dept. of Electronics, Carleton University, Ottawa, Canada
shakibmahmood@cmail.carleton.ca

Parneet Tethy
Dept. of Electronics, Carleton University, Ottawa, Canada
parneetthethy@cmail.carleton.ca

Richelle L. Smith
Department of Electrical Engineering, Stanford University, Stanford, CA 94305
smithrl@stanford.edu

Carl W. Werner
Rambus Lab, Rambus Inc., San Jose, CA 95134 USA
cwerner@rambus.com

Masum Hossain
Dept. of Electronics, Carleton University, Ottawa, Canada
masumhossain@cmail.carleton.ca

Abstract—**This paper describes inter-symbol interference (ISI) and crosstalk correction techniques for time domain signaling. The two novel digital equalization techniques introduced in this work can improve the data rate to 40 Gb/s compensating 40 dB loss, demonstrating 15 to 20 dB additional loss compensation capability compared to traditional equalization techniques. Similarly, crosstalk cancellation technique can achieve 20 dB reduction of crosstalk noise without any additional signal processing.**

Keywords—*Time domain, TDC (time-to-digital converter), ISI (inter-symbol interference), Crosstalk, Equalization*

I. INTRODUCTION

The bandwidth requirements driven by artificial intelligence (AI) and machine learning (ML) applications have motivated high-bandwidth interconnects. The introduction of chiplet and 3-D interconnects such as silicon interposer unveils the potential to meet Tb/s/mm bandwidth density and low latency connectivity. To take advantage of these low-loss interconnects, industry is adopting single-ended signaling standards such as UCIe (Universal Chiplet Interconnect Express), BoW (Bunch of Wires) etc. But the crosstalk and supply noise limit the achievable data rates in single-ended systems. Extending these systems to multilevel signaling causes significant reduction in signal-to-noise ratio (SNR).

An attractive alternative to multilevel amplitude-modulated signaling is to encode information in the time domain [2-5], such as the pulse width [1-7], phase, or edge position [8]. These modulation techniques have shown SNR advantage over PAM-4 [9]. Since zero crossings are less sensitive to amplifier nonlinearity and reduced voltage headroom, time-domain circuits have the potential to be better adopted in digital implementation. Therefore, time-domain circuits can achieve digital like energy savings through voltage and frequency scaling techniques (VFS). However, time domain signaling is also limited by random noise, inter symbol interference (ISI) and crosstalk.

Recently, Differential Edge Modulation Signaling (DEMS) has been proposed and demonstrated as an energy-efficient signaling scheme [8], where edge position is differentially modulated so that the common noise components are cancelled. This provides significant immunity of random jitter and power supply induced jitter. Although this work also demonstrated a time domain DFE technique, it was limited to 2-phase modulation scheme.

In this work, we take a more generalized approach towards multi-phase edge modulation technique compare the performance achievable using such techniques. The paper is organized in the following manner. In section II we discuss the bandwidth requirements and the nature of ISI in time domain signals. In section III we compare different equalization techniques followed by performance comparison. Section IV will discuss the crosstalk reduction.

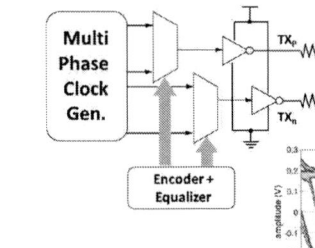

Fig. 1. Time modulated Link with example modulation.

II. ISI IN TIME DOMAIN SIGNALING

The power spectral density (PSD) of the modulated signal is the simplest way to estimate the required transmission bandwidth, and bandwidth efficiency [9]. The PSD of the quadrature phase modulated signal can be shown as:

$$S_u(f) = \frac{1}{T/8} S_b(e^{j2\pi \frac{fT}{8}})|G(f)|^2 \tag{1}$$

Here, T is the period of the carrier frequency and $G(f)$ frequency domain representation of the time domain pulse $g(t)$

$$|G(f)|^2 = \left(\frac{T}{8}\right)^2 \text{sinc}^2\left(\frac{fT}{8}\right). \tag{2}$$

As the modulated pulse width is a fraction of the carrier period, this modulation creates frequency tones that are multiple of the carrier frequency (f_c). For a 4-phase per edge modulation, there are high frequency tones at $7f_c$ and $9f_c$, but providing the bandwidth up to that frequency is challenging. If the system has less bandwidth, it shows inter symbol interference (ISI). Two key observations are: i) the magnitude of the ISI is much smaller and the polarity of the ISI is opposite in time domain compared to the amplitude domain. ii) ISI is a function of the modulation depth in phase

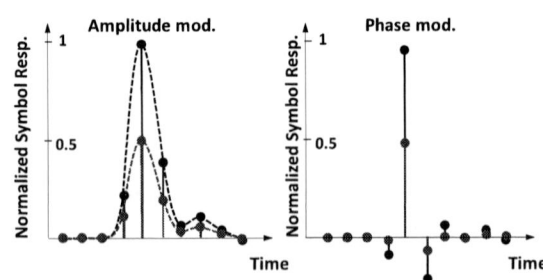

Fig. 2. Measured symbol response for amplitude and phase modulation for low loss chiplet channel.

modulation since modulation depth changes the bandwidth of the signal. These observations unlock the door for much simpler equalization and more spectrum-efficient modulation.

979-8-3503-5124-8/24 $31.00 © 2024 IEEE

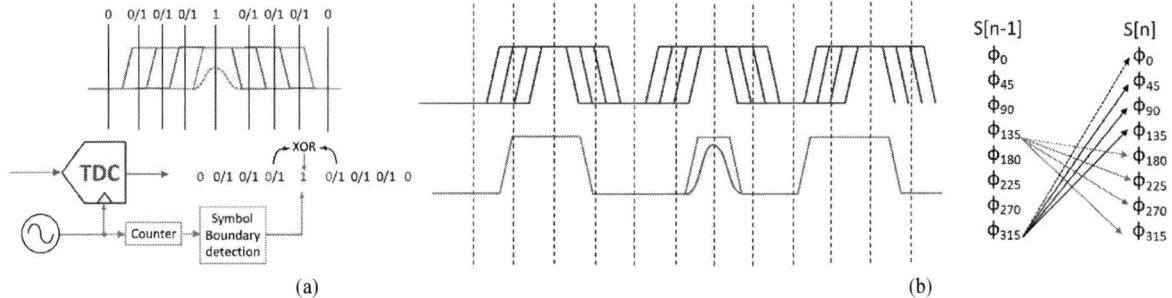

Fig. 3. Digital equalization techniques for time domain modulation (a) RX side TDC based and (b) TX side Encoder based equalization

III. EQUALIZATION FOR TIME DOMAIN MODULATION

Time domain equalization techniques are less explored in literature as their adoption is still in the research and exploratory phase. Potential equalization techniques include simplistic peaking similar to amplitude modulation and fully digital equalization that can be done from either the transmitter side or at the receiver side in the digital domain after time-to-digital converter (TDC). Particularly these digital equalization techniques are significantly different compared to traditional feedforward equalizer (FFE) and decision feedback equalizer (DFE) and therefore the main focus of this paper. However, for completeness, we will start with more known equalization techniques such as continuous time linear equalizer (CTLE) and pre-distortion from Tx etc.

A. CTLE and TX Predistortion

Given the channel attenuates the high-frequency signal components, by doing so introduces ISI. CTLE on the contrary, attempts to equalize by attenuating the low-frequency content and boosting the high-frequency content. This is similar to amplitude modulation except for multi-phase modulation carrier harmonics can be fairly high – so when the loss is high, providing the required boost at high-frequencies introduces analog-mixed signal challenge. From the time-domain view point, a narrow pulse becomes narrower when sent over a lossy channel. Therefore, pulses are widened before transmitting to compensate for the narrowing effect of the channel. Interestingly, this approach is also used in wireline equalization.

B. Digital Equalization of time domain modulation

The digital equalization approach is explained in Fig. 3. Given the maximum ISI in time domain mostly results from the narrowest pulse, this digital equalization approach mostly targets the narrow pulse. First, in the receiver side after the TDC, we can detect those missing narrow pulses by detecting the symbol boundaries. This is done by simply counting the number of edges. At the symbol boundary, we expect the value to be 1 or 0 based on the rising edge or falling edge. Once the symbol boundaries are detected, based on the previous and next symbol's TDC output values we can correct the boundary symbol.

The occurrence of the narrow pulse can also be predicted from the transmitted symbols sequence. Since there are transitions from the previous symbol (one of the 8 phases) to current symbols (also 8 phases) the narrowest pulse can occur only two ways: first from ϕ_{135} to ϕ_{180} and second, ϕ_{315} to ϕ_0. Therefore, from the transmitter side, we can exclude such transitions shown by dotted lines in Fig. 3(b). Note that there are possible 64 transition and only two transitions to be excluded. Therefore, by identifying those transition we can eliminate the narrow pulse and its associated transitions.

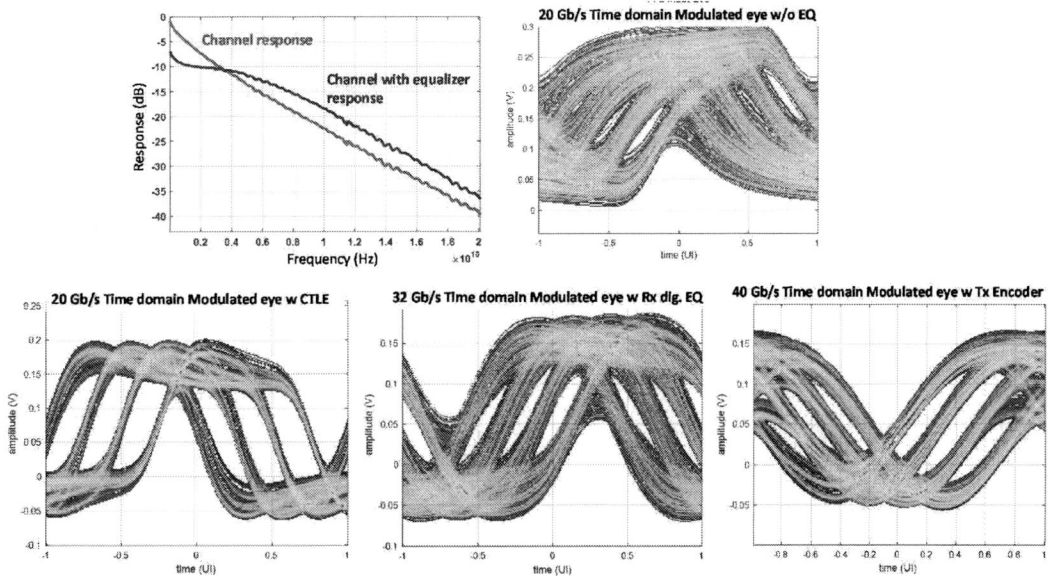

Fig. 4. (Top) Channel response with and without equalizer. Time-modulated eye without equalization @ 20 Gb/s (Bottom) Equalized Eye @ 20Gb/s, 32 Gb/s eye with RX Digital equalization and 40Gb/s encoded eye

979-8-3503-5124-8/24 $31.00 © 2024 IEEE

C. Performance Comparison

To compare the performance of different equalization techniques described above, we are using a typical die-to-die channel that has about 25 dB loss at 10 GHz and 40 dB of loss at 20 GHz. Despite their ISI advantages, it turns out that the data rate could not exceed more than 10 Gb/s. With transmitter pre-distortion, achievable data rate could be increased to no more than 15 Gb/s. Compared to that relatively simple CTLE such as 8 to 12 dB boost at moderate 10 GHz can exceed the bandwidth to 20 Gb/s (Fig. 4 bottom left). However, to further improve the data rate CTLE requirements become more and more stringent. But the digital equalization that is customized to restore the missing pulse can take it to 32 Gb/s. This approach also requires TDC that may be one of the bottlenecks for achieving higher energy efficiency. Transmitter side encoder is a better approach towards equalization due to two reasons – First, by eliminating the narrow pulse it also avoids the generation of the channel ISI. Second, it eliminates the need for higher resolution TDC and therefore, can achieve an energy efficient receiver architecture. This approach can achieve 40 Gb/s data rate and at the same time keeps a digital implementation that is scalable with technology.

IV. CROSSTALK CANCELLATION

Crosstalk is an important consideration in high-density interconnects, especially for single-ended signaling. Inserting ground traces between lanes is a common practice that improves crosstalk by few dBs at the cost of bandwidth density. In time domain modulation, choice of the phase vectors provides a better way to mitigate crosstalk. An example case is shown in Fig. 5 where for simplicity we are using only two phases for modulation as opposed to four phases. Here, the middle lane is the victim where the top and bottom lanes are the aggressors. But in the top lane we choose ϕ_{180}, ϕ_{225} as two possible phases and in the bottom lane we choose ϕ_0, ϕ_{45} as the two possible phases. Due to the randomness of the data patterns, the crosstalk noise from aggressors would not be exactly 180^0 out of phase all the time, but they mostly destructively interfere as shown in Fig. 6. This technique can provide ~20 dB rejection without sacrificing bandwidth density.

V. CONCLUSION

This paper describes two novel digital equalization techniques and a crosstalk correction technique applicable for time-domain signaling. These techniques have proven to be capable of addressing the uniqueness of the time domain ISI to achieve 40 Gb/s data rate compensating 40 dBs of channel loss. Compared to channel inversion approaches, these digital techniques can accommodate 15 to 20 dBs additional loss without significant complexity. These digital techniques along with crosstalk correction shown in this work can be combined to achieve the bandwidth density required for next-generation interconnect solutions.

REFERENCES

[1] W. Wang and J. F. Buckwalter, "A 10-Gb/s, 107-mW double-edge pulsewidth modulation transceiver," *IEEE Trans. Circuits Syst. I, Reg. Papers*, vol. 61, no. 4, pp. 1068–1080, Apr. 2014.

[2] M. Chae, H. Kwon, S. Bae, N. Kim, and H. Park, "A duo-binary transceiver with time-based receiver and voltage-mode time-interleaved mixing transmitter for DRAM interface," *IEEE Trans. Circuits Syst. II, Exp. Briefs*, vol. 68, no. 7, pp. 2409–2413, Jul. 2021.

[3] S. Jeon et al., "A framed-pulsewidth modulation transceiver for highspeed broadband communication links," *IEEE Trans. Circuits Syst. I, Reg. Papers*, vol. 67, no. 8, pp. 2825–2835, Aug. 2020.

[4] C.-Y. Yang and Y. Lee, "A PWM and PAM signaling hybrid technology for serial-link transceivers," *IEEE Trans. Instrum. Meas.*, vol. 57, no. 5, pp. 1058–1070, May 2008.

[5] M. H. Perrott, "Making better use of time in mixed signal circuits," in *Proc. 6th IEEE Dallas Circuits Syst. Workshop Syst.-Chip*, Nov. 2007, pp. 1–8.

[6] N. C. Sevuktekin, L. R. Varshney, P. K. Hanumolu, and A. C. Singer, "Signal processing foundations for time-based signal representations: Neurobiological parallels to engineered systems designed for energy

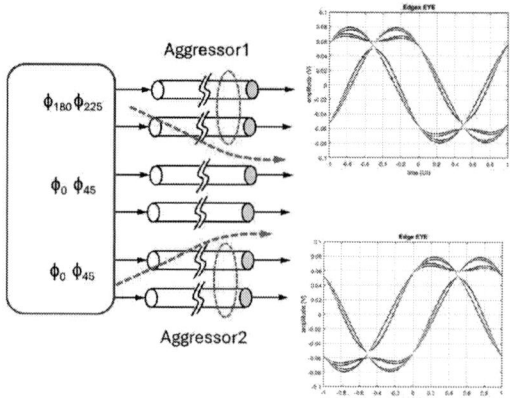

Fig. 5. Selection of the phase vectors for time domain modulation for 2 bits/cycle case

Red – Single aggressor crosstalk
Blue – two aggressor crosstalk

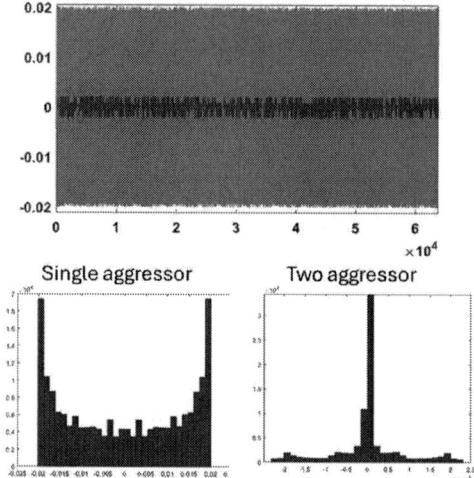

Fig. 6. Time domain and statistical distribution of crosstalk noise with and without cancellation technique.

efficiency or hardware simplicity," *IEEE Signal Process. Mag.*, vol. 36, no. 6, pp. 38–50, Nov. 2019.

[7] S. Ziabakhsh, G. Gagnon, and G. W. Roberts, "The peak-SNR performances of voltage-mode versus time-mode circuits," *IEEE Trans. Circuits Syst. II, Exp. Briefs*, vol. 65, no. 12, pp. 1869–1873, Dec. 2018.

[8] R. L. Smith et al., "Differential edge-modulated signaling with encoded clock and dynamic data rate scaling," *IEEE Solid-State Circuits Lett.*, vol. 4, pp. 84–87, 2021.

[9] R. L. Smith et al. "Differential Edge Modulation Signaling for Low-Energy, High-Speed Wireline Communication," *IEEE Trans. Circuits Syst. I, Reg. Papers*, vol. 70, no. 8, pp. 3359-3372, Aug. 2023.

A Highly-Scalable Parallel Boundary Element Method for the Full-Wave Electromagnetic Analysis of Large Interconnect Networks and Entire Packages

Damian Marek, Jasper Hatton, Yongzhong Li, Piero Triverio

Edward S. Rogers Sr. Dept. of Electrical and Computer Engineering, University of Toronto, Toronto, ON, Canada

damian.marek@mail.utoronto.ca, jasper.hatton@mail.utoronto.ca, yongzhong.li@mail.utoronto.ca, piero.triverio@utoronto.ca

Abstract—We demonstrate a scalable parallelization strategy for an advanced boundary element method suitable for the full-wave electromagnetic analysis of very large interconnect networks and packages. A new preconditioning technique and charge neutrality enforcement strategy show significant promise to overcome the scalability bottlenecks of existing methods. The proposed method is scaled up to 10,240 cores, extracting the scattering parameters of a complete electronic package in less than 2.5 minutes per excitation and frequency point.

Index Terms—boundary element method, high-performance computing, signal integrity, electronic packages

I. INTRODUCTION

Three-dimensional integration can bestow substantial performance and integration gains to integrated circuits (ICs), provided that several challenges are addressed. A key challenge is the proper design of the interconnect network that exchanges signals, power and ground levels within chiplets and between chiplets. This design phase requires extensive electromagnetic (EM) simulations to rule out signal and power integrity issues. At the IC level, these simulations have traditionally been performed with lumped models or quasi-static solvers. While scalable, these approaches are becoming inadequate for an increasing number of designs, as 3D integration gives rise to complex interference patterns which only full-wave simulations can reveal. Unfortunately, the full-wave analysis of even a word from a CPU memory bus in 3D ICs can be extremely time consuming, if not unfeasible, with integral equation (IE) and finite element (FEM) methods available to industry.

To sustain the nascent 3D IC industry, a new class of EM simulation algorithms must be conceived, with lower computational cost and ability to scale on large computing clusters. The efficient parallelization of EM solvers has been the subject of intense research. Most progress has been made for antenna-type problems involving perfect conductors in free space, using IE formulations and fast multipole methods [1]. For interconnect networks in layered substrates, some strategies have been proposed, based on parallel IEs [2]–[5] and FEM [6] algorithms, hybrid IE-FEM methods [7] and domain

This work was partially supported by: Advanced Micro Devices, Natural Sciences and Engineering Research Council of Canada, Digital Research Alliance of Canada, and CMC Microsystems.

decomposition [8]. Despite these advancements, the full-wave analysis of very large interconnect networks or entire packages remains very challenging.

In this paper, we present preliminary results from a new IE method for the full-wave EM analysis of complex interconnect networks which can scale efficiently on thousands of computing cores. The method is accelerated with Multi-AIM [9], which is a multilevel version of the adaptive integral method (AIM) [10], and uses a new algebraic multigrid (AMG) preconditioner to overcome the scalability barrier of factorization-based preconditioners. Finally, a parallelization-friendly approach to enforce charge conservation is devised. The proposed method is scaled up to 10,240 computing cores, enabling the full-wave analysis of an entire IC package in less than 2.5 minutes per excitation and per frequency.

II. PROPOSED METHOD

A. Formulation

We consider a structure made by many conductive objects with conductivity σ, immersed in a homogeneous medium of relative permittivity ε_r and relative permeability μ_r. The structure is excited by N_P ports, with the goal of computing its scattering parameters. Conductors are discretized with a triangular surface mesh. Our starting point is the augmented electric field integral equation (aEFIE) formulation [11]

$$\begin{bmatrix} jk_0\mu_r\mathbf{L}_A & -\frac{1}{\varepsilon_r}\mathbf{D}^\mathrm{T}\mathbf{L}_\Phi \\ \mathbf{D} & jk_0\mathbf{I} + \mathbf{C} \end{bmatrix} \begin{bmatrix} \mathbf{J} \\ c_0\boldsymbol{\rho}_\mathrm{c} \end{bmatrix} = \begin{bmatrix} 0 \\ \mathbf{I}_\mathrm{s} \end{bmatrix}, \quad (1)$$

where $\mathbf{J}$ and $\boldsymbol{\rho}_\mathrm{c}$ are the current density and the charge density on the conductors' surface, respectively. In the matrix, $\mathbf{L}_A$ and $\mathbf{L}_\Phi$ are the components of the EFIE operator related to vector and scalar potential, respectively. Coefficients k_0 and c_0 are the wavenumber and speed of light in free space, while $\mathbf{D}$ is an incidence matrix [11], and $\mathbf{C}$ is a matrix for coupling to the external lumped circuit [12]. Vector $\mathbf{I}_\mathrm{s}$ collects the port currents. Conductor losses are modeled using the Leontovich boundary condition, which is an accurate model when the skin depth is smaller than conductor thicknesses. For the structures considered in this work, the skin depth satisfies this condition when the frequency is above 100 MHz. System (1) is solved iteratively using the GMRES algorithm in PETSc. Accelerating the computation of the matrix vector product associated with (1) is critical to performance. We use

979-8-3503-5124-8/24 $31.00 © 2024 IEEE

a parallel implementation of MultiAIM that uses a hierarchy of multiresolution grids to quickly compute the EM interaction between mesh elements, and is significantly more efficient than AIM for the multiscale layouts typically found in 3D ICs [9].

B. Charge Neutrality Enforcement

In the original aEFIE [11], the total charge on each isolated conductor is set to zero by eliminating one charge density coefficient. Unfortunately, this solution reduces the sparsity of the matrix, and complicates the development of efficient preconditioners. We propose a better approach that enforces charge neutrality by setting the average charge density to zero without explicitly forming any sparse matrices, as is required in [11]. When using an iterative solver, this strategy can be implemented efficiently by subtracting the average of the charge density vector at each operation before multiplying by the system matrix in (1). Preliminary tests show that this approach achieves the same convergence rate as the method in [11], while being easier to implement and more scalable.

C. Algebraic Multigrid Preconditioner

The original aEFIE uses a constraint preconditioner based on the LU factorization of the Schur complement (1)

$$\mathbf{S} = jk_0\mathbf{I} + \frac{1}{jk_0\mu_r\epsilon_r}\mathbf{D}\text{diag}\{\mathbf{L}_A\}^{-1}\mathbf{D}^{\text{T}}\text{diag}\{\mathbf{L}_\Phi\}, \quad (2)$$

where $\text{diag}\{\mathbf{A}\}$ denotes the diagonal of $\mathbf{A}$. On a single machine, this sparse matrix can be factorized efficiently. However, our experiments revealed that this factorization does not scale on large computing clusters, due to its sequential nature and the communication caused by pivoting. We explore an alternative solution based on AMG preconditioners [13]. The form of (2) is similar to the 2D scalar Helmholtz equation discretized on a graph

$$-k^2\mathbf{I} + \frac{1}{h^2}\mathbf{D}\mathbf{D}^{\text{T}}, \quad (3)$$

where h is the characteristic length of the mesh. For this matrix, AMG preconditioners are very effective, especially when k is small (Laplace operator) [13]. When k is not small, AMG preconditioners remain effective as long as the coarsest grid used in the AMG preconditioner remains fine enough to resolve the smooth eigenfrequencies of the Helmholtz operator [13]. This is the strategy adopted in this work.

III. RESULTS

We compare the proposed method with two versions of the standard aEFIE [11] with its LU-based preconditioner factorized with MUMPS. One aEFIE code is accelerated with AIM [10], the other with MultiAIM [9]. Both aEFIE codes and the proposed method are parallelized with MPI, and use the new charge neutrality enforcement technique from Sec. II-B.

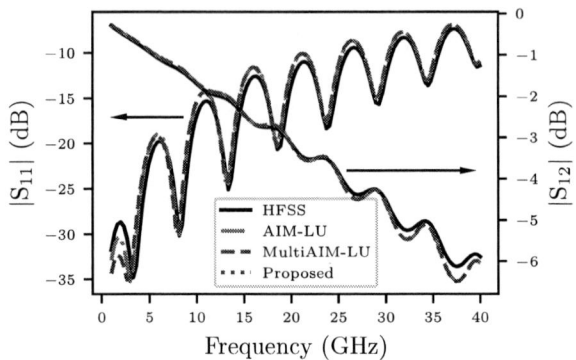

Figure 1. S parameters for the single-ended microstrip benchmark [14].

A. Single-Ended Microstrip Benchmark

We consider a single-ended microstrip, 14.85 mm long, with two probe landing pads and a ground plane [14]. Conductive objects are modeled as copper with RMS surface roughness of 0.3 μm. They are embedded in a homogeneous dielectric background material, with parameters obtained by averaging the properties of the material on the original layered substrate ($\epsilon_r = 3.4$, $\tan\delta = 0.023$). The structure is meshed with an average edge length of 41 μm, for a total of 59,536 triangles. Figure 1 shows the scattering parameters computed by the different methods, along with results from a commercial FEM solver (Ansys HFSS). The agreement validates the accuracy of the proposed solver.

B. IBM Plasma Package

IBM Plasma is an entire 3-2-3 IC package, with eight-metal-layer laminate and full signal fanout from the top bumps to a ball grid array underneath [14]. Despite dating back to 2006, its full-wave analysis remains very challenging, and has been attempted by relatively few groups, such as [2], [6], [8]. The package measures 32 mm × 32 mm × 0.7 mm. We consider a quarter cutout, a half cutout, and the entire package. All metal conductors are modeled as copper surrounded by a homogeneous medium with $\epsilon_r = 3.8$ and $\tan\delta = 0.035$. Four ports are defined at the two ends of the "DAT_02" and "DAT_03" signal nets. The full package resulted in a mesh with over 24 million mesh edges and 16 million triangles (average edge length of 45 μm). All codes were run on the SciNET Niagara cluster.

Fig. 2 reports the system generation time and the system solve time as a function of the number of cores P, for a single frequency (1 GHz). The system generation time accounts for generating all matrices in (1), forming the constraint preconditioner, and, when not using the proposed method, computing the LU factorization of the Schur complement in (2). The system solve time is the time spent solving (1) with GMRES. The upper panel of Fig. 2 demonstrates that both AIM-LU and MultiAIM-LU do not scale with P because of the poor scaling of the sparse LU factorization. From the lower panel, we see that the system solve time for both AIM-LU and MultiAIM-LU does not scale with P either. This issue

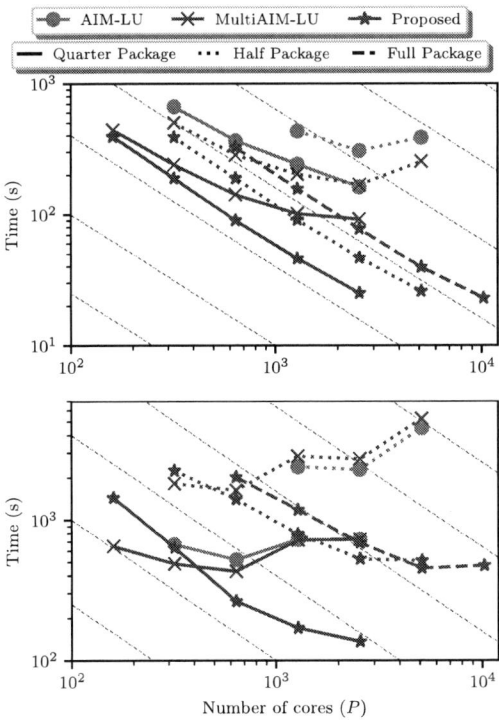

Figure 2. System generation time (upper panel) and solution time (lower panel) for the quarter, half and full Plasma package, as a function of the number of cores. Dashed lines represent ideal scalability slopes.

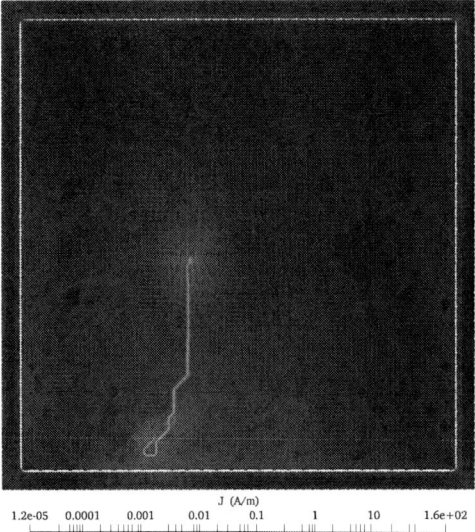

Figure 3. Surface current density computed for the Plasma package at 1 GHz.

is caused by the poor scaling of the forward and backward substitutions needed when applying the preconditioner in the solving step.

The entire Plasma package could only be simulated by the proposed method, as both AIM-LU and MultiAIM-LU ran out of memory. The proposed method achieves very good scalability for the entire package, up to 5,120 cores. The system generation phase scales almost perfectly up to 10,240 cores owing to the AMG preconditioner, in stark contrast to the LU-based preconditioner used in the AIM-LU and MultiAIM-LU algorithms. Overall, the proposed method is up to $9\times$ faster than AIM-LU and MultiAIM-LU. Fig. 3 shows the current density on the entire package computed by the proposed method at 1 GHz. With 10,240 cores, the proposed method performs the full-wave EM analysis of the entire package in less than 2.5 minutes per excitation and per frequency point. Although the current implementation assumes a homogeneous medium, extension to layered media is possible. Overall, the presented preliminary results show the merits of the proposed ideas and their potential to, in due course, overhaul the EM simulation capabilities made available to the microelectronic industry for the design of 2D and 3D ICs.

REFERENCES

[1] W.-J. He, Z. Yang, X.-W. Huang, W. Wang, M.-L. Yang, and X.-Q. Sheng, "Solving Electromagnetic Scattering Problems With Tens of Billions of Unknowns Using GPU Accelerated Massively Parallel MLFMA," *IEEE Trans. Antennas Propag.*, vol. 70, no. 7, pp. 5672–5682, 2022.

[2] J. Morsey, A. Deutsch, J. Libous, C. Surovic, B. J. Rubin, L. Jiang, and L. Eisenberg, "The use of accelerated full-wave modeling to analyze power island coupling in a HyperBGA SCM," *IEEE transactions on advanced packaging*, vol. 30, no. 2, pp. 288–294, 2007.

[3] C. Liu, K. Aygün, and A. E. Yılmaz, "A parallel FFT-accelerated layered-medium integral-equation solver for electronic packages," *Int. J. Numer. Model.: Electron. Networks, Devices and Fields*, pp. 1–17, Sep. 2019.

[4] D. Marek, S. Sharma, and P. Triverio, "A parallel boundary element method for the electromagnetic analysis of large structures with lossy conductors," *IEEE Trans. Antennas Propag.*, vol. 70, no. 11, pp. 10 736–10 750, Nov. 2022.

[5] D. Marek and P. Triverio, "An efficient parallel electromagnetic solver for extracting scattering parameters from large electrical interconnects with many ports," in *2022 IEEE 31st Conf. Elect. Perform. Electron. Packag. Syst.*, 2022, pp. 1–3.

[6] S. Sun, F. Zavosh, Z. Yang, Q. Liu, S. Cui, and L. Jiang, "Full Wave IBM Plasma Substrate Benchmark By Cadence Clarity," in *2023 IEEE 31st Conf. Elect. Perform. Electron. Packag. Syst.* IEEE, 2023, pp. 1–3.

[7] Z. Peng, Y. Shao, H.-W. Gao, S. Wang, and S. Lin, "High-fidelity, high-performance computational algorithms for intrasystem electromagnetic interference analysis of IC and electronics," *IEEE Trans. Compon., Packag., Manuf. Technol.*, vol. 7, no. 5, pp. 653–668, 2017.

[8] Y. Shao, Z. Peng, and J.-F. Lee, "Full-wave real-life 3-D package signal integrity analysis using nonconformal domain decomposition method," *IEEE Trans. Microw. Theory Techn.*, vol. 59, no. 2, pp. 230–241, 2011.

[9] Y. Li, D. Marek, and P. Triverio, "MultiAIM: Fast electromagnetic analysis of multiscale structures using boundary element methods," *IEEE Trans. Antennas Propag.*, 2024, (in press).

[10] E. Bleszynski, M. Bleszynski, and T. Jaroszewicz, "AIM: Adaptive integral method for solving large-scale electromagnetic scattering and radiation problems," *Radio Science*, vol. 31, no. 5, pp. 1225–1251, Sep. 1996.

[11] Z.-G. Qian and W. C. Chew, "Fast full-wave surface integral equation solver for multiscale structure modeling," *IEEE Trans. Antennas Propag.*, vol. 57, no. 11, pp. 3594–3601, Nov. 2009.

[12] Y. Wang, D. Gope, V. Jandhyala, and C. J. R. Shi, "Generalized Kirchoff's current and voltage law formulation for coupled circuit-electromagnetic simulation with surface integral equations," *IEEE Trans. Microw. Theory Techn.*, vol. 52, no. 7, pp. 1673–1682, 2004.

[13] Y. A. Erlangga, C. W. Oosterlee, and C. Vuik, "A novel multigrid based preconditioner for heterogeneous Helmholtz problems," *SIAM J. Sci. Comput.*, vol. 27, no. 4, pp. 1471–1492, 2006.

[14] IEEE EPS Technical Committee on Electrical Design, Modeling and Simulation, "Packaging Benchmark Suite," 2021. [Online]. Available: https://packaging-benchmarks.org/

Gradient-based method to find solution for Rational Polynomial Chaos coefficients for Uncertainty Quantification

Karanvir S. Sidhu and Roni Khazaka

Department of Electrical and Computer Engineering, McGill University, Montréal, Québec, Canada, H3A 0E9
Email: karanvir.sidhu@mail.mcgill.ca, roni.khazaka@mcgill.ca

Abstract—**Rational Polynomial Chaos (RPC) is an emerging method for forming the surrogate model for uncertainty quantification in circuit applications. However, the use of Rational Polynomial Chaos can be prohibitive as it generally requires a large number of samples to generate the surrogate models. In this paper, we propose the use of gradient-based algorithm to solve for the Rational Polynomial Chaos coefficients. Using the proposed approach we can compute the RPC coefficients with fewer samples than the traditional RPC method.**

Keywords— **Uncertainty Quantification, Machine Learning, Polynomial Chaos.**

I. INTRODUCTION

Uncertainty quantification and design for manufacturability are crucial aspects of the circuit design process, particularly as the device sizes decrease. The decrease in device sizes leads to process variations, which introduces circuit parameter variations. Traditionally, Monte Carlo-based techniques are used to quantify the effects of the parameter uncertainties on Quantity (or Quantities) of Interest (QoI). Monte-Carlo approaches are known for their accuracy and robustness [1]. However, the Monte Carlo approaches are computationally intensive. Therefore, significant efforts have been made to develop alternative techniques for computing QoI statistics arising from parameter uncertainties.

In recent years, Polynomial Chaos (PC) based methods have been proposed to form surrogate models of the QoI as a function of uncertain parameters [2]. The PC methods use uncertain parameters to form multivariate orthogonal polynomials, which are then used to create the surrogate model of QoI. The statistics of the QoI are computed by performing the Monte-Carlo analysis on the surrogate models. However, it has been shown in [3–5] that the PC method cannot model a class of problems where the QoI is highly sensitive to the circuit parameters. This has led to the introduction of the Rational Polynomial Chaos (RPC) method for uncertainty quantification [3,4].

One of the drawbacks of PC and RPC methods is that the number of polynomial basis functions increases rapidly with an increase in the number of random parameters. Due to this, the number of circuit evaluation data required to build the surrogate model also increases rapidly. The RPC method proposed in [3,4] uses the Sanathanan-Koerner(SK) approach

to compute the RPC coefficients. In this paper, we present a novel approach that uses gradient descent-based optimization techniques to find the RPC coefficients by minimising the loss function. Additionally, we use regularization to find the RPC coefficients with fewer circuit evaluation data than the polynomial basis. We use the backpropagation method to compute the gradient of the loss function with respect to RPC coefficients.

II. BACKGROUND

The Rational Polynomial models for uncertainty quantification were proposed in [4]. The rational polynomial model for the output stochastic QoI at sample $\boldsymbol{\xi}^{(k)}$, can be written as,

$$y(\boldsymbol{\xi}^{(k)}) \equiv \frac{N(\boldsymbol{\xi}^{(k)})}{D(\boldsymbol{\xi}^{(k)})} = \frac{\sum_{j=0}^{P-1} n_j \, \phi_j(\boldsymbol{\xi}^{(k)})}{1 + \sum_{j=1}^{P-1} d_j \, \psi_j(\boldsymbol{\xi}^{(k)})} \quad (1)$$

where n_j and d_j are the coefficients of numerator polynomials and denominator polynomials, respectively. The polynomials $\phi(\boldsymbol{\xi})$ and $\psi(\boldsymbol{\xi})$ are multivariate polynomial functions of random variables $\boldsymbol{\xi} = \begin{bmatrix} \xi_1 & \cdots & \xi_d \end{bmatrix}$, where $\xi_1, \ldots, \xi_d$ are the random parameters. The discrepancy between the model $y(\boldsymbol{\xi}^{(k)})$ and the observation, v_k at a given sample $\boldsymbol{\xi}^{(k)}$ can be written as

$$\epsilon_k = v_k - \frac{\sum_{j=0}^{P-1} n_j \, \phi_j(\boldsymbol{\xi}^{(k)})}{1 + \sum_{j=1}^{P-1} d_j \, \psi_j(\boldsymbol{\xi}^{(k)})} \quad (2)$$

Our goal is to compute the RPC coefficients $\{n_j\}_{j=0}^{P-1}$, and $\{d_j\}_{j=1}^{P-1}$ such that the sum of square of the error in (2) is minimised. One method to compute the RPC coefficients is to use the Sananthanan-Koerner (SK) approach, which is described in [4]. The polynomials $\phi(\boldsymbol{\xi})$ and $\psi(\boldsymbol{\xi})$ can be chosen either using the Full Tensor Order truncation [4] or using the Total Order truncation used in [5]. Despite the truncation style, the number of polynomial functions increases very rapidly with the increase in the number of parameters [1,4,5].

As the number of unknown coefficients increases, the number of circuit simulations, i.e., the QoI observations at different samples of random parameters required to build, also increases. The approach in [5] uses the derivative information with the SK method to reduce the number of circuit simulations for the frequency analysis problems. However, obtaining derivative information in the context of time-domain simulation can be costly. Furthermore, using the SK approach

979-8-3503-5124-8/24 $31.00 © 2024 IEEE

can also be computationally expensive for cases with a large number of polynomial bases.

III. PROPOSED APPROACH

We use the structure of a neuron as shown in Figure 1. As seen in Figure 1, the numerator polynomials, $\{\phi_j(\boldsymbol{\xi})\}_{j=0}^{P-1}$, and the denominator polynomials, $\{\psi_j(\boldsymbol{\xi})\}_{j=1}^{P-1}$, are the inputs to the neuron. The numerator and the denominator polynomials are evaluated at input training samples. In the next stage of our network, the numerator and the denominator polynomials are multiplied with the unknown numerator and denominator weights $\{n_j\}_{j=0}^{P-1}$ and $\{d_j\}_{j=1}^{P-1}$, respectively. The prediction,

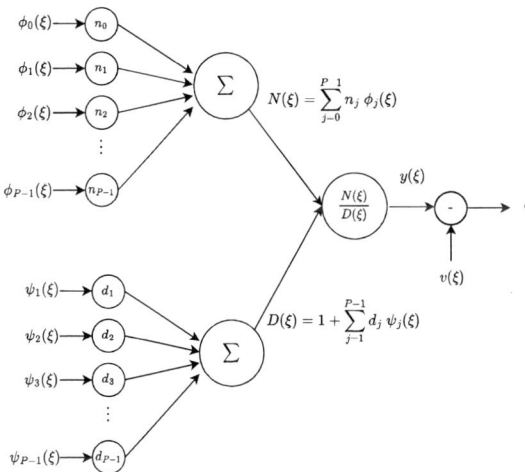

Fig. 1. The structure of Rational Neuron for a given input sample, $\boldsymbol{\xi}$.

$y(\boldsymbol{\xi})$, of the QoI at a given sample $\boldsymbol{\xi}^{(k)}$ is shown in (1) The unknown coefficients in the proposed scheme can be computed using the gradient descent-based optimization algorithm; to use this method, we need to compute the derivative of the error with respect to the unknown coefficients. To achieve this, we used the back-propagation algorithm. To fit the rational model to the data, we minimize the Mean-Square Error (MSE) between QoI observations, $v(\boldsymbol{\xi})$, at the samples of random parameters and the prediction, $y(\boldsymbol{\xi})$. The MSE, along with the regularization terms, can be written as

$$E = \frac{1}{2K}\sum_{k=1}^{K}\epsilon_k^2 + \frac{\lambda}{2}\|\boldsymbol{n}\|_2 + \frac{\lambda}{2}\|\boldsymbol{d}\|_2 \tag{3}$$

To avoid the tedious expressions, we start by performing the back-propagation for one training sample. Thus, the cost at a given sample can be written as,

$$E_k = \frac{1}{2}\left(v(\boldsymbol{\xi}^{(k)}) - y(\boldsymbol{\xi}^{(k)})\right)^2 + \frac{\lambda}{2}\|\boldsymbol{n}\|_2 + \frac{\lambda}{2}\|\boldsymbol{d}\|_2 \tag{4}$$

The derivative of the cost function without the regularization terms wrt the prediction is shown below

$$\frac{\partial E_k}{\partial y(\boldsymbol{\xi}^{(k)})} = -(v(\boldsymbol{\xi}^{(k)}) - y(\boldsymbol{\xi}^{(k)})) \tag{5}$$

Next, we compute the derivative of the prediction with respect to the numerator and denominator as

$$\frac{\partial y(\boldsymbol{\xi}^{(k)})}{\partial N(\boldsymbol{\xi}^{(k)})} = \frac{1}{D(\boldsymbol{\xi}^{(k)})} \tag{6}$$

$$\frac{\partial y(\boldsymbol{\xi}^{(k)})}{\partial D(\boldsymbol{\xi}^{(k)})} = \frac{N(\boldsymbol{\xi}^{(k)})}{D(\boldsymbol{\xi}^{(k)})^2} \tag{7}$$

Now that we have obtained the derivative of the prediction with respect to the numerator and the denominator at a given sample $\boldsymbol{\xi}^{(k)}$, we can use (6) and (7) to compute the derivative of the cost function with respect to the weights $\boldsymbol{n}$ and $\boldsymbol{d}$ using the data from all the K number of samples The derivative of the cost function and the regularization terms with respect to the numerator coefficient is given by

$$\frac{\partial E}{\partial \boldsymbol{n}} = \boldsymbol{\Phi}^T\left(\frac{\partial E}{\partial \boldsymbol{y}}\frac{\partial \boldsymbol{y}}{\partial \boldsymbol{N}}\right) + \lambda \boldsymbol{n} \tag{8}$$

where, $\frac{\partial E}{\partial \boldsymbol{y}}$ and $\frac{\partial \boldsymbol{y}}{\partial \boldsymbol{N}} \in \mathbb{R}^{K \times K}$ are diagonal matrices. The matrix $\boldsymbol{\Phi}^T \in \mathbb{R}^{K \times P}$ contains the numerator polynomials evaluated at K number of samples. Similarly, the derivative of the cost function and the regularization terms with respect to the denominator weights can be written as

$$\frac{\partial E}{\partial \boldsymbol{d}} = \boldsymbol{\Psi}^T\left(\frac{\partial E}{\partial \boldsymbol{y}}\frac{\partial \boldsymbol{y}}{\partial \boldsymbol{D}}\right) + \lambda \boldsymbol{d} \tag{9}$$

similar to the numerator case, $\frac{\partial \boldsymbol{y}}{\partial \boldsymbol{N}} \in \mathbb{R}^{K \times K}$ is a diagonal matrix and $\boldsymbol{\Psi}^T \in \mathbb{R}^{K \times P-1}$ contains the denominator polynomials evaluated at K samples. Now that we have obtained the derivative of the cost function with respect to the unknown parameters, we can use any gradient-based optimization algorithm. For our applications, we found that the Adam optimization algorithm [6] and the AdaGrad optimization algorithm [7] work well.

IV. NUMERICAL EXAMPLES

To demonstrate the accuracy and efficiency of the proposed method, we considered the following 3-port mixer shown in Figure 2. The local oscillator of this circuit is located at Port 1 and is excited with a 1GHz sinusoidal signal with an amplitude of 100mV. The input at Port 2 is excited with a sinusoidal signal with an amplitude of 1 mV and frequency of 100MHz. The third port is the output port indicated by V_{out} in Figure 2. All the ports of the example circuits are matched to the impedance of 50Ω. For this example, we used all $d = 24$ random parameters shown in Figure 2. All the parameters are assumed to be uniformly distributed with 10% relative standard deviation around their nominal values, provided in Figure 2.

We simulated the above circuit for 30ns using the Backward Euler approach with the time step of 0.1ns. To compute the Monte-Carlo results we simulated the above circuit 5000 times with different samples of random parameters. The statistical results obtained using the Monte-Carlo method are used as the ground truth to compare the results obtained with the other approaches used in this paper. We compared the results of the proposed approach with the RPC and the LS-SVM methods.

979-8-3503-5124-8/24 $31.00 © 2024 IEEE

For the RPC method, we used the Total order truncation of order 2; this resulted in a total of 651 polynomials. To build the surrogate model for the RPC method, we used $K = 655$ observations of the QoI. The Root Mean Square Error (RMSE) in standard deviation predicted by RPC method was 0.28mV with $K = 655$ simulations. The LS-SVM method required $K = 500$ simulations to reach an accuracy similar to that of the RPC method. The RMSE error in standard deviation predicted by LS-SVM with $K = 500$ simulations was found to be 0.13mV.

For the proposed method, we also used the Total order truncation of order 2; this resulted in a total of 649 polynomials, and we used only $K = 120$ samples to build the surrogate model. We used the Adam optimization algorithm to compute the RPC coefficients; we used the default hyperparameters for the optimizer as provided in [6]. The RMSE error in the standard deviation predicted by the proposed method was 0.19mV. As we can see, the proposed approach can compute the QoI statistics with accuracy similar to the RPC and LS-SVM methods but with fewer circuit simulations.

The mean and the standard deviation computed using the RPC approach, LS-SVM method, and the proposed approach are shown in Figure 3. As shown in Figure 3, the proposed approach can compute the mean and standard deviation as accurately as RPC and the LS-SVM method while using only $K = 120$ samples. The probability distribution of the V_{out} is shown in Figure 4; we can see in Figure 4 that the proposed approach accurately models the pdf of the V_{out}.

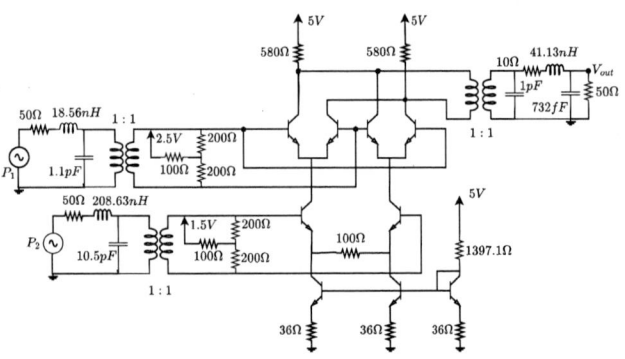

Fig. 2. An up-converter mixer circuit.

V. CONCLUSION

We proposed a technique to compute the RPC coefficients using the gradient-based method. The gradient of the error with respect to the RPC coefficients was computed using back-propagation, and we used the Adam optimization algorithm to find the RPC coefficients to form the surrogate model. We presented an example to showcase the accuracy and efficiency of our proposed method where we found that proposed method can compute the statistical properties of the QoI with fewer circuit evaluations.

REFERENCES

[1] A. Kaintura, T. Dhaene, and D. Spina, "Review of polynomial chaos-based methods for uncertainty quantification in modern integrated circuits," *Electronics*, vol. 7, no. 3, p. 30, 2018.

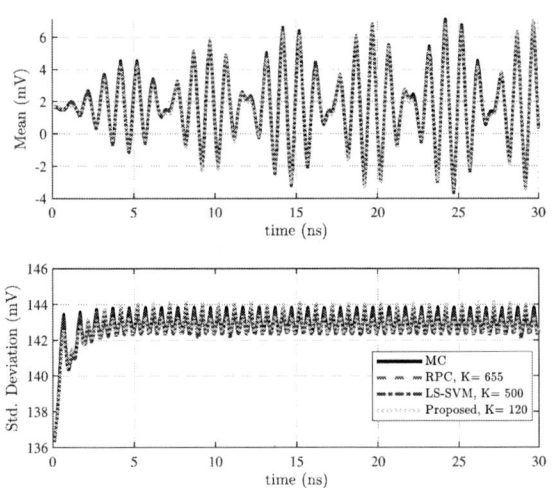

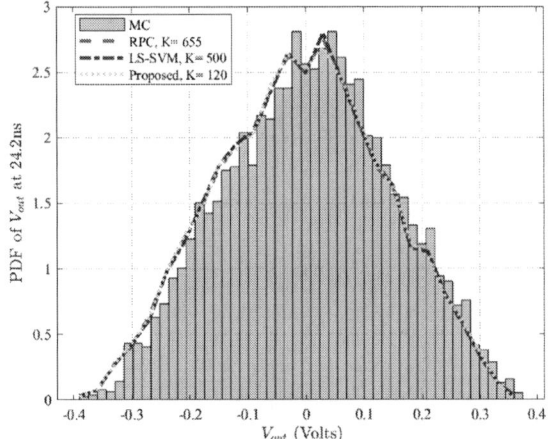

Fig. 3. Mean (upper panel) and the Standard Deviation (lower panel) of V_{out} from 0ns to 30ns.

Fig. 4. Probability distribution of the V_{out} at 24.2ns.

[2] D. Xiu and G. E. Karniadakis, "The wiener–askey polynomial chaos for stochastic differential equations," *SIAM journal on scientific computing*, vol. 24, no. 2, pp. 619–644, 2002.

[3] P. Manfredi and S. Grivet-Talocia, "Improved stochastic macromodeling of electrical circuits via rational polynomial chaos expansions," in *2019 Joint International Symposium on Electromagnetic Compatibility, Sapporo and Asia-Pacific International Symposium on Electromagnetic Compatibility (EMC Sapporo/APEMC)*. IEEE, 2019, pp. 511–514.

[4] ——, "Rational polynomial chaos expansions for the stochastic macromodeling of network responses," *IEEE Transactions on Circuits and Systems I: Regular Papers*, vol. 67, no. 1, pp. 225–234, 2019.

[5] K. S. Sidhu and R. Khazaka, "Derivative-enhanced rational polynomial chaos for uncertainty quantification," *IEEE Transactions on Circuits and Systems I: Regular Papers*, pp. 1–10, 2024.

[6] D. P. Kingma and J. Ba, "Adam: A method for stochastic optimization," *arXiv preprint arXiv:1412.6980*, 2014.

[7] J. Duchi, E. Hazan, and Y. Singer, "Adaptive subgradient methods for online learning and stochastic optimization." *Journal of machine learning research*, vol. 12, no. 7, 2011.

On the Parallelization of the MultiAIM Algorithm for the Fast Electromagnetic Analysis of 3D ICs

Yongzhong Li[1,2] and Piero Triverio[1]

[1] Edward S. Rogers Sr. Dept. of Electrical and Computer Engineering, University of Toronto, Toronto, ON, Canada
[2] Advanced Micro Devices, Markham, ON, Canada
yongzhong.li@mail.utoronto.ca, piero.triverio@utoronto.ca

Abstract—We propose a parallelization strategy for the MultiAIM algorithm, to accelerate the electromagnetic analysis of multiscale integrated circuit layouts. We devise a solution for shared memory architectures. With a careful choice of numerical libraries, we achieve a code that can be easily adapted to distributed-memory clusters. Preliminary tests on a commercial 3D integrated circuit show good scalability up to 32 cores, and outline a few directions for further improvements.

Index Terms—electromagnetic analysis, signal integrity, multiscale problems, multithreading

I. Introduction

The full-wave electromagnetic (EM) analysis of electronic interconnects and packages is critical for ensuring signal integrity (SI) and power integrity (PI) in modern integrated circuits (ICs). For packages and on-chip interconnects, simplified models based on lumped elements, transmission line theory, and quasi-static analysis have been extensively used for SI/PI assessments, with full-wave EM simulations advocated only where needed. However, the rise of 3D integration is disrupting the status quo in two ways. First, 3D integration substantially increases IC performance and density, leading to SI and PI phenomena of much higher complexity and extent that only full-wave simulations can reliably predict. Second, 3D integration makes IC layouts significantly more complex and multiscale, with a myriad of tiny microvias connected to much larger power and ground structures. Full-wave EM simulations, notoriously CPU intensive, can become extremely demanding under those circumstances, calling for both mathematical and algorithmic advancements.

While all algorithms for computational EM find application to interconnects, in this work we focus on the boundary element method (BEM). In presence of layered substrates, the BEM is often accelerated with the Fast Fourier Transform (FFT), through the adaptive integral method (AIM) [1] or pre-corrected FFT method [2]. Other acceleration strategies have been also proposed, based on domain decomposition [3], parallelization [4], [5], and hierarchical matrices [6], [7]. For multiscale problems, the cost of these methods typically grows

This work was partially supported by: Advanced Micro Devices, Natural Sciences and Engineering Research Council of Canada, Digital Research Alliance of Canada, and CMC Microsystems.

quite unfavourably, unless multiresolution or multigrid strategies are adopted. Recently, a multiresolution generalization of the AIM, called MultiAIM, has been proposed, and shown to be particularly effective for IC analysis, even in the presence of multiscale features [8], [9]. In this paper, we present a preliminary strategy for the parallelization of MultiAIM for shared-memory architectures, using multithreading. Particular attention is devoted to the choice of numerical libraries, in order to obtain a code that can be also used on distributed-memory clusters with minimal code changes.

II. Formulation

We consider a conductive network in a layered substrate, excited through lumped ports and discretized using a triangular surface mesh. We solve Maxwell's equations through the augmented electric field integral equation (aEFIE) [10]

$$\begin{bmatrix} jk_0\mathbf{L}^{(A)} + \eta_0^{-1}\mathbf{Z}_s & -\mathbf{D}^T\mathbf{L}^{(\phi)}\mathbf{B} \\ \mathbf{FD} & jk_0\mathbf{I} + \mathbf{C} \end{bmatrix} \begin{bmatrix} \mathbf{J}_S \\ c_0\boldsymbol{\rho}_S \end{bmatrix} = \begin{bmatrix} \mathbf{0} \\ \mathbf{I}_s \end{bmatrix}, \quad (1)$$

where the surface current density $\mathbf{J}_S$ and the surface charge density $\boldsymbol{\rho}_S$ are taken as unknowns. In (1), k_0, η_0 and c_0 denote the wavenumber, impedance, and phase velocity in free space, respectively. The dense matrices $\mathbf{L}^{(A)}$ and $\mathbf{L}^{(\phi)}$ arise from the discretization of the vector and scalar potential operators. Matrix $\mathbf{Z}_s$ implements the surface impedance boundary condition, while $\mathbf{I}_s$ and $\mathbf{C}$ apply a lumped port excitation. Expressions for $\mathbf{D}$, $\mathbf{F}$ and $\mathbf{B}$ can be found in [10].

System (1) is solved iteratively using GMRES [11]. For a fast solution, the matrix-vector products involving $\mathbf{L}^{(A)}$ and $\mathbf{L}^{(\phi)}$ must be accelerated. To this end, AIM and MultiAIM split $\mathbf{L}^{(A)}$ and $\mathbf{L}^{(\phi)}$ into near and far interactions as $\mathbf{L} = \mathbf{L}_N + \mathbf{L}_F$. The near component of the matrix is computed explicitly with numerical integration of the Green's function. The far component $\mathbf{L}_F$ is approximated as [9]

$$\mathbf{L}_F \cong \mathbf{W}^{(0,1)}\left(\widetilde{\mathbf{H}}^{(1)} + \mathbf{H}_c^{(1)}\right)\mathbf{P}^{(1,0)} - \mathbf{L}_c, \quad (2)$$

in order to enable to use of FFT. Matrix $\mathbf{P}^{(1,0)}$ projects the sources (charges, currents) from the original triangular mesh (level $l = 0$) to a uniform grid (level $l = 1$). On this grid, sources are propagated with FFT to determine the fields they produce at the grid points. This operation is represented by the convolution matrix $\widetilde{\mathbf{H}}^{(1)}$. Finally, fields on the original triangular mesh are computed with interpolation matrix $\mathbf{W}^{(0,1)}$.

979-8-3503-5124-8/24 $31.00 © 2024 IEEE

This is how AIM works [1]. Unfortunately, if the layout has tiny features, the uniform grid can be extremely fine, making FFTs very expensive. MultiAIM avoid this issue by introducing a hierarchy of grids, and defining the convolution matrix recursively as

$$\widetilde{\mathbf{H}}^{(l)} = \begin{cases} \mathbf{W}^{(l,l+1)}\left(\widetilde{\mathbf{H}}^{(l+1)} + \mathbf{H}_c^{(l+1)}\right)\mathbf{P}^{(l+1,l)} & l \in [1, L-2] \\ \mathbf{W}^{(L-1,L)}\mathbf{H}^{(L)}\mathbf{P}^{(L,L-1)} & l = L-1 \end{cases}$$

(3)

Matrix $\mathbf{P}^{(l+1,l)}$ projects sources from level l to level $l+1$. When level L is reached, matrix $\mathbf{H}^{(L)}$ propagates with FFT all sources to obtain fields on the level L grid, taking into account the surrounding layered substrate through its Green's function. In this way, fields can be propagated efficiently even over large distances, while still being able to resolve tiny features. Due to the singularity of the Green's function, correction matrices $\mathbf{H}_c^{(l)}$ and $\mathbf{L}_c$ have to be introduced.

III. PARALLELIZATION STRATEGY FOR MULTIAIM

The efficient parallelization of an accelerated BEM algorithm is not a trivial task, due to the complexity of the implementation and the variety of operations required, which typically imply the use of multiple numerical libraries. Furthermore, there are two competing paradigms for parallelization. Multi-threading is easier to implement, but restricted to single computing nodes with shared memory. Message passing works on both shared- and distributed-memory architectures, but is much more involved. In this paper, we focus on MultiAIM's parallelization for shared-memory machines. However, we pursue a solution based on a numerical library, PETSc, which is also suitable for large computing clusters, in an effort to reduce code differences between the shared and distributed memory scenarios.

A. Generation of Near-Region, Correction and Projection Matrices

The matrices $\mathbf{L}_N^{(A)}$, $\mathbf{L}_N^{(\phi)}$, $\mathbf{H}_c^{(l)}$, $\mathbf{P}^{(l,l+1)}$ and $\mathbf{W}^{(l+1,l)}$ are sparse. Their entries are computed by either integration [10] or interpolation [12]. These computations are independent, and can be easily parallelized with OpenMP.

B. Iterative Solver (GMRES)

We use PETSc's GMRES solver, as it is one of the few implementations that support complex data and can scale to large clusters. The challenge is that, while PETSc has extensive support for message passing (MPI) parallelization, its support for multithreading is limited. To avoid imposing MPI parallelization on the entire BEM implementation, we exploit PETSc's ability to offload most GMRES operations to an external library for linear algebra.

A first operation required by GMRES is matrix-vector multiplication between the matrices in (1) and the candidate solution vector. This operation requires sparse matrix-dense vector products such as $\mathbf{L}^{(A)}\mathbf{J}_s$. The default PETSc code for this operation does not have OpenMP support. However, one can link PETSc to an external BLAS library that supports both

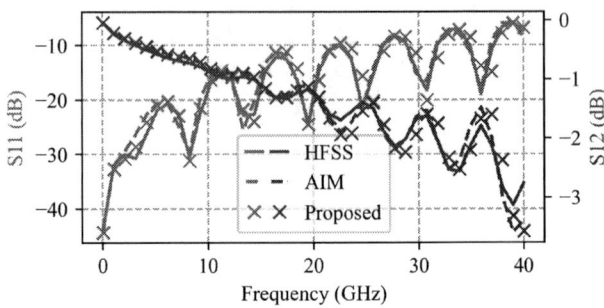

Fig. 1. Scattering parameters obtained for the microstrip benchmark in Sec. IV-A.

sparse matrices and multithreading. We chose the Intel's Math Kernel Library (MKL) [13] and set the sparse matrix type to `MATAIJMKL`. Other libraries, such as OpenBLAS [14], are unsuitable due to limited support for sparse matrices. This solution is also applied to the matrix multiplications in (2), which are coded inside a PETSc "shell matrix" object.

A second operation required by GMRES is the computation of the FFT associated to $\mathbf{H}^{(L)}$, for which we use the FFTW library [15], that has excellent OpenMP support.

The third operation required by GMRES is the inner product between vectors, as part of the orthogonalization process. By default, PETSc does not support multithreading for this operation. However, with the `vec_maxpy_use_gemv` flag, this operation can be offloaded to Intel's MKL.

C. Preconditioner

The AEFIE preconditioner [10] requires a sparse LU factorization and forward/backward substitutions. This preconditioner can be implemented in PETSc with user-provided code as a "shell" preconditioner. However, we could not find a direct way to enable multithreading in this section. Alternatively, PETSc supports offloading the LU factorization to an external, multi-threaded library, but this solution does not appear to be compatible with "shell" preconditioners.

IV. NUMERICAL RESULTS

A. Package Microstrip Benchmark

The proposed method was implemented in C++ with OpenMP parallelization, and ran on a dual-Xeon node of the SciNET Niagara cluster. As first example, we consider the first benchmark microstrip in [16], which is a microstrip with probe landing pads on both ends. The microstrip is placed in a single dielectric layer with thickness of 72.77 μm and permittivity of 3.4. The S-parameters were extracted at 40 frequency points from 50 MHz to 40 GHz and are depicted in Fig. 1. The results from the proposed method are in good agreement with those from AIM and from a commercial FEM solver (Ansys HFSS).

B. High-speed Bus from a Commercial 3D IC

The second example is a cut out of a high-speed communication bus from a commercial processing unit provided by Advanced Micro Devices. The structure is a dense network

979-8-3503-5124-8/24 $31.00 © 2024 IEEE

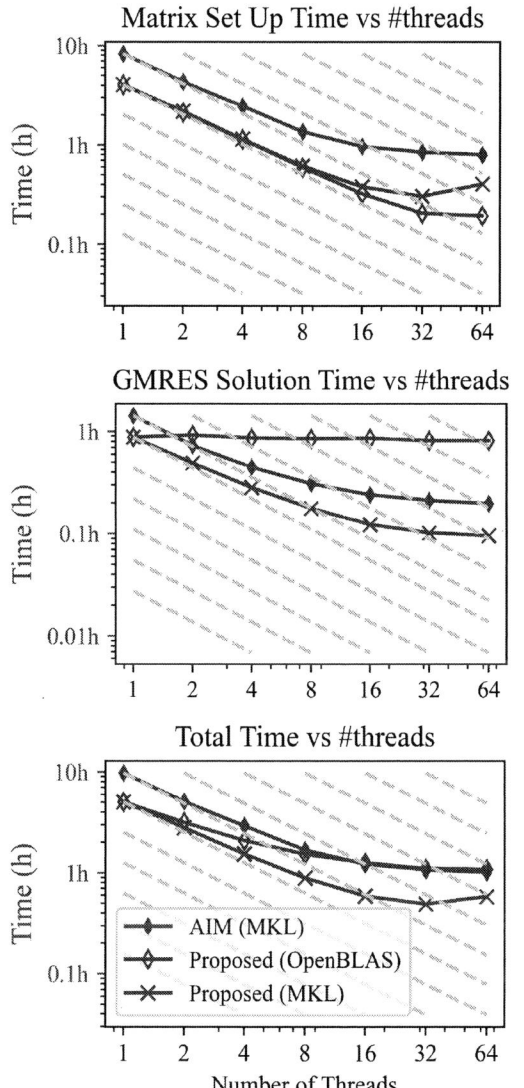

Fig. 2. Execution time as a function of the number of threads for the commercial layout in Sec. IV-B. Top panel: matrix filling time. Middle panel: GMRES. Bottom panel: total time.

not the case for the OpenBLAS implementation, as it lacks a multithreaded routine for sparse matrix-vector products. Finally, the bottom panel reports the total wall time for the three methods. Overall, the proposed parallelization strategy scales well up to 32 cores, and is faster than AIM while providing S parameters within 0.5 dB. Future work will focus on improving scalability beyond 32 cores, by finding a way to parallelize the preconditioner factorization and application in a PETSc-compatible way. Overall, these preliminary results show that the proposed parallelization strategy is effective and can significantly accelerate the full-wave EM simulation of large interconnect structures from 3D integration.

V. ACKNOWLEDGEMENT

We are grateful to Advanced Micro Devices (AMD) and to PETSc's developers for their support.

REFERENCES

[1] E. Bleszynski and J. T. Bleszynski, M., "AIM: Adaptive integral method for solving large-scale electromagnetic scattering and radiation problems," *Radio Science*, vol. 31, no. 5, pp. 1225–1251, Sep. 1996.

[2] J. R. Phillips and J. K. White, "A precorrected-FFT method for electrostatic analysis of complicated 3-D structures," *IEEE J. Technol. Comput. Aided Design*, vol. 16, no. 10, pp. 1059–1072, 1997.

[3] Y. Shao, Z. Peng, and J.-F. Lee, "Full-wave real-life 3-D package signal integrity analysis using nonconformal domain decomposition method," *IEEE Trans. Microw. Theory Techn.*, vol. 59, no. 2, pp. 230–241, 2011.

[4] C. Liu, K. Aygün, H. Braunisch, V. I. Okhmatovski, and A. E. Yilmaz, "A parallel iterative layered-medium integral-equation solver for electromagnetic analysis of electronic packages," in *Proc. IEEE Conf. on Elect. Perf. of Electron. Packag. and Systems (EPEPS)*, 2017, pp. 1–3.

[5] D. Marek, S. Sharma, and P. Triverio, "A parallel boundary element method for the electromagnetic analysis of large structures with lossy conductors," *IEEE Trans. Antennas Propag.*, vol. 70, no. 11, pp. 10736–10750, Nov. 2022.

[6] W. Chai and D. Jiao, "Linear-complexity direct and iterative integral equation solvers accelerated by a new rank-minimized $\mathcal{H}^2$-representation for large-scale 3-d interconnect extraction," *IEEE Trans. Microw. Theory Techn.*, vol. 61, no. 8, pp. 2792–2805, 2013.

[7] R. Gholami, S. Zheng, and V. I. Okhmatovski, "Surface-volume-surface efie for electromagnetic analysis of 3-d composite dielectric objects in multilayered media," *IEEE J. Multiscale and Multiphys. Comput. Techn.*, vol. 4, pp. 383–394, 2019.

[8] Y. Li, D. Marek, and P. Triverio, "Fast electromagnetic analysis of multiscale interconnect networks using MultiAIM," in *Proc. IEEE Conf. on Elect. Perf. of Electron. Packag. and Systems (EPEPS)*, 2023, pp. 1–3.

[9] ——, "MultiAIM: Fast electromagnetic analysis of multiscale structures using boundary element methods," *IEEE Trans. Antennas Propag.*, 2024.

[10] Z. G. Qian and W. C. Chew, "Fast full-wave surface integral equation solver for multiscale structure modeling," *IEEE Trans. Antennas Propag.*, vol. 57, no. 11, pp. 3594–3601, 2009.

[11] Y. Saad and M. H. Schultz, "GMRES: A generalized minimal residual algorithm for solving nonsymmetric linear systems," *SIAM J. Sci. Comput.*, vol. 7, no. 3, pp. 856–869, 1986.

[12] A. Brandt, "Multilevel computations of integral transforms and particle interactions with oscillatory kernels," *Comput. Phys. Commun.*, vol. 65, no. 1-3, pp. 24–38, 1991.

[13] E. Wang, Q. Zhang, B. Shen, G. Zhang, X. Lu, and Q. e. Wu, "Intel math kernel library," *High-Performance Computing on the Intel® Xeon Phi™: How to Fully Exploit MIC Architectures*, pp. 167–188, 2014.

[14] X. Zhang *et al.*, "OpenBLAS: An optimized BLAS library," https://www.openblas.net/, 2024, accessed: 2024-07-09.

[15] M. Frigo and S. G. Johnson, "FFTW: An adaptive software architecture for the FFT," Proceedings of the IEEE International Conference on Acoustics, Speech and Signal Processing, 1998, available at http://www.fftw.org.

[16] "Packaging Benchmark Suite," 2021. [Online]. Available: https://packaging-benchmarks.org/

of layers with many vias. Conductors are modelled as copper, and placed in a single dielectric layer with permittivity of 3.3, surrounded by air on both sides. Four lumped ports are placed on metal bumps, and excited at 10 GHz. Once discretized, the structure results in a mesh with 1,204,374 edges, with a multiscale factor of 14.5 (the ratio between the longest and shortest mesh edges). The proposed method is compared to a parallel implementation of the AIM. Fig. 2 reports the wall time as a function of the number of threads. The proposed method has been tested with both Intel's MKL and OpenBLAS as numerical libraries. The top panel of Fig. 2 shows that matrix generation time scales well for all three implementations up to 32 cores. The middle panel shows that, when MKL is used, GMRES' time also scales well up to about 32 cores. This is

979-8-3503-5124-8/24 $31.00 © 2024 IEEE

AUTHOR INDEX

Achar, Ramachandra 142, 193
Ahmed, Shameem 43
Ahn, Jungmin.................. 49, 79, 115, 148, 163
Ahn, Seungyoung 73
Ahn, Sung-Oh 7
Akinwale, Oluwafemi 34
Ali, Isaac 199
An, Hyunjun 46, 49, 79, 112, 115, 118, 148, 163
Ankamah-Kusi, Sylvester 4, 40
Araga, Yuuki 58
Araujo, Daniel De 175, 178
Araujo, Stefan De 175, 178
Arend, Paul Van Der 1
Aronsson, Jonatan 166
Asadi, Reza 43
Aygün, Kemal 184
Ayinet, Tiruye Mulat 55
Bae, Taeil 49, 109
Baissa, Gerba Olani 55
Bakir, Muhannad S. 100
Bandi, Sathvika 37
Bansal, Vinayak 28
Bao, Tina 13
Baydogan, Mustafa Gökçe 184
Bellaredj, Mohamed Lamine Faycal 196
Benner, Peter 28
Boef, Pascal Den 91
Bogaerts, Wim 145
Bracken, J. Eric 64
Bradde, Tommaso 124
Cai, Kevin 43
Carlucci, Antonio 121, 124
Carta, Corrado 136
Chang, Zhu-Chen 88
Chen, Jie 40
Chen, Xu 100
Chen, Xuan 190
Chen, Yen-Tung 16
Cheng, Chris 160
Cheng, Yu-Ying 16
Cho, Chulhee 85
Cho, Young-Chul 85
Choi, Dongho........................... 61, 67, 70
Choi, Hyun-Chul........................... 52, 106
Choi, Jung-Hwan 85, 118
Choi, Seonguk 46, 49, 79, 112, 115, 118, 148, 163
Choi, Sumi 112
Choi, Wooshin 85, 118

Choi, Yongjin 160
Chu, Xiuqin 22
Chun, Sunghoon............................61, 67, 70
Cocchini, Matteo 94
Delbue, Roger 178
Deschrijver, Dirk 145
Dhaene, Tom 145
Dounavis, Anestis 10
Duan, Xiaomin 127
Durgun, Ahmet Cemal........................184, 187
Elkin, Samuel T. 31
Eslampour, Hamid 7
Fang, Tao.................................133, 151
Fang, Yuan 133, 151
Feng, Lihong 28
Franzon, Paul D................................. 160
Fu, Greg 151
Gad, Emad 82
Ghaly, Germin 82
Ginste, D. Vande 172
Grivet-Talocia, Stefano 121, 124
Guo, Mengmeng 13
Haider, Clifton R. 19
Haider, Michael 31
Hatton, Jasper 205
Hickman, Alexander P. 19
Hong, Seokwoo............................61, 67, 70
Hossain, Masum 202
Hui, Teo T. 55
Huynen, M. 172
Hwang, Seunghyun 169
Iseini, Festim.................................136, 139
Jain, Ashish 1
Jang, Chorom.............................61, 67, 70
Javaid, Ahsan 142
Jeong, Jonghee 61, 67
Jeong, Sangnam 7
Jiao, Dan 169
Jin, Sungwoo..............................61, 67, 70
Jinesh, Anandajith 190
Jung, Haekang 49, 109
Kahmen, Gerhard 136
Kaller, Dierk 127
Kamath, Swathi 4
Kashyap, Priyank 160
Keegan, Ryan 178
Keuseman, Jordan R.............................19
Khazaka, Roni 208

Kikuchi, Katsuya 58
Kim, Donghyun Bill 43
Kim, Donghyun 37
Kim, Dongkyun 73
Kim, Haeyeon........................... 109, 115, 118, 148
Kim, Hyunsik.. 49, 109
Kim, Hyunwoo 61, 67, 70, 73
Kim, Jihun 112
Kim, Joungho.......... 46, 49, 79, 109, 112, 115, 118, 148, 163
Kim, Juneyoung 61
Kim, Kang-Wook 52, 106
Kim, Keunwoo.............. 46, 49, 79, 112, 118, 148, 163
Kim, Kwangho 85
Kim, Kyungsuk.. 61, 67, 70
Kim, Mun-Ju 52, 106
Kim, Nam Sung 100
Kim, Taesoo 49, 79, 115, 163
Kim, Woopoung 7
Kirchberger, Alexander 10
Ko, Youngjun 61, 67, 70
Koh, Cheng-Kok 169
Krishna, Ram 100
Kronenberg, Tom 4
Lee, Hyeongi 85
Lee, Jinan.. 61, 67
Lee, Jinwook 73
Lee, Junghyun................. 46, 49, 79, 112, 115, 118, 148, 163
Lee, Manho 85
Lee, Sanguk 73
Lee, Seonghi 73
Lee, Youngjae 85
Lena, Davide 154
Li, Aobo 22
Li, Chaofeng 43
Li, Guangxu 40
Li, Jingbo 13
Li, Shuxiang.. 133, 151
Li, Weizhe 13
Li, Yongzhong... 205, 211
Lin, Chien-Min 88, 103
Lin, Han-Ting 136, 139
Ling, Feng 166
Liu, Dan... 13, 34
Liu, Tong 157
Lu, Cheng-Yuan 103
Mahmood, Shakib 202
Malignaggi, Andrea 136
Mandredi, Paolo 124
Manfredi, Paolo 25
Manvelyan, Diana 91
Marek, Damian 205
Mathew, Manish K. 43

Mathur, Anuj 193
Matta, Kusuma 34
Maubach, Joseph 91
Mendez-Ruiz, Cesar 34
Min, Byung-Cheol 52, 106
Miskovic, Goran 196
Mousavi, Mehdi 43
Murugan, Rajen 4, 40
Mutnury, Bhyrav 175
Nakhla, Michel 82
Nascimento, Vinicius C. Do 169
Niazi, Alireza 181
Okhmatovksi, V. 172
Okhmatovski, Vladimir 181
Oreifej, Rashad.. 1
Ouchi, Shinichi.. 58
Özese, Doganay 184
Page, Andrew 94
Palermo, Samuel 157
Pappu, Santosh 37
Pareschi, Fabio 154
Park, Dongryul 73
Park, Eunkyeong 7
Park, Hyunah................. 49, 79, 109, 112, 115, 148, 163
Park, Hyunwook 118
Park, Joonsang 109, 112
Park, Junho 112
Park, Junyong 37, 43
Pelagalli, Nicola 136
Peng, Zhekun 37
Penta, Srujan 100
Qiu, Lu 97
Qiu, Qiang 169
Rahal-Arabi, Tawfik.. 1
Ranzaswamy, Granthana 37
Reinamendivil, David 199
Roh, Sungwon....................................... 61, 67, 70
Rosenbaum, Elyse 100
Roth, Thomas E. 31
Roy, Abinash 130
Rubia, Valentin De La 28
Ryu, Seunghun 73
Sahouli, Mohamed 199
Saidi, Mehdi... 1
Scearce, Stephen 151
Scharff, Katharina 127
Schilders, Wil 91
Schutt-Ainé, José E. 76
Seahra, Rajit... 1
Sen, Bidyut 43
Setti, Gianluca 154
Shim, Taehyun 61

Shimamoto, Haruo	58	
Shimomura, Soshi	58	
Shin, Jongchul	7	
Shin, Taein	49, 79, 115, 118, 148, 163	
Sidhu, Karanvir S.	208	
Sim, Taeyang	157	
Simone, Silvia	154	
Smith, Michael J.	169	
Smith, Richelle L.	202	
Smutzer, Chad M.	19	
Sohn, Young-Soo	85	
Son, Keeyoung	46, 49, 79, 109, 112, 115, 118, 163	
Son, Sanghyuk	112	
Song, Jinwook	61, 67, 70, 148, 163	
Song, Sangsub	7	
Subbarayan, Ganesh	169	
Suh, Haeseok	46, 49, 79, 115, 148, 163	
Sundaram, Sriram	1	
Suzuki, Yutaka	40	
Talbot, Gerry	199	
Tang, Xinlin	133, 151	
Tethy, Parneet	202	
Tian, Minzheng	13	
Travis, Blake	4	
Trinchero, Riccardo	25, 124	
Tripathi, Jai Narayan	142	
Triverio, Piero	205, 211	
Uematsu, Yutaka	58	
Ullrick, Thijs	145	
Ünal, Hasan Said	187	
Venkataraman, Srinivas	37	
Victor, Ashita	100	
Vojkuvka, Lukas	196	
Wang, Jun	22	
Wang, Kai	34	
Wang, Xu	37	
Wei, Bing	13	
Weisshaar, Andreas	136, 139	
Wen, Yeujiang	160	
Werner, Carl W.	202	
Wouw, Nathan Van De	91	
Wu, Ruey-Beei	88, 103	
Wu, Tzong-Lin	16	
Xu, Yan	22	
Yang, Xian-Long	97	
Yong, Seokbeom	7	
Yoon, Jiwon	46, 49, 79, 115, 118, 163	
Yoon, Sungjin	85	
Yuan, Felix	130	
Zhang, Kangkang	22	
Zhong, Yangfan	13, 34	
Zhou, Yi	76	

Zhu, Xiao-Wei	97
Zutter, D. De	172

IEEE
445 Hoes Lane
Piscataway, NJ 08854-4141

ISBN 979-8-3503-5124-8